The West Point Military Library

General Editors of the Series

COLONEL THOMAS E. GRIESS
Professor, United States Military Academy

and

PROFESSOR JAY LUVAAS
Allegheny College

THE

ARTILLERIST'S MANUAL,

COMPILED FROM VARIOUS SOURCES,

AND ADAPTED TO

THE SERVICE OF THE UNITED STATES.

ILLUSTRATED BY ENGRAVINGS.

JOHN GIBBON,
1ST LIEUT. 4TH ART'Y, U.S. ARMY

GREENWOOD PRESS, PUBLISHERS
WESTPORT, CONNECTICUT

Originally published in 1860 by D. Van Nostrand,
New York and Trubner & Company, London

Reprinted in 1970 by Greenwood Press,
a division of Congressional Information Service,
88 Post Road West, Westport, Connecticut 06881

Library of Congress catalog card number 74-84270

ISBN 0-8371-5007-8

Printed in the United States of America

10 9 8 7 6 5 4 3 2

TO

THE OFFICERS

OF

The United States Artillery

THIS WORK

IS

RESPECTFULLY DEDICATED.

PREFACE.

THIS work, originally designed as a book of instruction for the cadets of the Military Academy, has, since my separation from the department of artillery, been extended beyond the limits at first proposed, with a view of spreading information not popularly accessible, upon a subject of the first importance to our national defense.

It is submitted to my brother officers, trusting that many allowances will be made for its defects, and that some one more capable of doing justice to the subject will be induced to offer to the service, and to our militia,—on whom, in the event of war, the principal defense of our large fortifications must devolve,—a more complete system of instruction than I have been able to furnish.

Where translations have been made, it has been my endeavor to select such portions as are or may be applicable to our own service, leaving out those peculiar to the foreign.

J. G.

WEST POINT, N. Y., August 14, 1859.

CONTENTS.

ERRATA.

Page 36, 4th line from bottom, insert ϕ' between "wires" and "would."

Page 78, 12th line from bottom, for "GUIDE GUAGE" read GUIDE PLATE.

THE

ARTILLERISTS' MANUAL.

CHAPTER I.

GUNPOWDER.

THE inflammable property of a mixture of nitre, charcoal, and sulphur, was known long before its projectile force was discovered. Used in the form of dust, it is supposed by many to have been the substance of which the rockets and Greek fire of the ancients were made.

DISCOVERY.—Gunpowder is said to have been discovered in Europe somewhere about the XIII. century; but a somewhat similar compound is supposed to have existed many centuries before, in China. Until towards the close of the XV. century, it was used in the dust form; but about that time it was discovered that when formed into grains, its force was very considerably increased.

INGREDIENTS.—The ingredients used in the manufacture of gunpowder are charcoal, nitre (or saltpetre), and sulphur; and in order to explain the phenomena of explosion, a short description of each of these substances will be given.

CHARCOAL.—If charcoal is heated to a certain temperature, combustion takes place, and if perfectly pure, no ash will remain. The charcoal will be converted into an invisible gas. Placed under the receiver of an air-pump, and the air removed, it will

2

not burn. There is, therefore, needed for its combustion, an ingredient of common air—oxygen—and it will presently be shown how this element is combined with the charcoal so as to produce combustion in vacuo. The power of oxygen to support combustion is very great. This may be seen by a simple experiment. A piece of steel wire having any inflammable substance placed on the end of it and lighted, when dipped in a jar of oxygen gas is rapidly consumed. This being the case with regard to iron or steel, it may readily be imagined how much more rapidly charcoal immersed in the gas would disappear, from what is called combustion—really uniting with the oxygen—forming carbonic acid gas. Now, if instead of presenting the charcoal in a solid mass to the action of the oxygen, it, in the form of a minute powder, be brought in contact with a substance similarly prepared, and containing a large quantity of oxygen, the combustion, when fire is applied, will be much more rapid, producing deflagration.

Charcoal made from light, quick-growing wood is the most easily pulverized, and is the best for making gunpowder. In the United States willow and poplar are used. It should be sound, about an inch in diameter, and not more than three or four years old. The largest limbs are split up before burning; the smallest branches are used for fine sporting-powder. The wood should be cut in the spring, when the sap is running freely, and the bark at once taken off. It may be charred in the ordinary way, in pits; but the most usual way of preparing it for making gunpowder is by *distillation*. This process consists in placing the wood in iron cylinders, to which pipes are fitted to carry off the liquid and gaseous products. Heat is then applied to these cylinders and kept up for a number of hours. The fires are then allowed to go down, and when cool, the charcoal is taken out. This method has the advantage over the common way of making charcoal, of performing the operation of charring more completely, and of saving the ingredients, which by the common process are lost. Acetic or pyroligneous acid, tar, and wood naphtha are thus obtained. This charcoal is also said to be more free than any other from potash.

Another process has lately been introduced in France by M. Violette, and considered a great improvement. This consists in transmitting through the wood high-pressure steam, producing more effectually even than fire, the desired result. It should retain a certain degree of elasticity, have a brown color, present the fibrous appearance of the wood, and its fracture be irridescent.

Wood contains about fifty-two per cent. of carbon; but this process furnishes only about thirty or forty per cent. The coal is usually made in the immediate locality of the powder-factories, and, as it readily absorbs moisture, is made only as it is wanted. In sticks its specific gravity is about .300, which is increased by pulverizing to 1.380.

NITRE.—Nitre is the substance which, possessing such a large percentage of oxygen in a solid form, gives charcoal the ability to burn whether in contact with the air or not. Sulphur adds consistency to the mixture, and intensity to the flame, when lighted; and the result is gunpowder. Sulphur also renders the powder less liable to absorb moisture. The quality of powder depends very much upon the intimate mixture and proper proportions of the ingredients—or in other words, on our ability always to have in contact with the inflammable materials, charcoal and sulphur, the necessary oxygen in the form of nitre to support combustion, and change *all* the solid mass into gases. Charcoal by itself, *i. e.* in vacuo, will not burn, nor will nitre either in that way or in the open air; but together they burn in vacuo, and intimately mixed, deflagrate.

The finer the gunpowder and more intimately mixed the ingredients, the nearer it approaches a detonating powder. In large and compact charges, however, the reverse is the case, as then the powder itself forms an obstacle to the transmission of the flame, and it burns slowly and in successive layers. As a general rule, and before coming to the limit of dust, the smaller the grain the more rapid the combustion, and the greater the bursting force of the powder. This is the reason that in small arms the small-grained powder is used. It has but a short distance in which to act on the projectile, and being in small quantity, its bursting force is not such as to injure a gun; while in large pieces the charges are so great, that did they burn fast no metal could withstand the shock. In these, too, as the time which the powder has to act on the projectile is greater than in small arms, the whole force of a slow-burning powder will be produced before the projectile leaves the muzzle; and from the great heat evolved by the large charge, the largest grains will be consumed.

Nitre or saltpetre, the nitrate of potassa, is a natural salt found in many localities, especially in the East Indies, where it exists in great abundance on the earth's surface, and where it is natural to suppose its peculiar properties when combined with carbon were first observed. It is also found in other warm countries, and in limestone caves in the States of Virginia, Georgia, Tennessee, and Kentucky; and in this last in the form of rock ore, a sandstone containing a very large proportion of the salt. It may also be produced artificially in what are called "nitre beds," by exposing animal and vegetable substances together to the action of the air in moist situations. This method, although not first discovered, was perfected by the celebrated Berthollet, at a time when France, cut off from communication with other nations, was driven to these beds, or walls as they are sometimes called, to obtain supplies of an article so absolutely necessary to her defense.

Characteristics.—This salt when pure, has a clear, white, crystalline appearance, a specific gravity of 2.090, melts at 660° F., and decomposes at a red heat. It is soluble in water, much more so in hot than in cold, and has a cool, saline, and slightly bitter taste.

In the manufacture of gunpowder, it is of the greatest importance that all the ingredients, and especially the nitre, should be pure. It never appears so in commerce, but in the shape of grough saltpetre contains from six to twelve per cent. of foreign salts, earth, and water.

Analyzing.—To ascertain the quality of grough saltpetre, a certain quantity, say one lb., is washed in a saturated solution of pure nitre, and the liquor poured off; the washing is then repeated and poured on a filter, which is dried perfectly by placing it on glass imbedded in ashes or lime, and then evaporating. The saturated solution takes up only foreign salts, and what remains, allowing two per cent. for earthy matter, &c., will be the quantity of pure saltpetre. As a check and comparison, perform the same operation on the same quantity of pure saltpetre.

Purifying.—Great pains are taken to purify the nitre to

be used in the manufacture of gunpowder; and the process is much facilitated by the facts that some of the impurities are less soluble in boiling water than in cold; nitre is the reverse; and some are more soluble in hot or cold water than nitre. For purifying then, two tiers of vats are employed, one heated the other cold. For heating, steam is sometimes used, as at the celebrated establishment of Col. Hazard, at Enfield, Conn. The saltpetre of commerce is first placed in the vat "A," Fig. 1, and steam applied from below, producing complete ebullition. Particles of dust and other light materials rise to the top, and are skimmed off. The nitre being more soluble in boiling water than in cold, and such not being the case with chloride of sodium (common salt), this last will on the cessation of the boiling sink to the bottom, with some earthy matter, the precipitation of which is assisted by a solution of glue. The liquor, whilst hot, is then drawn off by a syphon into the vat below. Here it is allowed to cool slowly. Chlorides of calcium and magnesium, and the nitrates of lime and soda, being more soluble in water, hot or cold, than nitre, are taken up by the water, while the nitre is formed into crystals at the bottom. When the mass is reduced to about 104°, the water is drawn off, leaving the nitre behind pure. It is kept constantly stirred during this operation, to prevent the formation of large crystals, which would probably inclose water and some impurities. It may then be left to drain dry through holes in the bottom of the vat, or transferred to a kind of hopper specially made for the purpose,—Fig. 2. One of these operations is not always sufficient to rid the nitre of all its impurities; a matter of the first importance, since the presence of the most injurious of these, nitrate of lime and impure common salt having a powerful affinity for water, will very materially affect the quality of the powder made from it, more especially in regard to its preservation. The importance of this will be seen when the effect of moisture on gunpowder is observed. It rapidly impairs

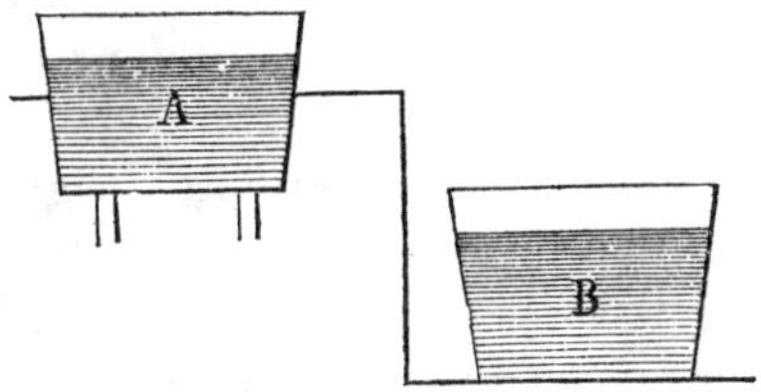

Fig. 1.

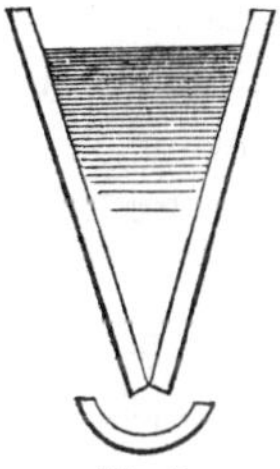
Fig. 2.

its strength, and in time produces a marked change in the structure of the grain; the nitre efflorescing on the surface in the form of a white powder, and the whole mass being formed into cakes and lumps.

FUSING.—The nitre is now in small, perfectly white grains; but before being used for powder it is *fused*, for the purpose of driving off any water which may be retained among the particles, so that it may be accurately weighed. It should never be fused at a higher temperature than 600°, as beyond that it is apt to be injured by partial decomposition. It should not contain more than 1-3000 part of chlorides; but for powder which is to be used at once, this degree of purity need not be required; and this fact gives the means of reducing considerably the proportion of nitre in a powder designed for immediate use. When fused, the nitre is cast into flat cakes and allowed to cool. In the manufacture of powder in this country, the nitre is not fused after it is reduced to fine pure grains, unless it is to be kept for some time before being used. But in Europe it is looked upon as an important step; though objectionable on some accounts, as with too high a heat it is partially decomposed, oxygen being driven off.

SULPHUR.—Sulphur is found in all volcanic countries in great abundance. Sicily supplies the principal part of the market. It is found there in large masses, very nearly pure. The greater part of that used in the United States comes from the French refineries.

REFINING.—Before being used in the manufacture of powder, this substance has also to be refined. Two methods are used, one by simply fusing it, when the heaviest impurities sink to the bottom of the vessel, and the lightest rise to the top. The intermediate portion being left pure, may be withdrawn. The other method is by sublimation, by which the sulphur is vaporized at about 170°, and again condensed in the form of a fine powder called "flowers of sulphur." When melted and run into moulds, it is called "roll sulphur."

CHARACTERISTICS.—Pure sulphur is of a citron-yellow color, has a shining fracture, and crackles when pressed in the hand. Its specific gravity is 2.033, which is reduced by sublimation to 1.900, and still further by trituration. It melts at 220°, but at 320°

takes the consistency of paste, and sublimes or vaporizes at 680°. Insoluble in water, but in oils and alcohol it is soluble. As when pure it is entirely consumed by combustion, a simple test of its purity is to burn some on a clean surface, and see if any residue remains.

MANUFACTURE.—Having described the different substances which enter into the composition of gunpowder, we will now proceed to the manufacture of this last, which consists in pulverizing the ingredients, incorporation, compression, granulation, drying, glazing, and dusting.

POUNDING-MILL.—Two methods of making gunpowder are used,—by the pounding-mill and by the rolling-mill. In the pounding-mill the ingredients are pulverized, incorporated, and compressed, at the same time, by the pestles: twenty or twenty-four mortars and pestles are placed in two rows in the mill; the mortars are nearly spherical, and dug out of a piece of oak, having a piece of harder wood in the bottom. Each mortar receives about twenty lbs. of composition. The charcoal in small pieces is first placed in with a quantity of water, and pounded for half an hour, after which the saltpetre and then the sulphur, previously pulverized and sifted, are put in, and the whole well mixed with the hand; and, after being pounded for an hour, it is transferred to the next mortar, and so on, changing every hour. At the sixth or eighth change add a half-pint of water, to guard against explosion. During the last two hours no change is made, in order to allow the composition to form into cake. It is then taken from the mortar, the moisture reduced to four per cent., and then grained; for which purpose it is partially dried and placed in a graining-sieve of parchment pierced with holes, which is moved by hand or machinery, whilst a lenticular disk of hard wood, weighing five lbs. breaks up the powder, which passes through the holes as soon as sufficiently reduced. The cakes may also be grained by being passed between wooden rollers. The grains are then sifted, to separate those which are too coarse and too fine, as also to separate the different kinds—cannon, musket, and rifle—from each other.

GLAZING.—About five hundred lbs. of it, containing about three per cent. of moisture, is then placed in a large glazing-barrel, and glazed by revolving the barrel fifteen or twenty times a minute for twenty-four hours or less, according to the degree of glazing required. Glazing is necessary in order to prevent the absorption of moisture, and breaking up of the grains in transportation—forming dust, which sifting through the bags reduces the charges, and retards the inflammation by choking up the interstices between the grains.

DRYING.—The powder is then dried, either in the open air or in a drying-house. In the first case it is spread on sheets placed on tables, and allowed to remain ten or twelve hours, being stirred frequently to expose it well to the sun. When the sun is too hot, it is sometimes necessary to cover the powder over to prevent the loss of the sulphur. In a drying-house, it is exposed, in layers of from one to four inches, to a current of air heated to 140° by means of a furnace or hot-water pipes.

DUSTING.—It is then dusted by being sifted in fine sieves or bolting cloths, in order to clean it thoroughly and cool it before being barreled. The dust may be worked over again to make inferior powder, or mixed with other composition in the pounding-mill.

TIME.—It requires from eleven to fourteen hours to make powder by this process. This time may be somewhat reduced by previously pulverizing and mixing the ingredients in rolling-barrels formed of strong leather or hides stretched over a frame-work, the slats of which, nine inches apart, project half an inch inwardly. Each barrel contains 100 lbs. of zinc or composition balls which serve to pulverize the materials when the barrel is revolved around its axis. The charge for a barrel is fifty lbs., and the elasticity of the leather prevents the powder from adhering to the sides, as it would do with wood.

ROLLING-BARRELS.—The charcoal and sulphur are first placed in and rolled for two hours, to pulverize them; the saltpetre (refined) is then added, and the rolling continued for two hours longer; when ten per cent. of water is added, and it is pounded in the mortars for three hours; or it may be spread in thin layers, moistened with the same quantity of water, distributed equally with a fine watering-pot, and brought to a state of cake by pressure in an hydraulic or screw-press; and this method of forming the powder into cakes is sometimes adopted after the ingredients have been incorporated in the pounding or cylinder mills.

The pounding-mills are used altogether in France, for the military service; but the charcoal used is made in open pits (*à l'air libre*) where it is more thoroughly burnt and becomes more friable than cylinder coal, which last is too hard to be pulverized sufficiently by the action of the pestles.

CYLINDER OR ROLLING MILL,—Fig. 3. The cylinder or rolling mill used in England and this country consists of a circular bed of cast iron or marble, the inner diameter of which is about three feet, and two cylinders or wheels weighing about five tons each, of the same material, which run on this bed, followed by a wooden plow which throws the materials towards the centre of the track. The cylinders revolve about ten times a minute, and run from one half to three hours on fifty pounds of composition. The powder is therefore made much quicker, and it is found that it exhibits a greater degree of strength in cannon, than powder incorporated in any other way; but this superiority, though uniform, is not so great as to give the rolling-mills an absolute preference over all other methods of incorporation, so far as regards the strength alone; and the choice between them must be decided by the relative economy, and the other qualities which each imparts.

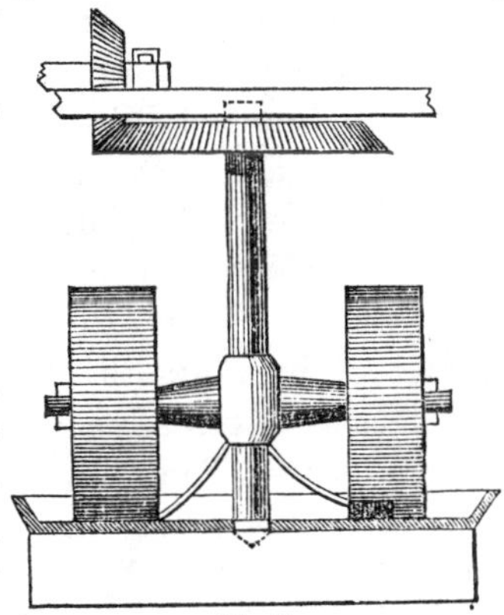

Fig. 3.

Experiments in this country, as well as in France, seem to give the preference to the rolling-mill powder; but the pounding-mill is still retained in France; where it is claimed,—1. That the pounding-mill makes powder better adapted to the promiscuous service of all arms. 2. That it is less injured by exposure to moisture; and 3. That it is less injurious to the gun. The first is of no advantage in the United States, as we use different kinds of powder for cannon and small-arms; and rolling-mill powder may be so worked as to render it equally resistant to moisture. In the third place, by a simple alteration in the cartridge (reducing its diameter), the destructive force of dense powder on the gun is reduced without diminishing its projectile force; and this may be further reduced by the reduction in the charge which we are enabled to make in dense powder, from its greater force. In small-arms, the barrels are so strong that the destructive effects of the small charges used constitute no objection, to the use of a powder more violent even than the strongest rifle-powder proposed to be made. The pounding-mill powder, in order to be of a force nearly equal to rolling-mill powder, must be worked not less than fourteen or sixteen hours; and even then the grain is hardly firm enough to resist the jolting of ammunition wagons. The force retained by some Waltham powder after being kept for thirty years without special care, speaks strongly in favor of the English mode of manufacture.

It is a characteristic of gunpowder, that the greater the opposition offered to the expansion of its gases whilst the powder is burning, the greater becomes the force developed. This fact is taken advantage of in various ways in the science of artillery. Thus, for instance, a gun is made smaller at the seat of the charge by making use of a *chamber*, because the powder is thereby more confined, and a greater force developed; and it is found that the *form* of chamber which delays the escape of the gas the longest, is productive of the greatest force, though for other reasons it is not adopted. In firing shells from *mortars*, where the angle of elevation is 45°, and the *weight* of the projectile almost directly opposed to the effort of the powder, we are enabled to reduce the amount of the powder used, on account of the increased force developed. Thus, too, in blasting rocks, the amount of execution by a small charge of powder is apparently out of proportion to its size, on account of the manner in which the powder is confined. Good powder should be perfectly free from dust. Smooth, brilliant grains are more solid and fit for war purpose than those which are not so. When the grains are porous, combustion takes place more rapidly, and the powder approaches more nearly a *bursting* powder. Thus a powder which has absorbed a certain quantity of moisture, and been redried, may become more destructive to guns than when first taken from a magazine.

Gunpowder, in order to be of good quality, should have the

ingredients carefully prepared and thoroughly incorporated, should be perfectly dried and highly glazed; all of which are favorable not only to the production of the greatest force, but to the quick combustion of the grains and the rapid transmission of the flame through the whole mass of the powder. The injury caused to powder by moisture, is too apparent to need any further comment on the necessity of drying it well. Glazing does not affect the strength of powder; and though it lessens somewhat the rapidity of burning, this is more than compensated by the advantages resulting from the process. The preparation of the materials having been previously noticed, it now remains to refer to the other conditions.

Density.—The more a powder is worked, the more dense it becomes, and up to a certain point, the greater is the force of the powder, but beyond this limit it is found that no increase of force follows the continued working. The density of gunpowder may be approximatively determined by taking the weight of a cubic foot in ounces. This is called the *gravimetric* density, and is obtained by means of a *gravimeter*, a brass cylindrical vessel holding 1-27 of a cubic foot (64 cubic inches). By taking the weight of the contents when loose and after being shaken, the relative irregularity and size of the grain will be indicated by the difference. Experiments show that the gravimetric density of cannon powder should not be less than 870, nor greater than 920.

Specific Gravity.—Gunpowder being but a mechanical mixture, water readily dissolves out the nitre, and its specific gravity cannot be determined directly. The French use the following method: A cylindrical glass vessel of uniform diameter, the edges well ground, to which is adapted a polished glass cover so made as to hermetically seal the vessel, is filled with distilled water, carefully closed by sliding on the cover so as to exclude the air, and after being wiped dry on the outside is accurately weighed. The weight of the vessel and cover having been previously ascertained, take the difference, which will give the weight of water which the vessel holds (W). Find in the same manner the weight (W') of a saturated solution of nitre which the vessel holds. Pour out a portion of this, and pour in 1500 grains of gunpowder *slowly* so as to exclude the air; fill up with the satu rated solution, put on the cover, wipe the vessel, and again weigh it. From this weight subtract the weight of the vessel and cover and the powder. The remainder will be the weight of the saturated solution occupying the vessel with the powder. Subtract this from the weight of the solution before found, and the remainder will be the weight (w') of the solution which occupies the same volume as the powder. Then $W' : W :: w' : w =$ the weight of an amount of water equal in volume to the powder; and $w : 1500$ grs. : : spec. gravity of water (or 1) : specific gravity of powder. Or, specific gravity of powder$=\frac{1500 \text{ grs.}}{w}$. This operation is repeated three times, and the mean taken. Alcohol may be used instead of the solution; though neither of them gives perfectly accurate results. The principal reason for this seems to be the difficulty of expelling all the air in the powder. Another cause of error in the use of nitre-water is the difficulty of keeping it in a state of saturation, the heavier portion settling to the bottom; and a still greater cause of error may result from a slight change of temperature during the experiment, causing the solution to deposit some of its nitre, or take up some of that of the powder. By placing gunpowder in any liquid, such as rectified alcohol, sufficiently thin to penetrate all the pores of the grains, it must be in a great measure disintegrated, and furnishes thus, not the specific gravity of the mixture which forms the powder, but the combined specific gravity of the ingredients themselves; and the results would depend less on the intimacy of the mixture than on the trituration of the ingredients, which increases the specific gravity of charcoal, while it decreases that of sulphur.

Packing.—United-States powder is packed in barrels of 100 lbs. each; the barrels being large enough to allow vacant space sufficient for the powder to move when rolled, to prevent its caking. They are made of well-seasoned white oak, hooped with hickory, cedar, or copper. Cedar is less liable than hickory or oak to the attack of worms. The hoops should cover two thirds of the barrel, and when made of copper are fastened with copper nails. A screw-hole one and a half inches in diameter is made in the head of the barrel, and closed with a wooden screw with an octagonal head, which renders the removal of the head of the barrel unnecessary, and the hoops can be secured with copper nails. To protect the powder from moisture, a thin washer of leather steeped in a solution of beeswax in spirits of turpentine is placed under the head of the screw; and, for transportation, a piece of cloth should be glued over the head. Dimensions of powder-barrels, $20\frac{1}{2}$ in. long by 14 in. in diameter at the head.

Inspection.—The powder received into the United States service is inspected and proved; a sample for which is taken from each barrel by means of an extractor, a copper tube 18 in. long and 1 in. in diameter, pointed at the end, and having an opening about 9 in. from that end, by covering which with the hand the powder can be poured from the mouth of the tube. Each sample is placed in a tin canister marked with a number, a corresponding number being placed on the barrel from which it is taken. These samples are inspected, and the charges weighed from them when the strength is tested.

Characteristics.—The powder should have an even grain, angular and irregular in form; such being the form which exposes most surface to the flame, and therefore burns most rapidly: round grains burn slowest. It should be so hard as not to be easily crushed by pressure with the finger, should leave no dust when poured on the back of the hand, and should leave no bead or foulness when flashed, ten grains at a time, on a polished copper plate. The size of the grains is tested by means of sieves having holes of the maximum, minimum, and medium size, for each kind of powder. Ten grs. troy, contain of

CANNON POWDER.	MUSKET POWDER.	RIFLE POWDER.
150 grains.	2,000 or 2,500 grs.	12,000 or 15,000 grs.

Usually, the uniformity and size of grain is judged of by mere inspection.

Proportions.—Before mixing, the ingredients are carefully weighed in the proper proportions. These depend upon the kind of powder to be made. When it is remembered that on the quantity of oxygen present depends the rate of burning of the charcoal, it will readily be perceived why it is that to produce a quick burning powder we must increase the amount of nitre; as the same amount of consistency, however, is required, the sulphur, which produces this, remains the same, and the nitre is increased at the expense of the charcoal. Different nations vary slightly the proportions for military powders. In the United States two sets are used, as follows:

	NITRE.	CHARCOAL.	SULPHUR.
(1)	76	14	10
(2)	75	15	10

This also is the English. In France the proportion is—

	75	12½	12½
and in Prussia,	75	13½	11½

For sporting powder, which, of course, burns much quicker than other kinds, France and the United States have adopted—

	78	12	10

which England has varied slightly by taking—

	78	14	8
and	75	17	8

In blasting-powders, which it is not necessary should burn fast, France has adopted—

	62	20	18

Gunpowder explodes when suddenly submitted to a temperature of about 600°. It will also explode when struck a violent blow with a hard substance. This may be demonstrated satisfactorily by firing from a rifle a hollow pointed ball having a few grains of powder in the aperture near the apex, against an iron target: on striking, the powder explodes. It is more than probable that many premature explosions of shells, and more especially of spherical-case shot, in the bore of guns are due to this fact, instead of to the driving in of the fuze, as has been sometimes supposed. Moderate heat applied to the constituents of gunpowder separately, produces no chemical change. If applied to sulphur and charcoal mixed, a substance (bi-sulphuret of carbon) is produced which may be condensed to a liquid. If charcoal and nitre be reduced to fine powders, mixed and heated, the two combine and deflagrate, as has been stated, but not violently, forming carbonate of potassa, nitrogen, carbonic oxide, and carbonic acid, of which the first only is solid—combined in such proportions that none of them are left unconsumed; the volume of the gas resulting is, of course, a maximum, and is about 1,000 times the volume of the ingredients. This combination would give but small projectile force in a gun, as, burning but slowly, it would only be partly consumed, and throw the rest of the charge with the ball from the gun, producing on the ball but little effect. *Sulphur* and *nitre*, powdered, mixed, and having heat applied so as to first melt them, explode with great violence, though the gas evolved is somewhat less in volume than in the other case. The greater violence is probably due to the more intimate mixture of the two substances in consequence of their liquid state, and the greater heat resulting from the burning of the sulphur. The substances formed in this case are *sulphuret of potassium*, *nitrogen*, and *sulphurous acid*. A mixture of this kind would, in its effect, resemble too strongly a fulminating powder to be used as a projectile force, and would probably burst the gun. Now, a powder, to give a good projectile force, must neither burn too slow as in the first case, nor too fast as in the last; so by adding *sulphur* to the first, we increase the intensity of the flame and rate of burning, and by *charcoal* in the second, produce the contrary effect, thus striking the medium and producing gunpowder. The proportion of the ingredients has been arrived at by calculation (by means of their chemical equivalents); and repeated experiments give results almost identical.

Thus,—

	NITRE.	CHARCOAL.	SULPHUR.
Atomic theory,	74.64	13.51	11.85
Practice	76	14	10

Prussia has followed the theory even more closely. Thus,

75	13.5	11.5

COMBUSTION.—The substances resulting from the combustion of gunpowder, are nitrogen and carbonic-acid gas—both highly elastic gases; sulphuret of potassium, which produces the sulphurous smoke, and in part adheres to the sides of the bore. Brought in contact with water, sulphureted hydrogen is produced, causing the offensive smell noticed in washing out a gun.

It has been stated that burning a small quantity of powder is a good test of its quality; leaving no residue when burnt on white paper, therefore, shows the proportions are correct and the powder properly made. Should it, however, blacken the paper, it contains too much charcoal; while a yellow stain shows too much sulphur. Its color should be a dull brownish black or slate, not shining. If black, too much charcoal has been used; and this being the principal absorbent in powder, such a one will soon deteriorate from moisture. The grains should be sharp and angular, as these present a greater surface to the action of the flame, and ignite quicker, than round grains. When thrown from the hand it should leave no dust, which fills up the intervals between the grains, and forming a compact mass retards combustion very much. When the grains are very large, a mixture of fine grains (not dust) is advantageous, as they serve to transmit the flame from one grain to another, and thus assist combustion.

Ignition and *combustion* of powder must not be confounded. Ignition takes place when it begins to develop light and heat, and this takes place from grain to grain with great rapidity, assisted as it is by the gas first produced forcing the flame through the other portion of the grains. Combustion takes place more slowly, and refers to the total decomposition of all the grains, and the total evolution of all their gases.

The larger the grains the more rapid is the ignition, but the slower the combustion. With small grains the ignition is slower, but the combustion much faster; and hence, in small-arms, when small grains are used, the projectile gets the full force of the powder in a smaller space than in large guns.

Powder is a poor conductor of heat; and hence, in a compact mass, as a rocket filled with meal-powder, or a cake, if one end is lighted, combustion will take place very slowly. This shows the importance of graining powder.

Ramming.—In small-arms the charge is rammed; because, besides increasing slightly the distance the ball has to pass over before leaving the gun, it retards slightly the ignition of the powder, which, from the rapid combustion of small grains, has still time enough to develop its full force before the ball leaves the bore. Ramming is not advantageous in large guns; since, as it decreases the ignition, the large grains would not all be consumed by the time the ball reached the mouth of the piece.

Pressure.—Supposing a powder to be formed according to the theory of chemical equivalents, which is nearly in accordance with the practical formula, it is found that 135 grains occupying a space of about 3-10 of a cubic inch, will furnish gases which occupy about 623 times that space. But in an explosion these gases are evolved at a very high temperature, which, of course, increases immensely their bulk, besides converting the sulphuret of potassium into vapor. The volume has been estimated as high as 5,000 times the original bulk; and Gen. Piobert estimates the ordinary maximum pressure of these gases on a projectile at between 2,000 and 3,000 lbs. to the square inch. Figures, however, are merely approximations, and no definite idea can be given of the absolute force of gunpowder. The expansion of powder, however, like everything else, has its limit; and the idea once entertained, that a thimbleful of powder completely confined at the center of the earth would, if ignited, produce a general earthquake, is now exploded. During some recent experiments with concussion fuzes, an 8-inch shell was charged with ½ lb. of fine rifle-powder, and a bronze fuze-plug screwed in. The shell did not explode, but on being recovered and an attempt made to unscrew the fuze plug, it was found impossible to remove it. An attempt to remove an iron screw which stopped a smaller hole used for charging the shell, met with a similar result. By means of cold-chisels and punches a hole was now drilled through; but no sooner was an entrance made than the tool was hurled from the workman's hand across the room, he himself was knocked backward with a feeling on his hand and arm as if burnt, and the escaping gas made a noise like the escape-pipe of a small steam-engine. This powder, therefore, in a gaseous form, had been bottled up for several hours in the shell. The same thing occurred several times, and the workmen soon became more careful in opening recovered concussion shells. A simple remedy was soon found, however, by increasing the charge to 1 lb.

Ignition.—Powder does not inflame instantaneously. If it did, no gun could be made strong enough to resist its force. Its inflammation is gradual and progressive, and in a gun the projectile commences to move, probably, before all the charge is ignited. The development of its force is affected by the proportion of the space which it occupies, to the space around it in the gun. It is found that as this vacant space increases, the projectile force of the powder greatly decreases, while its absolute force or *shock* seems to increase. Thus, for instance, when a ball is not rammed *home*, it is not moved by the first portion of the gas evolved, but it remains stationary until nearly the whole force of the charge is developed, when, acting by its inertia against the immense shock of this force, the violent reaction of the fluid takes effect before the ball moves, and bursts the gun. In the same way, a musket whose muzzle has been stopped with mud, or even snow, has been known to burst just behind the obstruction. As a gun is usually loaded, the ball moving with the first gas generated, its inertia is gradually overcome, it furnishes space by degrees as the development of the gas requires it; no shock is felt, and nearly the whole force of the powder is transferred to the ball instead of the gun.

TESTING.—Besides the tests of the external senses applied to powder, analysis is sometimes resorted to. The simplest method is to dissolve out the nitre with pure water, and the sulphur by the aid of a solution of potash, thus isolating the charcoal. These, when dry, are weighed. Should the solution of nitre precipitate nitrate of silver, the presence of common salt or carbonate of soda would be shown. If it blacken a solution of acetate of lead, hydro-sulphuric acid is present; other chemical tests may be applied to determine the presence or absence of other impurities.

RATE OF BURNING.—Its rate of burning may be determined by noting the time taken for the flame to travel from one end of a uniform groove, in a plank or piece of metal filled evenly with powder, to the other, protecting the flame from the action of winds, &c. The relative rates of two similar kinds of powder may be determined by filling such a groove, circular in form, one half with each kind of powder, and applying a light at one of their points of meeting. Of course, the one which burns past the opposite point of meeting, is the fastest powder. This test is the most satisfactory when applied to small-grained powder, which can be placed evenly in the groove. All the powder used in the United States' service is made in private factories; but before being accepted, is examined and tested by officers of the army, who apply all the foregoing tests, besides applying those in regard to strength. These last have, until late years, been very unsatisfactory and imperfect, being the results of experiments with the *mortar eprouvette*, the inaccuracy of which will be presently shown. Since the adoption of the "gun-pendulum eprouvette," much more satisfactory results have been obtained; but it remained for Capt. Navez, of the Belgian army, to attain the highest degree of accuracy in determining this important element of gunpowder, by the invention of the wonderful electro-ballistic apparatus.

TESTING WITH THE MORTAR EPROUVETTE.—Powder was formerly tested by means of the mortar eprouvette, a small piece used for this purpose alone, and cast with an iron bed-piece (or sole), the plane of which makes an angle of 45° with the axis of the piece. An iron bed-plate, countersunk to receive the piece cast on the mortar, is solidly placed in a level position, so that when the mortar is in its place, its axis has an elevation of 45°. It is carefully loaded with the powder to be tried, and a ball, which is first accurately weighed, and then fired. Being elevated at 45° the range will be the greatest possible for that *quantity* of that kind of powder, and for that piece; but will it be indicative of its strength in any other quantity or piece? It is found that the force of powder does not increase at all regularly as the quantity in the charge is increased, but irregularly, and in a much greater proportion. The result then obtained by using a small charge gives no indication of the effect with a large one, when the heat is so much greater, and consequently the tension of the gases so much increased. The eprouvette mortar being very short, if the powder tried is a very quick one, of fine grain (rifle-powder, for instance), the whole force will be expended on the ball before it leaves the piece, and a greater range will be obtained than with a large-grained, slow-burning powder such as is used in cannon. But experiment has fully demonstrated that this last is the very kind of powder which in practice gives the greatest ranges; so that the mortar eprouvette gives in this case the very opposite of the truth. It has, therefore, been discarded as a proper test for the strength of powder; though, for want of a better, it may be used for determining the relative strength of different kinds, when they are all obtained from the same factory, and it is known that the same ingredients and the same manner of making it have been employed.

The Gun and Ballistic Pendulum.—The only true manner of determining the relative strength of powder is, to measure its effects when fired under the same circumstances as it is to be employed in service, using the kind of piece, the quantity of powder, and weight of projectile, that is to be used in the field. A great improvement has, therefore, been made in the adoption of the pendulum eprouvette; which consists of a gun, usually a 6 or 12 pdr. slung on a horizontal axis in such a way that the axis of the gun passes through the center of oscillation of the pendulum thus formed, and is, when the gun is fired, perpendicular to the vertical plane passing through the axis of suspension. The axis then receives no shock, and knowing the time of oscillation of the pendulum, a formula is easily deduced, by which the *initial velocity* of the shot, or the space passed over in the first second of time after it leaves the bore, is determined. This instrument is called the gun pendulum. The ballistic pendulum consists of a hollow cast-iron frustrum of a cone, slung in the same way as the gun, with its axis in the same horizontal plane as the axis of the gun when at rest. This cone is filled with sand packed in baskets destined to receive the ball fired from the gun pendulum, or from a stationary gun. A graduated arc under each of these pendulums shows, by means of a sliding-pointer, the angle through which they recoil when the first is fired, or the second struck by the ball. This angle is, of course, the element from which the initial velocity is determined. These pendulums may be used together or separately; but when used together they serve as checks on each other, and great accuracy has been attained in determining the force of powder, as well as many other points important to the science of artillery, such as the proper size of the charge, its form, the proper position of the vent, the best length of bore, thickness of metal, form of gun, the effect of wads and sabots on the force and accuracy of the shot. But these data, after all, are true only for the gun which is used in the pendulum, and it by no means follows that points which have been determined for a 6 or 12 pdr. will hold good for a 32 or 42 pdr.; so that for any gun below or above calibers used, the results are simply approximate calculations.

It is evident that by using the gun-pendulum, too large an arc will be obtained, giving, consequently, too great a result, whilst with the ballistic, too small an arc and velocity will be shown; as in the latter case the ball by its weight, will tend to retard the recoil of the pendulum, whilst in the former it passes to the front of the center of gravity, and the gas is still acting on the bottom of the bore to increase the recoil of the gun after the ball has left its mouth, and received its initial velocity. A mean, then, between the two processes should give the truest result. Great pains are taken in weighing and adjusting the powder and ball, measuring the spaces occupied by them in the gun, in adjusting the axis of the gun and pendulum block in the same line, so as to strike the block in as near the same place as possible every time.

To Benjamin Robins is due the invention of these pendulums, which, under the improvements of Hutton, have made such a great stride towards perfection in the determination of the strength of gunpowder. Could the process, with any sort of economy, be applied to guns of all sizes, little else would apparently be left to be desired by the most enthusiastic artillerist; but this great improvement seems, like most others, to have had its day, and to be obliged to give place to one destined, it is thought, to supersede it entirely, and of such accuracy and minuteness in its results as seem to preclude the idea of any farther improvements in the measurement of the flight of projectiles.

NAVEZ'S MACHINE.—It is evident that if any machine can be produced for measuring the flight of projectiles, and can be indiscriminately applied to every gun used as it is in actual service, and be made to exhibit accurately the time occupied by the ball in passing over different parts of the trajectory, the whole problem of measuring the force of gunpowder is solved, to say nothing of the many other important results which follow the possession of such knowledge.

Such an instrument has been invented in the electro-ballistic apparatus of Capt. Navez, of the Belgian army; and one cannot read the accounts of its wonderful operations without being struck with astonishment that such an instrument should have been invented ten or twelve years ago, and adopted, since, into almost every European army, without finding its way into this country, or, at least, being heard of by more than one or two of our most fortunate officers. One of these instruments has already been obtained for the Military Academy, and two more have been ordered for the government; so that it is to be hoped we will before long be enabled, with the rest of the world, to reap some of the advantages flowing from this new and wonderful application of electrical power.

The apparatus consists of three distinct parts:—

1. The pendulum.
2. The conjunctor, or establisher of currents; and
3. The disjunctor, or interrupter of currents.

The first consists, Fig. 4, Plate 1, of a strong vertical plate of brass, L L, to which is attached a pendulum, P, the disk of which is also of brass, but has inserted in its side, at *p*, a small piece of wrought iron. The rod of the pendulum is of steel, and is inserted into a piece of very hard bronze, which serves as an axis of suspension for the pendulum. It is sustained in position by two cylindrical pivots of cast steel, forming the extremities of two screws, one of which is seen in the figure at *v*. The other, which is opposite on the other side of the plate, is held in position by what clock-makers call a *bridge*. The pendulum is hung with great delicacy; the pivots being very small, the screws on which they are placed being very fine-threaded, and counter-screws are provided to fix their position when once adjusted.

The bronze axle passes through another piece terminated by a circle of iron R, and the two are so nicely fitted that when the axis moves, R, by friction, is carried around with it; and this friction may be further regulated by the spring *r*, one branch of which is fixed to the shoulder of the axle *a*, and the other presses against a corresponding shoulder, *b*, on the piece through which the axle passes; by means of the screw *s* these two shoulders are pressed together, and the friction increased, or the reverse. This arrangement is shown in section by figure 5, Plate 1. By this arrangement the pendulum in its movement carries around with it the iron circle R, together with the pointer I, which is fixed to it. The screw *u*, which passes through the pointer and rests against the plate R, serves to regulate the distance of the vernier, V, from the large plate L L. A pin, T, stops the pointer in such a position that the zero of the vernier coincides with the zero of the graduated arc, A B, divided into 150°. The vernier enables us to read to the twentieth part of a degree, or three seconds. An opening is made near the edge of the plate L L, through which passes the end of an electro-magnet, shown at Q. This is mounted in such a way that it can be moved up and down by the screw K. At the center of the plate L L, is placed a circular opening equal in diameter to the small plate R, allowing the entrance of the two extremities of a strong horse-shoe electro-magnet which is placed behind the plate L L. The two ends of it approach each other sufficiently to allow them to fit accurately into the opening in the plate. The interval between them is filled by a plate of brass, pierced to allow the passage of the axle of the pendulum. This brings the back-face of the plate R directly in front of and near the extremities of the magnet.

The instrument is fastened to a solid piece of wood, which is leveled by means of spirit-levels and screws, seen at N, C, D, &c. The whole should be inclosed in a glass case.

THE CONJUNCTOR. *Fig.* 6, *Pl.* 1.

An electro-magnet, E, moves along the column C; its movements being regulated by the screw V. Two strips of copper, bent in zigzags, serve to establish communication between the magnetic wire and the screws 5 and 6.

Under the electro-magnet is a small iron cup, M, in which some mercury is placed. The screw H passes through the side of the cup, and serves to regulate the height of the mercury. Around the cup, and extending a short distance above it, is placed the brass cylinder, O. A copper ribbon or wire, R, connects the cup of mercury with the screw 7. From the screw 8 a blade of tempered steel, L, projects, the end of which being directly over the cup, has an iron point projecting downwards towards the mercury.

A leaden weight, P, surmounted by a piece of wrought iron, is kept in the position shown in the figure by the attraction of the electro-magnet, E, when active. The wooden base to which this apparatus is fixed is leveled by means of three leveling screws, A, B, and C, and a plumb-line hung in the inside of the column, and seen through slits cut along the length of the column.

THE DISJUNCTOR. *Fig.* 7, *Pl.* 1.

Two blades of copper, L, L, separated by a piece of ivory, and kept in position by a stirrup lined inside with ivory, are connected, the one with screw 9, the other with 10 by copper bands running on the under side of the board on which this instrument is built.

Two other similar blades, L′ L′, which are also separated by ivory, form a movable system whose extremity may be introduced between the blades L and L, rubbing them slightly as they enter. Each of these movable blades is connected by the zigzag copper ribbons B B, with a screw on its own side of the platform (11 and 12).

A steel rod jointed to the piece of ivory separating the movable blades, passes through the cylinder, C, and having a thread cut on the end, screws into the knob, E. In the cylinder C is placed a strong spring, which, acting on the movable blades, keeps them separated from the fixed ones. By pressing on the knob E, the spring yields and the movable blades are pushed up between and in contact with the others. The nose of a seer beneath the platform, and pushed up by a small spring, catches in a notch made in the steel rod, and holds it and the blades in that position until by pressing on the trigger at D, the nose of the seer is disengaged and the blades L′ L′ fly back.

These constitute the apparatus of Capt. Navez; which, for use, is placed securely under shelter, where it will be completely protected from the weather.

The pendulum and conjunctor are placed on a heavy table, the feet of which rest on solid ground, and no part of which touches the walls of the building, in order that no disturbance may result to the instruments from the firing of the gun.

The disjunctor is placed upon a small table not touching the other. The working of the disjunctor causes a jerk which might interfere with the operation of the conjunctor if placed on the same table with it.

Figs. 4, 6, and 7, Pl. 1, represent the three instruments in their proper positions.

It is *very important* to place the pendulum in such a position that the oscillating system between its initial point and that of stable equilibrium shall have passed over an angular distance of 75°, corresponding, from the construction of the instrument, to one half the graduated arc. This is effected in the following manner: The pendulum bed is first leveled by means of the leveling-screws. The disk is then raised until the small piece of iron, *p*, comes in contact with the end of the electro-magnet at Q. The pointer carried around by the movement of the pendulum will rest against the pin T, and the zero of the vernier coincide with the zero of the limb. Let go the pendulum, and allow it to come to a state of rest. If in that position the zero of the vernier coincides with the seventy-fifth degree of the limb, the instrument is properly adjusted. If it does not, the electro-magnet is lowered or raised by the screw K, and the operation is repeated until the zero does coincide with the seventy-fifth degree. Before raising the pendulum, see that the pointer is situated somewhere between 0 and 75°. The moment the pointer strikes the pin, the axis of suspension will commence to turn in the washer which supports the pointer, and continue to do so till the disk meets the magnet, Q.

To regulate the *suspension* of the pendulum, the screw behind the plate L L is withdrawn far enough to allow the plate R to come in contact with the electro-magnet behind it. It is then screwed slowly in again until the pivot of suspension acting against the axis of suspension pushes back the oscillating system far enough to separate the plate R from the magnet. The space between the two should not be greater than .004 of an inch. The extremity of the other point of suspension should not touch the bottom of its aperture by about .02 of an inch. The tension of the spring *r* is regulated by the screw *s* in such a manner that the pointer shall be carried around with the pendulum. The vernier is placed very near the graduated limb, by means of the screw *u*.

The conjunctor being placed along-side of the pendulum, its column is placed in a vertical position by mean of its plumb-line and the leveling screws. The mercury in the cup should be pure and brilliant, and have its surface just below the end of the iron point fixed to the steel blade L. This is regulated by the screw H.

The disjunctor is placed on its table by the side of the conjunctor; and the distance the nose of the seer is to enter the notch in the steel rod is regulated by the assistance of a small screw placed under the bed of this instrument. The pressure of the fixed blades on the movable ones may be slightly increased by moving the screw *i*, which acts upon the ivory of the stirrup; but this should not be much used, as it is designed to rectify the effects of use on the instrument. Two galvanic batteries are needed to work the apparatus. Bunsen's are generally used (Fig. 8); and to avoid corroding the instruments from the gases evolved, the batteries are placed outside of the building which contains the instruments. Two target frames, C and C′, Pl. 2, are placed along the line the shot is to travel, and at such distances as to comprise between them that part of the trajectory the time in describing which it is designed to measure.

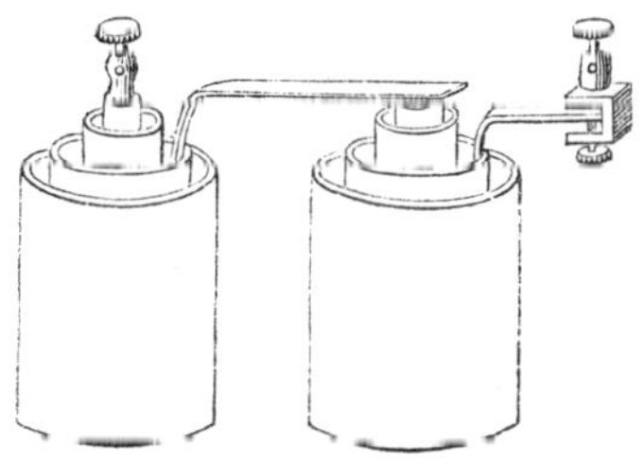

Fig. 8.

The size of these frames will depend upon the distances at which they are placed from the piece, and the accuracy of fire. These frames are covered with copper wire in parallel lines, the spaces between being about two-thirds the diameter of the shot to be used. The wire, which must be annealed, is supported on nails covered with gutta percha (Fig. 9), which are driven along the uprights of the frames. The size of the wire will depend upon the size of the gun used. If cannon is used, it should be about 1-10 of an inch in diameter. If small arms, not more than half that size will suit better.

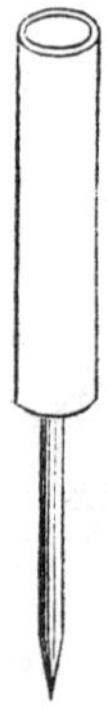

Fig. 9.

With very large target frames it is better to place the lines of wire vertically, in order to avoid injury from their being thrown around horizontally when struck by the projectile; when it is necessary to make these lines very long, a second set, which should be *non-conductors*, is woven into the other in a direction perpendicular to them. Cotton cord covered with a coating of varnish, is very good for this purpose.

The apparatus, batteries, and targets, are connected by means of the smaller wire (1-20 of an inch), which is held in its position by the press-screws of the different instruments. Posts ten or fifteen yards apart are used to support the wire running to the targets, being provided for this purpose with nails covered with gutta percha, each one of which to protect it against moisture has over it a small rectangular piece of zinc, curved downward and fastened in the post as represented in Fig. 10.

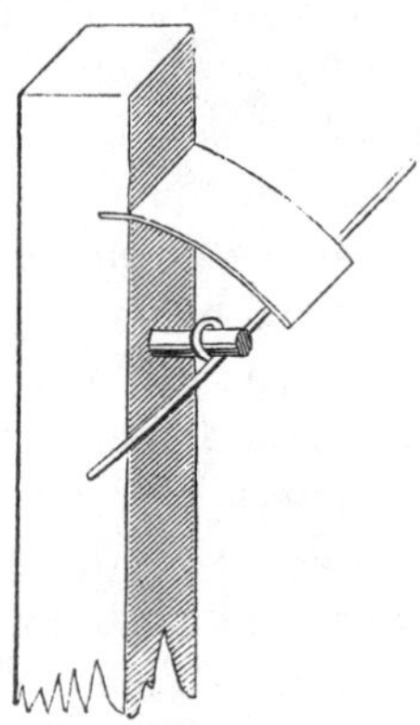

Fig. 10.

Where the conductors pass through the walls of the building, they should be protected by gutta-percha tubes, through which they run.

For the conductor between the different parts of the apparatus, the wire should be covered with silk or varnished cotton. This precaution is necessary to avoid accidental deviations of the currents by the wires touching each other; but outside of the building in which the instruments are placed, it is not necessary.

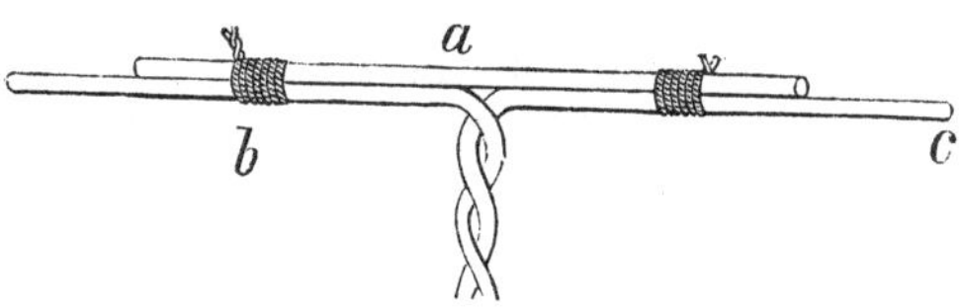

Fig. 11.

The connection between two pieces is made as represented in Fig. 11, and the short splice, *a*, is tied to the main wires, *b* and *c*, with fine wire.

Three currents are established for the working of the apparatus, whose courses will be designated on Fig. 12, Pl. 2.

No. I., leaving the battery, P, reaches the coil of the electro-magnet, Q, through the press-screw, 2, magnetizes that magnet and passes out through the screw 1 to the target C, from whence it returns to the screw 11 of the *disjunctor*, and passes into the *left* movable blade; and as this is in contact with one of the fixed blades (the knob E having been pushed up) the current passes on to the screw 9, and returns to the battery from which it proceeded.

No. II., leaving the battery P′, proceeds through the conductor to C′, returning to the screw 5 of the *conjunctor*, from whence it passes into the coil of its magnet, magnetizing that and passing out at the screw 6, it reaches the screw 12 of the *disjunctor*, passes through the *right* movable and fixed blades and the screw 10 to return to its battery P′.

No. III., leaving the same battery as No. I. (P), passes through the coil of the large magnet of the pendulum, magnetizing it, and passing out by the screw 4, proceeds to the screw 8 of the *conjunctor*. The steel blade carries it to the cup of mercury, from whence it goes to the screw 7, and so back to the battery P.

We will now describe the manner in which the instrument works. The gun is loaded, the disjunctor is not in gear, *i. e.*, the movable and fixed blades do not touch each other, and in the conjunctor, the point on the steel blade does not touch the mercury. None of the currents are established. The operator seated before the instrument, with the right hand puts the disjunctor in gear to establish the currents I. and II. With the finger of the left hand, or the small handle shown behind the disk, he raises the pendulum until the iron *p* in the disk touches the magnet Q, which, being magnetized, retains it there. The zero of the vernier coincides with the zero of the limb. He then suspends the weight P to the magnet of the conjunctor by placing its iron end in the brass tube fixed by friction on the end of the magnet. The iron part of the weight being slightly conical, it is released entirely from the tube by raising this last up a little on the magnet. This leaves the weight suspended by magnetic attraction alone. This manner of putting the weight in position is for the purpose of getting its axis coincident with the vertical axis of the magnet, without which it would not fall accurately in the cylinder O.

The operator now presses upon the trigger, D, of the disjunctor, releasing the movable blades, which, flying back, rupture the currents I. and II. *simultaneously*. The pendulum and weight commence falling as soon as their respective magnets become sufficiently unmagnetized. As soon as the weight strikes the end of the steel blade, this bends, putting its point in contact with the mercury, which, establishing current No. III., the large magnet of the pendulum becomes active, acts on the iron plate R, fixing that, and, consequently, the pointer attached to it. The pendulum, however, continues to oscillate, the axis turning in the muff. See Fig. 5.

Having noted the arc passed over by the pointer, which we will call φ, withdraw the weight from the cylinder O. This releases the steel blade, which, rising, breaks the current No. III., and the large magnet no longer attracts the plate R, which is pulled away from its position against the ends of the magnet by seizing the muff with the thumb and forefinger.

The disjunctor is immediately put in gear, the pendulum replaced in its old position, and the weight suspended again to its magnet. The operator now gives the signal to fire, and the projectile as it passes through the targets C and C′, cuts the wire, breaking in *succession*, first current I. and then II. The apparatus performs as before, and the pointer is found fixed after having passed over a certain arc, φ', which will be greater than the one first noted.

Had the projectile cut simultaneously, instead of in succession, the two wires, would have been equal to φ. The difference, then, between these two arcs $(\varphi' - \varphi)$ will correspond exactly to the time employed by the projectile in passing over the distance between the two targets.

This method, then, consists in making the apparatus work two successive times under *exactly the same circumstances.* The first operation, effected by means of the disjunctor, is, in fact, the same as if the projectile cut both wires at the same time, or the space passed over was nothing, whilst, in the second, effected by means of firing, the space passed over by the projectile is that comprised between the two targets.

Both operations are performed in a few seconds. During this short space of time the intensity of the currents cannot sensibly vary, and consequently the electro-magnetic action should be the same in both cases. All the resistances opposed by the apparatus are the same in the two cases, and the heights from which the weight falls are equal. In the second operation, the time to be measured has simply been added to the time which in the first was consumed by the pointer in passing over the arc φ. It follows then that the arc $(\varphi'—\varphi)$, by which the arc φ is increased, corresponds to the time to be measured.

It becomes useless in this method to regulate the currents, since the times necessary for magnetizing and unmagnetizing the magnets being the same in both operations, can have no influence upon the accuracy of the results shown by the apparatus.

Besides the advantage of ridding the results of the apparatus of the inaccuracies to which all other electro-magnetic pendulums are liable, this method allows the operator to choose the point of the arc corresponding to the commencement of the time to be measured. The arc φ is increased or diminished by raising or lowering the magnet on the column of the conjunctor, since the falling time of the weight P is varied in consequence. The most advantageous position of the arc $(\varphi'—\varphi)$ is evidently that one where the middle of the arc corresponds to the greatest velocity of the pendulum, since it is in this position that the arc corresponding to any given time is the greatest. It can generally be so arranged in the experiments that the center of this arc $(\varphi'—\varphi)$ will be about the middle of the pendulum's course. To show the importance of this advantage, we will take an example. If we wish to measure the 1-1000 part of a second, taking for a starting point the initial position of the pendulum, the arc corresponding to that time will be only the 0.27 of a degree, whilst the arc corresponding to the same time, taken near the middle of the limb, will be 7.53 degrees. *The process, then, in the case supposed, has the advantage of giving, as a measurement of the time, an arc 28 times greater than would be obtained by counting from the initial position of the pendulum.*

The results obtained will be so much the more accurate as the nicety with which the apparatus is constructed increases, and as the skill of the operator improves by practice.

In order to be satisfied of the accuracy with which the apparatus works, it is only necessary to employ it for the measurement of a known space of time comprised between two ruptures made in the currents destined to be broken by the projectile. Suppose we choose the time 0, which corresponds to a simultaneous rupture, that is, an infinite velocity of the projectile. All that is necessary is to operate with the apparatus twice, and the difference between the two arcs indicated by the pointer gives us in time the degree of precision of the apparatus. This gives us an easy method of proving at any moment the proper adjustment of the instrument for use.

The manner of forming the table made use of for determining the time corresponding to an arc $(\varphi'-\varphi)$ will now be described.

The time of a small oscillation of the pendulum will be determined by counting the number of oscillations in a given time by means of a good seconds watch or chronometer. In order to obtain a sufficient number of oscillations, the pendulum must be let fall from a sufficient height. The oscillations will diminish then in size until the movement ceases altogether. But we must find the time of an oscillation so small as to be confounded with the cycloidal arc,—*i. e.*, that the time of oscillation shall be constant, or independent of the size of the arc. It is therefore necessary to take into account the influence of the variable size of the circular oscillations.

By comparing the expression $t=\pi\sqrt{\frac{l}{g}}$, which gives the time of a very small oscillation of a simple pendulum whose length is l, or of a compound pendulum of which l is the distance from the axis of suspension to the center of oscillation, with this,

$$t=\pi\sqrt{\frac{l}{g}}\left(1+\left(\frac{1}{2}\right)^2\frac{a}{2}+\left(\frac{1.3}{2.4}\right)^2\left(\frac{a}{2}\right)^2+\left(\frac{1.3.5}{2.4.6}\right)^3\left(\frac{a}{2}\right)+, \&c.\right.$$

which represents the time corresponding to the circular oscillation, having a for the versed sine of half the arc, we find the increments of duration which depend on the amplitude of the arc of vibration. The time of an infinitely small oscillation being taken as unity, the increments of duration will be—

For an amplitude of 10°, . . 0'.00012
" " 20°, . 0.00190
" " 30°, . . 0.00426, &c.

Classifying the oscillations into groups according to their amplitude, and representing by $n\ n'\ n''$ the number of oscillations in each group; by $a\ b\ c$, the increments of duration, and by T, the whole time of observation; the time of a very small oscillation will be—

$$t=\frac{T}{n\left(1+\frac{1}{a}\right)+n'\left(1+\frac{1}{b}\right)+n''\left(1+\frac{1}{c}\right)+\ \&c.}$$

which will serve to determine the distance from the center of oscillation to the axis of suspension. This distance is found by making use of the formula

$$l=\frac{g}{\pi^2}t^2.$$

Knowing the position of the center of oscillation, the velocity of the pendulum at any point of its course can be determined by using the formula—

$$v=\sqrt{2gy.}$$

y representing the vertical descent of the center of oscillation.

If a, Fig. 13, represents the constant angle of half oscillation, and x the angle variable with y; this last value may be expressed in terms of the first two.

$y=l(\cos(a-x)-\cos a)$; and by substituting

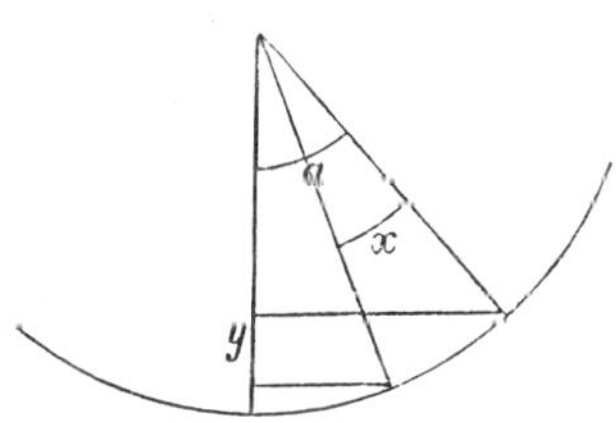

Fig. 13.

$$v=\sqrt{2gl(\cos(a-x)-\cos a)}$$

the velocity of the pendulum will be sought in a series of positions so close together that the movement may be considered as uniform between two consecutive positions. Then by dividing the length of each small arc comprised between two neighboring points by the velocity corresponding to that arc, we will obtain the time employed by the pendulum in passing over respectively each of the small arcs the sum of which will form the part of the oscillation made use of.

As, from the construction of the instruments, the arcs to which the times to be measured correspond, never include the first part of the oscillation, it is not necessary to commence the table of times from the initial position of the pendulum. The positions of the pendulum at which its velocity is calculated, may be the farther apart as the small arcs comprised between them fall nearer the lowest point of the oscillation. Towards the middle of the oscillation the movement of the pendulum becomes almost uniform. These different calculations are applied to the establishment of a table used to facilitate the employment of this apparatus. The elements will, of course, be different with different pendulums and in different latitudes. We will take as an example those employed by Capt. Navez.*

The time of a very small oscillation is found to be 0''.3234.

The distance l of the center of oscillation from the axis of suspension=4.00323 in. (0^m10168).

The time consumed by the pendulum in passing over each degree from the fortieth is calculated. The constant angle of half oscillation =75°.

The time employed by the pendulum to pass over each degree is given by the formula—

* Capt. Navez has since adopted a much more simple method of determining the length of the pendulum. This consists in a much more delicate suspension of the pendulum on knife edges, which, in the practical operations of the instrument, would be quickly destroyed.

By this arrangement the pendulum oscillates through a very small arc about 2,000 times before the motion ceases.

By means of a good chronometer the time corresponding to 500 double oscillations of the pendulum is noted twice in succession, and the mean taken. The time of a very small oscillation is then determined by noting the time for 2,000 oscillations, and correcting one operation by the other; the length of the pendulum becomes known by making use of the expression—

$$l=\frac{g}{\pi^2}t^2.$$

$$t=\frac{2\pi l}{360v}=\frac{2\pi l}{360\sqrt{2gl(\cos.75^\circ-x)-\cos.75^\circ}}$$

in which are substituted the various values of a: 40°, 41°, 42°...75°.

In this way the table on the opposite page is formed. The numbers in the third column of this table are obtained by adding all the times which precede in the second column together.

To show how this table is used, an example will be given. To facilitate the calculations, the arcs passed over by the pointer will be noted in degrees and decimal fractions of a degree. As the vernier indicates to the 20th of a degree, two of the subdivisions will represent a tenth (0.1), and where an odd division occurs it will be noted as half of this, or (0.05).

Let us suppose that the operation with the disjunctor gives $\varphi=44^\circ.65$, and that by firing the piece immediately afterwards we obtain $\varphi'=101^\circ.35$. We wish to find the time corresponding to the difference between these two arcs $(\varphi'-\varphi)$. Find in the table the "sum of the times" corresponding to

the 101st degree = ... 0″.09234

To this must be added the time corresponding to the remaining 0.35. The time corresponding to the 101st degree is 0″.00157, or 0″.0000157 for 0.01; and for 0.35, . . 0″.00055

The sum = . . . 0″.09289

The "sum of the times" corresponding to the 44° is . . . . 0″.00663

The time corresponding to the 44° is 0″.00162 for 0.01, the time will be 0″.0000162, and for 0.65; . . . . 0″.00105

The sum = . . . 0″.00768

The difference = . . 0″.08521

will be the time corresponding to the arc $(\varphi'-\varphi)$ or the time consumed by the projectile in passing between the two targets.

Degree.	Time taken by the pendulum to pass over the de ree.	Sum of the times.	Degree.	Time taken by the pendulum to pass over the degree.	Sum of the times.	Degree.	Time taken by the pendulum to pass over the degree.	Sum of the times.
40	0″.00168		64	0″.00148	0″.03742	88	0″.00149	0″.07260
41	.00166	0″.00168	65	.00147	.03890	89	·00149	.07409
42	.00165	.00334	66	.00147	.04037	90	.00149	.07558
43	.00164	.00499	67	.00147	.04184	91	.00150	.07707
44	.00162	.00663	68	.00147	.04331	92	.00150	.07858
45	.00161	.00825	69	.00146	.04478	93	.00151	.08007
46	.00160	.00986	70	.00146	.04624	94	.00152	.08158
47	.00159	.01146	71	.00146	.04770	95	.00152	.08210
48	.00158	.01305	72	.00146	.04916	96	.00153	.08462
49	.00157	.01463	73	.00146	.05062	97	.00154	.08615
50	.00156	.01620	74	.00146	.05208	98	.00154	.08769
51	.00155	.01776	75	.00146	.05354	99	.00155	.08923
52	.00154	.01931	76	.00146	.05500	100	.00156	.09078
53	.00154	.02085	77	.00146	.05646	101	.00157	.09234
54	.00153	.02239	78	.00146	.05792	102	.00158	.09391
55	.00152	.02392	79	.00146	.05938	103	.00159	.09599
56	.00152	.02544	80	.00146	.06080	104	.00160	.09708
57	.00151	.02696	81	.00146	.06230	105	.00161	.09868
58	.00150	.02847	82	.00147	.06376	106	.00162	.10029
59	.00150	.02997	83	.00147	.06523	107	.00164	.10191
60	.00149	.03147	84	.00147	.06670	108	.00165	.10355
61	.00149	.03296	85	.00147	.06817	109	.00166	.10520
62	.00149	.03445	86	.00148	.06964	110	.00168	.10686
63	.00148	.03594	87	.00148	.07112	111		.10854

The manner of determining the degree of accuracy of the apparatus has already been described. The variations may be limited so that the difference between two successive operations by means of the disjunctor will not be greater than one-fourth of a degree, which from the table corresponds to 0″.00036. When using the instrument, then, for timing the flight of projectiles, the distance between the two targets must be great enough to allow an accidental variation of 0″.00036.

The principal advantages of the apparatus over all others yet invented, consist in its accuracy, the ease with which it is worked, and its cheapness as compared with the unwieldy and expensive machines used at present, in this country, for the testing of gunpowder. But its principal advantage is its applicability to every kind of piece by which the force of the powder as it is actually used in every gun in service can be determined.

Established near a battery, it may be applied to any piece, by a slight variation in the direction of the targets, and the communicating wires.

INITIAL VELOCITIES.—*In proving cannon powder*, the initial velocity of a ball of medium weight and windage, with a charge of *one-fourth* its weight of powder, should be:

From a twenty-four-pounder garrison gun, not less than 1,600 feet.

From a twelve-pounder field gun, not less than 1,550 feet.

From a six-pounder field gun, not less than 1,500 feet.

In proving small-arm powder, the initial velocity of a musket ball, with a charge of 120 grains, should be:

With musket powder, not less than 1,500 feet.

With rifle powder, not less than 1,600 feet.

With fine sporting powder, not less than 1,800 feet; and with the new rifled musket, sixty grains of musket powder, and the elongated ball, about 1,000 feet.

HYGROMETRIC QUALITIES.—The susceptibility of powder to absorb moisture may be judged of by exposing any quantity, as one pound, to the air, in a moist place (such as a cellar which is not too damp), on a glazed earthen dish for fifteen or twenty days, stirring it sometimes so as better to expose the surface: the powder should be previously well dried at a heat of about 140°. Well-glazed powder made of pure materials, treated in this way, will not increase in weight more than five parts in 1000, or a half of one per cent. Such powder kept in casks in a dry magazine will absorb about eight-tenths of one per cent. of moisture. A sample thus kept for fifteen years in a common barrel, was found to lose but nine-tenths of one per cent. in drying.

A more accurate and expeditious method of comparing the hygrometric qualities of different samples of powder, is to expose them to air saturated with moisture. For this purpose, samples of about 1,500 grains weight may be placed in a shallow, tin pan, nine inches by six inches, set in a tub, the bottom of which is covered with water; the pan of powder should be placed about one inch above the surface of the water, and the tub covered over. In this manner any sample of powder may be compared with another of known good quality. Good powder.

made of pure materials, will not absorb more than two and a half per cent. of moisture in twenty-four hours.

RESTORING UNSERVICEABLE POWDER.—When the quantity of water absorbed by gunpowder does not exceed seven per cent., the powder may be restored by drying; this may even be effected in the magazine, if it is dry, by means of ventilation, or by the use of chloride of lime for twenty or thirty days. Quick-lime may be used; but the use of it is attended with danger, on account of the heat evolved in slaking.

When powder has absorbed from seven to twelve per cent. of water, it may still be restored by drying in the sun or in a drying house, but it remains porous and friable, and unfit for transportation; in this case it is better to work it over. In service, it may be worked by means of the rolling barrels, as described for making round powder.

When powder has become mixed with dirt or gravel, or other foreign substances which cannot be separated by sifting, or when it has been under water, or otherwise too much injured to be reworked, it must be melted down, to obtain the saltpetre by solution, filtration, and evaporation.

STORING.—Gunpowder is preserved in magazines specially constructed for the purpose, made fire-proof and dry, and protected by lightning rods, which are attached to *masts* or poles planted from six to ten feet from the walls of the building; the *stem* of the rod need not be thicker than the conductor, nor more than six feet high; but the mast should be of such a height that the point of the stem may be about fifteen feet above the building. The barrels are generally placed on the sides, three tiers high, or four tiers if necessary; small skids should be placed on the floor, and between the several tiers of barrels, in order to steady them, and chocks should be placed at intervals on the lower skid to prevent the rolling of the barrels. The powder should be separated according to its kind, the place and date of fabrication, and the proof range. Fixed ammunition, especially for cannon, should not be put in the same magazine with powder in barrels, if it can be avoided.

In a room thirteen or fourteen feet wide, the barrels may be arranged in a double row in the centre, two alleys two and a half feet wide, and two single rows, six to twelve inches from the walls: in this way the marks of each barrel may be seen, and any barrel can be easily reached. In a room twelve feet wide, an equal number of barrels may be placed in two double rows, with a central alley of three feet, and two side alleys, next the walls, of about ten inches each; there should be an unincumbered space of six or eight feet at the door or doors of the magazine.

Should it be necessary to pile the barrels more than four tiers high, the upper tiers should be supported by a frame resting on the floor; or the barrels may be placed on their heads, with boards between the tiers.

Besides being recorded in the magazine book, each parcel of powder should be inscribed on a ticket attached to the pile, showing the entries and the issues.

Ventilating Magazines.—For the preservation of the powder, and of the floors and lining of the magazine, it is of the greatest importance to preserve unobstructed the circulation of air, under the flooring as well as above. The magazine should be opened and aired in clear, dry weather; the ventilators must be kept free; no shrubbery or trees should be allowed to grow so near as to protect the building from the sun. The moisture of a magazine may be absorbed by chloride of lime suspended in an open box under the arch, and renewed from time to time; quick-lime, as before observed, is dangerous.

The sentinel or guard at a magazine, when it is open, should have no fire-arms, and every one who enters the magazine should take off his shoes, or put socks over them; no sword, or cane, or anything which might occasion sparks, should be carried in.

Transportation.—Barrels of powder should not be rolled for transportation; they should be carried in hand-barrows, or slings made of rope or leather. In moving powder in the magazine, a cloth or carpet should be spread; all implements used there should be of wood or copper, and the barrels should never be repaired in the magazine. When it is necessary to roll the powder for its better preservation and to prevent its caking, this should be done, with a small quantity at a time, on boards in the magazine yard.

In wagons, barrels of powder must be packed in straw,

secured in such a manner as not to rub against each other, and the load covered with thick canvas.

In transporting powder an escort should always be sent with it, the number of men depending on circumstances. With wagons, a man is attached to each one, and the commander of the escort frequently inspects them. Pavements are avoided as much as possible, and the train is marched in single file, and usually at a walk. No smoking is allowed near the train. Towns and other inhabited places are avoided if it can be done without making too great a détour. If this cannot be done, the pavements, if the weather is warm and dry, should be watered. If forges are with the trains their fires should be put out; and in case powder is found sifting from any of the wagons, they are placed in the rear, and fifty yards apart.

In camping, care should be taken to place the wagons by themselves, and away from the camp fires. A special guard is placed over them, and strict orders given to allow no unauthorized persons to approach.

GUN-COTTON.

In 1846 the important announcement was made of the discovery of a substitute for gunpowder, possessing, it was said, all the advantageous properties of the latter, with none of its defects.

The announcement was first made by Prof. Schönbein of Bâle, Switzerland, who attempted to keep the preparation a secret; but the same or similar compounds were soon after discovered by different chemists in Europe, and the whole military world was set to work making experiments on this new-found material, which was destined, according to some, to supersede entirely the use of gunpowder, thrusting this last into the class of things which, having had its day, must give way to the march of improvement and later inventions.

Preparation.—To prepare it, well-cleaned, ordinary cotton is steeped for about half a minute in highly concentrated nitric acid. It is then washed several times in pure water and dried, when it is ready for use.

Characteristics.—Gun-cotton thus prepared explodes like ordinary powder, when struck a sharp blow, or when brought in contact with a coal of fire. It burns at about 380° F.; and will, therefore, not set fire to gunpowder when burnt in a loose state over it. By varying the mode of preparation, it may be made to explode at a much lower temperature; and great care should, therefore, be taken in drying it.

In the first experiments made, its projectile force was found to be so great as to favor the idea that for military purposes it was far superior to gunpowder; it being found that in moderate charges its force was equal to that of about twice, or according to some experimenters three times, its weight of the best powder. But by further experiments it was found that its explosive or bursting force is much greater than that of ordinary gunpowder, resembling more the action of fulminates. It is, therefore, in its effects on guns much more injurious than powder; though for mining purposes it is well adapted, especially in sieges, as, burning without any smoke, the workmen in the galleries, etc., would be less incommoded, though the acid vapors resulting from the combustion of gun-cotton may prove as objectionable as smoke. From the rapidity of its action, too, it is suitable for loading shells, which are burst into a much greater number of pieces than when loaded with five times the amount of ordinary powder.

When compressed by hard ramming, as in filling fuses, it burns slowly.

It is more liable than powder to absorb moisture, by which its force is rapidly diminished; but by drying it is immediately restored with but slight diminution of strength, possessing thus one great advantage over ordinary powder, which is very difficult to restore.

When well prepared, it leaves no perceptible stain when a small quantity is burnt on white paper. The principle residua of its combustion are water and nitrous acid; which last is very injurious to the gun, soon corroding it if not wiped after firing.

Velocity.—In the experiments made at Washington in 1846, it was determined that sixty grains was the proper charge to give the requisite force to the musket ball for service; and to test its applicability to our muskets, in which two or more cartridges are frequently inserted by mistake, a barrel was loaded with 120 grains, a ball and wad. On being fired it burst at the breech.

Another barrel was then loaded with two charges of sixty grains each, one ball, and a wad. This barrel also burst, blowing out of the opening the upper charge unburnt.

A third barrel was loaded with a charge of sixty grains, two balls, and two wads, and by the discharge was somewhat swelled at the seat of the shot, and on being fired again with the same load burst.

These barrels had all borne the regular proof charges at the armory; and the results fully demonstrate that the use of gun-cotton is unsafe in our present small arms; nor is it probable that the advantages resulting from its use will compensate for the increased weight necessary to be given the piece to guard against the accidents of service.

The experiments in this country with cannon have not been quite so extensive as those with small arms; but it is reported that very extensive ones have been lately carried on in Austria, that they have been very favorable, and that the arsenals of that country are now engaged in casting short, thick howitzers for the express purpose of using gun-cotton.

In Major Mordecai's experiments with a twenty-four-pounder, it was found that two pounds of gun-cotton gave the same initial velocity (1,422 feet) as four pounds of good cannon powder; and by analogy it is calculated that to produce the same initial velocity as is furnished by one pound of cotton, would require two and a quarter pounds of common powder, which is about the same proportion as in the musket.

Gun-cotton, if not designedly colored, has nothing to distinguish it in appearance from ordinary cotton; it would, therefore, to avoid accidents, be well to dye it, which can be done without injuring its ballistic properties. Compressed in the hand it produces a peculiar crackling sound, resulting from the fact that the oily matter which imparts to raw cotton the same kind of adhesion as exists in wool, has entirely disappeared by the action of the acids.

Space.—Ordinarily, gun-cotton occupies double the space of the same weight of powder, though by compression it may be made to occupy only about one-third that occupied by the same weight of powder.

Heat.—The influence of heat on gun-cotton is much more marked than on gunpowder. Sulphur commences to evaporate at 35° R. (110.75 F.), but so slightly that ordinary powder may be submitted to that temperature without any loss in weight or effect being perceived. Powder inflames at 240° R.; whereas gun-cotton exposed to from 10 to 12° R. (54.50—59° F.) for a few days is decomposed; the gas (oxide of nitrogen) escapes; and combining with the oxygen of the air forms gaseous nitric acid; and this takes place in a few hours at a temperature of 35°R. or more. Gun-cotton inflames at a temperature only about one-half that necessary to inflame gunpowder; and in one particular case, it has taken place at a point much below the boiling point of water. It is also more easily exploded by a blow than gunpowder.

COMPRESSED.—If gun-cotton is compressed so as to diminish its volume more than one-fifth its ordinary bulk, its combustion is retarded very much, and its projectile effect may be entirely destroyed. Pressed into a metallic tube, suspended as a pendulum, it burns very slowly and without imparting any motion to the pendulum, whilst under the same circumstances gunpowder causes considerable oscillation. A piece of gun-cotton held tightly between the fingers, or wrapped closely with paper and lighted, burns up to the fingers or paper and then goes out.

COST.—It is estimated that the cost of manufacturing gun-cotton, including the cost of the ingredients, is three times as great as that of ordinary gunpowder; but before it can be made up into ammunition it has to undergo another operation, which will increase the cost considerably. The charges of ordinary gunpowder are not *weighed*, but *measured* with instruments prepared for the purpose; but as gun-cotton cannot be so measured, the charges have to be all weighed, which will take ten times as long, require a large number of accurate balances and other utensils which must be carried into the field, to say nothing of the necessity for the employment of reliable and skillful men.

The principal advantage claimed for gun-cotton is that a charge of it produces as much effect as three times the amount of powder; but this claim loses much of its force when it is remembered that the two charges occupy about the same amount of space, the volume of one pound of gun-cotton being equal for instance to that of three pounds of gunpowder, so that in regard to *space* nothing is gained by this reduction in weight, which, moreover, in the whole armament of an ammunition wagon would be very little.

CHAPTER II.

ARTILLERY.

By the term *artillery*, is meant all fire-arms of large caliber, together with the mechanical machines and implements used with them. It also refers, in a technical sense, to the art of constructing, preserving; and using, all kinds of machines and munitions of war, as well as to the particular troops which perform these different duties. In the United States' service, these duties are distributed between two different corps, the artillery and the ordnance; the former being entrusted with the use of the arms and munitions, and the latter with their construction and preservation.

Ordnance.—The term *ordnance* is applied to the guns themselves; and, in the United States' service, the ordnance is divided into *guns*, *howitzers*, and *mortars*.

Form.—The form originally given to ordnance was that of a truncated cone, both within and without, similar to the common household *mortar*, which is supposed to have first suggested it. This was first used as our present mortars are, stone balls being employed. It gave, however, but little velocity or accuracy, from the fact that as soon as the ball moved from its position, the greater part of the force of the powder was lost by the gas escaping around its sides, and no certain direction could be given to it in the bore of the piece. The escape of gas was so much greater, too, as the powder then used was in the form of *dust*, and burnt very slowly.

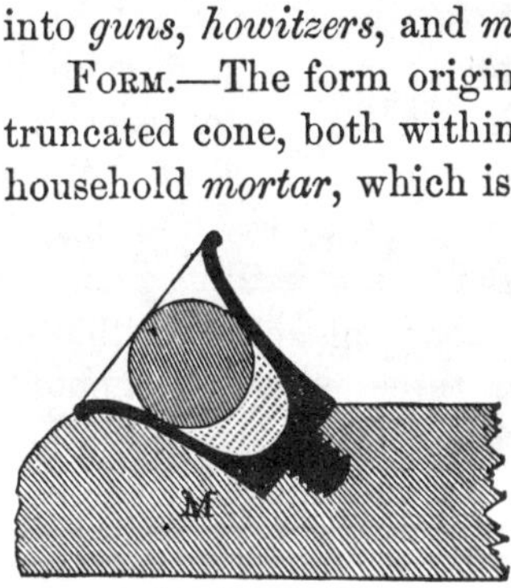

Fig. 14.

To get the full force of the powder, the bore was made nearly cylindrical, from four to eight times its diameter in length, and terminated by a long narrow *chamber*, the object of which was to make it more difficult for the gas to escape; and this was further increased by stopping the mouth of the chamber with a wooden tompion driven down against the powder. The bore was finally made perfectly cylindrical, the piece called a *stone mortar*, and used to fire nearly horizontally like our present guns and howitzers. Fig. 15. From the earliest times, the diameter of the projectile, or that of the bore of the piece, has been used as a unit in the construction of the pieces. These ancient pieces were originally made by placing together bars of iron, and hooping them like the staves of a barrel. They were then made of *forged*, and finally of *cast iron*. *Bronze* pieces were used as early as 1350.

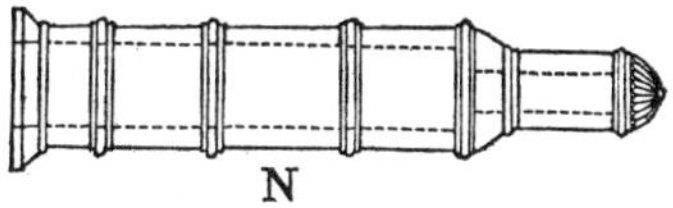

Fig. 15.

BREECH-LOADING.—Loading at the breech was early attempted; and for this purpose a rectilinear enlargement was left in the piece, in which was inserted a box containing the powder, kept in its place by a wedge and key, Fig. 16. But this construction was found wanting in solidity, and was soon abandoned.

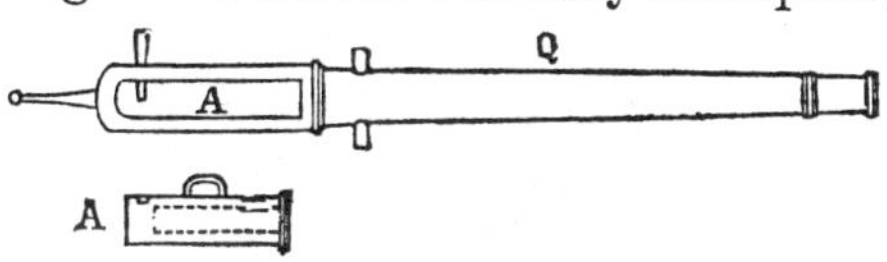

Fig. 16.

LENGTH.—It was thought, reasoning from the analogy between small-arms and cannon, that the longest guns were the most effective; but the question of mobility and the difficulties of fabrication caused it to be laid down as a principle, that the largest calibers should be proportionally the shortest and lightest, so that the small calibers approached more nearly in dimensions the portable arms. Some, however, were made very long, and one is mentioned 26 feet in length, carrying an 18 lb. ball.

When powder instead of being used in the form of dust was *grained*, its rate of burning being much increased, the very long pieces lost their advantages; for the nearer the combustion of a powder approaches to being instantaneous, the less necessity there exists for having the gun long in order that the ball shall receive the full effect of the powder before it leaves the mouth.

Under Charles V. an enormous piece was cast at Genoa, which

was fifty-eight calibres long, and carried a ball thirty-six pounds in weight. It had less range than an ordinary twelve-pounder, and was about being re-cast, when the expedient was adopted of cutting off from the muzzle, first, eight calibres, then six, and then one; and the range was found to increase as the piece became shorter; which shows that there is *for each piece a maximum length, which should not be exceeded.* The reason of this seems to be, that for any increase of length over what is necessary to insure the total combustion of the powder before the ball leaves the muzzle, a loss of velocity follows from the friction of the ball against the sides of the bore. This limit, from experiment is found to be about twenty-four times the diameter of the bore. In the United States' service the length of the pieces varies from fourteen to twenty-three calibres.

CALIBRE.—Each power at first adopted a series of calibres or weight of ball, decreasing in geometrical progression. The first was 32, 16, 8, 4, 2, forming the French calibres, and the second, 48, 24, 12, 6, 3, 1½, forming the German. In time, each nation adopted one or other of these series. This was found necessary from the immense multiplicity of calibres, and the want of uniformity among them.

This large number of calibres being found very disadvantageous in war, the number was reduced to five by Charles IX. In 1732 France adopted her first regular system of artillery, but it was too heavy for field service, and the teams being harnessed in single file, it was very difficult to move at a rapid pace.

In 1765 this system was replaced by that of Gribeauval, who introduced many improvements, especially in the carriages, and formed a distinct artillery for service in the field, lighter than the old, and drawn by double teams. Since that period the French calibres have varied but little, the precarious condition of politics in Europe rendering a total change at any time dangerous, to say nothing of the immense expense.

MORTARS AND HOWITZERS.—Short *chambered* pieces, fired under large angles, similar to our present mortars, are of a very early date, and were used in Flanders in 1478 to crush and set fire to buildings in sieges. The attempt was also early made to throw hollow projectiles filled with powder, and carrying with them a lighted match to explode them. According to Villaret such projectiles were made use of at Bordeaux in 1452; but it is probable that the accidents consequent upon their imperfect construction, caused them to be very soon abandoned. The fuze was lit by a man standing near the muzzle, at the same time that the charge was fired. General Cotty says the Turks made use of them at the siege of Rhodes in 1522. A work of Tartaglia (1537) represents an inflamed shell thrown from a short piece similar in every way to our mortars. It was not until 1634 that these projectiles were introduced into France, giving birth to the present mortar.

Efforts were also made to throw hollow projectiles from stone mortars and guns; but the loading was so difficult, and so many accidents happened, that the attempt was abandoned, until the Dutch, having reduced the pieces to such a length that the charge could be placed in by hand, again brought them into use. The pieces from which they were fired were called *howitzers*, from the German haubitz, and were introduced into France in 1749. They were very similar in appearance to the present eight-inch siege howitzer of our service.

The Russians continued to make use of long howitzers, after finding means to load them without danger, calling them *Licorns*. Almost all European powers adopted them about 1826.

Construction.—The general principles upon which cannon of the present day are constructed, are the results of experiments and calculation. In place of the extremely heavy pieces, the service of which was so difficult, and the small pieces scarcely more formidable than small arms, and the employment of which did harm by decreasing the *moral* effect of artillery, a certain number of mean calibres has been adopted, regard being had to their destination. The maximum effect to be produced under every circumstance of war has been determined, by which was obtained the maximum calibre for each service.

For the inferior limit, a calibre large enough to act powerfully, both physically and morally, is taken. Hence results, that all pieces possess the necessary efficacy and degree of mobility required in the different circumstances under which they are used, and that the number of calibres is reduced as much as possible; which facilitates the equipment, and render replacements in the field easy.

Materials.—Pieces for service in the field are generally made of *bronze*, or gun-metal, a mixture of one hundred parts of copper and eleven of tin. This mixture possesses great tenacity; which allows a diminution in thickness and an increase in length for a given weight, a matter of considerable importance in the manufacture of guns.

By increasing the proportion of tin in bronze it becomes harder, but more brittle and fusible. If the tin is decreased, the metal becomes too soft and loses a part of its elasticity. Between eight and thirteen per cent. is generally used.

Bronze is almost entirely free from any ill effects resulting from exposure to air and moisture. It has, however, the disadvantage of being a very dear material for guns, being seven or eight times more costly than cast iron; the price of bronze guns in this country being about forty-five cents per pound, whilst that of iron is only about six or six and a half cents per pound.

The bore in bronze guns is more apt to be injured by the shot than in iron guns; as, the iron being much harder than bronze, when the shot strikes against the sides of the bore the indentation is greater than in the iron gun.

Naval guns, and those used in garrisons and sea-coast defenses, are generally made of cast iron; which, besides being cheaper, is not furrowed and cut up by the shot, and the pieces preserve their accuracy of fire to the last. Iron has the disadvantage of being less tenacious than bronze, rendering necessary increased thickness for strength, and decreased length for a given weight. It has no elasticity, in consequence of which the metal gradually loses its force of cohesion from firing, and without giving any sign of danger the piece bursts unexpectedly, even when fired with light charges. Iron is also very liable to rust.

These disadvantages become greater as the quality becomes worse. In Sweden, where the material is excellent, all pieces are made of iron. So also are all United-States guns and howitzers, except those used in field artillery. In France, both field and siege and garrison guns are made of bronze, which is also generally the case in Mexico.

Attempts have been made to make bronze pieces with cast or wrought iron bores, but have failed; the difference in tenacity, elasticity, and expansion causing separation between the envelopes. Various efforts have been made to use forged iron; but the difficulty of welding it completely, especially with large masses, the high price of manual labor, the want of elasticity of the metal, and the want of hardness relative to cast iron, render very doubtful the advantages claimed for guns made of it by some authors; at the head of whom stands the celebrated Mr. Mallet, whose valuable work on cast iron contains a mass of interesting and instructive information for the artillery student. Forged guns usually fail from defective welding, especially after use and exposure to the air and moisture; the least opening gives rise to rust, which of course soon weakens the union between the different parts of the metal. This peculiar weakness of wrought iron will be better understood when it is stated that there is no iron of so close a texture as to prevent gases or even fluids from passing through its fibres. Grease will penetrate through the sides of an iron pot; and the finest steel blade, if oiled and bent, and the oil wiped off clean, will, when allowed to resume its natural position, exude particles of oil, thus showing that the fluid has penetrated the pores of the metal.

The Bore.—The bore of guns is smooth, though many attempts have been made, by cutting grooves in them, to produce rifle motion in the shot; a much more difficult problem than in small arms, since the metal of cannon balls is too hard to be forced into the grooves as in a common rifle.

Richochet firing plays an important part in warfare, since by the bounding along of the ball on the surface of the ground, points are reached not otherwise attainable by direct fire. The best form for richochet firing being the spherical, a ball of that shape having a girdle around it on one of its great circles similar to the English rifle-bullet, would present some of the advantages of a rifle-ball, without losing all those of one fired from a smooth bored gun. But the great difficulty is in compelling the girdle to follow the groove in the gun. If it becomes in the least jammed the progress of the ball is arrested, and with the immense force of the expanding gases behind it, the gun is very apt to burst. Though, in this particular method of rifling *balls*, the danger is less than with elongated projectiles (as in a sphere the centres of gravity and figure are but slightly separated), the force of rotation required to retain the ball in the vertical plane of fire need not be so strong as with elongated projectiles, and the inclination of the grooves may be reduced sufficiently to allow the girdle to follow them, even with strong charges.

When elongated projectiles were used, buttons were placed on each side of the cylindrical portion, which fitted into the grooves of the gun and produced rotation. But great care had to be taken in placing them in the gun; and, during some experiments at West Point, in 1858, a gun of the kind was burst, probably either from the shot getting a little out of its proper direction, and becoming jammed in the bore, or from too great a reduction in the windage.

No difficulty is found in cutting the grooves in the gun, thanks to the marvelous improvements made in machinery, and they can be made of any number, size, shape, or curve, with ease. The only difficulty being in making the shot follow them,—a subject which will be more fully discussed under the head of "*Projectiles.*"

The friction of iron against bronze being one of the lightest known, bronze pieces would make the best rifles for firing the kind of projectiles just mentioned.

Exterior Form.—Cannon are in general of a truncated conical form, with a cylindrical opening along the axis to a certain depth. The strongest part surrounds the seat of the charge. The form is one adapted to the effects of powder, the gases of which act equally in all directions, and with a decreasing force as the projectile moves towards the mouth of the piece. The rate of this decrease, however, was comparatively unknown until the late Col. Bomford, of the Ordnance, succeeded in demonstrating an approximate law by an ingenious device. Commencing near the muzzle of an ordinary piece, a hole was bored perpendicular to the axis. In this a pistol-barrel was securely screwed, and a bullet inserted in it. The velocity of the bullet blown out by the discharge of the gun was measured, and similar experiments being made in succession at different points along the piece, a series of velocities increasing toward the breech was obtained, showing the relative force of the powder at the various points, and giving an indication of the requisite thickness of metal. On the data thus obtained, our first shell-guns, or columbiads, were formed.

Weight.—The weight of a piece is determined by that of its projectile, and the maximum velocity necessary to give this, according to the service in which it is to be used. It is evident that in cannon, as in small-arms, the piece is acted on by a greater effort than that which throws the projectile forward; the difference in the result being due to the weight and friction in the gun. The recoil of the gun will, therefore, become greater as it becomes lighter, and the reverse.

The velocities vary inversely as the masses ($V : v :: m : M$); therefore by making the piece 300 times heavier than the ball, the recoil will be scarcely 1-300 of what it would have been had the piece and ball been of equal weight. Take the case of an iron 12-pdr. (300 times heavier than its ball) which, giving with 3 lbs. of powder an initial velocity of 1742 ft., would recoil $\frac{1742}{300} = 5.8$ ft., a velocity not too great for the preservation of the carriage, which, by its friction on the platform or ground, decreases still farther the recoil. Had the piece been but 100 times heavier than its shot, as is about the case with the 12-pdr. *field*-gun (121 times), the velocity of recoil would have been three times as great, or 17.4 ft., and the carriage would be destroyed unless the charge was diminished; and in field artillery this is done.

The weight of the ball and the velocity it is to have, being given, the weight of the piece becomes known; and then the metal is distributed along its length, in the way that experiment proves is most suitable to resist the force of the powder. The piece is made thicker as the tenacity of the metal decreases.

FORM.—If the tension of the gas diminished at a uniform rate, the surface of the piece might be continuous; but it is not so, and the piece presents several truncated cones, forming offsets, which are finished off with mouldings. These cones are called *reinforces;* the one next the breech being the first, the next one the second reinforce, and the third the *chase.* The different thicknesses are so calculated as to offer sufficient resistance to the force of the powder and the shock of the projectiles.

If the action of the gas alone was considered, it would be necessary to give but little thickness towards the muzzle. But the shocks of the projectiles glancing along the bore would soon injure the accuracy of fire of a piece so constructed, and a thickness greater than is required merely to resist the force of the powder is necessary.

The *chase* is terminated by the *swell of the muzzle* in guns, and a *muzzle band* in howitzers, an increase of thickness which is found necessary to prevent the mouth from splitting from the shocks of the projectiles.

The interior of guns, called the *bore*, is perfectly cylindrical, and has a diameter greater than that of the shot; the difference between the two is called the *windage*, the object of which is to insure the service of the piece, and prevent any difficulty in introducing the projectile, either from the formation of rust, or from heat, as in firing *hot shot.*

WINDAGE diminishes the accuracy of fire, weakens the effect of the charge by allowing an escape of gas, and is the principal cause of deterioration in cannon. It is therefore of importance to make the windage as small as possible compatible with ease and efficiency in loading.

LENGTH.—The weight of a piece and its thickness being given, its length is determined; yet in some cases a certain length is indispensable, as in the case of embrasure guns, where the muzzle must project so far into the mouth of the embrasure as not to injure the cheeks. Then the question of length is more difficult to solve.

At first sight it would seem, that the longer the piece the greater would be the velocity impressed on the ball; but the shocks and friction of the latter counterbalance, and might end with exceeding, the action of the motive force. For this reason *bronze* pieces are not made longer than 24 or 25 calibers. Iron dieces, in order to render them movable, and in consequence of the want of tenacity in the metal are made generally shorter, though our 12-pdr. *iron* gun is longer than the 12-pdr. bronze field-piece, an anomaly which probably results from the necessity for length in the first piece, and lightness in the second. The limit of length appears to be more extended for small calibers than for large; and for very dense projectiles than for those of little density.

CHAMBER.—When a light piece is intended for throwing large and heavy projectiles, the effect of recoil may be lessened by employing small charges of powder; but as they would be difficult to manage, and would form but a small mass retained with difficulty in its proper place in the gun, a cavity called a *chamber* is made in the bottom of the bore, designed to keep the powder together in its proper place, and by keeping it more more confined, increasing its efficiency.

There are three kinds of chambers: the *cylindrical* (A), Fig. 17, conical (B), and spherical (C).

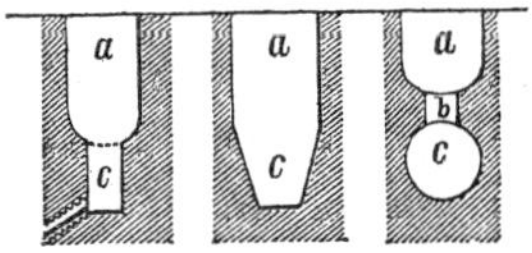

A B C

Fig. 17.

Cylindrical.—In the first, the bottom of the bore at the mouth of the chamber is formed of a portion of a sphere, so that the projectile closes the mouth of the chamber. This is the chamber used in our different howitzers, and in the eprouvette mortar. It is, however, joined to the bore in the way above described only in the eprouvette, and in one of the howitzers (the 8-in. siege). In the other howitzers it is connected by means of a conical surface, the junctions being rounded off to prevent being worn away by the action of the powder. Cylindrical chambers, when narrow and deep, give greater ranges than shallow wide ones, which do not confine the powder so much, but as in the former the gas acts on but a small segment of the projectile (usually hollow), it sometimes breaks it; and, for this reason, too great a depth in cylindrical chambers is avoided.

The *Gomer* chamber (after its inventor) consists of the frustrum of a cone connected with the bore by a portion of the surface of a sphere. This kind of chamber is considered the most advantageous for mortars, and is used in most of ours. Being large at the mouth, it allows the powder to act on an entire hemisphere of the projectile, and no risk is run of breaking it. It, however, gives a less range than either of the others, but its capacity is greater.

The *spherical* chamber consists of a sphere, joined to the bore of the piece by means of a small cylinder which serves as a channel to the gases. As this cylinder decreases in diameter, the gas finds more difficulty in escaping, and greater force is developed. This chamber, however, is even more objectionable than the cylindrical one, from its liability to break the projectiles, although it gives the greatest range of the three chambers. This chamber is, in the first place hard to make, and, in use, soon becomes foul, and is very difficult to clean out. It is now entirely given up.

With pieces which are fired nearly horizontally, and loaded

with a rammer, the cylindrical chamber is the best, and being connected with the bore by a conical surface, the charge slides up into the chamber without any difficulty. In the 8-in. siege howitzer, which is short enough to be loaded by hand, this connection is not necessary, and it is made spherical for the purpose of using the shell without a *sabot*. The advantages of which will be mentioned hereafter.

For small charges a chamber is always advantageous, no matter what the length of the piece; but the advantages are more marked the shorter the piece is made. In proportion as the piece is lengthened and the charge is increased, the influence of the chamber becomes less marked, until with a bore of 10 or 12 calibers long and a charge 1-7th the weight of the projectile, it ceases to be appreciable. In *guns* where the charges are large, a chamber would diminish the velocity by lengthening and confining the charge, thus retarding its inflammation.

In all our guns and howitzers, the bottom of the bore is a plane surface perpendicular to the axis of the piece, and joined on to the sides of the bore by an annular surface whose cross-section is part of the circumference of a circle described with a radius of 1-4th the diameter of the bore or chamber at that point. This construction, which facilitates the boring and cleaning of the gun, adds somewhat to the solidity at the breech.

The vent is the cylindrical channel which serves to convey fire to the charge. It is, in our service, 2-10ths of an inch in diameter, and is calculated in such a manner that the escape of gas shall not be too great, and that the priming-tubes, wires, &c., used with the pieces, may be large enough for effective service. In general, the vent enters the bore at a distance from the bottom equal to 1-4th the diameter of the bore; that is, just where the curved surface joining the two commences. It usually makes an angle of 80° with the axis of the piece, which, besides giving a greater thickness of metal to pass through, makes it more convenient to prime.*

In bronze pieces, the heat and current of gas through the vent melts out and carries away the *tin*, rapidly widening the

* This inclination of the vent is objectionable when friction-tubes are used, by rendering it easier to pull the tube out of the vent. The vents of the new guns, 12, 64, and 128 pdrs., have in consequence been placed perpendicular to the axis.

aperture. For this reason the vent is made in a *vent-piece* of pure copper screwed into the gun, which enables us, when the vent becomes too much enlarged, to replace it by another. When the charges used are very small, and the projectile not very heavy, no vent-piece is used. In iron pieces the vent is made in the metal itself.

With the usual charges, the influence of the position of the vent on the force of the shot or recoil does not appear to be great, whether it enters at the middle or the bottom of the charge. In firing small charges, however, the latter position is the best, so that every portion of the cartridge-bag may be expelled. Should the vent enter too far from the bottom of the bore, the priming-wire and fire might miss the cartridge by passing to the front of it.

TRUNNIONS.—Pieces are mounted on their carriages by means of *trunnions*. These are cylinders cast on the gun, having their common axis perpendicular to that of the gun. Their size should be proportional to the force of recoil, and their position is determined in such a way as to weaken or favor the recoil according to need. In long pieces they are placed slightly in front of the center of gravity of the piece, in order that the breech may always remain resting upon the elevating apparatus, for convenience in sighting, &c. In mortars, they are now placed entirely behind the piece, so that the chase of these rests upon their elevating quoins, which are placed in front.

It is on the position of their axis in reference to that of the gun that the recoil depends to a certain extent. If this is *below* that of the piece, A, Fig. 18, the recoil will be diminished; for, being transmitted along the axis of the piece, *b b*, and the carriage resisting, by virtue of its inertia, the tendency would be to turn the piece around the axis of the trunnions, with a force proportional to the arm of the lever, *a b*, and this, pressing down the breech, forces the trail to the ground, and diminishes the recoil.

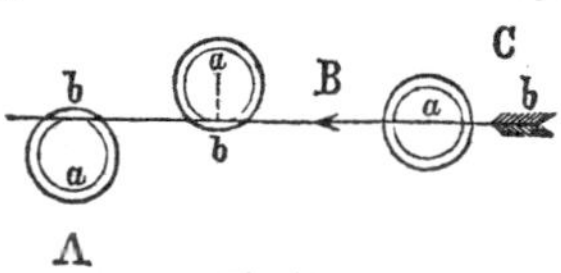

Fig. 18.

If the axis is *above* that of the piece, B, the tendency to turn around the trunnions is in the opposite direction. This relieves the trail and favors the recoil, but the carriage, and especially the axle-tree, experience less shock. These two positions are used by some nations, according to circumstances.

In all our pieces, the axis of the trunnions is in the same plane with the axis of the piece, C, which transmits the recoil directly to the trunnions.

RIMBASES.—The trunnions are joined to the piece by larger cylinders called *rimbases* which serve to consolidate them and keep the piece in the same position on its carriage.

KNOB.—Guns and howitzers are terminated at the breech by a knob or ball, called the *knob of the cascable*, which, placed at the end of a contraction called the *neck*, gives a point of attachment or support in manœuvering the piece. Without this arrangement the piece would offer but little hold for the action of forces.

HANDLES.—Some pieces, as our 12-pd. field-gun, and 24-pd. and 32-pd. field-howitzers, have *handles* placed over the center of gravity, to assist in manœuvering them. The pieces mentioned have *two*, which are cast with them, and are rounded off on the inside so as not to cut a rope, when used as a means of suspending them. Mortars, when any are placed on them, have but *one*. They are not usually placed on iron pieces, as, unless they are very large and clumsy, the metal is too fragile to support the weight of the gun. Small bronze pieces, as the 6-pd. field-gun, 12-pd. field-howitzer, and 12-pd. mountain-howitzer, have none.

Formerly guns were covered with all sorts of ornaments, coats of arms, devices in relief, &c., which, besides increasing the cost of them, were the cause of weakening the pieces. They are now very plain, and marked after their reception into the service.

UNITED STATES' ORDNANCE.

ORDNANCE for the land service is divided into three classes, according to form and the kind of projectile used with each, viz.: *Guns*, *Howitzers* (including *Columbiads*), and *Mortars*.

I. GUNS are the largest and heaviest pieces in regard to their projectiles, which, acting simply by their force of percussion, are always fired with large charges, ¼ and sometimes ½ the weight of the ball. They are used for striking objects *direct* and at a distance; or, by their ricochet fire, for attaining objects not accessible by direct fire, and overthrowing or breaking them down by the force of percussion. They are also used in battering down the walls of fortifications. They are always designated by the weight of the solid shot which they carry.

DIVIDED.—We have seven different calibers, which are divided into three classes, according to the positions in which they are used.

1. *Sea-Coast*............	32 and 42-pdrs.	Cast iron.	For dimensions, see Appendix, p. 1.
2. *Siege and Garrison*..	12, 18, and 24-pdrs.	Cast iron.	
3. *Field*..................	6 and 12-pdrs..	Bronze.	

The *sea-coast*, Fig. 19, being the heaviest, are used principally in permanent fortifications on the sea-board. They are neither as long nor as heavy, *in proportion to the size and weight of their projectiles*, as the 2d Class.

The *line of metal*, or *natural line of sight*, is the line joining the highest points of the breech and muzzle, but in these two pieces the intervening metal prevents the highest point of the muzzle from being seen when the eye is at the highest point of the breech. In firing, therefore, at what would be *point-blank* distance, or within that, a *muzzle-sight* has to be used, the top of which is at a distance above the highest point of the muzzle equal to the difference between the semi-diameter of the muzzle and the semi-diameter of the largest circle of the breech. This difference is called the *dispart*. The above remark applies also to the sea-coast *howitzers*.

The *natural angle of sight* is the angle which the natural line of sight makes with the axis of the piece.

The *dispart* is therefore the tangent of the natural angle of sight to a radius equal to the distance from the rear of the *base-ring* to the highest point of the muzzle, measured on a line parallel to the axis of the gun.

The *base-ring* is a projecting band of metal encircling the gun next to the base of the breech, and is joined to the body of the piece by a concave moulding.

The *base of the breech* is the part immediately in rear of the breech, and in these guns is a *spherical* segment.

The *cascable* is all that part of the gun in rear of the base ring, and is composed of the *knob*, *neck*, *fillet*, and *base of the breech*.

The *breech* is that part of the solid metal between the bottom of the bore and the rear of the base-ring.

The *neck* is the smallest part of the chase of the piece. Around it is placed the *chase-ring.*

The *swell of the muzzle* is the largest part of the piece in front of the neck. It is terminated by the muzzle mouldings, which in these guns consist simply of a *lip.*

The *face* of the piece is the terminating plane perpendicular to the axis.

The *bore* of the piece includes all the part bored out. It is terminated, in *all* guns, at the bottom, by a plane perpendicular to the axis, joined to the bore by an annular surface, as before described.

A *lock-piece* is usually cast on iron guns near the vent for the purpose of attaching a lock, but is now never used since the introduction of friction tubes.

The *preponderance of the breech* is the excess of weight in rear of the trunnions over that in front, and is measured by the weight which must be applied in the plane of the face to balance the gun when suspended freely on its trunnions.*

The *siege and garrison* guns, Fig. 20, are used in permanent fortifications, as well as in field works, and in *sieges* to batter down the walls of the besieged, forming breaches, to dismount their guns, &c. As when used in a siege they have to be fired through earthen embrasures, they are made long enough to project into the mouth, so that the blast from the gun will not destroy the cheeks.

The *natural angle of sight* in these guns is $1\frac{1}{2}°$. The *muzzle* mouldings consist of a *lip* and *fillet.* The *base* of the breech is of a *conical* shape.

The *field*-guns, B, Fig. 21, are used in the field as light artillery. They are made of *bronze,* and are our lightest pieces. The natural angle of sight is 1°. The 12-pdr. has over its center of gravity two handles. The base of the breech is a part of a cone, like the preceding guns.

Fig. 21, A represents the new 12-pdr. gun recently adopted

---

* A more accurate method is to determine the *lifting* force in pounds necessary to be applied at the base-ring to balance the piece in its trunnion-beds; and this is the plan now pursued.

for field service. As it is designed to use with this piece both hollow and solid shot, it will probably replace in our field artillery both the guns and howitzers now in use, except where very light pieces are required. For dimensions, &c., see Appendix, p. 1.

In all our guns the diameter of the trunnions is the same as that of the bore of the piece. Their axis is in the same plane as the axis of the piece, and perpendicular to it. In the French guns it is below the axis of the piece.

A very good way to remember the diameter of the bore of any particular piece is by means of the following table, showing a common difference with all the pieces but one.*

	42	32	24	18	12	6	
Diameter of the bore.....	7	6.4	5.82	5.3	4.62	3.67	In inches.
Common difference.......	..	.6	.6	.6	.6	.6	
Which would give.......	7	6.4	5.8	5.2	4.6	1.*	Very nearly the same.

By an examination of the table at page 1 of the Appendix, it will be seen that the windage also in these guns decreases very regularly.

A *howitzer* is a *gun* with a *chamber* in it. It is used principally for firing hollow projectiles or shells; and as these require for efficiency a certain velocity, and are liable to break if too much powder is fired behind them, it becomes an important object to use a small charge, and give it all the power of which it is susceptible. These points are gained by the use of the chamber, which in all our howitzers is cylindrical in shape.

The howitzers are divided into four classes, viz :—

1. *Sea-Coast*..............	8-and 10-in.	Cast-iron.	For dimensions, see Appendix, p. 2.
2. *Siege and Garrison*....	8-in. and 24-pdrs.		
3. *Field*................	12, 24, and 32-pdrs.	Bronze.	
4. *Mountain*...........	12-pdrs.		

The calibre of the howitzer is designated either by the weight of the *solid* shot it would carry, or the diameter of the bore in inches.

The *sea-coast howitzers*, Fig. 22, are used in permanent fortifications principally on the sea-coast. They are similar in appearance to the sea-coast guns, except being shorter and thicker, though with a less thickness of metal, as the charges are so much less. The parts are designated in the same way, and in addition we have the chamber. It is sometimes stated as a reason for not making these chambers conical in form, that the cartridge is apt to slide forward, instead of lying against the bottom of the bore as it ought; but many of the naval howitzers are made with conical chambers and found to work well.

Howitzers are used generally to throw shells and other hollow projectiles, which act by their explosion as well as by their force of percussion, and set fire to towns and other shelters of an enemy. The piece being shorter and lighter in proportion to its projectile than the gun, and the charges used being smaller, the accuracy of fire is much less. This is compensated for, however, in the greater execution of the shell when it bursts.

The system of shell-guns first practically brought into use in 1822 by the French general Paixhans, was adopted soon afterwards by us, and the number of pieces has been steadily increasing ever since; a special piece called the *Columbiad*, differing slightly from the ordinary howitzers, having within late years been added to our armament. The propriety of this *addition* is questionable. The columbiad requires a special carriage, drill, and method of management, thus increasing the number of these in the service, which is objectionable. As the columbiads carry the same-sized projectiles as the sea-coast howitzers, these last should be discarded from our armament at once, if the superiority of the columbiad is demonstrated. The form of this last is objectionable in several particulars, which will be noted under another head.

The *siege and garrison* howitzers, Fig. 23, are used in the trenches at sieges, and in the defenses of permanent fortifications. The 8-in. usually in sieges, the 24-pdr. generally in garrison, more especially in flank casemate defense, in which position a special carriage is appropriated to its use.

The 8-in. is a very short piece, and is loaded by hand. The chamber, instead of being joined to the bore by a conical surface, has a spherical one, thus allowing the shell to fit down close over the mouth of the chamber, without using a *sabot* or block of wood, as with other howitzers. This is a matter of considerable importance, as the piece being used in siege batteries, is frequently compelled to fire over the heads of the men in the advanced trenches. A sabot so fired, and flying to pieces, would be dangerous to one's own party. The breech of this howitzer is made as heavy as possible, so as to throw the centre of gravity to the rear and make the muzzle project as far as possible to the front, to save the embrasures.

The 24-pdr. having but one reinforce, shows no band similar to the one on sea-coast howitzers at the junction of the 1st and 2d reinforces. The swell of the muzzle is replaced by a *muzzle band*, as is the case also with the 8-in. and the field and mountain howitzers. The base of the breech is conical, as in all guns and howitzers, except the sea-coast.

The *field* howitzers, fig. 24, are used in the field in light batteries, the 12-pd. with the 6-pd. gun, and the other two with the 12-pd. gun, which fits on the same carriage as the 32-pd.

The chamber increasing the thickness of metal near the breech, enables us, where greater lightness is necessary, to reduce the weight by taking off a portion of the metal on the outside around the chamber. This is done in the 24 and 32 pdrs., and in the 8-in. siege howitzer, as also in the mountain howitzer; forming what is called the *recess*.

The 24 and 32 pdrs. have each two handles.

The *mountain howitzer*, fig. 24, is a very light 12-pdr. a trifle over a yard long, without any reinforce, and used for service in countries so rough as not to admit the passage of wheeled vehicles. It and its carriage, taken to pieces, are carried on the backs of mules, which, with more favorable roads, may be used to draw the carriages (two wheeled) with the piece mounted ready for use. A special carriage with four wheels has been invented for use with this piece on the prairies, where the two-wheeled carriage is inconvenient on account of its liability to upset.

Columbiads, A, fig. 25.—These are of two sizes, the 8 and 10 inch. They are longer, thicker, and much heavier than the sea-coast howitzers, and are used in similar positions. They have no band at the junction of the 1st and 2d reinforces, like the howitzers, and the breech is not so full and rounding. The first, we will show hereafter, strengthens the gun; and the last feature weakens it. The bottom of the chamber is *hemispherical.* The knob of the casable is also different from other pieces, and is cut through for the purpose of forming on the base of the breech a system of steps or notches, which are of very great assistance in elevating the piece, the mass of metal being so great.

New Columbiads. B, fig. 25, represents the new columbiads, or 128 and 64 pdrs.

MORTARS.

A *mortar* is a short-chambered piece used for throwing at a great elevation, shells, which, acting by their velocity acquired in falling, serve to crush the vaults and shelters of the enemy, and set fire to them. The chamber is particularly necessary, as the piece, being very short, small charges have to be used. The chamber usually employed in our service, is the *conical* or Gomer.

Mortars are designated usually by the diameter of the bore in inches. They are cylindrical in shape, terminated in the rear spherically, and have, of course, no *angle of sight.*

In our service there are seven, divided into the following classes, viz.:

Heavy, Sea-Coast	10 and 13 in.		Iron.	For dimensions, see Appendix, p. 3.
Light, Siege and Garrison.	8 and 10 in.			
Stone	16 in.		Bronze.	
Coehorn	5.82 in.	(24 pdr.)		
Eprouvette	5.655 in.	"	Iron.	

The last is the only one with a cylindrical chamber.

The *heavy sea-coast mortars,* fig. 26, are used principally for sea-coast defense. Their chamber is terminated at the bottom by a hemisphere, and joined on to the bore by a portion of another in which the shell fits. This connecting surface in the siege and garrison mortars, is so small a portion of a sphere as to be scarcely perceptible, being, in fact, simply a rounding off of the metal where the chamber and bore meet. All sharp angles at this junction, as well as at the mouth of the chamber, in all the mortars and howitzers, are rounded off to prevent their being worn off by the action of the powder.

In place of handles, the heavy mortars have an *ear* cast over the center of gravity, by means of which, and a bolt, a *clevis* is attached, to be used in handling the piece. The stone mortar is the only other one having a similar arrangement.

The *light siege and garrison mortars*, fig. 27, are for use in the trenches at sieges, and in the defense of fortifications of all kinds. They are made light enough to be easily transported with a marching army. The *stone* mortar, fig. 28, is used under the same circumstances for throwing baskets of stone or small shells for very short distances. It is used by the besiegers to clear the breach just before the advance of the storming party, or by the besieged to repulse this last as it advances up the breach.

The divided state of the projectiles thrown from it, renders it unnecessary to have the sides very strong, which are therefore thin ; and the piece is made of bronze. It is further decreased in weight by removing a large portion of metal in rear of the piece, which changes it from the spherical form of the other mortars. The bottom of its chamber is hemispherical.

The *Coehorn*, Fig. 28, named after its inventor, is used like the preceding, and is especially adapted for sieges, where on account of its lightness, it can be carried from trench to trench by two men. It is strong enough for the use of large charges, and with 8 oz. gives a range of 1,200 yards.

The *Eprouvette*, Fig. 28, can scarcely be considered as a part of our armament, since it is used simply for testing powder ; and even for this, it is now almost entirely superseded by the pendulum.* It is cast with a sole, the plane of which makes an angle of 45° with the axis of the piece. This sole slides into a recess in the cast-iron bed-plate, which last is bolted to the platform.

Although called a 24-pdr., its bore is less (5.655), for the purpose of reducing the windage as much as possible, and preventing the escape of gas. For the same reason the vent is only 1 10th inch in diameter. The chamber is cylindrical, with a hemispherical bottom.

Such are the pieces, as they now exist, in our service ; and although it is not likely that any very great changes will be made

* It is sometimes used at the powder factories for determining the *relative* strength of powders of the same kind.

in the calibres, improvements are constantly taking place as experiment and use show them to be necessary. The 10-in. sea-coast howitzer has fallen into disuse; and experiments are being carried on with a columbiad of 12-in. bore. The form of the columbiad, too, has changed, becoming more that of Dahlgren's gun, with little or no swell of the muzzle, and no chamber. They are now designated as 64-pdrs. and 128-pdrs.

Manufacture.—All the ordnance for the United-States service is made at private foundries, and afterwards inspected and proved by officers of the ordnance detailed for the purpose. The foundries where the most of our pieces are made, are the following:—The West Point Foundry, near Cold Spring, N. Y.; the Fort Pitt Foundry, near Pittsburgh, Pa.; The Tredegar Foundry, near Richmond, Va.; the Algers Foundry, near Boston, Mass.; and the Ames Foundry, near Chicopee, Mass. The last two furnish the bronze ordnance, and the others the iron.

MANUFACTURE OF ORDNANCE.

The "*model*" or form of a gun, is made of hard wood or iron turned to a size a little larger than the gun to be made, in order to allow metal enough to turn the outside smooth. The model is generally of wood, and composed of four or five pieces. The first, Figure 29, forming the body of the piece, extends from the chase ring to the base ring. The second, Fig. 29, forms what is termed the "sprue" or "dead-head" and extends from the top of the first piece to the end of the model, which is longer than is actually required for the gun, in order to furnish metal to supply shrinkage, and to produce greater pressure and density in the lower part of the gun.

For reasons which will be explained in discussing the manner in which the gun cools, these two pieces are now made in one at some foundries, the upper part being formed of one unbroken cone, as represented by the broken line in figure 29.

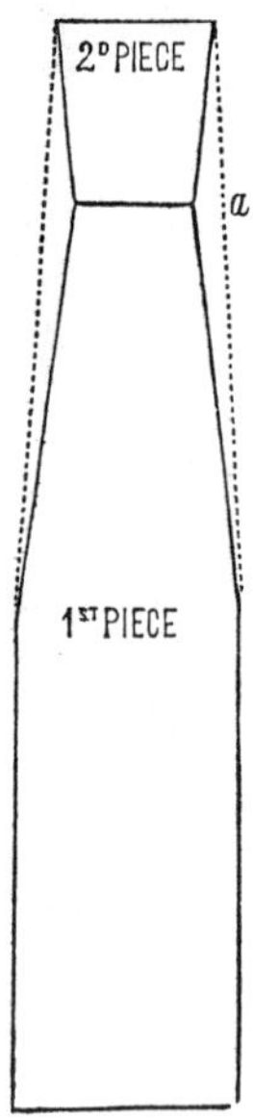

Fig. 29.

The third piece, Fig. 30, forms the breech and cascable of the gun, having a square projection, longer than that necessary to form the knob, which is used in attaching machinery to give the piece a rotary motion in turning and boring.

Fig. 30.

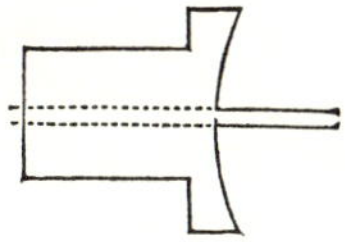
Fig. 31.

The models for the trunnions and rimbases form the 4th and 5th pieces, as represented in Fig. 31, and are connected with the body of the piece at the proper place by means of bolts passing through their axes.

The "*mould*" is made of hard, angular, and refractory sand, moistened with water in which clay has been stirred, to make it adhere and retain the form given to it. This mould is formed in a jacket or case made of sheet iron in four or five different parts, each part being formed of several segments fastened together by bolts and nuts passing through flanges on the edges, and the different parts are fastened together in the same way before pouring in the metal. Fig. 32.

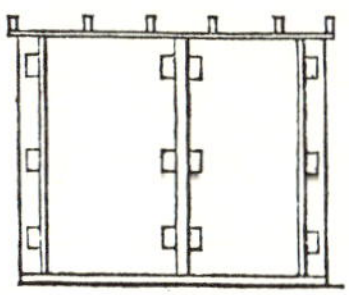
Fig. 32.

To form the mould, piece No. 2, or, in the case of the improved form, the body of the model, is placed on a plain surface, "sprue head" down, inside of the first part of the jacket, and adjusted in a vertical position, the axes of jacket and model coinciding, Fig. 33. The moulding-sand is then packed tightly around this until the top of the jacket is reached, when it is leveled off, made perfectly smooth and even with the top, *a b.*, and sprinkled with fine dry sand, to prevent the succeeding portion from sticking to it. Another portion of the jacket is now fitted around the model on top of the other portion, and the mould completed in the same way; and so on until the position for the trunnions is reached. The end-plates of the trunnion jackets *a* are then taken off as represented Fig. 34; the trunnion models are run in and bolted to the body of the model, and the sand is packed in around them until

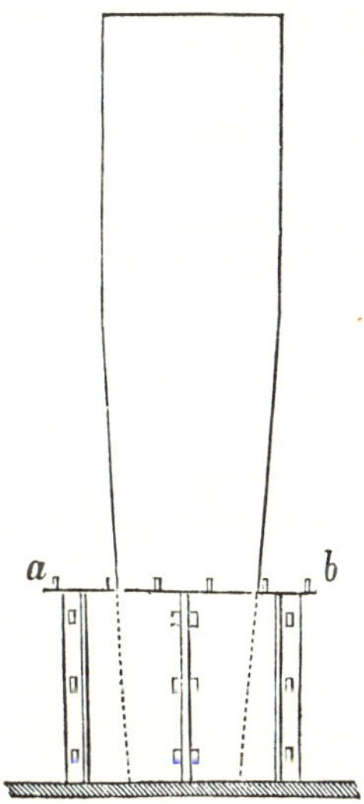

Fig. 33.

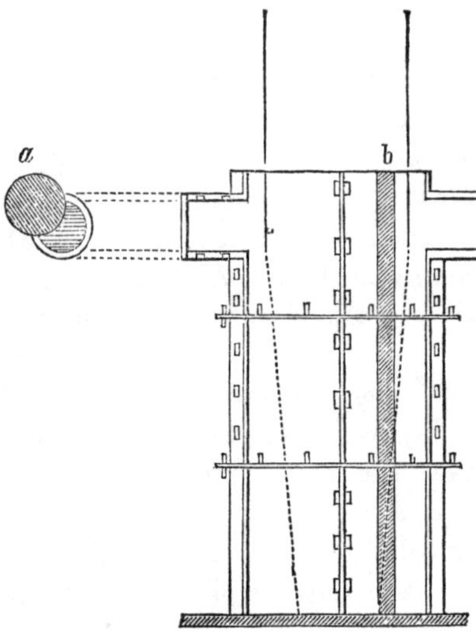

Fig. 34.

the trunnion jackets are nearly full, when the bolts are withdrawn, the remaining portion of the trunnion jackets filled in even with the faces, the end plates placed on and fastened. This portion of the mould is then completed by filling in sand from above as before; another portion of the jacket is placed on; and so on till the mould is finished; taking care to sprinkle the top of each portion of the mould and the model with dry sand, to prevent sticking between the parts.

The top portion, containing the model and mould for the breech, is now taken off, and the breech model removed; the body of the model is drawn out, the trunnion bolts having already been removed; and when the upper portions of the jacket are taken off, the trunnion models are removed by hand.

Such portions of the mould as have been broken are now repaired by hand, with instruments used for the purpose; and the whole interior covered with a *cokewash* made of water and powdered coke, put on with a fine brush, in order to make a smooth surface and prevent the sticking of the metal to the mould. All the parts are then placed in an oven, and baked hard.

In most of the castings made in this country the metal is introduced at the *bottom* of the mould, through a channel (*b*, Fig. 34) made in the same way and of the same material as the mould, in order to prevent the falling metal from injuring the form, and also with the idea that by giving the entrance a position oblique to the axis and imparting a circular motion to the fluid iron, the casting is more homogeneous, and such scoria as enters the mould is kept in the centre and rises to the top. In some countries, however, the metal is allowed to fall in at the top, the advantage being claimed for this method that by the concussion of the falling metal the formation of air-bubbles is prevented.

Sand is used in making the mould, because it is considered that in casting in an iron mould, the metal cools too quickly, and forms on the surface hard projections which it is very difficult to remove; but it is contended that in casting bronze guns, the metal of which cools much more slowly than iron, the more rapid cooling in iron moulds is advantageous to the strength of the guns.

The superfluous metal in the "dead" or "sprue" head is for the purpose of making the casting more dense by the pressure, and to feed the casting, required by shrinkage.

The metal cools in layers, commencing on the outside of the casting, and contracting as it cools; the fluid mass in the center supplies the metal, causing a current down from the dead-head to fill the vacancy. It is found that the narrow part at *a*, Fig. 29, cools so fast as to obstruct the supply from the dead-head, and the model is now made of an unbroken cone, as represented by the dotted line in the figure, the surplus metal being turned off afterwards.

The mould, having been baked hard, is put together again, *breech down*, in a pit, the sprue-head being a little below the level of the point of the furnace from which the metal is to flow; the different parts of the jacket being bolted firmly together, to prevent their separation when the metal is poured in. A trough is now made from the furnace to the top of the mould, the valve of the furnace opened, and the metal run in, in as short a time as possible; holes being left at different heights to allow the metal to pass into the mould. As it rises it is agitated with a pine stick, and the scoria kept in the middle to prevent its entering the trunnion moulds; which are enlarged slightly *on top*, so that any defect there may be removed by the cutter when the trunnions are turned.

When full, a quantity of charcoal is thrown on the top of the dead-head to absorb gases and prevent oxidation from taking place; and the casting is then left for several days to cool.

BORING AND TURNING.—It is then taken out of the mould, the sand cleaned off, and when perfectly cooled, is taken to the boring-room to be bored and turned. These operations are performed by fixing the gun on a track, in a horizontal position, supporting it near the muzzle by a collar in which it turns, and at the breech by the square projection in rear of the cascable to which the machinery is fastened to give it the rotary motion about its axis. Fig. 35, Plate 6.

When accurately adjusted, and the machinery put in motion, a cutter is applied at the proper position near the muzzle, and the *dead-head* cut off; after which the bore-cutters are adjusted in the prolongation of the axis, and pressed by means of a weight and system of cog-wheels against the gun. Several kinds of cutters have been used; but the one now generally adopted consists of a hollow cylinder with cutters projecting from the base sufficiently to cut out a cylinder somewhat larger than the one to which the cutters are attached. This allows the cutter cylinder to pass down into the mouth of the piece, and the iron cuttings to pass out of the gun along the outer surface of the cylinder. A solid mass of iron is thus left inside the cutter cylinder, which is broken off by wedging, after the cutter has gone down as far as it can (between two and three feet), and been withdrawn. The cutter is then run in again; and so on, until the boring is finished. A smaller cylinder for the chamber in boring howitzers, or a chamber-cutter, is used; and the whole bore is reamed out and finished by a standard reamer of the proper size.

While the piece is being bored, cutting instruments are applied to the exterior, which is turned down to the proper size. That portion of the gun situated between the trunnions cannot be so removed; and is taken off in a planing machine, in which the piece moves backwards and forwards under a cutter. Such portions of the surface as cannot be reached by these two machines, are removed by the chisel.

The piece is then placed in the trunnion-lathe, where centers are pressed against the trunnion faces at the proper points; a rotary motion is given by machinery around this axis, and the trunnions and rimbases are turned. Fig. 36, Pl. 6. While in this lathe, the axis of the piece is placed at a proper angle with the horizontal, a borer applied at the proper point, leveled, and the vent bored. In bronze pieces, the gun is placed in a proper position, and a hole bored through for the vent-piece; after which the thread of a screw is cut in it, and the vent-piece, made of well-hammered copper with a corresponding thread on it, screwed into its position, the vent bored through this as before, and the part of the vent-piece projecting into the bore, cut off by a cutter inserted in the bore.

The square projection at the end of the knob is not removed until after the piece is inspected, in order to enable the founder to replace the piece in the machinery to correct any irregularities in the form. It is then removed by boring small holes in it, and splitting off the metal with wedges. The piece is then ready for inspection and proof.

For Inspection, the piece is placed on skids, for the purpose of being easily moved. It is examined carefully on the exterior to see there are no cracks or flaws in the metal, whether it is finished as prescribed, and to judge, as far as practicable, of the quality of the metal. The gun must not be covered with paint, lacker, or any other composition, before it is inspected. Any attempt discovered to fill up flaws or cavities with plugs or cement, causes the rejection of the piece without further examination. The exterior diameters of the piece are now measured by means of *callipers*, Fig. 37, Pl. 6, constructed for the purpose. The lengths of the different portions are also measured with a long *rule* of hard wood.

A common *looking-glass* is now held so as to reflect the sun's rays into the bore, which can be seen with great distinctness to the bottom. In case of the absence of the sun, a short piece of candle, lighted and placed on the end of a pole, is inserted to the bottom of the bore.

The Searcher, Fig. 38, Pl. 6, is then used. This is an instrument for determining the presence of small cracks or flaws in the bore not visible to the eye. It consists of four flat springs turned *out* at the ends and sharpened, the other ends being fastened in a socket into which a handle is screwed. The handle used with this and several other of the inspecting instruments is—

The Cylinder-staff, Fig. 39, Pl. 6, a round staff of mahogany or other hard wood, in two parts, joined together by brass sockets and screws. The other instruments used with this staff are the *cylinder-guage*, Fig 40, Pl. 6, *guide-plate*, Fig. 39, Pl. 6, and *measuring-point*, Fig. 39, Pl. 6, all of which can be screwed on the end. The staff is graduated in inches and tenths, on a strip of brass let into the side. The graduation is arranged to read the distance from the end of the *measuring-point*, when screwed into its place. The searcher is about fifteen inches long, and the blades twelve. To insert the instrument in the muzzle of a gun, the blades have to be pressed up together, and the points then move along the surface from the stiffness in the blades. The searcher is pushed slowly to the bottom of the bore and withdrawn, turning it at the same time. If one of the points catches, its distance from the muzzle is read from the staff, its position in the bore noted and marked on the exterior of the gun. The size and figure of the cavity is then determined, by taking an impression of it in *wax* placed on the end of a hook.

To keep the staff in the axis of the piece whilst using the different instruments, it is supported at the muzzle by a *half-tompion* of wood, having in the center a groove to fit the staff. The *rest* for the *star guage* may be also used for this purpose.

THE CYLINDER GUAGE is a hollow cylinder of cast or wrought iron turned to the exact minimum or true diameter of the bore. Its length is equal to its diameter, and it has at each end cross-heads perpendicular to each other. In one of the cross-heads, and in the axis of the cylinder, a smooth hole is left of the size of the cylinder staff. At the other end the hole is threaded for the reception of the end of the cylinder staff, on which the guage is now screwed, and this pushed slowly to the bottom of the bore. It must go all the way down, or the bore is too small; but if it goes down, the bore may still be too large, and irregular in its dimensions.

STAR GUAGE, Figs. 41 and 42, Pl. 6.—To ascertain this, a more complicated and delicate instrument is used, called the *star* guage from the shape of its head, which is of brass, with four steel sockets, two movable and two stationary, for the *measuring points*. There are four measuring points for each calibre; and when two of these are screwed into the fixed sockets, the distance between their points is equal to the true diameter of the bore. The movable sockets rest against the inclined sides of a slider or wedge whose sides incline 0.35 in. in a length of 2.2 in., so that by pushing the slider the 35th part of this distance (about 0.06 in.) the distance between the two sockets or the measuring points, if screwed into their places, is increased .01 in.

The slider is fastened to a square steel rod consisting of three parts, which are screwed together according to the length of bore to be measured. This rod passes through a brass tube which is also made in three parts, and to screw together. This tube is graduated into inches and quarter inches, commencing at the plane of the measuring points, so as to indicate the distance of these from the muzzle of the gun.

The handle, Fig. 42, Pl. 6, is of wood, attached to a brass cylinder or socket through which the rod passes into the handle. The socket of the handle slips over the end of the brass tube made smaller for the purpose, and has a slit in it allowing the brass tube to be seen through. On the side of this slit a scale is constructed, to indicate the movement of the measuring points. Each joint of the long tube has a mark on it, to show the position for the zero of the scale when the instrument is properly adjusted for any particular calibre. In this position the handle is fixed to the sliding rod by means of a screw clamp.

A Ring Guage, or ring of metal, for each calibre, is used for adjusting the instrument for use. The handle is loosened, the proper measuring points are screwed in, the ring guage placed on them, and the slider pushed out until all the points touch the inner circumference. The zero of the scale is then made to coincide with the mark on the tube, and the handle clamped, when the instrument is ready for use.

A Rest, Fig. 43, Pl. 6, in the form of a T, is placed in the mouth of the gun to keep the instrument in the axis of the piece. It has three sliders which can be adjusted on the different limbs, to suit any sized bore.

Commencing at the muzzle, the diameter of the bore is measured at intervals of a calibre, as far as the trunnions. From that point to the seat of the shot, a diameter is measured at every inch, and for every quarter of an inch, for the rest of the bore. No variations over 0.03 of an inch, are allowed.

The Trunnion Square, Fig. 44, Pl. 6, consists of a horizontal piece of wood with two perpendicular limbs projecting from it, the distance between which is equal to that between the rimbases of the gun. The bottom edges of the limbs are shod with iron, and are in the same plane, parallel to the upper edge of the connecting piece, so that when the square is placed with its feet resting on the trunnions, the upper edge of the connecting piece is parallel to their axis. At the middle point of the horizontal piece a pointed slider projects down, with a thumb-screw to fasten it in any position. Each of the limbs has an iron plate projecting from its side, the lower edge of which is perpendicular to the limb. It is placed on top of the trunnions, whilst the edges of the feet press against the side to determine whether the trunnions have the same axis perpendicular to that of the piece.

To find whether the axis of the trunnions is in the same plane with that of the piece, the feet are placed on the *top* of the trunnions, and their edges should touch throughout. The slider is pushed down till its point touches the surface of the gun, and is then fastened with the thumb-screw. Turn the gun over, and apply the square in the same way to the other side. If the feet now rest on the trunnions, and the pointer touches the surface of the gun, the two axes are in the same plane. Should the point of the slider not touch, the axis of the trunnions is below that of the piece; but should that touch and the feet not, it is above. If the alignment of the trunnions be accurate, the edges of the feet will fit on them when applied to different parts of them; and if their axis is perpendicular to that of the piece, the edges of the feet will touch throughout the trunnions while the iron perpendicular projection will rest on the top of them.

TRUNNION GUAGE.—The size of the trunnions is determined by the *trunnion guage*, which is an iron ring which must fit closely on the trunnion, its outside circumference being of the same diameter as the rimbases, and thus serving to verify them at the same time.

THE GUIDE GUAGE, Fig. 45, Pl. 6, is a thin circular iron plate of the minimum diameter of the bore, for directing the measuring point to the center of the bottom of the bore. It has a hole through the center, with a thread by which it is screwed on to the end of the cylinder staff.

THE MEASURING POINT, Fig. 46, Pl. 6, is a short pointed piece of iron, with one end cylindrical, and having a socket into which the end of the cylinder-staff is screwed after the guide-plate is placed on. The scale on the cylinder staff is made to commence from the end of this point.

To find the length of the bore push the staff, with the guide-plate and measuring point in their positions, to the bottom of the bore, resting the staff on a half-tompion or the star-guage *rest*. With a straight-edge against the face of the piece, read off the length on the staff.

THE TRUNNION RULE, Fig. 47, Pl. 6, is a long graduated rule, having on it a piece of metal in the shape of an L, one leg of which rests on top of the trunnion, while the other rests against its side, and the distance of the trunnion from the base-ring is read off from the staff. Other external dimensions of the piece are measured by a wooden rule, or verified by means of an accurately cut profile.

A RAMMER-HEAD, or simply a profile cut to fit the bottom of the bore, is used to determine the point at which the vent enters the bore, by thrusting in a priming-wire and marking where it makes an impression on the wood. The position of the exterior orifice of the vent is also verified.

THE VENT GUAGES, Fig. 48, Pl. 6, are two pieces of pointed steel wire, greater and less than the true diameter of the vent by .005 inch. They are used to determine the size of the vent.

THE VENT SEARCHER, Fig. 48, Pl. 6, is a hooked steel wire about half the size of the vent, used to detect flaws or cracks.

The dimensions and form of chambers are verified by means of forms cut from wood or metal. After the powder-proof, the bore is washed and wiped clean, the bore and vent again examined, and the bore measured. The results of both measurements and examinations are noted on the inspection reports opposite the number of the gun.

Figure and letter stamps are required, to affix the necessary marks.

INSPECTION OF BRASS ORDNANCE.—Brass cannon are measured, and their dimensions recorded, as prescribed for iron cannon. *The exterior form and dimensions* are verified by the application of a *profile* cut out of sheet iron, of the exact shape of a longitudinal section of the piece. All brass ordnance, except stone mortars, should be bored under size from .04 to .05 inch, and after proof, reamed out to the exact calibre. When the powder-proof is finished, the bore should be cleaned and examined; the vent should then be stopped with a greased wooden plug, the muzzle raised, and the gun filled with water, to which pressure shall be applied to force it into any cavities that exist; or, the water shall be allowed to remain in the bore about twenty-four hours. *The bore* must then be sponged dry and clean, and viewed with the *mirror* or *candle*, to discover if any water oozes from cracks or cavities, and also if any enlargement has taken place. The quantity of water that runs out of a crack or honey-comb will indicate the extent of the defect; and if it exceeds a few drops the gun should be rejected, although the measured depth of the cavity may not exceed the allowance. If the water oozes out between the vent-piece and the metal of the gun, a new vent-piece must be inserted, and the gun again proved with one charge, and the water-proof repeated. After the bore has been reamed out to the proper size, its dimensions are again verified, and an examination of the bore and vent is made, to detect any defects which may have been caused or developed by the proof. *Whitish spots* show a separation of the tin from the copper, and, if extensive, should condemn the piece. A *great variation from the true weight*, which the dimensions do not account for, shows a defect in the alloy. Any attempt to conceal cavities by filling them with screws, or by any other methods, should cause the rejection of the piece.

Brass cannon should be rejected for the following cavities or honey-combs:

Exterior.—Any hole or cavity 0.25 inch deep in front of the trunnions, and 0.2 inch deep at or behind the trunnions.

Interior.—From the muzzle to the reinforce, any cavity 0.15 inch deep. Any cavity from the reinforce to the bottom of the bore.

The specific gravity of the metal of brass ordnance should be occasionally ascertained, by taking that of some of the heaviest, and some of the lightest pieces, at each inspection.

The exterior surfaces of all ordnance are turned, or dressed smooth.

Proof.—Gunpowder for proving ordnance, should be of the best quality. It should be proved immediately before being used, unless it has been proved within a year before; and there is no reason to suppose it has become damaged.

The cartridge-bags are made of woolen stuff or paper, the full diameter of the bore or chamber. They are filled by weight; and if not filled where the proving takes place, each bag should be enveloped in a paper cylinder and cap, marked with the weight and the proof range.

The shot must be smooth, free from seams and other inequalities that might injure the bore of the piece; and they must be of the true diameter given in the tables.

The *wads* are made of junk.

Proof of Iron Ordnance.—Guns and howitzers are laid with the muzzle resting on a block of wood, and the breech on the ground, or on a thick plank, giving the bore a small elevation.

Mortars are mounted on strong wooden frames or beds, at an elevation of 45°, supported by the trunnions.

In proving iron ordnance, after pricking the cartridge, prime with powder, or a tube, and place over the vent a piece of portfire, set in clay or putty, long enough to permit the man who fires it to reach a place of safety before the charge explodes.

Proof Charges for Iron Guns.—*First and Second Rounds.*—A charge of powder equal to one half of the weight of the shot; two shot and one wad.

Third Round.—A charge of powder equal to one third of the weight of the shot; one shot and one wad.

In proving new guns, a compound shot, or a cylinder with hemispherical ends, of the true diameter of the shot, and equal in weight to the two shot, shall be used instead of them.

The wad is placed over the cylinder or the upper ball; the whole being well rammed.

Should any of the guns proved at one time, fail to sustain the above proof, the remainder shall be again fired twice with a charge of powder equal to one half of the weight of the shot, one shot and one wad; and if, in either or both of these trials, one fourth of the whole number of guns should fail, the whole shall be rejected.

Other iron ordnance are fired with the following charges:

Columbiads.—

10-*inch.*—1st round: 20 lbs. powder, one 10-inch strapped shot, and one wad over the shot.

2d round: 24 lbs. of powder, one 10-inch shell strapped.

8-*inch.*—1st round: 12 lbs. of powder, one 8-inch strapped shot and one wad.

2d round: 15 lbs. of powder, one 8-inch shell strapped.

HOWITZERS.—

Sea-coast.	10-*inch.*—2 rounds, with 15 lbs. powder, one 10-inch strapped shot, and one wad over the shot.
	8-*inch.*—2 rounds, with 12 lbs. of powder, one 8-inch strapped shot, and one wad over the shot.
Siege and Garrison	8-*inch.*—2 rounds, with 4 lbs. of powder, one 8-inch shot, and one wad over the shot.
	24-*pdr.*—2 rounds, with 3 lbs. of powder, one 24-pdr. strapped shot, and one wad over the shot.

MORTARS.—

Heavy ..	13-*inch.*—2 rounds, with 20 lbs. of powder, and one 13-inch shot.
	10-*inch.*—2 rounds, with 10 lbs. of powder, and one 10-inch shot.
Light . . .	10-*inch.*—2 rounds, with 5 lbs. of powder, and one 10-inch shot.
	8-*inch.*—2 rounds, with 2½ lbs. of powder, and one 8-inch shot.

Should any columbiad, howitzer, or mortar, fail to sustain the above proof, the remainder of those offered at the same time shall be again fired twice with the same charges; and if, in either or both of these trials, one fourth of the whole number should fail, the whole shall be rejected.

THE WATER-PROOF, as described for brass cannon, must also be applied occasionally to iron cannon, at the discretion of the inspector.

The bore and vent, and the exterior surface of every piece which is approved, should be well covered with sperm oil immediately after the inspection.

PROOF OF BRASS ORDNANCE.—They are mounted on appropriate carriages or beds, and fired three times; guns and howitzers at an elevation of 5°, mortars at an elevation of 45°, with the following charges:

FIELD GUNS.—A charge of powder equal to one third of the weight of the shot, one shot and one wad.

HOWITZERS.—

Field . . .	32-pdr.—3¼ lbs. of powder, one strapped shot and one wad.
	24-pdr.—2½ lbs. of powder, one strapped shot and one wad.
	12-pdr.—1½ lbs. of powder, one strapped shot and one wad.
Mountain,	12-pdr.— ¾ lb. of powder, one strapped shot and one wad.

MORTARS.—*Stone Mortar.*—2½ lbs. of powder, covered by a wooden tompion 2 inches thick; a basket filled with alternate layers of stones and earth, weighing 100 lbs.

Coehorn, 24-*pdr.*—$\frac{3}{4}$ lb. of powder, and one 24-pdr. shot.

In proving brass cannon in service, or after they have been bored to the proper calibre, the shot should be wrapped in cloth or strong paper to save the bore as much as possible from injury.

Marking.—All cannon are required to be weighed and to be marked, as follows, viz: the *number of the gun*, and *the initials of the inspector's name*, on the face of the muzzle; the numbers in a separate series for each kind and calibre at each foundry; the initial letters of the *name of the founder* and of the foundry, on the end of the right trunnion; the *year of fabrication* on the end of the left trunnion; the *foundry number* on the end of the right rimbase, above the trunnion; the *weight of the piece in pounds* on the base of the breech; the letters U. S. on the upper surface of the piece, near the end of the reinforce.

The natural line of sight, when the axis of the trunnions is horizontal, should be marked on the base ring and on the swell of the muzzle, whilst the piece is in the trunnion lathe.

Cannon rejected on inspection are marked X C, on the face of the muzzle; if condemned for erroneous dimensions which cannot be remedied, add X D; if by powder-proof, X P; if by water-proof, X W.

Injuries.—Brass cannon are little subject to external injury, except from the bending of the trunnions sometimes after long service, or heavy charges.

Internal injuries are caused by the action of the elastic fluids developed in the combustion of the powder, or by the action of the shot in passing out of the bore. These effects generally increase with the calibre of the piece.

Of the first kind, which exhibit themselves in rear of the shot, are: *the enlargement of the bore* by the compression of the metal, which is seldom a serious defect; *corrosion of metal*, particularly at the angles, such as the inner orifice of the vent, or the mouth of a cylindrical chamber; *cracks*, from the yielding of the cohesion of the metal; *cavities*, cracks enlarged by the action of the gas, and by the melting of the metal, observable especially in the upper surface of the bore.

Injuries of the second kind, which appear in front of the charge, are: *the lodgment of the shot*, a compression of the metal on the lower side of the bore, at the seat of the shot, caused by the pressure of the fluid in escaping over the top of the shot. There is a corresponding *burr* in front of the lodgment; and the motion thereby given to the shot causes it to strike alternately on the top and bottom of the bore, producing other *enlargements*, generally *three* in number; the first, on the upper side, a little in advance of the trunnions; the second, on the lower side, about the astragal; the third, in the upper part of the muzzle.* It is chiefly from this cause that brass guns become unserviceable; the extent of the injury varies according to the length of the bore. *Scratches*, caused by the fragments of a broken shot, or the roughness of an imperfect one. *Enlargement* of the muzzle, by the striking of the shot in leaving the bore. *Exterior cracks*, or longitudinal splits, caused by too great a compression of the metal on the interior.

The durability of brass cannon may be much increased by careful use, and by the precautions of *increasing the length of the cartridge*, or that of the *sabot*, or using *a wad over the cartridge*, in order to change the place of the shot; by *wrapping the shot in woolen or other cloth*, or *in paper*, so as to diminish the windage and the bounding of the shot in the bore. In field *guns*, the paper cap which is taken off from the cartridge should always be put over the shot.

Iron cannon are subject to the above defects in a less degree than brass, except the corrosion of the metal, by which the vent especially is rendered unserviceable from enlargement. The principal cause of injury to iron cannon is the *rusting* of the metal, producing a roughness and enlargement of the bore, and an increase of any cavities or *honey-combs* which may exist in the metal.

Iron has no elasticity, and less tenacity than bronze when subjected in guns to the violent shocks of gunpowder; it does not yield like bronze, but losing gradually its cohesion, finally bursts at the breech suddenly and without any external indication of the

* These shocks are sometimes so violent as to break the projectile, the fragments of which form furrows and ridges which prevent the introduction of the shot afterwards. Bronze pieces, after long use, show these indentations at the mouth, and become then altogether inaccurate in their fire. They ultimately fail in the *chase*.

danger; the fracture generally passing through the vent. Iron guns are more injured by exposure to the weather than bronze.

It is considered that after 1200 rounds iron guns are not safe; but bronze guns have been fired as high as 2400 times without failing. An important discovery was made by Gen. Piobert, by which the strain on guns was reduced, and their durability consequently increased. This consists in reducing the *diameter* of the cartridge, which, giving a greater length of bore between the bottom and the projectile, allows more room for the gas to expand; and a part of the force which was before exerted on the piece is transferred to the projectile, giving an increased velocity. This discovery is of the more importance, from the fact that guns under some circumstances have been rendered unserviceable after fifty shots.

The injuries to mortars and howitzers, progress more slowly than in guns. The trunnions in mortars are sometimes bent from using heavy charges.

The service to which an iron cannon has been subjected may generally be determined by the appearance of the vent.

SPIKING AND UNSPIKING CANNON, AND RENDERING THEM UNSERVICEABLE.—To spike a piece, or to render it unserviceable, drive into the vent a jagged and hardened steel spike with a soft point, or a nail without a head; break it off flush with the outer surface, and clinch the point inside by means of the rammer. Wedge a shot in the bottom of the bore by wrapping it with felt, or by means of iron wedges, using the rammer or a bar of iron to drive them in; a wooden wedge would be easily burnt by means of a charcoal fire lighted with the aid of a bellows. Cause shells to burst in the bore of brass guns; or fire broken shot from them with high charges. Fill a piece with sand, over the charge, to burst it. Fire a piece against another, muzzle to muzzle, or the muzzle of one to the chase of the other. Light a fire under the chase of a brass gun, and strike on it with a sledge to bend it. Break off the trunnions of iron guns; or burst them by firing them with heavy charges and full of shot, at a high elevation.

When guns are to be spiked temporarily, and are likely to be retaken, a *spring spike* is used, having a shoulder to prevent its being too easily extracted.

To *unspike a piece*:—If the spike is not screwed in or clinched, and the bore is not impeded, put in a charge of powder of $\frac{1}{8}$ the weight of the shot, and ram junk wads over it with a handspike, laying on the bottom of the bore a strip of wood with a groove on the under side containing a strand of quick-match by which fire is communicated to the charge; in a brass gun, take out some of the metal at the upper orifice of the vent, and pour sulphuric acid into the groove for some hours before firing. If this method, several times repeated, is not successful, unscrew the vent piece, if it be a brass gun, and if an iron one, drill out the spike, or drill a new vent.

To drive out a shot wedged in the bore:—Unscrew the vent piece, if there be one, and drive in wedges so as to start the shot forward, then ram it back again in order to seize the wedge with a hook; or pour in powder, and fire it after replacing the vent piece. In the last resort, bore a hole in the bottom of the breech, drive out the shot, and stop the hole with a screw.

PRESERVATION OF ORDNANCE.—Cannon should be placed together, according to kind and calibre, on skids of stone, iron, or wood, laid on hard ground, well rammed and covered with a layer of cinders, or of some other material, to prevent vegetation.

Guns and long howitzers.—The pieces should rest on the skids in front of the base ring and in rear of the astragal; the axis inclined at an angle of four or five degrees with the horizon, the muzzle lowest; the trunnions touching each other; or if space is wanting for that arrangement, the trunnion of one piece may rest on the adjoining piece, so that the axes of the trunnions is inclined about 45° with a horizontal line; the vent down, stopped with a greased wooden plug, or with putty or tallow. If circumstances require it, the pieces may be piled in two tiers, with skidding placed between them, exactly over those which rest on the ground; the muzzles of both tiers in the same direction and their axes preserving the same inclination.

Short howitzers and mortars.—On thick planks, standing on their muzzles, the trunnions touching, the vents stopped.

Iron ordnance should be covered on the exterior with a lacker impervious to water, the bore and the vent should be greased with a mixture of *oil and tallow*, or of *tallow* and *beeswax* melted together and boiled to expel the water. The lacker should be renewed as often as requisite, and the grease at least once every year.

The lacker and grease should be applied in hot weather, in order to enter the pores of the metal.

The cannon should be frequently inspected, to see that the moisture does not collect in the bore.

In England, France, Belgium, and Sweden, howitzers and mortars take their denominations, as with us, from the diameter of the bore, or from the calibre of a gun of corresponding bore; in Austria and Prussia, from the weight of a stone ball of the calibre of the bore; in Russia, from the true weight of the shell.

CHAPTER III.

FORM OF CANNON, MATERIALS, &c.

Allusion has been made to the defects in the form given to cannon, which will now be treated of.

The ancient method of forming guns from a number of pieces, and hooping them together, after the fashion of a barrel, no doubt gave rise to the use of *bands* and *mouldings* on the exterior, with an idea that they were strengthening the piece. Latterly, these have been modified in order to produce a certain *finish* in the appearance of the piece, it being presumed that even if they did no good, they certainly could be productive of no harm. We will try to prove that this is altogether a mistake, and that these *bands*, *mouldings*, &c., are actually the cause of weakness to the gun.

Fracture.—In Robert Mallet's work "On the Construction of Artillery," many important facts bearing upon this subject, are laid down; and among the rest is the general direction and course of the lines of fracture in cast-iron guns which have burst under fire. It is generally laid down that such guns burst through the vent, that being a weak or starting point for the action of the powder. From there the line of fracture passes along the axis to the front of the trunnions, where it turns off to the right or left, or both, leaving the rest of the chase entire. That this is by no means always the case it is not necessary to state; but the heavy dotted lines in figures 49 and 50 will show the most usual course for the fracture. The eye is supposed to be above the gun, placed vent upward.

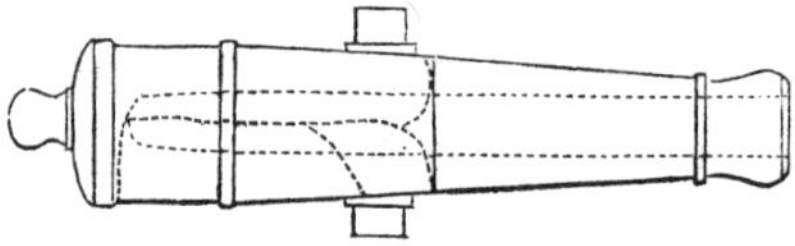

Fig. 49.

On examining a burst piece, the dividing plane in the direction of the axis will be found suddenly curved on one or other of the sides near the exterior of the gun, as shown in section in Fig. 51, Pl. 7; showing that the fracture commenced at one side, *s*, opposite the inflected portion, and spread from that, the parts dividing and turning from each other upon the point of inflection at *f*.

Fracture, therefore, appears to commence at the interior, rending the metal apart from within to without, a result agreeing with mathematical investigations, since the metal must yield first where the pressure per square inch is greatest upon its resisting unit of section, and this is on the interior.

Planes of Weakness.—The planes of fracture follow the track, with almost unerring precision, of all *re-entering angles* on the exterior of the gun. Thus, although the longitudinal fracture often passes through the vent, it more frequently passes along the re-entering angle formed by the lock-piece with the gun; whilst the transverse fractures follow the edges of the base-ring and reinforce mouldings, and those near the trunnions, the re-entering angles formed by these with the body of the piece. These general directions of the fracture cannot be the result of accident; there must be some cause to place the fracture there in preference to other parts of the gun.

Mallet says, "It is a law (though one I do not find noticed by writers on physics) of the molecular aggregation of crystalline solids, that when their particles consolidate under the influence of heat in motion, their crystals arrange and group themselves with their principal axes in lines perpendicular to the cooling or heating surfaces of the solid,—that is, in the lines of direction of the heat-wave in motion, which is the direction of least pressure within the mass; and this is true whether in the case of heat *passing from* a previously fused solid in the act of cooling and crystallizing on consolidation, or of a solid not having a crystalline structure, but capable of assuming one upon its temperature being sufficiently raised by heat applied to its external surfaces, and so *passing into* it.

If one of these crystalline bodies be cast in the form, say of an ingot, and "broken when cold, the principal axes of the crystals will always be found arranged in lines perpendicular to the bounding planes of the mass; that is to say, in the lines of direction in which the wave of heat has passed outwards from the mass in the act of consolidation."

The same effect is produced by applying heat far below that of fusion, to the surface of solids of similar substances.

If a cylinder of lead four or five inches long, and as many in diameter, be cast around a cylindrical bar of iron one and a half inches in diameter, and two or three feet long, the lead, cooling rapidly in contact with the cold iron, will have a perfectly homogeneous structure, entirely devoid of crystals. If the end of the iron bar be now placed in a furnace and heated red hot, and time be allowed for the heat to pass along the bar and into the lead until the temperature is raised to about 550° Fahr., it will, when struck several blows with a hammer, all fall to pieces, and show a complete crystalline structure with the principal axes of the crystals radiating from the center of the cylinder. See Fig. 52. Pl. 7.

If a flat, thick piece of malleable or rolled zinc which, if not homogeneous, has its fibres lying in the plane of the plate, be laid flat upon a cast-iron plate, heated to within a few degrees of the melting-point of the zinc, it very soon becomes crystalline, the axes of the crystals being now all arranged perpendicular to the sides of the plate; the planes of internal structure being actually turned round 180°.

Any one who has ever remarked upon the formation and melting of ice, will have seen the same principles exemplified on a grander scale. The little sharp crystals, as they form, are lying flat upon the surface of the water. When, in the spring, the ice becomes rotten from the heat it has absorbed from the water beneath and the air on top, these crystals will all be found arranged vertically, and are easily pushed through by the weight of the foot or a stick. Large floating cakes of ice have been known to disappear on being struck by a vessel, or on striking the shore, called sometimes the *sinking* of the ice. The shock destroys the little cohesion remaining between the crystals, and the whole mass falls apart.

Cast iron is one of the substances which, in cooling, obeys more or less perfectly this law, so that it may be laid down as a fact, that the crystals of an iron casting arrange themselves perpendicularly to the exterior surfaces. The crystals being small, this arrangement is not very apparent to the eye. Their development depends upon the character of the iron and the size of the casting; the largest casting presenting the largest and coarsest aggregation of crystals, but not the most regular arrangement of them, which depends upon the rate at which the mass is cooled, and the regularity with which heat has been carried off by conduction from its surfaces to the surrounding mould. Hence, in casting by what is called the "*chilling*" process, where the mould is cold, thick iron, whose high conducting power carries off the heat rapidly, the most complete crystalline structure occurs perpendicular to the chilled surface.

The arrangement of the crystals in castings of different forms, can be better seen by referring to the figures which represent sections of the different forms.

Fig. 53, Pl. 7, is a section of a round bar where the crystals all radiate from the center.

In the square bar, Fig. 54, they are arranged perpendicular to the four sides, and hence in the diagonals of the square, the terminal planes of the crystals abut or interlock; along these lines the crystalization is always confused and irregular.

In the flat bar, Fig. 55, the crystals are arranged as in the square, with an extension in one direction.

In Fig. 52 is shown the arrangement in a hollow cylinder.

Fig. 56 represents a portion of the closed end of the cylinder of an hydraulic press which broke under a great pressure, the end of the cylinder coming out in the form of a flat frustrum of a cone, showing that the planes where the crystals perpendicular to the different surfaces join confusedly together, are *planes of weakness*, along which the cohesion of the metal is less than in any other part.

In consequence of the failure of this cylinder, the form was changed to that represented in Fig. 57, where the direction of the crystals, changing gradually, formed no planes of weakness; and this stood the pressure without breaking. The similarity of these forms to the breeches of guns will be readily recognized.

Figs. 58 and 59 are sections of different re-entering angles found on guns as now cast; and show the planes of weakness resulting therefrom. Fig. 58 is through the lock-piece; Fig. 59, through a trunnion; and Fig. 60, through a reinforce band, or chase-ring; and all show planes of weakness to exist exactly where the planes of fracture pass in burst guns, as will be seen by referring back to Figs. 49 and 50. "The conclusion, therefore, seems inevitable, that however incapable the unaided eye may be to discern any difference in the crystalline arrangement of one part of the gun more than another, such planes of weakness do exist, in the positions, and from the causes here pointed out."

It follows, then, that in casting our guns, all *sudden* changes in the direction of the surface should be avoided; that all unnecessary projections, such as the chase and base rings, reinforce bands, mouldings and muzzle bands, should be dispensed with; and that the piece when finished should present as unbroken a surface as possible, and with a rounding (if possible spherical) breech; so as to avoid the great planes of weakness now formed there, which there is little doubt were the cause of the great number of failures among the columbiads of the old pattern. With such a form it is worse than useless to add *more* metal. The defect increases with the size of the casting, and the strength gained is not enough to compensate for the increased weight. This fact is beginning to be appreciated in our service; for very great changes have been made in the new columbiads, by omitting the chase and base rings, rounding off the breech, and reducing the swell of the muzzle. The rounding of the breech has not yet been carried sufficiently far, probably on account of its interfering with the elevating notches; though these have not as yet been placed on the new pieces. This, however, can be no serious objection, since if once the *strongest form* is obtained, mechanical ingenuity will very soon invent a means of manœuvering the piece.

It may be objected that these bands, ornaments, &c., are not cast on the piece, but turned from it after the casting is cool, and can therefore have no effect upon the arrangement of the crystal. Not certainly *at first;* but it must be remembered, that the law quoted applies as well to the body when receiving heat as when giving it off; and that after a gun becomes heated by repeated firing, the heat, having passed through from the inner to the outer surface, has produced its effect upon the crystals, and arranged them normal to the surfaces by which the heat goes out. Hence, old guns which have been a great deal used take a crystalline form, and are liable to fall to pieces at any time from the shock of gunpowder, a fact not often appreciated until too late at our popular 4th of July and other celebrations.

This is by far the most satisfactory way of accounting for the weakness and bursting of old guns. Cast-iron guns have not the tenacity of bronze, which commence splitting long before they burst. But the change gradually takes place in the *metal*, and when, after long use the shock becomes too great to be borne, the mass flies apart like a huge cake of ice in the spring, which in the winter, before its crystalline structure was *upset*, was capable of supporting thousands of tons.

In testing columbiads within late years, it has been found that it is almost impossible to make them strong enough to withstand the proper number of charges, notwithstanding that the very first quality of iron was used, and one which, as a metal, had borne the very highest trials for tenacity and strength. After attributing the failures to quick powder, and various other reasons, the expedient of changing the form of the piece has been finally resorted to, and it is to be expected with very beneficial results. These trials generally take place as soon as a gun is finished. But it is more than probable that if left for years (during which time a constant change is going on in the structure of the gun), the metal would assume a permanent state or *set;* just as a wagon bow does, the fibres of which, after a time, accustom themselves to the new direction, which then becomes the natural one.

In casting guns, some parts which are smaller than others, will cool the fastest, forming strains in the piece which are liable to act injuriously. Thus, for instance, the neck of a piece will cool much before the body, thus cutting off the supply of liquid metal from the dead head; and a vacant space or cellular portion of metal is left in the interior of the gun, having very little tenacity. On this account the models for guns are not now made so sloping towards the muzzle, but approach more nearly a cylinder in shape, and the gun is afterwards turned down by machinery. This is found to render the piece very much stronger.

The exterior of course cools first, and successive layers become solidified and contract until the centre is reached, the particles of liquid metal being drawn out towards the cooling surfaces. The tendency of this is to produce a state of internal tension in the molecules of the metal, tearing away the external portions from the internal nucleus, and producing along the axis a line of weakness, where the metal is soft, porous, and with coarse and separated crystals, leaving, in spite of the constant feed of liquid metal through the dead-head, actual cavities in the centre of the casting.

Fortunately, in pieces of artillery this portion is nearly all bored out; but where the boring does not extend well back to the exterior of the breech, a portion of this soft spongy mass remains, forming the bottom of the bore. This is more especially the case in very large mortars; and the defect increases, of course, as the mass of the casting becomes larger. Figure 61, Pl. 7, represents a large mortar with the dead-head still on, and the bore marked out in black lines. The weak point is evidently at the bottom of the chamber.

A remarkable exemplification of this weakness was exhibited in the English mortars (13-in.) used in the bombardment of Sweaborg during the last war with Russia. Three of these mortars burst after firing from 100 to 300 rounds. They split into two almost equal halves, in a plane passing through the axis and vent, and exhibited no defect or injury except just at the bottom of the chamber, where "a small, irregular cavity was found, with jagged sides and bottom, as though burrowed into by some corroding agent." The metal of which they were made appears to have been of a very indifferent quality, but pieces taken out near the muzzle, inside the bore, showed a mixed metal of the very coarsest kind; and could the part bored out have been examined, it would no doubt have shown a much worse texture. Most of the other mortars showed similar defects.

Hollow Casting.—To avoid these defects, it has been proposed to cast guns, more particularly heavy mortars, hollow, or on a core, and in this way to have in cooling a hollow cylinder, which we have shown to be the strongest form. The piece may be cast on a core placed in the centre of the mould, just as a shell is cast.

Capt. Rodman, of the ordnance department, has for some time been making experiments on a plan of his for casting guns hollow, which promises very important results.

A core is formed on a water-tight cast-iron tube, closed at the lower end. By means of an interior tube in the center of the other, and open at the lower end, a stream of water is conducted to the bottom of the larger tube, and rising through the circular space between the two, flows out at the top. A fire is built around the jacket at the bottom of the casting pit, and the gun-mould kept at nearly red heat. Twenty-five hours after casting, the core is withdrawn, and the flow of water continued through the space left by it for forty hours longer.

The amount of water used is about fifty times the weight of the casting, and the heat imparted to the water, and carried off by it, is equal to 60° on the whole quantity used. This is in casting an 8-inch columbiad. For larger pieces, the amount of water and time of cooling are greater.

All the guns cast in this way present a marked superiority in endurance, over those cast solid in the ordinary way; and the 8-inch columbiad mentioned above, sustained 1,500 rounds, including proof charges, without bursting; whilst another of the same calibre, cast solid, from the same metal, at the same time, and under precisely the same circumstances, failed at the seventy-third round.

In the ordinary method of casting, all the cooling takes place on the outside. The temperature of the metal being much above the melting point, cools down until the exterior surface reaches the freezing point, when a thin layer is formed by congelation on the exterior, inclosing within it the liquid metal.

Another layer is now formed within this and upon it, and in the meantime the first layer loses a portion of its heat, and its temperature falls below the freezing point and below that of the next layer. Next, a third layer is formed, and so on until the whole becomes solid; at which time, if we suppose the outside layer to have arrived at the temperature of the surrounding bodies, the difference between its temperature and the center of the casting will be about 2700°.

The outside layer has now ceased to contract, whilst the interior ones are just beginning; and the effect of such a contraction must be to effect a very injurious strain upon the exterior, and on the interior to form a very porous, weak metal, especially if the supply of liquid metal is cut off, as in the case of casting guns with a narrow part at the neck.

The case supposed where the difference between the temperatures of the interior and the exterior is a maximum, is, of course, an extreme one, which will never occur in practice; but it serves to explain the law relating to the contraction of iron in cooling.

The more equally and uniformly a casting cools, the less liable it is to be weakened by these strains; and the merit of Capt. Rodman's invention evidently consists in retarding the cooling of the outside by the application of heat, and hastening that of the inside by the application of cold, thus rendering the cooling of the whole mass slower and more uniform, and preventing the formation of those successive layers of different temperatures; or if they are formed at all, making them commence on the interior, by which the strain is *reversed* and actually made to add to the strength of the piece.

The method can only be perfected by experiments by which the proper degrees of heat and cold to be applied may be determined; for to cool the piece *too* rapidly on the interior, would be even worse than too rapid exterior cooling.

Capt. Rodman has also suggested an improvement to Col. Bomford's method of determining the thickness of metal at different points. This consists in substituting for the pistol-barrel and bullet, a piston having a punch at one end. This is put in the hole made in the gun, and a block of copper so placed that as soon as the piston is acted on by the inflamed charge, the punch is forced into the surface of the copper, making a certain indentation which is afterwards compared with one made in the same block of copper by the same or a similar punch in a machine where any amount of pressure can be given by the application of weights. This, although not an accurate process, tne forces applied in the two cases being so different in their nature, gives probably the nearest approximation yet obtained.

Capt. Rodman has met with such success in his plan of hollow casting, that he is now about to cast by direction of the government, a gun of 15 inches in calibre, which will probably be a complete test of the system; since, if such an immense piece can be formed with a corresponding strength, there will remain no doubt of the advantages of this method of casting.

Trunnions.—Another great cause of the want of strength in guns as at present formed, is the *position* of the trunnions, as regards the *point* to which the recoil of the gun is transmitted.

The existence of these trunnions, by forming great re-entering angles on the surface of the gun, is of itself a great cause of weakness, as has already been explained; but their position on all pieces except *mortars*, has a weakening effect in a different way, which will now be explained.

Let M be the mass of a shot, and D its diameter, v=its initial velocity, and M' the mass of the gun neglecting the mass of the powder. The "work done" by the explosion will be$=Mv^2+M'v'^2$; the quantity of work of the recoil being$=M'v'^2$, which, transferred against the interior of the breech, tends to *stretch* the metal from the section y to the muzzle. Fig. 50.

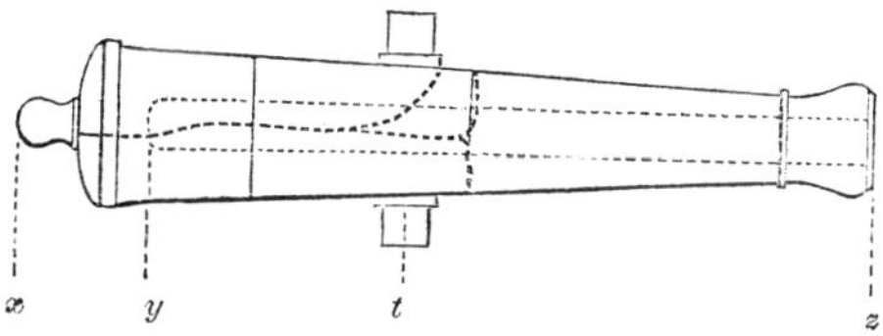

Fig. 50.

The rate of propagation is very rapid, so much so as to present to the eye the appearance of a simultaneous recoil of the whole gun. The first effect, however, is to elongate the gun, successively straining each portion. The gun resumes its original length, provided the elongation is far within the elastic limits of the metal.

If the gun, instead of resting on its trunnions, abut firmly against a fixed object at the breech, x, the segment in rear of the cartridge will be *compressed* by a force equal to the whole recoil in the direction yx, while the other portion of the gun will be *extended* by a force in the direction yz, which, at the section y, is equal to the recoil, and at the muzzle is$=0$. If F, then, be the work done to tear the gun in two at y, and t=the time the shot takes to reach the muzzle,

$$\frac{M'v'^2}{2}t=Ft,$$

will be the longitudinal strain on the gun.

If the breech be unsupported at x, the strain tending to tear it off at

$$y=\frac{M'v'^2}{2}t.$$

diminished only by the inertia due to its small mass between y and x, or

$$Ft=\left(\frac{M_{\prime}v'^2}{2}-\frac{mv'^2}{2}\right)t.$$

If the gun be fixed rigidly on trunnions placed in the usual position at t, the strain tending to tear them off is equal to the whole work done by the recoil. The bearings always yield some, however.

If M'' be= the mass between x and t, the tendency to tear the gun in two at the trunnions, is

$$\left(\frac{M'v'^2}{2}-\frac{M''v'^2}{2}\right)t.$$

This extension at t at the first moment is, on the principle already stated, followed by a compression in the direction zt, of equal amount; and so for any other position of the trunnions.

The amount of extension or compression—assuming the gun of equal transverse section of metal throughout its length, and its elasticity perfect—being proportionate to the extending or compressing forces, it follows that the longitudinal strain on the gun, due to recoil, will be a minimum *if the trunnions be placed at the farthest point to the rear*, as in mortars, or if the breech be rigidly fixed there, and supported, without trunnions, against the recoil.

The metal of a gun is, therefore, subjected at the same time to three different strains, which act at right angles to each other. 1st. The tangential or bursting, which is extensional, and accompanied by—2d, compressive or radial strains, which are normal, and—3d, the longitudinal and re-active strains of the recoil, which are chiefly extensional. There are good grounds for presuming that the existence of the latter two tend, to a certain extent, to weaken the resistance of the metal to the former.

Could, therefore, a convenient method be invented for supporting guns upon their carriages without the use of trunnions, the recoil being received by a support entirely in rear of the piece, the trunnions could be dispensed with, and the gun very much strengthened in two ways. Muskets and other small arms arms are, therefore, arranged in the way best suited to their strength, and they can therefore be made much lighter than would otherwise be safe.

Whether designedly or not, the ancient guns possessed this element of strength, for they were without trunnions, and transmitted the recoil to supports placed behind. Being of immense size, this might have been the result of calculation, provided trunnions had ever been used on guns in early times; but it is probably the case that they were an invention of a later day.

It is, therefore, concluded, that were these causes of weakness in guns obviated in the way proposed,—1st, by forming the piece perfectly plain, and with no re-entering angles; and 2d, by dispensing with the trunnions,—they would become much stronger, and could be made to support the same amount of strain by using less metal, thereby reducing the weight.

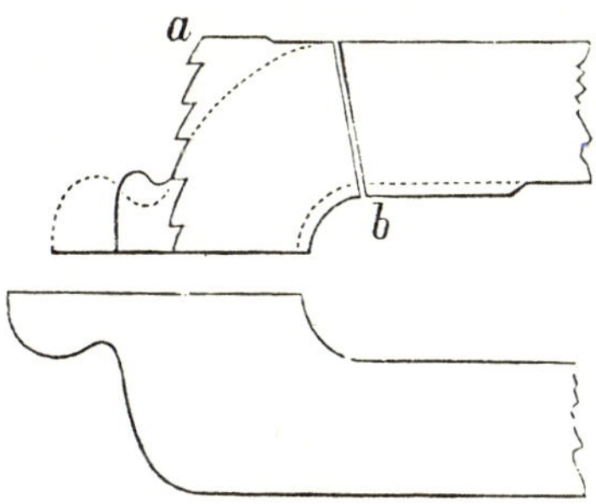

Fig. 62.

By looking at Fig. 62, it will be seen what, according to the preceding remarks, is regarded as the weak point of our present columbiads. The sharp angle at the breech forms a plane of weakness in the direction *a b*; and the additional metal outside of the dotted circle, instead of strengthening the piece actually makes it weaker.

This fact is beginning to be appreciated by our authorities; for finding it impossible to obtain the proper strength with the present form of the columbiad, they have been compelled to change it to that represented on the lower half of the figure. But this, although an approximation to the theory of Mallet, stops short of it; and I think a gun formed as represented by the dotted line on the upper half of the figure, would give greater strength, at the same time that less metal would be used and a lighter piece be obtained.

The principle that sharp angles in castings do cause weakness being acknowledged, it is to be presumed that other reasons, besides a disinclination to make such a complete change in the form of the piece, exist to prevent our authorities from following the theory more strictly; and among these is probably the fact that a gun so formed would require a great change in the present carriage, and possibly an entirely new one.

The breech of the "Dahlgren" gun, now so well known in our navy, seems to have been formed in accordance with the principles just laid down; and a paragraph in a letter from its distinguished inventor is so directly to the point in the matter, that it is inserted. He says,—

"It would be an egregious error to suppose that the resistance demanded for the gun in its different parts, is to be obtained solely by the thickness of metal; other controlling considerations come into question, as I think you will admit upon noticing the dimensions of different models of guns, and the leading fact that in scarcely any instance has a weak gun been wanting in *quantity* of metal."

The iron used in the manufacture of these defective pieces was of the very best quality, so good in fact, that it began to be suspected that it was *too good*, and that it was necessary to mingle with it some of inferior quality, in order to produce those peculiar qualities of metal most suitable to resist the force of gunpowder.

As if to completely demonstrate that in the form lay the defect of the pieces, an incident occurred some time ago at Boston where an opportunity was offered of comparing the two forms of the army columbiads and navy shell guns.

Some of each were cast from the same metal at the same time and by the same persons, and when completed were proved. The columbiads 8-in. failed under proof, two bursting by service charges at the 180th and 45th rounds. The navy guns 9-in. succeeded completely, enduring 1500 rounds without bursting.

The cause of the failure of the columbiads being supposed to be due to the high bursting quality (*brisante*) of the powder used, the navy department caused another gun of the same lot to be tested with powder similar to that used with the columbiads. The gun sustained 1500 service rounds, appeared then to be in good order, and was finally burst with proof charges at the 23d round, the shot having been gradually increased to 7, weighing about 650 lbs.

The pieces were about the same weight, about 9,000 lbs. A 68-lb. shot and 10 lbs. of powder were used with the columbiads, and a shell weighing 70 lbs. and 10 lbs. of powder, with the navy guns.

In connection with this subject of endurance, an important fact was noticed in making some experiments in 1852. It was found that the length of time that a piece had been cast had a very great influence upon its endurance.

Three 8-in. columbiads of the same form and dimensions, and cast in the same way, were tried. One of them had been cast only a few days, the other two were six years old, and of metal of inferior strength to the first. The one tested immediately after casting failed at the 72d round; of the other two, the weaker failed at the 800th round, and the other sustained 2582 fires without yielding.

Here is an apparent anomaly which appears to be satisfactorily explained by the supposition that the strain which has been referred to as existing in the gun when cast is limited in duration, and that, like many other substances, iron possesses the property of accommodating itself to an unnatural position, and finally of adopting this as its natural one, and of actually being as strong or even stronger in that than in the original state; just as a wagon-bow or barrel-hoop after becoming used to the new direction, will not only not return, when released, to the old, but require force to bring it back, showing that the fibres possess the power of accommodating themselves in accordance with the solicitation of external forces. There is nothing in the character of iron which precludes this idea of a new arrangement of the fibres; on the contrary, as has been shown, it does take place under certain circumstances.

It would therefore seem to be advisable and of benefit to the strength of ordnance, that it should be allowed to remain for a certain length of time undisturbed after casting and before being tested; and, most probably, the longer this period the greater becomes the strength of the piece.

MATERIALS FOR ORDNANCE.

Iron.—It is laid down that iron for our cannon must be of the best quality of *charcoal* iron, made in a smelting-furnace, with a cold blast, and should be selected especially with regard to its strength. There are, however, many important details necessary to be observed in making the casting, and without which defects are almost certain to result: among these may be mentioned the temperature of the metal at the time the casting is made, the number of times it has been melted, the method of making the casting, &c., all of which vary according to the particular kind or "make" of metal which is used. An accurate knowledge of all these details is necessary to a successful founder; but to enter deeply into them would scarcely be within the scope of this work.

Gray.—The iron suitable for cannon should be soft, yielding easily to the file or chisel; its fracture an uniform dark gray, brilliant appearance, with crystals of medium size.

This iron is distinguishable from the other principal division (*the white*), by being softer and less brittle, slightly malleable and flexible, and not sonorous. The color of its fracture is lighter as the grain becomes closer, and the hardness increases at the same time.

A medium-sized grain, bright gray color, lively aspect, fracture sharp to the touch, and a close compact texture, are the characteristics of a good quality. If the grain be very small or very large, or the iron present a dull, earthy aspect, loose texture or dissimilar crystals mixed together, the metal is of an inferior quality. If the iron is too soft, loose and coarse, its strength and elasticity may be increased by remelting it once or twice, and by continuing it in fusion several hours under a high heat.

Gray iron melts at a lower temperature than white, becomes more fluid, and preserves its fluidity longer; it runs smoothly; the color is red, and deeper in proportion as the heat is lower; it does not stick to the ladle; it fills the mould well; contracts less, and contains fewer cavities, than white iron; the edges of a casting are sharp, and the surface smooth and convex.

The mean specific gravity of pig-iron is 7.0, and its tenacity, about 16,000 lbs. to the square inch, which are both increased in the gun, the former slightly, the latter to between 25,000 and 30,000 lbs. The tenacity and specific gravity of gray iron are diminished by annealing; whilst in white iron the former is increased and the latter diminished by the process.

White iron is very brittle and sonorous; it resists the file and chisel, and is susceptible of high polish. The fracture presents a silvery appearance, and, compared with gray iron, is comparatively smooth. The differences between the two, when placed side by side, are very marked. When melted, it is white, and throws off a great number of sparks, but is seldom used by itself, without being mixed with a superior quality of iron. A mixture of the two kinds (white and gray) is generally used in making shot and shell, where strength is not of the same importance as in guns.

Besides these two divisions, manufacturers distinguish more particularly the different varieties of pig-iron by numbers, according to the hardness, No. 1 being the gray, and No. 6, the white iron.

The qualities of these different kinds of iron depend very much upon the proportion of carbon, and the state in which it exists in the metal. In the darker kinds, where the proportion is sometimes 7 per cent., it exists partly as graphite or plumbago, which softens the iron. In white iron the carbon is thoroughly combined with the metal, as in steel.

Swedish Castings.—The tenacity of gray cast iron is much lessened by using intense heat in the blast-furnace; and, where castings are required to offer great resistance, it should not be employed. That obtained from less refractory ores, and in furnaces of lower temperature, gives greater strength, provided it is not too gray, *i. e.*, too graphitic, and does not expel too large a quantity of graphite (in cooling).

In certain cases, neither gray iron nor white should be used for cannon. Gray iron produced from a mixture of refractory ore and flux (any material used to assist the reduction of the ore, as limestone or clay), is unsuitable, because it then contains a large proportion of earthy metals; but even with readily fusible ores it is extremely difficult so to work the smelting-furnace that the pig-iron shall neither be too gray nor too white, either of which is equally injurious for gun-founding.

The plan pursued in Sweden, where they have the finest cast-iron guns in the world, is to charge the furnace partly with roasted and partly with raw ore, and the blast is so regulated that the yield shall be regular, and the slag, or refuse, nearly colorless. These two kinds of ores, having different degrees of fusibility, are reduced after different periods in the furnace, and, hence, afford, one a gray and the other a white iron. If the minerals are properly proportioned, there is obtained a very finely mottled gray iron, which is said to be less porous, harder, and more tenacious than the gray iron obtained in the ordinary way.

Cast iron frequently retains a portion of foreign ingredients from the ore, such as earths, or oxides of other metals, and sometimes sulphur and phosphorus, which are all injurious to its quality. Sulphur hardens the iron, and, unless in a very small proportion, destroys its tenacity.

Testing.—These foreign substances, and also a portion of the carbon, are separated by melting the iron in contact with air; and soft iron is thus rendered harder and stronger. The kind of iron most suitable for casting guns should be ascertained by trial in the furnace in whichit is to be used; and a trial-gun is made and tested to extremity. The piece adopted for this proof, is a long 9-pdr., which is fired with the following series of charges, and is required to sustain the first four without breaking.

1st.	20	rounds,	3	lbs.	of powder,	1 ball.	Without	wads.
2d.	20	"	4.5	"	"	2 balls.	"	"
3d.	10	"	4.5	"	"	3 "	"	"
4th.	5	"	9	"	"	6 "	"	"
5th.	—	"	18	"	"	13 "	"	"

Samples of the iron are also taken from the castings, and tested by an ingenious machine invented by Major Wade, in which the tenacity of the iron and its capacity to resist compression, a bursting force, and transverse and torsional strains, are all measured in pounds.

This method of applying the strains to iron can, of course, give but an approximate result with regard to the strain resulting from the use of powder in the gun. It would appear that a bursting force, where a steel cone is forced through a cylinder of the tested metal, ought to represent more nearly than any of the others, the real action of powder in a gun. But even there, the force applied is by a steady, regularly increasing pull, whereas the force of the powder acts suddenly, and, as it were, by a shock; so that the one power cannot accurately represent the other.

The nearest approach which has been made to the real action of powder in a gun, comes to us from one of the European arsenals, where very slightly conical steel punches are forced into conical openings of the same inclination formed in small cylindrical specimens by the action of a given weight falling from a known height.

Alloys.—Attempts have been frequently made to improve the quality of cast iron for guns, by mixing with it foreign metals in minute proportions. Copper, tin, lead, and manganese have thus been tried, but with no favorable results. The affinities of iron, in forming alloys, are very slight, and its previous combination with carbon, for which it possesses so powerful an affinity, seems to reduce the former to so low a point that its alloys are little better than heterogeneous mixtures, which separate on cooling after being heated.*

All attempts, too, to form guns compounded of different metals, placed one within the other, with the idea of gaining strength, have failed. In this way bores of cast iron, surrounded with rings or jackets of bronze or wrought iron, to obviate the interior defects of bronze, and strengthen the outside, or protect it from the weather by a better metal, have been formed, but soon abandoned.

Wrought Iron, &c.—The yielding of the ball when a piece is fired, gives the inflamed gases an opportunity to expand, and prevents their effort being expended upon the piece; but for this, a piece made of unyielding material, and loaded with a tight-fitting heavy projectile, must give way to the force of the powder. If a piece could be made of elastic material which, yielding to the first force of the powder without breaking, would give time for a sufficient force to be developed to expel the ball, the same injurious effects upon the piece would not result.

Were a tube of glass and one of india-rubber to be made of the same strength, as determined by the ordinary methods, it may be readily imagined how much more easily one of these would fail from the explosion of a charge of gunpowder than the other, from its utter inability to yield to the first force of the powder.

In the manufacture of guns, this fact was early appreciated, and efforts were made to so construct them as to yield a certain amount to the force of the powder. It is mentioned that guns were made by placing over each other tightly-fitting layers of leather; but such pieces, as may be imagined, did not long retain

* With a good deal of mystery, the production of a metal "infinitely superior to bronze," for field pieces, is confidently predicted by a recent English writer. Such a metal, or probably *mixture*, could not long remain a secret. It is, probably, one of those hopeful chimeras of which the science of artillery affords so many instances.

their places before the rapidly advancing improvements in fire-arms.

Up to the present time, insuperable obstacles have presented themselves to the formation of cast-iron guns of a larger calibre than 10-inch. Beyond that, so many defects are liable to occur in the casting, it is not considered safe to go; though with mortars the calibre has been extended to 13-inch. As, so far, all attempts to make castings larger than this have failed, attention has naturally been directed to wrought iron, with the hope of increasing the calibre.

Modern artillery is here referred to; for in the construction of ancient artillery the reverse of this process seems to have been adopted, and the wrought iron tried first, or at least cotemporary with the cast. These ancient pieces, some of them as large as 33 inches in calibre, were made of longitudinal bars, hooped round with heavy wrought iron, shrunk on while hot. With these pieces, however, were used stone balls and the slow-burning, ungrained powder of early times, which did not require the same amount of resistance as iron balls and grained powder. The carriages too, from which they were fired, were very different from those now in use; and the pieces being without trunnions, the recoil was transmitted directly to the breech, which probably increased considerably the resistance and endurance of the pieces.

The great objection to wrought-iron guns, especially those of large size, is the great difficulty of forming a perfect weld between the parts, the effect of the blow decreasing very rapidly as the distance from the point struck increases. This imperfect welding, besides causing weak points in the piece, which are constantly made weaker by service, allows the penetration of moisture, causing oxidation which is rapidly destructive to the metal.

Wrought-iron pieces are usually made by placing together, in the form of a faggot, a number of bars of the metal, which are then welded together under a heavy hammer, other bars being added as the welding progresses, until the proper size is reached. During this time, the mass of iron has to be kept more or less heated. Experiments made on the metal taken from the gun, and compared with those made on the metal before forging, prove that this method of manufacture is highly injurious to the strength of the iron.

In the manufacture of the "Peacemaker," the bursting of which on board of the steamer Princeton in 1843, was so disastrous to the country, the above plan was pursued; and although the metal of which the piece was made was not of the best quality, it was found to have decreased in strength, from the intense heat applied in forging it, from 46.950 lbs. to the square inch, to 33,586, or more than 13,000 lbs.

Such a result ought alone to be conclusive evidence against the practicability of making wrought-iron guns of large calibre, by welding in a mass. For small calibres, the case is not so clear, and pieces have been made giving results which would seem to lead to an opposite conclusion. But from the fact that the principal powers of Europe, where these experiments have been made, still retain bronze as the metal for their field-pieces, it would seem that the question is far from being decided in favor of the wrought iron. It is stated that a 12-pdr. steel gun has been made in France which has been fired more than 2,000 times, without showing any signs of injury, except an enlargement of the vent, and that after bouching it with copper, the firing was to be continued.

How is it that ancient cannon, some of which are of immense calibre, and made of wrought iron, have shown such strength, lasted such a length of time, and now present an appearance so free from those defects which are stated as characteristics of the material? The reason is, that these guns, instead of being forged in one mass, like the "Peacemaker," were *built up* usually of longitudinal wrought-iron bars, beveled and fitted at the edges; only a small amount of heat and welding force being applied. On this hollow cylinder were driven heavy rings, applied when hot, and shrunk into position.

Such pieces would, of course, be liable to the attacks of rust; but the joints appear to be remarkably well fitted, and on examination, are found to be filled with rust, converted, by time, into crystalline hematite, as hard as the iron itself, the whole presenting the appearance of one solid mass.

Others of these ancient pieces are formed of a number of segments, which, put together when required for use, form the piece, and which can be taken apart for the purposes of transportation, &c. The carriage for such pieces was simply a large wooden trough, in which the piece was laid, abuting at the breech against a solid structure which received the recoil, the force of which was sometimes lessened by the interposition of a pad or cushion; and in case of trunnions being dispensed with on modern cannon, this arrangement might furnish a useful hint for forming a structure by which the recoil of the piece might be almost entirely overcome, by bringing into play the elastic force of material. This method of receiving the recoil, is of much more importance with these *built-up* guns than with any other, as the constant tendency of the forces exerted on a gun mounted in the ordinary way would be to separate the rings and parts, and this in a much greater degree than where the recoil is all transmitted directly to the breech.

Mallet's Mortar.—On a principle somewhat similar to this last one, Mr. Mallet has constructed his monster mortar, which has a calibre of *one yard.*

It consists of an immense breech of cast iron, into which is let a wrought-iron chamber; of a chase which consists of three large rings, all built up of smaller ones of wrought iron, rabetted into each other. These rings, jointed together, fit on top of the breech, and are strengthened and held firmly together and to the breech by six wrought-iron staves, which run along the outside of the rings and, passing through openings in the muzzle ring, and a projecting ring on the cast-iron breech, connect the whole system together. The binding power of these staves is regulated by quoin-like wedges driven through the ends of the staves beneath the projecting breech-ring, thus tightening the longitudinal binders by a blow, when necessary.

This immense structure, with its bed, weighs about fifty tons, and carries a shell weighing some twenty-five cwt. It is said to have cost $40,000, and is denominated by a recent English writer "a monster failure;" though, judging from the fact that they are now engaged in making experiments with the piece at Woolwich, and from the description given of its execution by one of our officers who saw it fired, this condemnation appears to be premature. There can be no question of the immense execution attending the fall of such a mass of metal, and the explosion of its charge of powder (480 lbs.); and should the present experiments determine favorably the practical questions involved, the effect upon the present system of engineering must be very great. No bomb-proof shelter or masonry vault, as now constructed, would be any protection against such shells.

Military men, however, may well be incredulous with regard to the practical success of such a system of mortars. It is true that such a piece at Sebastopol, *when once established* in position, might have materially shortened that memorable siege, by the devastating effects of a few of its shells. But such sieges as that do not often take place; and it may be considered as questionable in the great majority of those which do occur, if a piece of this kind, taking all things into consideration—the transportation, setting up of the piece, its slow firing, &c.—could not better be replaced by a number of smaller pieces, the rapidity of firing of which is much greater, and the points struck much more numerous.

Bronze, or brass, for cannon, is made of 100 parts of copper and about 10 of tin; a variation of 1 part of tin, more or less, being allowed. Bronze is more fusible than copper, much less so than tin. It is more sonorous, harder, and less susceptible of oxidation, than either copper or tin; less ductile than copper, more so than tin. It is harder than either. Its fracture is of a yellowish color, with little luster; a coarse grain, irregular, and often exhibiting spots of tin of a whitish color, which indicate defects in the metal. The specific gravity of bronze is about 8.700, which is greater than the mean of the specific gravities of copper and tin.

Copper, when pure, is of a red color, inclining to yellow, and has a fine metallic lustre. Its fracture exhibits a short, even, close grain, of a silky appearance. Copper is very ductile and malleable. The greater the purity of it, the more malleable it becomes, and the finer its grain. It has a specific gravity of from 8.600 to 9.00.

Tin, when pure, is of a white color, a little darker than silver. It is malleable, but not very ductile, though it can be rolled into sheets. It is very soft; and, when bent backward and forward, gives a peculiar crackling sound, the distinctness of which is in proportion to the purity of the metal. Specific gravity, from 7.290 to 7.320.

Analysis.—To analyze bronze, put 10 parts (say 100 grains) of it in small particles in a small glass matrass, with 80 parts of very pure nitric acid, at 22° Beaumé's hydrometer; heat it gradually to ebullition, and continue the heat until red vapors cease to come over. The acid dissolves the copper, and converts the tin into an insoluble peroxide. Let it settle; pour off the liquid, add to the oxide of tin 20 parts of nitric acid, and let it boil ten minutes; decant the liquid again, and repeat the same operation; dilute the first portion decanted with two or three times its volume of water, and pass it through a filter. Do the same with the second and third portions. Place the oxide of tin on a double filter, the two parts of which are equal, washing it until the water that comes off no longer gives a blue color when heated with ammonia, and does not change the color of litmus paper. Spread the filter on paper, and dry it perfectly in a stove or sand bath. Weigh it, adding the exterior filter to the weights, which will give the quantity of peroxide of tin which remains on the upper filter: 127 parts of peroxide give 100 parts of pure tin.

If the bronze contains lead, it will be dissolved by the acid. To detect it, add sulphate of soda after the solution is cool, in order to precipitate the lead in the state of an insoluble sulphate, 145 parts of which contain 100 of lead.

Composition.—It is a singular fact that the ancient composition of bronze, as in the case of gunpowder, has remained almost unchanged to the present day; and the metal from cannon is found to be almost identical in the proportions of copper and tin with the rude weapons of Scandinavian, Celtic, Egyptian, Greek, and Roman warfare. It has never been satisfactorily demonstrated, however, as in the case of powder, that this composition is the best that can be formed from the two metals for the manufacture of guns. For, strange as it may appear, it is asserted by Mallet that, after five hundred years' habitual use of the material, not a single series of systematized and accurate experiments on the physical properties of gun-metal has been undertaken; and that such a series is entirely beyond the reach of private means, or the demands of commerce, and can be carried on to successful and satisfactory results only as a national undertaking.

The amount of tin in bronze varies in different countries from 9 to 12.5 parts to 100 of copper: 12.5 parts are used in this country; and in France 11 parts is fixed by law as the proper amount.

As with many other metallic alloys, the combination between the two metals in bronze is so imperfect, that very slight forces are sufficient to cause its separation into two or more different alloys, which, on cooling, are found to occupy different portions of the mass. In casting a gun, for example, the outside which cools first has a constitution different from that assigned by the proportions of the metals, as fixed for fusion. The interior, which cools last, has another different from both, and always richer in tin. On being examined after cooling, portions at different heights are found to differ from each other; and this difference varies along the exterior and interior portions, so that no two adjacent portions have strictly the same chemical constitution; the maximum of copper being found in the exterior and breech of the gun, making these portions less flexible, and the maximum of tin in the interior and higher parts.

Specimens taken from the top and bottom of the casting, show also a very great difference in density and tenacity; the density at the breech being much greater, and the tenacity, in one instance cited, more than double. The dead head, in consequence of these facts, is made much longer in casting bronze guns than is other wise necessary.

It is found, that the constitution of the alloys changes not only in cooling, but in melting, by the reduction of the quantity of tin, which oxidizes much faster than the copper; and this takes place at such a rate that after six successive meltings the amount of tin is reduced one half.

The alloy used in France appears to result after cooling into one more nearly coincident with the atomic constitution, the small excess of tin being oxidized during the process of melting, and leaving the solid metal almost precisely in accordance with the atomic theory (17 Cu + S N). This more staple alloy is therefore considered the preferable one.

Cooling.—Very rapid cooling is considered most advantageous for bronze pieces; and as the cooling will be slower in proportion as the temperature of the metal when poured into the mould is higher, it results that, the lower the temperature at which the metal remains sufficiently fluid to fill the mould perfectly, the better will be the gun when cast.

Gun metal, when suddenly cooled, as when the red-hot metal

is plunged into cold water, is rendered softer, tougher, and more malleable. The more quickly, then, a gun is cooled the better; and it has been proposed to employ iron moulds of nearly the true dimensions of the gun in place of the ordinary composition, thus producing a casting more nearly of the exact size required, economizing in metal, and reducing the expense and time in turning down the exterior surface.

By this method, too, a greater amount of pressure could be given by increasing the height of the "dead head," which is found very materially to increase the density and other good qualities of the gun.

Major Wade suggests that the iron moulds be lined with a thin layer of clay or other suitable material, and that a free circulation of air be kept up around the mould. As this method of rapid cooling would soon cause a large difference of temperature between the exterior and interior portions, resulting in cavities in the interior, a division in the alloys, and a want of homogeneousness in the mass of the casting, he suggests the application of Rodman's method of hollow casting and cooling by water, by which an equality of temperature in all parts may be maintained. The core to be so arranged as to discharge the water horizontally near the lower part of the dead head, so that the latter will be unaffected by the flow of water, leaving it liquid metal, which by its own gravitation will supply the void spaces created by the contraction of the solid.

Re-heating the solid soon after it is formed, and then cooling slowly, is found to greatly improve the tenacity of the metal; and Major Wade therefore suggests, that as soon as the gun has become solid the flow of water should be stopped, and the free circulation of air interrupted by covering over tightly the casting pit.

Tenacity.—In testing the tenacity of bronze taken from the dead-heads of guns, near the junction with the muzzle, it is found that the tenacity of those samples of which the planes of fracture passed through the axis of the gun, was much greater than in those in which the planes were perpendicular to the axis or parallel to the face of the muzzle, although the densities of both were about the same. The cause of this is not explained, though probably a result of the same general law regarding the formation of crystals as laid down by Mallet for cast-iron, and most other crystalline bodies. The crystals being formed normal to the cooling surface are more difficult to fracture by a force acting in the direction of their axes than by one acting perpendicular to them.

The principal injuries to which bronze guns are subject in service being the indentations formed in the bore, rather than any want of tenacity or endurance in the metal itself, the relative qualities of different alloys to resist a force of compression is a matter of the greatest importance, and should be the subject of extensive investigation; which is accordingly recommended by Major Wade.

MATERIALS FOR CARRIAGES, IMPLEMENTS, &c.

TIMBER.

The kinds of wood principally used in ordnance constructions are the following:—

White Oak.—The bark is white, the leaf long, narrow, and deeply indented; the wood is of a straw color with a somewhat reddish tinge, tough and pliable. It is the principal timber employed for all kinds of artillery carriages.

White Beech—Red Beech—are the most suitable for fuzes and mallets; also for plane-stocks and various other tools.

White Ash, is straight-grained, tough, and elastic, and is therefore suitable for light carriage-shafts; in artillery, it is used chiefly for sponge and rammer staves; sometimes for handspikes, and for sabots and tool-handles.

Elm, is well-suited for fellies and for small naves.

Hickory, is very tough and flexible; the most suitable wood for handspikes and tool-handles, and for wooden axletrees.

Black Walnut, is hard and fine-grained; it is sometimes used for naves, and the plank for ammunition boxes; it is used exclusively for the stocks of small-arms.

White Poplar, or Tulip Tree, is a soft, light, fine-grained wood, which grows to a great size; it is used for sabots, cartridge blocks, &c., and for the lining of ammunition boxes.

White Pine, is used for arm-chests and packing-boxes generally, and for building purposes.

Cypress, is a soft, light, straight-grained wood, which grows to a very large size. On account of the difficulty of procuring oak of suitable kind in the Southern States, cypress has been sometimes used for sea-coast and garrison carriages. It resists, better than oak, the alternate action of heat and moisture, to which sea-coast carriages are particularly exposed in casemates; but, being of inferior strength, a larger scantling of cypress than of oak is required for the same purpose; and on account of its softness, it does not resist sufficiently the friction and shocks to which such carriages are liable.

Bass-Wood, or American Lime, is very light, not easily split, and is excellent for sabots and cartridge blocks.

Dog-Wood, is firm and hard-grained, suitable for mallets, drifts, &c.

SELECTION OF STANDING TREES.

The principal circumstances which affect the quality of growing trees are *soil*, *climate*, and *aspect.*

Soil.—In a moist soil, the wood is less firm, and decays sooner, than in a dry, sandy soil; but in the latter the timber is seldom fine: the best is that which grows in a dark soil mixed with stones and gravel. This remark does not apply to the poplar, willow, cypress, and other light woods, which grow best in wet situations.

Locality.—In the United States, the climate of the Northern and Middle States is most favorable to the growth of the timber used for ordnance purposes, except the cypress.

Trees growing in the center of a forest, or on a plain, are generally straighter and freer from limbs than those growing on the edge of the forest, in open ground, or on the sides of hills; but the former are at the same time less hard. The aspect most sheltered from the prevalent winds, is generally most favorable to the growth of timber. The vicinity of salt-water is favorable to the strength and hardness of white oak.

Time.—The selection of timber trees should be made before the fall of the leaf. A healthy tree is indicated by the top branches being vigorous and well covered with leaves; the bark is clear, smooth, and of a uniform color. If the top has a regular, rounded form; if the bark is dull, scabby, and covered with white and red spots, caused by running water or sap, the tree is unsound. The decay of the uppermost branches, and the separation of the bark from the wood, are infallible signs of the decline of the tree.

FELLING TIMBER.

TIME.—The most suitable season for felling timber is that in which vegetation is at rest, which is the case in midwinter and in midsummer. Recent experiments incline to give preference to the latter season, say the month of July; but the usual practice is to fell trees for timber between the first of December and the middle of March.

The tree should be allowed to obtain its full maturity before being felled; this period in oak timber is generally at the age of from seventy-five to one hundred years or upwards, according to circumstances. The age of hard wood is determined by the number of rings which may be counted in a section of the tree.

CUTTING.—The tree should be cut as near the ground as possible, the lower part being the best timber; the quality of the wood is in some degree indicated by the color, which should be nearly uniform in the heart-wood, a little deeper towards the centre, and without sudden transitions.

Felled timber should be immediately stripped of its bark, and raised from the ground.

DEFECTS OF TIMBER TREES (ESPECIALLY OF OAK).

SAP, the white wood next to the bark, very soon rots, and should never be used, except that of hickory. There are sometimes found rings of light-colored wood surrounded by good hard wood; this may be called the *second sap*; it should cause the rejection of the tree in which it occurs.

BRASH WOOD, is a defect generally consequent on the decline of the tree from age; the pores of the wood are open, the wood is reddish-colored, it breaks short, without splinters, and the chips crumble to pieces. This wood is entirely unfit for artillery carriages.

WOOD WHICH HAS DIED BEFORE BEING FELLED should in general be rejected; so should *knotty trees*, and those which are covered with tubercles or excresences.

TWISTED WOOD, the grain of which ascends in a spiral form, is unfit for use in large scantling; but if the defect is not very decided, the wood may be used for naves and for some light pieces.

SPLITS, CHECKS, AND CRACKS, extending towards the centre, if deep and strongly marked, make the wood unfit for use, unless it is intended to be split.

WIND SHAKES are cracks separating the concentric layers of wood from each other: if the shake extends through the entire circle, it is a ruinous defect.

All the above-mentioned defects are to be guarded against, in procuring timber for use in artillery constructions. The *center heart* is also to be rejected, except in timber of very large size, which cannot generally be procured free from it.

SEASONING AND PRESERVING TIMBER.

As soon as practicable after the tree is felled, the sap should be taken off and the timber reduced, either by sawing or splitting, nearly to the dimensions required for use. Pieces of large scantling, or of peculiar form, such as those for the bodies of gun carriages and for chassis, are got out with the saw; those of smaller dimensions, such as spokes, are split with wedges. Naves should be cut to the proper length, and bored, through the axis, with a 1½ in. auger, to facilitate their seasoning and to prevent cracking as much as possible. They should be cut from the butt of the tree.

Timber of large dimensions is improved by *immersion in water* for some weeks, according to its size, after which it is less subject to warp and crack in seasoning.

DRY SEASONING.—For the purpose of seasoning, timber should be piled under shelter, where it may be kept dry, but not exposed to a strong current of air; at the same time, there should be a free circulation of air about the timber,—with which view, slats or blocks of wood should be placed between the pieces that lie over each other, near enough together to prevent the timber from bending. In the sheds, the pieces of timber should be piled in this way, or in square piles, and classed according to age and kind. Each pile should be distinctly marked with the number and kind of pieces, and their age, or the date of receiving them. The piles should be taken down and made over again at intervals, varying with the length of time which the timber has been cut. The seasoning of timber requires from two to eight years, according to its size.

KYANIZING.—Gradual drying and seasoning in this manner is considered the most favorable to the durability and strength of timber; but various methods have been proposed for hastening the process. For this purpose, *steaming* timber has been applied with success; and the results of experiments with Mr. Kyan's process of saturating timber with a solution of corrosive sublimate, have been highly satisfactory: this is said to harden and season the wood, at the same time that it secures it from the dry rot and from the attacks of worms. The process of Mr. Earle, which consists in saturating the wood with a hot solution of copperas and blue vitriol mixed together, has been tried by the Ordnance Department; but the results have not been favorable, as regards its effect on the strength or preservation of the timber.

KILN DRYING is serviceable only for boards and pieces of small dimensions, and is apt to cause cracks and to impair the strength of wood, unless performed very slowly. *Charring* or *painting* is highly injurious to any but seasoned timber; as it effectually prevents the drying of the inner part of the wood, in which, consequently, fermentation and decay soon take place.

Oak timber loses about *one fifth of its weight* in seasoning, and about *one third of its weight* in becoming perfectly dry.

MEASURING TIMBER.

Sawed or hewn timber is measured by the cubic foot; or more commonly by *board measure*, the unit of which is a superficial foot of a board 1 in. thick. Small pieces, especially those which are got out by splitting (such as spokes), and *shapes*, or pieces roughed out to a particular pattern (such as stocks for small arms), are often purchased by the piece.

Usual rule for measuring round timber:

Multiply the length by the square of one fourth the mean girth, for the solid contents; or, $\frac{LC^2}{16}$; L being the length of the log, and C half the sum of the circumferences of the two ends. But when round timber is procured for use in the Ordnance Department, it should be measured according to the square of good timber which can be obtained from the log.

TABLE, *showing the Superficial Feet in one lineal foot of Boards of various widths.*

WIDTH.	AREA.	WIDTH.	AREA.	WIDTH.	AREA.
In.	Sup. ft.	In.	Sup. ft.	In.	Sup. ft.
0.25	0.0208	4.25	0.3542	8.25	0.6875
0.50	0.0417	4.50	0.3750	8.50	0.7083
0.75	0.0625	4.75	0.3958	8.75	0.7292
1.00	0.0833	5.00	0.4167	9.00	0.7500
1.25	0.1042	5.25	0.4375	9.25	0.7708
1.50	0.1250	5.50	0.4583	9.50	0.7917
1.75	0.1458	5.75	0.4792	9.75	0.8125
2.00	0.1667	6.00	0.5000	10.00	0.8333
2.25	0.1875	6.25	0.5208	10.25	0.8542
2.50	0.2083	6.50	0.5417	10.50	0.8750
2.75	0.2292	6.75	0.5625	10.75	0.8958
3.00	0.2500	7.00	0.5833	11.00	0.9167
3.25	0.2708	7.25	0.6042	11.25	0.9375
3.50	0.2917	7.50	0.6250	11.50	0.9583
3.75	0.3125	7.75	0.6458	11.75	0.9792
4.00	0.3333	8.00	0.6667	12.00	1.0000

To find the number of feet, *board measure*, in any piece of timber of a given width, multiply the tabular *area*, for that width, by the length in feet and the thickness in inches.

CHAPTER IV.

RIFLES.

Origin.—Rifled small arms date from a very early period (1498),—in France, from about 1515; though the principle from which their accuracy results does not seem to have been well understood.*

A *rifle* is a gun having in its bore a number of grooves, helical in form, which, as the projectile passes out of the bore, give it the rifled or rotating motion. In rifles loaded at the muzzle, the projectile must be made to "take" the grooves, by being forced into them. It will then turn in the piece as often as the entire curve is repeated in the length of the bore, and continue the same motion after leaving the muzzle. The rate at which it revolves will depend, for any given velocity, upon the inclination which the grooves have to the axis of the piece.

Theory.—The deviation of a shot is caused by the resistance to its passage offered by the air. The effect of rifling a ball is to make it keep the same part always to the front. This is shown by marking the top part of a rifle ball as it is placed in the gun. That part always strikes the target first. This renders the retarding effect of the air on the ball much more uniform than it would otherwise be.

Suppose that the point *M*, Fig. 63, of the front surface, be the point of application of the resultant P M of the resistance of the air acting obliquely to the vertical plane of fire of which A B is the trace. As the ball revolves, the point M takes the position *m*, and the force M P assumes the posi-

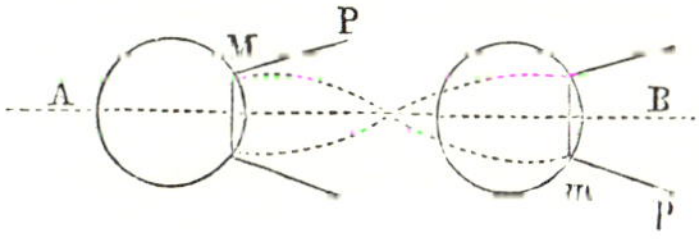

Fig. 63.

* The grooves of the first rifles made were perfectly straight, the metal being cut out to receive the dirt accumulated in firing a gun, as well as to allow the escape of the air in forcing down the ball, made with an increased diameter to diminish windage.

tion m p, directly opposed to the first position, and tending to throw the projectile back to the left with the same force it before exerted to throw it to the right, thus neutralizing this last motion. The force M P describes, as the ball revolves, a surface which we may take for a cone. For any two positions of the force M P opposite to each other in this cone, the deviating influences mutually counteract each other.

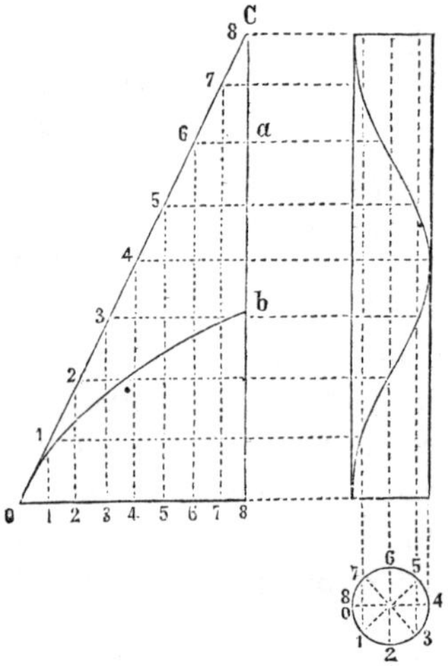

Fig. 64.

A helix is a curve on the surface of a cylinder which meets the elements of the cylinder constantly under the same angle. To trace one practically upon a cylinder, take a paper triangle, 8 0 8, whose base is equal to the circumference of the cylinder, their altitudes being the same. Place the line 8 8 on one of the elements of the cylinder, and roll the latter in the triangle; the hypothenuse 8 0 will describe the helix. Should we desire the helix to make but the $\frac{1}{4}$ of a revolution instead of an entire one, the triangle a 8 6 would be used, in which a 8 is $\frac{1}{4}$ of 8 8.

Cutting Grooves.—The ordinary ones are the easiest described and the best. To describe the curve on the bore of the piece, it is only necessary to give the barrel a certain motion of translation and rotation whilst a stationary point presses upon it. Or the barrel may rotate while the point is given a motion of translation. Or finally, the barrel may remain stationary and the point have both motions, and this last, is the usual way of rifling in practice. The cutter is placed on the end of a shaft moved back and forth by machinery, and having on it a projection which is forced to move in a groove cut in the proper form on the surface of a metal cylinder. A groove of the same form, consequently, is cut in the bore of the piece. This is the plan usually adopted in rifling large guns; but with small arms the barrel is fixed as before, and the grooves are made by a cutter which moves back and forth, turning at the same time by the action of a ratchet, moved up and down by a wheel at the bottom moving on an inclined plane or curve, according to the kind of helix to be cut, and which works into teeth placed on the frame, which support the rod of the cutter. This method has recently been applied also to rifling large guns.

These grooves are sometimes made with an increasing curve or a "twist," when the angle which they make with the axis increases as they approach the muzzle. The inclination at any point, as at the muzzle, is determined by the angle made by the tangent at that point with the axis. It may be described like the ordinary helix, by substituting for the hypothenuse of the triangle a parabola, or any other regular curve, as 8 4 or 6 0, Fig. 64, tangent to it at the extremity. This will give a curve less developed than the right line, and its mean inclination will be less. The friction, consequently, with a rifle of this form, will be less.

LENGTH.—Friction is a matter of considerable importance in rifles, and its retarding effect seems to increase in a greater ratio than the velocity. Hence, the inclination of the grooves should increase as the gun decreases in length, and the reverse. Other considerations will have a bearing upon this same point. The velocity of rotation should be great enough, otherwise the tendency of the ball to turn round its shortest axis overcomes the rotary motion, and the firing is inaccurate. It should not be too great; as in that case the velocity of translation is reduced. When the charge is increased, the inclination of the grooves should be decreased, otherwise the ball is forced *across* them without being rifled; and it is in this way that the velocity of rotation takes from that of translation.

If the groove makes one complete revolution in the length of the gun, it is supposed to impart that rate to the ball when it leaves the muzzle, and that it makes one of these revolutions in the same space in air. This, of course, is not strictly true, from the friction of the air, which will commence to reduce that velocity as well as the velocity of translation, from the moment the ball leaves the piece.

The curve, however, generally makes but a *part* of a revolution in the piece, the ball perfecting the revolution after it gets out—that is, after it passes over one *complete* helix. Let us suppose the length of the helix to be 16 feet, and the initial velocity of the ball to be 1000 feet. It will take $\frac{16}{1000}$, or $\frac{1}{100}$ part, of a second to pass over one length of the helix; or, the ball makes 100 revolutions in a second. Knowing, then, the initial velocity of translation of the ball, and the length of a completed helix of the gun, *the initial velocity of rotation* is obtained by dividing the former by the latter.

Lead being a softer substance than iron, the projections on the ball will be the more liable to be torn off, or what is termed "*stripped*," as the inclination of the groove to the axis and the charge of powder are increased. To avoid this, therefore, there must be a certain relation established between the dimensions of the grooves, their number, and inclination,—and the velocity of ball.

INCLINATION.—Experiment demonstrates that the inclination should increase as the spherical form is departed from in the projectile. Thus, an elongated ball having a greater bearing surface than a sphere, will have more metal forced into the groove, take a better hold of it, and be less liable to "strip." Such a ball, too, will stand a larger charge of powder before stripping, than a spherical ball.

The first rifles were made with grooves like teeth, and a great number of them, as in Fig. 65; but the sharp edges soon became worn by the rammer, and the piece became inaccurate.

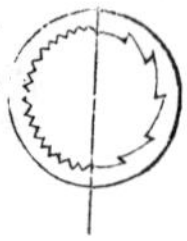

Fig. 65.

A screw, when used without oil, is soon destroyed. Grease or oil reduces the friction, and renders its movement more uniform. Grease applied to the grooves of a rifle, produces the same effect, and it is for this reason that the patch of a rifle ball is greased.

As the power of powder is increased by offering a greater resistance to it, the friction of the ball in the grooves requires a greater thickness of metal in a rifle than in a smooth-bored gun.

SLUGGED.—It is believed that the first attempt to use the rifle principle, was in *breech-loading* arms. In breech-loading pieces, the bore consists of two separate parts, viz., the long part or barrel, which is rifled, and the chamber immediately behind it, the bore of which is larger than that of the barrel. In this part, the charge and ball fit. When the ball is forced into the barrel by the ignited powder it enters the grooves and becomes what is termed *slugged*.

The advantages of this mode of loading, are, that it is safer;

that no soldier, however awkward, can get more than one charge in the gun; that no rammer is required in loading (a great advantage, especially on horseback); that the ball remains in its place, no matter if the gun is carried muzzle down and at a rapid pace; and that the piece can be loaded much more rapidly and easily.

The disadvantages are, that no matter how perfect the mechanism at the joint, *the powder being exploded at that point*, the gas is sure to penetrate it. The parts become rusty and foul, work with difficulty, close badly, and allow an escape of gas injurious to the marksman. These disadvantages are increased by the rapidity with which the piece can be fired.

The most noted arms of recent date in this country, loaded in this way, are those of Colt, Hall, Sharp, and Burnside. All of these have their peculiar advantages and defects, which will not be discussed at any length. The last-named, however, having been recommended by two successive boards of officers, as peculiarly adapted for the *military* service, it may not be out of place to give a short description of it.

Burnside's.—Its great advantage in a military point of view, is the protection of the cartridge from moisture. This is effected by means of a metallic case either of brass or steel, capped at the larger end with a lead ring.* By means of a steel mould, this lead, after the back end of the ball is placed in it, is forced around the ball so as hermetically to seal the joint. The case is slightly conical, and at the smaller end is a small hole closed when the powder has been poured in and the ball forced in its position, with a little tallow or wax.

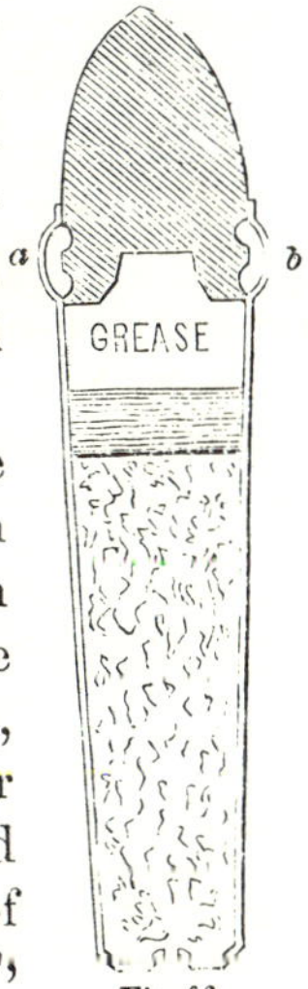

Fig. 66.

The chamber, of the *same form* as the cartridge case, is in a piece of metal, the back end of which drops down, being connected at the front end with the barrel, by means of a hinge *below* it. The chamber being down, the cartridge is dropped in, entering as far as the line *a b*. When the chamber is raised the ball enters the barrel, which is reamed out to receive the enlarged part, *a b*. This ring of lead being directly at the joint, effectually "*packs*"

* This lead is now left off.

it and prevents the escape of gas. When the charge is fired, the ball is driven forward through the barrel, leaving the *case in the gun.* Being conical in form and fitting the chamber, the sides of this support the case and prevent its being destroyed. The chamber being opened, the empty case is taken out, and a loaded one put in again.

Should the case stick after the gun is fired, an ingenious contrivance releases it. Being a conical surface fitting into a similar one, the least movement forward separates the two surfaces and releases the case. This movement is effected by means of a *cone seat* placed in rear of the chamber, and against it the cartridge-case rests. On touching this seat, the case is moved forward.

The gas was found to be very effectually cut off by the cartridge, and there being no necessity for the chamber-piece to fit tightly, either to the barrel or the side pieces which connect this with the stock, there is no danger of rust preventing the mechanism from working. The piece fired accurately, and passed successfully through some very severe tests with water.

Since the great improvements made in the muzzle-loading rifle, in regard to range and facility of loading, the breech loading principle is not a matter of so much importance. When it is remembered that soldiers in action, even when well disciplined, expend from 10,000 to 30,000 cartridges for every man disabled, it becomes a self-evident fact that they fire too fast already, and that it is only adding to the evil to give them the means of firing four or five times as fast, by placing breech-loading guns in their hands. It is, then, only special corps to which this arm should be supplied, and not to the main body of an army.

English.—Another plan for forcing the ball was as follows: The ball, very little smaller than the bore, was forced down on the powder with a heavy rammer, and driven into the grooves by several sharp blows. This method was not found very effective, and was injurious to the powder by mashing it. After a few rounds the piece becomes foul and works with great difficulty. Projections were then formed on the ball, corresponding to grooves in the gun, and no forcing was requisite; but this method was very inconvenient when firing and loading rapidly in succession, or in the dark, as the ball had to be placed in a particular position before it would enter the bore. These are the objections to the English or rather Brunswick rifle, which has but two grooves, and on the ball a projection in the form of a girdle, which fits into them. Fig. 67.

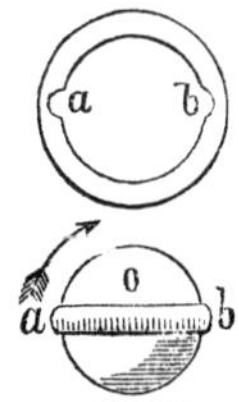

Fig. 67.

A third plan was by means of the common greased patch of linen or thin leather which envelopes the ball.

DELVIGNE, a French infantry officer, suggested an improvement in forcing, which, although soon given up, was the cause of another of much greater importance. He proposed to place at the bottom of the breech, a small chamber having an abrupt connection with the bore, Fig. 68. The powder nearly filled the chamber, and the ball, having been inserted like an ordinary musket ball, was arrested at the mouth of the chamber. To force it into the grooves, it was rammed several times with a heavy rammer having a concave head. This forced it also, into the chamber, by which the powder was injured. The ball, however, was forced very much out of shape, and being flattened in front, increased the resistance of the air. This rendered an increased twist necessary, in order to keep the ball from turning around its shorter axis. To prevent the ball being driven into the chamber, a *sabot* of hard wood was placed below it, which, resting on the offsets at the mouth of the chamber, prevented the ball from entering.

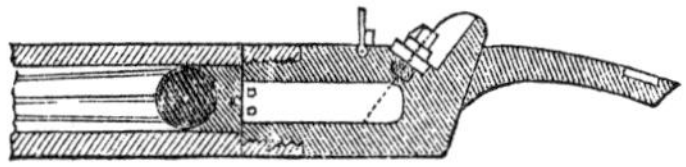
Fig. 68.

CARABINE À TIGE.—The invention which has been alluded to as the result of the preceding one, seems to be a very simple and natural consequence of it. It consists in substituting for the chamber a small cylinder or pin, Fig. 69, projecting into the lower part of the bore, and on the top of which the ball rests; thus furnishing the celebrated "*carabine à tige*," or stem-rifle. The powder now lies in a ring-shaped chamber around the stem, which supports the extreme point of the ball, allowing it to be forced more easily and with less detriment to its shape. This piece, the inventor of which was Col. Thouvenin, seems to have been the first military arm in which an *elongated* projectile was used.

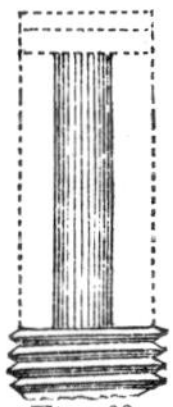
Fig. 69.

ELONGATED BULLETS.—It has long been known that the spherical form was not the one to which the air offered the least resistance; but all attempts to keep the projectile from turning around its shortest axis had failed.

The arrow, among elongated projectiles, has, with equal velocity, the most flattened *trajectory*, or path through the air. This is due: 1st, To the little resistance offered to the point of the arrow by the air, and to its great mass compared to that of a ball. 2d, To the volume which the arrow occupies, which prevents its falling as rapidly as a body of greater density; and, 3d, To the resistance offered by the air to the feathers placed on the rear part, which assist to prevent the arrow revolving on its shortest axis. These feathers are sometimes placed obliquely on the shaft so as to produce the rifle motion in the arrow. The arrow, then, is the type of rifle-balls, and the elongated projectile which can be made to approach nearest its trajectory in form will be the most advantageous.

Metal which must be used in connection with powder in order to resist its effect and have the proper penetration when it strikes, is too heavy and unmanageable to be used in the arrow form; but the proportions may be approximated to, and the proper amount of rotation be given by rifling.

The first form used was the cylindro-spherical, Fig. 70; then the cylindro-conical, Fig. 71; and, finally, the ogee-shaped ball, Fig. 72, having circular grooves cut upon the cylindrical part, which, acting like the feathers of an arrow, aided in maintaining the point to the front. This ball having been used with the chambered pieces, and found not to take the groove well, the chamber was suppressed, and a "*tige*" or stem substituted, which was found to force the ball better. The size of the stem is so calculated that the space around it is still large enough to contain the necessary charge after the piece has been fired fifty times. It was found necessary to increase the twist of the groove so as to make one revolution in about six feet, it having been before one in about twenty feet.

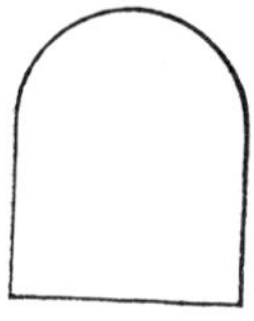

Fig. 70.

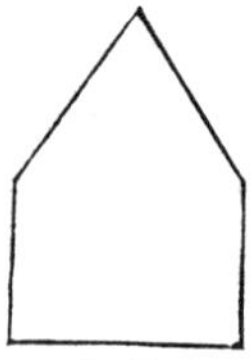

Fig. 71.

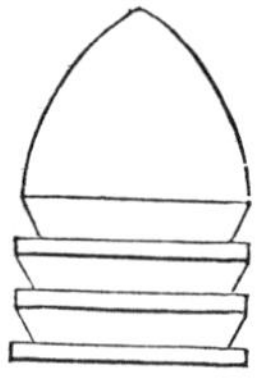

Fig. 72.

The difficulty of cleaning and keeping this gun in order was a serious objection to it, and means were sought of forcing the ball without the aid of the stem. This led to the adoption of the hollow *Minié* ball, having in the conical opening an iron *cup*, which, moving before the inertia of the lead is overcome, forces the ball into the grooves. This was found to force the ball quite as well, if not better, than the *tige*, at the same time that the ball was not deformed. When using very large charges, the cup is sometimes forced entirely through the ball. It is now discovered that the ball is forced as well without as with the *cup*, which has accordingly been suppressed.

A projectile may be so formed as to fly point first, like an arrow, even when fired from a smooth-bored gun; but in order to do so, it must fulfill the same conditions as the arrow, and have its center of gravity near the point, and such an arrangement on the rear part of the ball as will be acted on by the air to prevent its turning. Suppose, for instance, we have a projectile formed as in Fig. 73, the front being made of lead or iron, and the back part of hard wood, throwing the center of gravity well forward of the center of the figure. Such a projectile, although perfectly useless in practice, would travel point foremost when projected through the air. If the center of gravity be moved back farther and farther, a point will be reached where the projectile begins to show a tendency to turn over. It is evident, then, that a very slight rotation around its longest axis would at first maintain the point to the front; and this velocity of rotation would have to be increased as the center of gravity receded from the front.

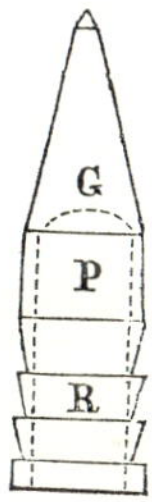

Fig. 73.

The causes of inaccuracy in firing spherical projectiles from smooth-bored pieces, are: 1st, Windage; 2d, A want of coincidence between the centers of gravity and figure; and, 3d, A rotary motion around an axis oblique to the plane of fire; this motion being partly a consequence of the formation of the bullet, partly in consequence of its shocks in the bore. Rifling nullifies all these causes of error. It destroys the windage, prevents the bullets from oscillating in the bore, and by substituting for its old motion of rotation another, around an axis lying in the vertical plane of fire, keeps the same part to the front.

The air retards the movement of projectiles, changes their velocity, and diminishes their accuracy.

When a spherical bullet is fired from a musket, its form is not changed at all. But if fired from a rifle, after having been rammed down to fit the grooves, it leaves the bore flattened in front, and somewhat cylindrical in shape. This form increases the resistance of the air, at the same time that the curve of the trajectory becomes flatter, on account of the air *supporting* the projectile. Hence result less velocity and range, but not less accuracy. The rotary motion of a *disk*, to which the flattened bullet approximates, is better maintained around the shortest axis than any other, and when the motion is around any but the shortest, the constant tendency is to return to this. This is also the case with elongated projectiles, unless some means are used to prevent it.

When moving front foremost, these experience less resistance from the air than spherical balls of the same caliber, especially if these last be flattened by ramming. By its greater mass, the elongated projectile has greater power to overcome the resistance of the air. Hence, an elongated bullet may leave the bore with less velocity than a spherical one, and yet, at a short distance from the piece, have a greater. This superiority of velocity will cause greater penetration, which will be still farther increased by the pointed form.

In order to have a rapid rotary motion, without too great an increase in the inclination of the grooves, a great velocity of translation is indispensable. But in attempting to impress a great velocity of translation on a spherical bullet, it is apt to "strip," not having so great a hold on the groove as the elongated one; moreover, as this last loses its velocity less rapidly than the spherical, less initial velocity will be required in order that the bullet may have the necessary velocity of rotation.

The motion of a rifled elongated projectile has been likened to that of a *top;* and the analogy is great. Could this toy be so constructed that its centers of figure and gravity accurately coincided, it would stand alone when placed vertically on its peg; but as this construction is mechanically impossible, it must incline to one side or the other. If, however, a rapid rotary motion be given to it, it will, whilst its axis is inclined, describe spirals on the ground, and finally assume a vertical position, until the rotary motion is exhausted, when it falls. When in inclined positions, the air on the under side being of the greatest density, tends to force the top up until it reaches the vertical, when the density of the air being equal on all sides, and the inequalities of the top being neutralized by the rotary motion, it stands in that position until the rotary motion becoming too small, to fulfill its object, the top falls.

The same is the case with the rifled projectile; and if its longest axis should not coincide at first with the direction in which it is traveling, its point will describe spirals like the top until those lines do coincide, if, in fact, they ever do so accurately. This spiral motion is made distinctly evident by the whistling noise made by a projectile not well rifled. The irregularity in the motion is easily detected by the ear.

The projectile cannot move through space in a right line, but in consequence of the action of gravity and the resistance of the air, it must describe a curve, the elements of which change direction at every instant. In order that the point of an elongated projectile shall strike first, the axis around which it revolves must be continually tangent to the curve of the trajectory. This will be apparent by reference to Fig. 74.

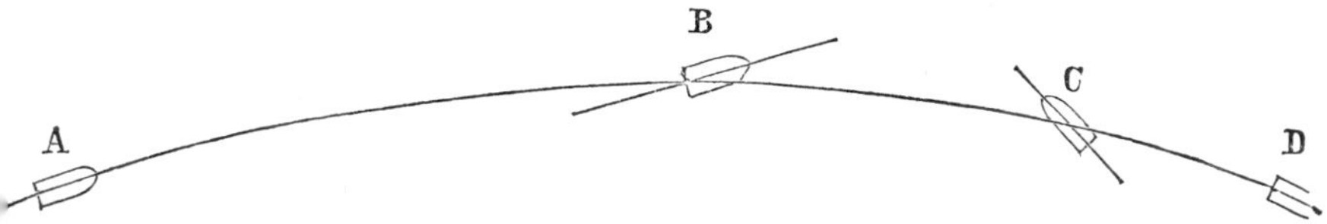

Fig. 74.

Supposing A to be the direction in which the projectile leaves the piece, if its axis should remain parallel to itself in all its positions, it would at length reach one as at B, where its side being presented to the pressure of the air, it would turn over, and might assume any other position, as C. Rifling such a projectile will keep it in the vertical plane of fire, but it still may turn over in that plane. To prevent this, circular grooves are cut upon the cylindrical part of the projectile, as shown in 72. The moment the axis fails to coincide with the tangent to the trajectory, the air acts on these grooves on the side of the bullet opposite to the one on which the point is deviating, and forcing this back to its proper position (thus acting like the feathers of an arrow), keeps the axis coincident with the tangent. Such being the way in which the elongated bullet travels, the resistance to its passage through the air is evidently a minimum for that sized and shaped body; the loss of velocity is likewise a minimum; and the velocity when it strikes is the greatest.

From what precedes, it appears there are three principal things to be considered in making a rifle.

1st. The charge of powder.

2d. The inclination of the grooves; and

3d. The diameter and form of the projectile.

In connection with the grooves, should be considered—

1st. The number.

2d. The depth and width; and

3d. The form.

The number of grooves should not be less than two; but an uneven number is the best, as there is then a *land* opposite each groove, and the bullet is not so much deformed by forcing as it would be were two grooves opposite to each other.

The number adopted in our service is three, of a rounded form, and equal in breadth to the lands; and instead of being of uniform depth throughout, they decrease as they approach the muzzle, the *slope* being the same in all arms; commencing at the muzzle with .005 inch in depth, and increasing in the musket at the breech to .015. They are uniformly spiral, and make in the long pieces (muskets), one turn in 6 feet, and in the short ones (pistols) one turn in 4 feet.

Bullets for altered Muskets.

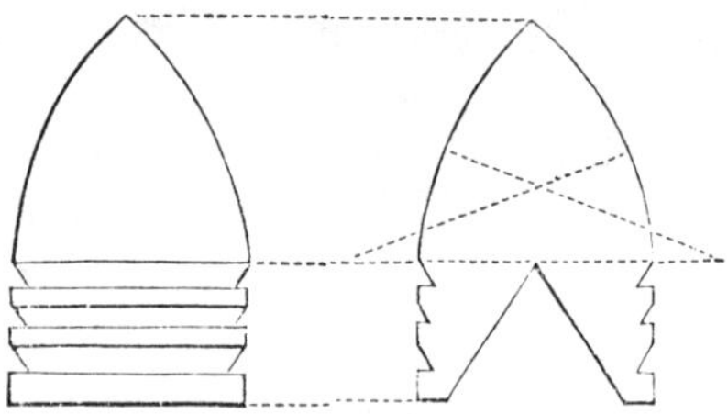

Fig. 75.

Figures 75 and 76 represent the forms and dimensions of the projectiles found most suitable, and adopted, in the United States' service.

Bullets for new Rifle-Musket and Pistol Carbine.

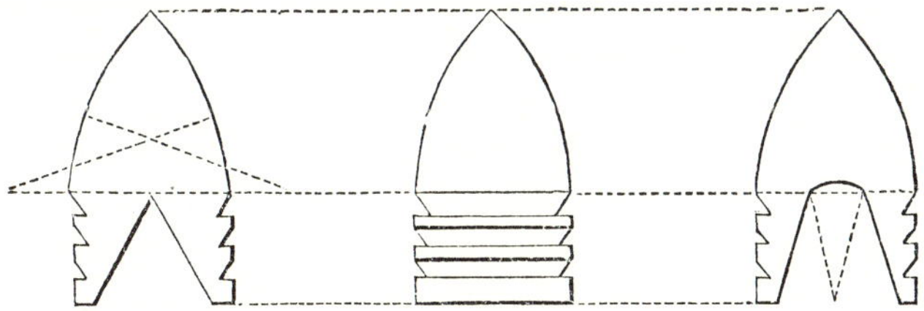

No. 1. Fig. 76. No. 2.

The No. 2 bullet, Fig. 76, is necessary for use with the pistol, on account of the less charge used; the force of the powder not being great enough to force the thick sides of No. 1 into the grooves. The two, however, differ only in the size of the cavity and of course, weight, and may be used indiscriminately if necessary.

By placing a percussion cap on the point of these elongated projectiles, they are very formidable against caissons, &c., for blowing them up.*

Almost every European army has adopted its own kind of rifle, the different arms differing from each other, sometimes very essentially in their construction, sometimes only in the lesser details. Nearly all have the elongated projectile; but the iron culot of Minié has been for some time abandoned as useless, though still retained by some. The English use a wooden plug in place of the culot with their projectiles which are transported to any great distance, as to India, to keep them from becoming deformed.

GERMANY.—The small German States use a great many different rifles, with bullets of almost every conceivable shape and size,—some round, some oblong; and on the latter the grooves vary in number from one to four, whilst the grooves in the rifles vary from two to eight. Some of these States use the common

* It is somewhat remarkable that during the experiments carried on in various countries with a view of perfecting the form, &c., of the elongated bullet, the discovery that the culot could be dispensed with, and the bullet forced simply from the action of the powder in the cavity, was made about the same time in several different places, and without the experimenters having communicated the results of their investigations to each other. Thus, in England, Belgium, and the United States, the discovery was made about the same time. In this country it was a result of a suggestion of Mr. Burton, master armorer at Harpers' Ferry.

rifle, with the ordinary expanding ball; while others have the rifle à tige; and others still the chambered rifle.

PRUSSIA still retains her national breech-loading needle gun, the complicated machinery of which will yet have to give place to the more simple constructions of the age. The bullet of this piece is peculiar in shape, being ogeval-shaped in front, and terminated in the rear by a hemisphere, a little less in diameter than the bullet, which has no grooves. Prussia has also in use the *carabine à tige*, with a cylindro-conical ball; and a rifled musket, which carries a heavy three-grooved ball, in which the culot is used.

BRUNSWICK.—The two-grooved rifle and belted ball introduced into England, and of which the Lancaster rifle is a modification, is the invention of Capt. Berner; and when adopted by Brunswick was of a somewhat more complicated structure, under the name of the oval musket. The grooves, which were broad at the breech, decreased in width as they approached the muzzle, where they finally disappeared in the oval bore, precisely like that of the Lancaster rifle. Spherical balls were generally used, and of two sizes; the smallest being used with a patch, the largest without.

HANOVER at first adopted the Berner system, but has now two models; one a rifle-musket à tige, the other a rifle à tige; both of the same caliber (0.62 in.). The musket has 7 grooves, the rifle 8; and the barrel of the latter is the shortest. Both pieces appear to be in use at the same time in the same corps. In the musket the grooves are of uniform depth; in the rifle they vary in depth, but have the same inclination as in the musket. The projectile used with these pieces is of a peculiar form, and is represented in Fig. 77.

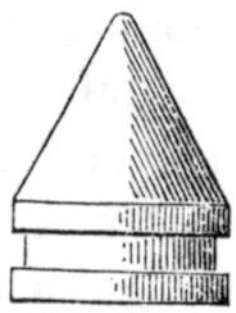

Fig. 77.

SPAIN adopted in 1852 a four-grooved rifle-musket, 0.60 in. in caliber, and firing a culot ball, which was exchanged three years afterwards for a ball similar to Peeters', Fig. 81. This arm has been issued only to a part of her troops.

RUSSIA originally adopted the two-grooved rifle and girdled ball, like the English; and has since adopted a very sharp-pointed oblong ball, with a button on each side which fits into the grooves, as the girdle did formerly. She also uses the carabine à tige and a solid, sharp-pointed ball, with one large groove near the base;

and of late years has been busily engaged in rifling her muskets, with which the expansive ball is used.

SARDINIA has a chambered eight-grooved carbine, on Delvigne's principle, which fires a short, thick-set ball, with a cylindrical body, and *conical* front. She has also, since 1854, been using a four-grooved rifle-musket, with a stem, which fires an oblong ball similar to ours, but solid. She has lately offered a large reward for the best military arm which may be brought forward, and does not limit the inventors to the citizens of her own territory.

SWEDEN uses an ordinary carbine transformed into the "à tige," with a bullet like that the French originally used with their à tige rifle. The piece has eight grooves; but only a portion of the men in each company are armed with it. She has also a complicated breech-loading rifle-musket; as also Norway, whose military administration remains distinct from that of Sweden, notwithstanding the union of the two countries under the same crown. The bullets for both these last are somewhat similar, being elongated with a girdle cut out near the base. Sweden is also preparing a large number of rifled muskets with the culot ball.

ENGLAND.—The rifles in use and experimented upon in England are of a great many different kinds and calibers. The principal one adopted for the army is the Enfield rifle-musket, manufactured in the government establishment. It is very similar to our present rifle, with three wide grooves. With it is used the Pritchett bullet, which is more rounding and full in front than ours, and has no grooves. A culot of hard wood is used with the bullet, to protect its form. A large number of rifled muskets have also been prepared for government use.

THE WHITWORTH rifle, one of the principal competitors of the Enfield, has a bore with a hexagonal cross-section, to which a twist of two turns in the length of the barrel (39 inches) is given. This formation imparts a very rapid rotary motion to the projectile, which was first made of a hard alloy of lead, tin, and manganese, very long and of the same cross-section as the bore, only slightly reduced of course, and with a very shallow cavity at the base, to aid expansion. The results obtained with this rifle (which is made of steel), both in regard to accuracy and penetration, are very remarkable. The velocity of rotation imparted to the bullet is immense, and is said to reach 15,000 revolutions in a minute.

This must necessarily create a large amount of friction in the piece, and a great development of force in the powder, necessitating more strength in the barrel; which is 1 lb. heavier than the Enfield, both being of the same length, whilst the bore is much less, being, from angle to angle, 0.49 in., and from side to side 0.438. The diameter of the Enfield bore is 0.577 in. Strong objections are urged against the Whitworth as a military arm, on account of its small caliber; though if no greater fault can be found than this, it ought not to be condemned, as will be demonstrated in discussing the rifles used in Switzerland. Another bullet, made of lead, and circular instead of hexagonal in its cross-section, and hollow at the base, has been adapted to this rifle, and is now being experimented upon. It is proper to state, in reference to this rifle, that the composition bullets used with it were very accurately cut in a lathe, to fit the gun. The rapid twist in the bore (once in 20 inches) would render it very difficult to load after becoming foul, much more difficult to keep in repair and free from dirt and rust, than a more smooth-bored piece. Its superiority over the Enfield rifle in point of penetration is said to have entirely disappeared on using in the Enfield a similar composition bullet.

The artillery carbine is a piece similar to the Enfield rifle-musket, of the same bore, but shorter (2 feet), and is used for arming the artillery. The Whitworth method of rifling dates as far back as the time of Louis XIII.

The Lancaster rifle is constructed on the same principle as the Lancaster cannon, but is not by any means so signal a failure. The bore is smooth; but in order to impart the rifle motion to the projectile, the cross-section of the bore is made elliptical or oval, the major axis being only .005 of an inch greater than the minor; so that the bore being given a certain amount of twist, imparts it to a projectile of the same cross-section, or one cylindrical in shape and expanded in the bore on the usual principle. The "gaining twist" is used with this rifle, and it has been reduced from one turn in 3 or 4 ft. to one in 6 ft. 6 in. The grooves, so to speak, offer less opposition to the action of the powder than in an ordinary rifle, and the ball is less liable to "strip;" but the irregular form imparted to it is objectionable, although the accuracy obtained with this rifle is said to be very satisfactory.

The bullets first used were of a form to fit the bore; but they have since been made like the ordinary expanding projectiles, and depend for the rifling upon the expansion acquired on the explosion of the powder.

FRANCE, originally one of the most backward powers in arming her troops with the rifle, stands now probably the first in experience and efficiency in that arm.

Delvigne, by placing a chamber at the bottom of the bore of an ordinary rifle, and making use of it to force the ball, dropped loosely into the bore, did away with *the* great objection to the use of rifles in war, the difficulty of loading them, and gave an impetus to investigations with regard to the arm, which has created a perfect revolution in the system of arming infantry, by leading to the present efficient weapon. This improvement led to the adoption, in 1842, in the French army, of the chambered carbine and rampart rifle-musket firing spherical balls.

Thouvenin soon followed with his improvement of the stem (tige) in place of the chamber, which led to the adoption of the *carabine à tige* for the chasseurs in 1853, in which elongated projectiles were early tried (solid at the base and with three grooves), at the suggestion of Delvigne, though the bullets were found to succeed better in this arm than in his, in which they had been tried some years before in Africa. This officer, it appears, was the first of recent date, in France, to announce the fact that such bullets, hollowed at the base, were expanded by the action of the gas, and forced into the grooves; though no practical results seem to have flowed from the discovery until some time in 1847, when Capt. *Minié* suggested his iron culot in the base of the bullet, which was found to force it as well, if not better, than the "tige;" and soon afterwards the culot was left out altogether, and the projectile was forced on the principle announced as a discovery four years before, by Capt. Delvigne. It would appear, therefore, that Minié is not without competitors for the honor of this important invention; for Col. Thiroux claims to have stated the same method for forcing rifle-bullets in 1845; and, in England, Mr. Greener, after in vain besieging the ordnance department for a hearing, announced the same thing in 1841, six years before Minié, and two before Delvigne. Mr. Greener's priority of invention has, he says, been admitted both by the Emperor Napoleon, and his own government; though there is no reason to suppose, as Mr. Greener does, that MM. Delvigne and Minié *copied* from him; for the real fact is, that the idea of elongated expansive bullets was no new one in France, or even in England, as mention is made of them in both countries at a much earlier date. Minié, an energetic and observant lieutenant of chasseurs, was fortunate in making a well-founded suggestion at a time when the arm and the different elements of a system were in the most favorable condition for deriving benefit from such an idea; and Capt. Minié is certainly entitled to all the credit which attaches to the success of his suggestion in the adoption of the now universally denominated Minié bullets.

This discovery caused the stem to be at once dispensed with as useless, the Minié rifle (*i. e.*, an ordinary rifle of any kind firing a Minié bullet), became the favorite of the new arms, and the smooth-bored musket was, by the simple operation of rifling, transformed into the long-range rifle (fusil-rayé) of the imperial guard, and is, as in most other armies, rapidly extending itself as the arm of the troops of the line.

SWITZERLAND.—Of all the countries of the old world, this is pre-eminently the country of the rifle. With an unrestricted right to bear arms, her citizens have long been celebrated for their shooting-matches and proficiency in the use of what may be called their national arm; and although we may not expect to find among them a superlative *military* rifle, it is natural to look for one possessed of very superior qualities for the use of a detached marksman.

Carbine. Rifle-musket.

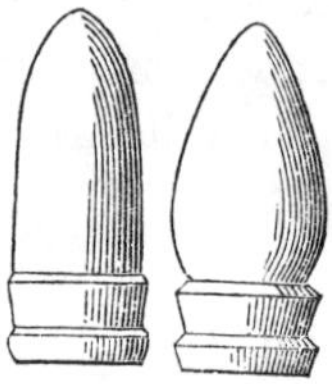

Fig. 78.

The new federal carbine is of a quite recent origin, having been adopted in 1851, after a series of experiments by a military commission, and is remarkable for its range and accuracy, its small calibre (0.4 in.), and the peculiar form of its bullet, which weighs but 257 grains, and is formed as represented in Fig. 78. The remarkable results of this rifle fired with 62 grains of powder will be seen by the following table.

DISTANCES, IN PACES.	DIMENSIONS OF THE TARGET IN FEET.	PER CENT. OF SHOTS STRIKING.	
200	2 × 2	100	
400	2 × 2	100	
600	4 × 6	97	Hard wind blowing.
600	8 × 8	100	" "
800	4 × 6	78	Wind.
800	4 × 6	90	"
800	8 × 8	100	
1000	4 × 6	from 58 to 56	
1000	8 × 8	" 86 to 96	
1000	13 × 10	" 92 to 100	

This arm was compared with many other rifles, and found to have a marked superiority over all of them in accuracy; the trajectory with the Wurstemberger bullet being much more flattened than with either the French rifle with a culot ball, the Bavarian rifle à tige, or the Swiss carbine when a spherical ball was used. The grooves, eight in number, were narrow and rounded; and so satisfactory were the results that the piece was at once adopted by the Swiss government. The ball, which is used with a patch, is the invention of Col. Wurstemberger, of the Swiss artillery.

The rifle-musket adopted soon afterwards, is of the same caliber as the carbine, but has only four grooves, and carries a ball of the same weight, though somewhat different in form, Fig. 78. This piece was compared with a number of others of different countries, and still maintained its superiority, though slightly less in accuracy and penetration than the carbine first adopted.

To determine whether the bullet was large enough for practical purposes, experiments were made by firing at the body of a horse, knocked on the head just before the firing commenced so that the blood and warmth of the body were retained during the course of the experiments; which resulted in completely demonstrating the perfect capacity of the rifle for disabling an enemy to the degree required on the field of battle.

Austria, like most of Germany, has always been very much in favor of the use of rifles; and as soon as the invention of Del-

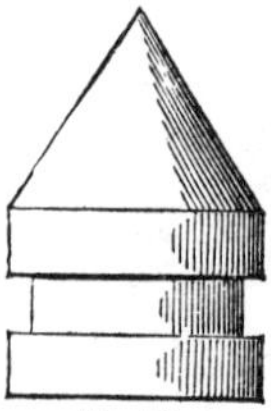
Fig. 79.

vigne became known, she adopted it somewhat modified by Baron Augustin, using with this Delvigne-Augustin rifle a cylindro-conoidal ball with a girdle cut out of the cylindrical part, Fig. 79. It is a very large calibre (0.7 in.), and has 12 grooves; and though the barrel is only about 26 inches long, the arm, without the bayonet, weighs 8½ lbs.

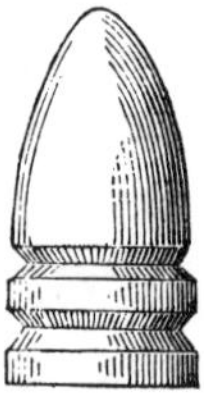
Fig. 80.

Rifled-Musket.—In 1855, she adopted for her infantry of the line the rifled-musket, and one cartridge for all small arms, the ball being the Lorenz, or modified Wilkinson ball, weighing 450 grains, Fig. 80. The barrel has four wide grooves equal to the land, making one turn in 83 inches, and is 37 inches long. Fig. 80.

The *rifle à tige*, which has now replaced the Delvigne in the chasseurs, is used in those corps at the same time with rifles from which the tige is removed, the former being given to the non-commissioned officers, 3d rank, and the best marksmen in the battalion, and have hausses regulated to 1200 paces. The others, which differ only in the absence of the tige, are regulated only to 1000 paces. These rifles differ from the rifled-musket only in being shorter and having a different bayonet. They are all of .55 calibre. They have been found very accurate; but, strange to say, the rammer is carried separated from the piece, which is very inconvenient.

Bavaria adopted, in 1848, a rifle à tige, or rather three, the bores of which differed slightly from each other, and all are used in the same corps. Three different sizes of balls were required. When the grooves of rifle No. 1 became worn, the piece was repolished so as to increase it to caliber No. 2; and this operation performed on No. 2 produced the caliber No. 3. The rammer had at its upper end a wooden knob, to assist the men in forcing the ball on the tige; and the rammer, as in the case just mentioned in the Austrian service, was carried separate from the piece, tied to the powder-flask.

In 1852, a new model of the tige rifle was adopted, carrying a ball very similar on the exterior to our present rifle-ball. This piece was provided with a hausse, so arranged as to correct the error arising from the "drift," an invention of Capt. Minié. The slider, as it rises, is made to follow a groove formed in the left hand upright of the hausse, so that at each elevation the sighting notch is carried to the left a distance corresponding to the deviation to the right resulting from drift.

The firing of this piece was very accurate.

BELGIUM, so celebrated in the military world for her thorough and practical investigations into questions relating to the military art, possesses, in her manufacturing establishment at Leige, where many European states have arms made, great facilities for taking advantage of every improvement which appears. She adopted the Delvigne rifle almost at the same time as France, and has ever since kept up her experiments in such a way as to remain second to no nation in her improvements, and in some things far ahead of all.

Her rifle à tige differs but little from the one adopted in France, and carries a bullet similar, on the exterior, to that used in our service, the caliber being the same as our altered musket, 0.69 in.; but the ball, being solid, is somewhat heavier than ours (756−730=26 grs). The grooves are 4 in number, as in the French rifle.

The rifled-musket, adopted soon after Minié's invention, does not differ essentially from the corresponding French arm, and is of the same caliber as the preceding piece. The bullet, however, is not Minié's, but the invention of one of the workmen in the Liége arsenal (Peeters), and is somewhat lighter than the bullet for the rifle à tige, in consequence of the cavity. (See Fig. 81.) This arm is somewhat superior in accuracy to the rifle, and is adopted for all the Belgian infantry.

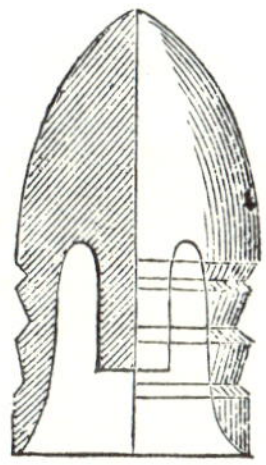

Fig. 81.

DENMARK had, as early as 1829, some of her troops armed with rifles, firing spherical balls. But little is published in regard to her present armament, though she has used cylindro-conoidal bullets in considerable quantities.

RIFLED CANNON.—The importance of applying the rifle principle to guns of large calibre, is too evident to need any explanation; and many and various have been the attempts made to succeed in it, but up to the present without any degree of certainty, although much progress has been made. The importance of the question is much enhanced by the fact that the moment a successful plan is discovered, the great problem of concussion and percussion shells is solved. That a way will sooner or later be discovered, there can be but little doubt.

Large projectiles, being made of iron, cannot of course be forced into the grooves of a gun like the leaden ball of small arms. All projectiles designed for rifled cannon are elongated in shape. Some have been made with spiral grooves, or projections, on the "after" surface, through which the gas as it rushed produced the rotary motion; others have these grooves on the fore part of the ball, and the rotation is produced by the action of the *air* on them as the ball moves forward; but although the effect was produced by these two methods in a slight degree, the force of rotation developed was not sufficient to insure the direction of the ball with any certainty.

Attempts have been made to cast on the outside of the cylindrical part of the shot some softer metal, such as lead, or composition, to "*take*" the groove and give the necessary rotation; but it has invariably been found that, although these metals take the groove at first, they are immediately torn to pieces and off the iron part of the shot by the force of the powder. The increase of force in powder cannot be calculated upon like any other motive power; and because a leaden projection of 0.01 in. will hold in its position a common rifle-ball, say of 2 oz. in weight, when acted on by 60 grs. of powder, it does not follow that the same effect will be produced when these elements are increased a hundred times, much less when they are increased a thousand times. This is the common mistake in a great many inventions of the day. Large breech-loading guns are constructed, with no other protection against the escape of gas than is used with a breech-loading rifle. Such inventions, many of them very ingenious, must fail, from the simple reason that they are not so calculated as to withstand the tests to which they must be submitted in service. Breech-loading cannon, however, have been made to work successfully for a time, like that of Col. Cavalli; but they are generally too complicated for active service.

Several were recently cast at the West Point Foundry for the English government, the breech-piece of which alone weighed several tons, and was moved in and out by means of machinery. No favorable report has ever been received of them.

The French have long been making experiments upon a cannon with two grooves, having an elongated projectile with buttons on the sides to fit into these grooves; and similar pieces have been tried in this country at Fort Monroe and West Point; the piece at the latter place bursting after a few rounds. Increased range and accuracy seem to be attained, but no dependence can be placed on them for any length of time.

The English Lancaster gun, which was brought prominently into notice some time ago, seems now to have proved a failure and fallen entirely into disuse. To convey some idea of the form of its bore, let us suppose a right cylinder, with elliptical bases fastened at one end so it cannot move with the longest axis of the bases in a horizontal position. Twist the free end of the cylinder around until the longest axis of that base is in a vertical position. This will represent the bore of the Lancaster gun, whose cross-section will show two grooves, so to speak, but so gradually merging into the bore as to be scarcely discernible. The projectile was elongated and ellipsoidal in shape, to fit the bore. The rifle-motion is produced, but not always with certainty; and the pieces are very liable to burst, probably from the projectile becoming jammed by a little obliquity in the bore. The same principle has been applied to small arms, with much better success.

The most satisfactory results yet obtained in this country are from a projectile invented by Dr. Reed, of Alabama, in which the Minié principle is used, by attaching a cup of wrought iron to the back part of an elongated projectile. By the explosion of the powder the malleable iron is forced into the grooves of the gun, and the shot is rifled perfectly, if the connection between the wrought and cast iron holds. This has been the great difficulty; but recent experiments seem to show that it has been overcome.

In the first shot made, the cup was of the form represented in Fig. 82, having a round hole in the bottom. This was placed upside down in the bottom of the *mould*, and the melted metal poured in, but prevented by a core from filling any more of the cup than to a certain distance, as *a*. The wrought iron was thus firmly imbedded in the cast; but when fired, it was found that the cast iron in front of the cup was

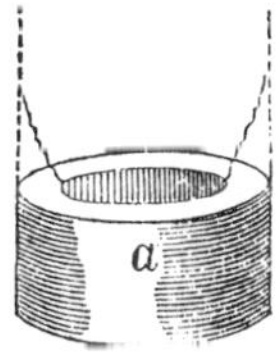

Fig. 82.

split off as represented by the broken lines in the figure, releasing in a measure, the bottom rim of the cup, and separating it from the shot. It was found, too, that where the cup was welded on the side, the parts were forced apart by the powder. In one case, when the cup had been made very shallow, and its rim projected only about $\frac{1}{2}$ in. behind the cast iron, the whole cup came off in the gun, sticking in the grooves; showing enough metal had not been allowed to run in to hold it, but showing also how well the rifling was effected.

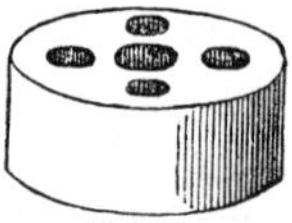

Fig. 83.

To remedy these defects, the cup was made with an entire bottom. and then several holes bored in, through which to allow the passage of the metal. The cast iron was cut away around the shot in front of the cup, so as not to allow a bearing surface to push off the pieces. This was a great improvement, but still does not succeed as perfectly as it should; and further experiments are being made, to work out the little details, such as thickness of iron, &c., to make it as perfect as possible. The cup, instead of being welded as before, is punched by machinery out of a flat piece of sheet iron, and is found to answer much better. The shot, as improved, is represented in Fig. 84.

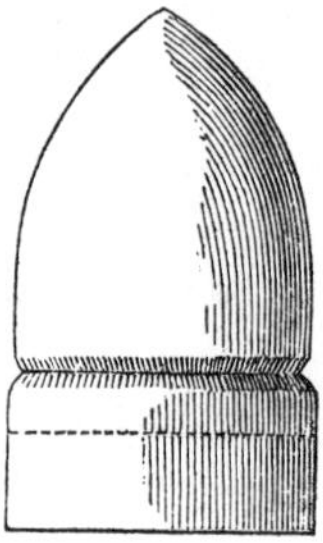

Fig. 84.

A little further improvement has been made, on which experiments are now taking place. This consists in forming the projectile and wrought iron cup as represented in Fig. 85, which shows the form of a 12 pdr. shell. This construction gives even more satisfactory results than the preceding; and it is thought that with some changes in the method of rifling the gun, more perfect ones will follow.

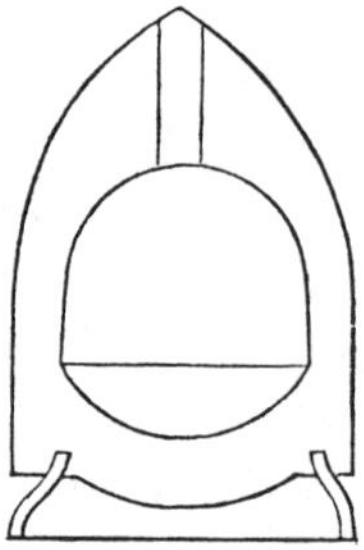

Fig. 85.

It is found, in using these shells, that they do not at first take the grooves, but appear to slide sideways, sometimes as much as an inch before the inclination to go straight out is overcome by the resistance offered by the part of the cup expanded into the grooves, which are broad and three in number. It is thought, from this, that more grooves are necessary, and that they should be increased in number as the caliber of the piece increases.* The projectiles vary in form somewhat for different sized pieces.

Experiments have recently been carried on with an oblong iron ball having a recess cut from the cylindrical part. In this recess a coating of lead is cast, on the outside of which is secured a covering of canvas designed to prevent the grooves from leading. Channels lead from the rear of the shot obliquely into the recess, so that when the charge is fired the gas passes through them under the lead coating, swelling this out and forcing it into the grooves, and imparting the rifled motion. This projectile is said to have given very accurate firing with a 12-pdr., the largest piece, it is believed, with which it has as yet been tried. It has the same objection that has been urged against all compound shot, viz., that the sheathing will break loose either before or after leaving the gun, and the projectile will not, therefore, be suitable for firing over the heads of our own men.† It is more than probable that when this projectile is applied to pieces of large caliber, the lead will not prove strong enough to produce the necessary rotation. In connection with this subject, it may be well to remark that shells made on Dr. Reed's plan, with cups of *lead*, succeed very well with pieces of small caliber.

THE ARMSTRONG GUN.—Great importance has recently been given in England, to this newly invented piece, which has, so far, been made of only two calibers, both small, viz., 2½ and 3¼ inches. From such information as has reached this country, these pieces seem to have met with remarkable success both as regards range and accuracy. No very detailed and authentic description of the gun has, however, as yet, been received; and but from the the fact of so much importance being attached to the invention, and the knighting of the inventor, it might appear premature and out of place to venture any opinion or description of it.

It appears, from the imperfect descriptions received, to be a breech-loading, "built-up" gun; the central portion, or part around the bore, being formed of a steel tube bored out and rifled

---

* This change has since been effected, and the shells are now no longer found to slip, but take the grooves perfectly, and in consequence, the accuracy of fire is correspondingly improved.

† Projectiles similar to these have been experimented upon both in France and England, and rejected principally on this account.

with forty grooves, making one turn in twelve feet. This tube is wrapped spirally with wrought-iron overlapping and welded together, and outside of this is another wrapping in the opposite direction and welded like the other; these outside tubes being, in fact, formed exactly as are the barrels for small arms. The objections to this kind of a barrel for large cannon, have been noticed elsewhere, and it is not deemed necessary to repeat them here, with the exception of remarking that the more the mass of metal is increased, the greater will be the difficulty of welding properly, and although the plan may succeed well for light guns, it by no means follows that it will be equally successful for large ones.

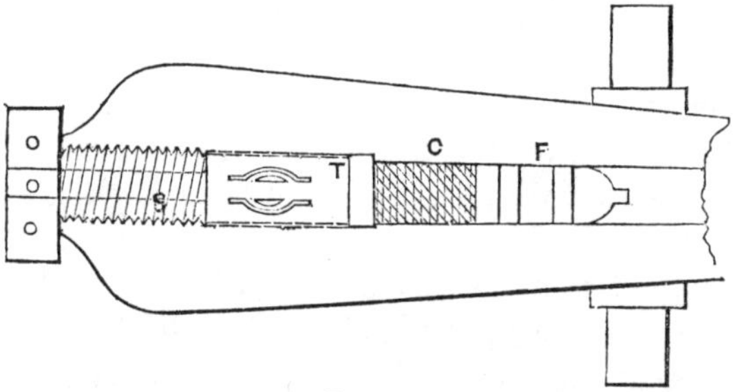

Fig. 86.

By referring to Fig. 86, the method of loading at the breech will be seen. In rear of the bore, and extending to the end of the piece, is a threaded opening into which fits a screw with its axis perforated, and an arrangement at its rear end for turning it. In front of this screw, the metal of the gun over the bore is cut away so as to form a recess to receive a block of metal, which is lifted out by handles attached to it. This block is represented in Fig. 87, and is solid, with the exception that the *vent* T is perforated in it, opening into the front end. This block being removed from the gun, the projectile is placed in the recess and rammed forward into its position by a bar of iron passing through the opening in the breech-screw. The cartridge is placed in in the same way. On account of the crooked vent, the cartridge cannot be pierced as in an ordinary gun; and to fire it, a small cartridge is placed in a recess formed in the front end of the block (as shown at D, Fig. 87),

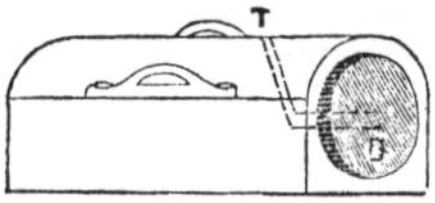

Fig. 87.

and fired in the usual way with a friction tube. These breech-blocks are furnished in duplicate, one being with the gun, the other carried in the limber.

If the breech-block is now placed in position, and forced against the front part of the recess by the breech-screw, it is evident that although the gas will be partially cut off, its escape will not be entirely prevented. To prevent entirely the escape of gas, advantage is taken of the expansibility of a softer metal, such as *copper*. A ring or short cylinder of this metal is let into the rear end of the bore, and forms a part of the chamber for the main charge of powder. By the explosion of the powder, this ring is forced into the joint and cuts off the escape of gas.

Another description of this gun makes no mention of the recess and breech-block, and merely describes a *solid* breech-screw which is supported after it is withdrawn on a slide which passes under the gun. But the above is believed to be a correct description of it, as mention is made of an accident which occurred in firing one of the pieces, from the breech-screw being left loose, and the block blown from its position.

It is supposed that, like all breech-loading pieces, the chamber is larger than the bore. The shot originally used, was covered throughout with lead; but this is now replaced by two projecting bands (as shown in Fig. 86) which are forced into the grooves as the shot moves through the bore.

It is stated that two fuzes are used with these projectiles: one contained wholly within the shell, and essentially a concussion fuze; the other communicating from the outside, and partly a concussion, partly a time fuze, which does not depend upon the flame from the discharge for its ignition. In the latter, fire is communicated to the composition column by means of a small quantity of percussion powder, which is exploded by the impact of a pointed steel bar let loose by the force of the explosion; the point at which fire is communicated to the column of composition being regulated by a cap and index, probably in a similar manner to that laid down for the Breethaupt fuze.

The results said to have been obtained with the Armstrong gun are very remarkable. It is stated that a range of 9,600 yards has been reached; and that shot after shot may be depended on for striking a target two feet square, at 1,000 yards. The loading of the piece is, however, necessarily slow.

It is scarcely fair to condemn an invention before the inventor has had full opportunity to test it thoroughly; but if the above proves to be a correct description of the celebrated Armstrong gun, it is doubtful if, taking all things into consideration, it will ultimately be found to possess such marked superiority over all other cannon as is claimed for it, and more than doubtful whether, when applied to guns of the largest calibers, its success will be any thing like as great as it has been for small guns.

Cavalli Gun, Fig. 88.—This piece, invented by Col. Cavalli, of the Sardinian service, has become somewhat noted in Europe as a *breech-loading* rifle of some success.

The chase of this gun does not differ essentially from the usual form of cannon; but at the breech the piece, instead of being round, has the four sides planed off so as to present from the rear an appearance of a square with the corners rounded off. It is bored throughout its length, and rifled with two flat grooves with rounded edges. The rear of the chamber is enlarged, and these grooves being continued through it, although shallower than in the chase, are deep enough to receive the wings or projections on the shot and hold it up till it reaches its seat in the gun.

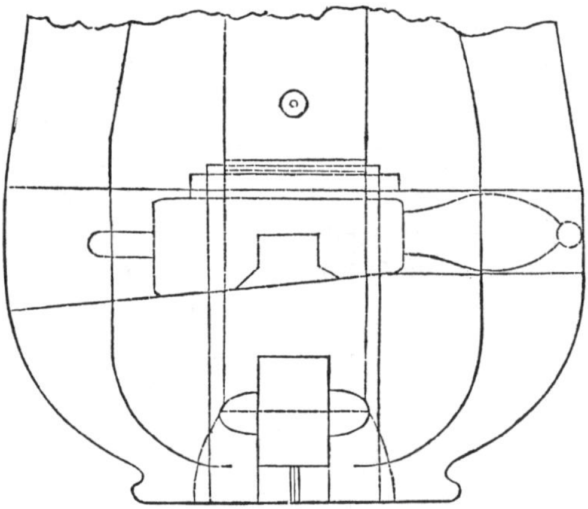

Fig. 88.

Crossing the bore at right angles, with its front face perpendicular to the axis of the piece, a wedge-shaped opening with a rectangular cross-section is cut. It is for a 32-pdr. 9.4 in. deep, 5.4 in. wide at the large end, and 3.7 in. at the small. This opening receives the quoin or wedge destined to close the breech in rear of the charge.

The wedge, Fig. 89, is made of hardened iron or steel, and of the same shape as the part of the opening which it is to fit. The wedge shape enables the bottom of the bore to be more perfectly closed, and prevents the escape of gas, whilst it also enables the breech-piece to be more easily moved after firing. The front face

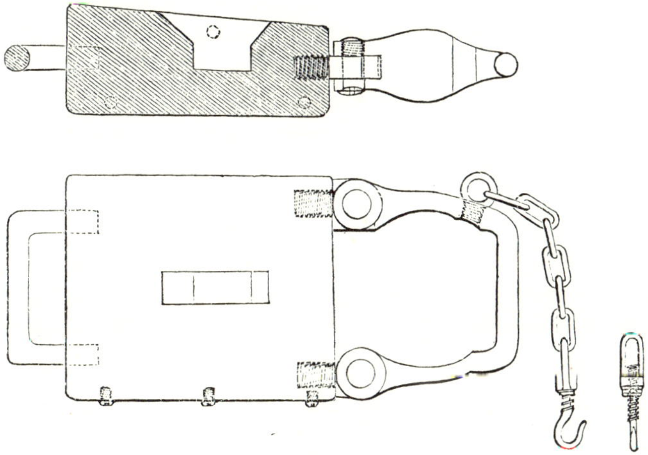

Fig. 89.

of the breech-piece is perpendicular to the axis of the piece, whilst the rear face makes with it an angle whose tangent is 1-8th, the co-efficient of friction of the hardened iron wedge against the cast iron of the piece. By means of this disposition and keeping the surfaces in contact well greased or moistened according as required, the breech-piece is found after firing to be more or less moved, at the same time that there is no danger of its being pushed too far or thrown out of its place.

The breech-piece is provided with two handles, which serve to handle it in pushing in and withdrawing it from its position. The large handle placed on the right, is of such a size that, when the breech is open and ready for the charge, the projectile can be passed through it, the lower part of the handle supporting the projectile, and guiding it through the breech opening into the chamber. To the large handle is attached a small chain and hook, the latter fitting into an eye screwed into the top of the opening on the right. This chain, when the breech is being opened, arrests the quoin when the opening in the large handle comes opposite the bore. When the charge is introduced, the cannoneer on the left pushes, and the one on the right pulls each by his handle, and the quoin comes back to the proper position for firing. In case any forcing is necessary, either for this or to move it after the piece is fired, a mortise is formed in the rear face of the quoin to receive the square end of a large iron handspike. It is not generally found necessary to use this lever except for the last-named purpose. In order to diminish the adhesion of the parts, three points are placed under the lower face of the quoin, which keep the quoin at the proper height in the cut.

CUT-OFF.—To cut off the escape of gas, a ring of hammered copper is used, similar to the one in Armstrong's gun. The cross-section of this ring is about one inch square. A recess for it is cut out of the gun at the rear part of the bore, the inner diameter of it being greater than the outer one, in order to retain it in position. The ring projects about $\frac{1}{4}$ in. behind the bore, and is pressed against by the quoin when in position. The interior diameter is the same as that of the chamber, and corresponding recesses are cut in it to allow the passage of the shot-flanges.

THE PROJECTILE.—Two projectiles have been used with this piece, both oblong, but differing in the front part, one being conical, the other ogee-shaped. They are made entirely of iron, and have two flanges or wings, properly inclined to suit the grooves, and extending from the rear to a short distance in front of the cylindrical part, for the purpose of preventing as much as possible the oscillations of the projectile, resulting from the windage, which is not destroyed in this piece. With the same object two other short flanges are placed at the junction of the cylindrical and conical portions, just thick enough to allow their entrance into the bore, leaving *for them* a windage of .015 in.

LOADING.—The quoin having been pushed to the left until stopped by the chain, the projectile is introduced, with the flanges in the grooves, and pushed in with a rammer until a stopper on the handle of the latter strikes against the breech, and shows the shot is in the proper position. The charge, inclosed in strong paper, is then pushed in, raising it slightly to pass it over the connection between the large and small parts of the bore. After this a *culot* is put in. This culot is a block of wood, the outer surface of which is turned with a shoulder, so that the front part of the culot enters the small part of the bore, immediately in rear of the cartridge. The object of this culot is principally to receive the residue of the powder, and keep it from soiling the front face of the quoin. It has a recess and thread of a screw in the rear, and is placed in the gun, and taken from it, by screwing into this a handle, which has, near the end, a sponge used to wet the parts when loading. The culot being in position and the handle withdrawn, the quoin is jerked back to its position by the cannoneers acting on both the handles. After the piece is fired, the quoin is forced back with the handspike, the culot withdrawn by screwing in its handle, and plunged into water without detaching the handle, when it is ready for the next charge. The parts should all be well washed and greased after firing.

Cavalli designs this piece more particularly for casemate battery, or positions where it can be protected by blindages and covers of different kinds. He has suggested a method of covering them from an enemy's fire by revetting the front of the battery with guns of the old pattern, with their muzzles planted in the ground and sloping backwards. For field artillery it is recommended to use rifled guns, but not with the breech-loading apparatus. The French rifled gun now in use in Italy is supposed to be something of this kind, with two or four grooves of a form similar to those used in the Cavalli gun, but merging into the bore more gradually.

FRENCH.—The French have a rifled 30-pdr. (corresponding to our 32-pdr.), with two grooves. The projectile, which is oblong, has, instead of wings or flanges, two *buttons*, as before mentioned. These allow too much oscillation in the bore, causing the shot to wedge, and either break it, or burst the piece. These oscillations of the shot in the bore probably were the cause of the deviations observed in the firing; and it has been estimated that with a windage of 0.079 in., the angle of departure of the shot may be changed 1°7′, and the lateral deviation reach $\frac{2}{103}$ of the range. On the other hand, too great a reduction of the windage, by not allowing sufficient play to the shot, is apt to cause it to break.

BELGIUM.—Recent experiments have been made in Belgium, which, from modifications in the form and curve of the groove, and in the wings of the shot, have indicated great improvements.

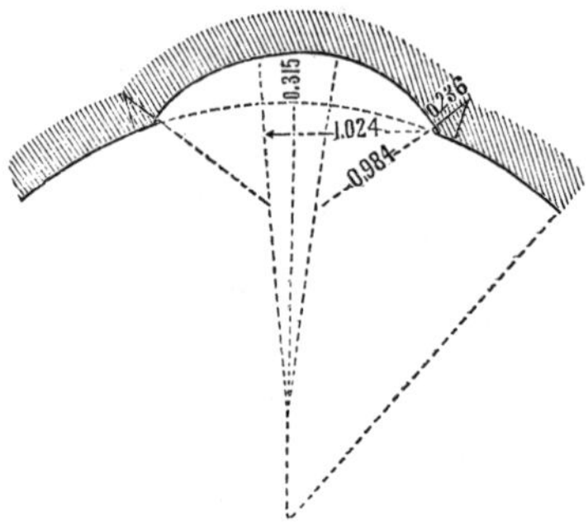

Fig. 90.

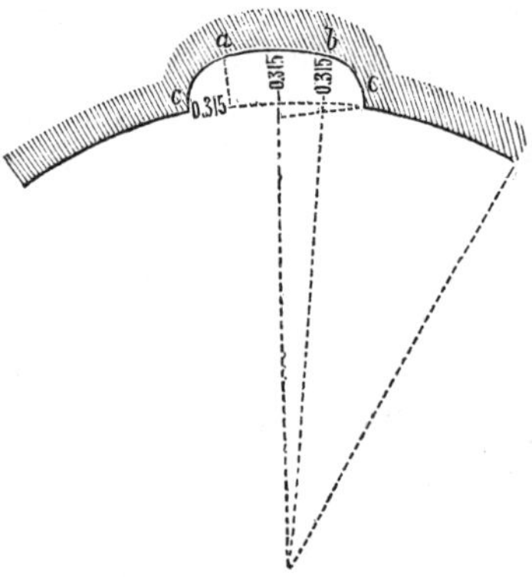

Fig. 91.

The form of groove first used, Fig. 90, was too open, allowing but little support to the buttons, and permitting them to twist out and break. The dimensions in the figure were those used in an 18-pdr.

The form used by Cavalli, Fig. 91, is considered preferable to the other, if the edges where it joins the bore were somewhat rounded off. The depth of this groove is 0.315 in., and its outer opening 1.26 in. The bottom of the section is an arc of a circle concentric with the bore, and is joined on to the bore with arcs of 0.315 in. radius. The width of the grooves is so calculated as to give sufficient thickness to the wings of the shot to resist the force of the charge. The width stated (1.26 in.), was used in a 30-pdr. The Belgium 18-pdr. groove, although greater in width (2.048 in.), caused the shot to break, in spite of the less twist given, probably from its defective form. By simply increasing the width of the Belgium groove, Fig. 90, and merging it gradually into the bore, a gun similar to Lancaster's would be produced; and it is more than probable that the same cause which is assigned for the breaking of the projectile in the Belgium gun existed, though to a greater extent, in the Lancaster. The groove in the latter gun, however, was a progressive helix, whilst that in this is a uniform one.

PROJECTILE.—During Cavalli's experiments with the 30-pdr., the wings of his shot were partly cut off, so as to leave them only 0.118 in. long, and it was found that these buttons sustained the force of the powder as well as the entire wings. The Belgians

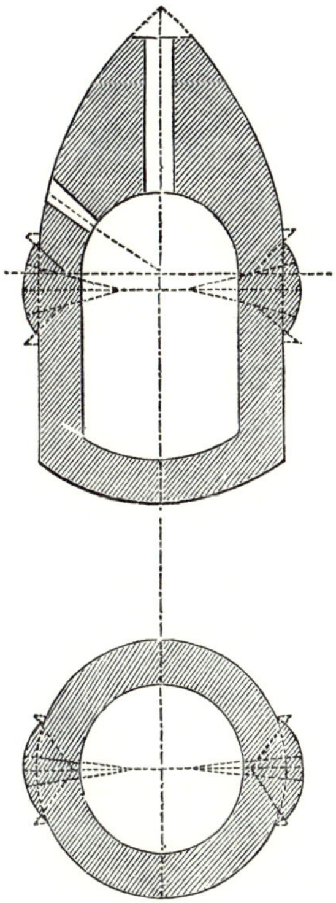

Fig 92.

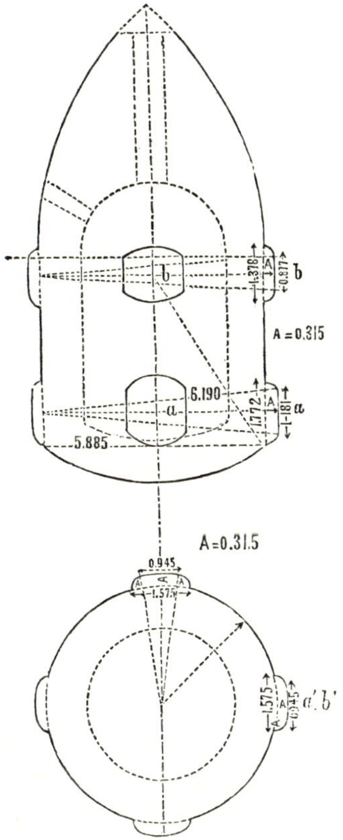

Fig. 93.

formed their first shot on this plan, constructing it as shown in Fig. 92, but making the outer side of the button a curved surface in both directions, so that but a few points in each came in contact with the gun, the windage allowed the buttons being 0.079 in. The shot moved freely about in the gun, and did not give as good results as Cavalli's shot.

In order to steady the shot in the bore, it is suggested by Capt. Gillion, of the Belgian artillery, to construct it as represented in Fig. 93, with two or four *pairs* of buttons placed on the front and rear ends of the cylindrical part of the shot. Two pairs would give more stability than two single buttons; and by placing four grooves in the gun, and four pairs of buttons on the shot, still greater steadiness would be attained, and the pressure against the buttons more uniformly distributed. The form and dimensions of the buttons are sufficiently explained by the figure.

Grooves.—The experiments in Belgium have developed certain facts in regard to the kind and length of helix most appropriate for rifles, which are of great importance. The velocity of rotation is directly proportional to the caliber and the velocity of translation, and inversely as the length of the helix. The velocity of rotation in the bore increases progressively from the action of the velocity of translation, which makes the shot, in equal times, pass over portions of the helix which become more and more extended as it approaches the mouth of the piece, at which point it has its maximum velocity. After the shot leaves the piece, the velocity of translation diminishes very fast, whilst that of rotation does so much more slowly, and this fact may enable both direct and ricochet firing to be successfully made from the same piece; in which case the inclination of the groove should be so regulated that the smallest charges would produce the necessary amount of rotation, and the largest ones not be destructive either to the piece or its projectile. For a piece to be used only in direct firing, the helix should be such as to insure only the amount of rotation necessary to direct the shot, in order that the grooves may be as little injured as possible.

There can be no doubt that each piece has a certain length of helix with which it will give more favorable results than with any other; but experiment goes to show that a considerable variation can be made from this particular length without producing any great differences in the results. It was first laid down as a law that the length of helix for different calibers should be proportional to the calibers, and consequently that when once the best length for any particular caliber was determined, the length for any other piece could be calculated; but this law, as it leaves out of consideration two important circumstances, is not exactly correct.

1st. The difficulty which the projectile experiences in following the grooves, increases in proportion as the calibers increase, since there is at the same time, with equal initial velocity, an increased moving force both of translation and rotation.

2d. The moving force of rotation will be preserved in direct proportion to the size of the caliber, since the resistance of the air acts in direct proportion to the surfaces, and in inverse proportion to the volumes of the projectiles.

Hence it follows:—

1st. That the velocities of rotation of similar projectiles, of different calibers, will be to each other inversely as the calibers; and

2d. That the inclination of the helices will be proportional to the preservation of the velocities of rotation.

The lengths of the helices will be to each other as the *squares* of the calibers. This appears to be more correct than the hypothetical law first stated.

To demonstrate this fact, call H and h the lengths of helices for two pieces, whose calibers are D and d; a a' the angles of inclination of the helices upon the developed surfaces of the bores. Suppose V to be the initial velocity, in both cases. $\frac{V}{H}$ and $\frac{V}{h}$ will represent the number of turns per second made by the two projectiles, and $\frac{V}{H}\pi D$ and $\frac{V}{h}\pi d$ the velocity of rotation on the surface of the cylinders.

By the first proposition we have $\frac{V}{H}\pi D : \frac{V}{h}\pi d :: \frac{1}{D} : \frac{1}{d}$

and hence $H : h :: D^2 : d^2$.

By the second proposition,

$$\text{tang. } a : \text{tang. } a' :: D : d \text{ or } \frac{H}{\pi D} : \frac{h}{\pi d} :: D : d$$

and hence $H : h :: D^2 : d^2$.

The first law would give the same inclination of the helix for all pieces, whereas this one shows that the length of helix should increase in a greater proportion than the caliber, which would make the inclination less for the large calibers.

Cavalli's experiments in Sweden, with a 30-pdr. (6.5 in.), gave good results with two lengths of helix—one 33.69 feet; the other 12.37 feet long. By calculating from the preceding formula, the length of helix corresponding to our 24-pdr. (5.82 in.), 18 pdr. (5.3), 12-pdr. (4.62), and 6-pdr. (3.67), we have—

30-pdr.	24 pdr.	18-pdr.	12-pdr.	6-pdr.
Feet.	Feet.	Feet.	Feet.	Feet.
H=33.99	27.24	22.59	17.17	10.6
H=12.37	9.91	8.22	6.29	3.92

which we may take as the limits between which the length of

helix should be comprised. It will, however, be better, for firing with large charges, to approach the larger limit in preference to the smaller, in order to lessen the drift and preserve the grooves.

The experiments with the 18-pdr. (5.41.), and a helix of 19.685 feet in length, gave more favorable results than with either 16.40 feet or 13.12 feet. The lengths corresponding to this for our 24-pdr., 12-pdr. and 6-pdr., would be 22.77 feet, 14.34 feet, and 9.06 feet. Various lengths, from 25 to 100 feet, have been used in the United States; but as yet no systematic experiments have been made to determine the best.

With regard to the kind of helix best adapted for rifles, it is now well established that the uniform is the most suitable.

The progressive helix acts favorably only with small charges. This kind of helix appears at first sight more favorable than the uniform one, for the passage of the projectile along the grooves; but as the velocity of rotation increases with that of translation, and the latter continues to increase until the mouth of the piece is reached, it is evidently easier for the projectile to enter an inclined groove at the commencement of its movement, when its projectile force is small, than to follow one whose inclination is greater in proportion as the velocity of translation increases.

This reflection caused the trial of one of these grooves reversed, that is, with its greatest inclination at the bottom of the bore, but the results obtained were very indifferent, and caused its prompt abandonment. Both the progressive and retrogressive groove is found to deform the indentations on a bullet; whilst in passing over the uniform groove the impressions remain undamaged, and the projections follow the grooves. So far, the progressive groove in cannon gives indifferent results.

Whilst experiments are going on in this country with regard to rifled cannon, it would be well to find the best length of helix between the limits laid down on page 153, for any one gun, and then, by experiments on other pieces, show whether the rule in regard to the lengths varying in proportion to the square of the caliber, holds good.

CHAPTER V.

PROJECTILES.

The projectiles first used in artillery were irregular in form, and consequently very inaccurate in their fire; and it was not long before the advantages of the spherical form were demonstrated.

The Sphere presents the minimum surface for a given volume; and the wind, which causes so much inaccuracy in elongated projectiles, has comparatively but little effect on the round one, which, having its centers of gravity and figure more nearly coincident than any other, presents, when it rotates, an equal surface always to the action of the air. If it strikes any object in its flight, it is less deflected from its course than one of any other form;—an important fact, since ricochet firing is of great importance in war, it being sometimes the only means of reaching an enemy behind obstacles.

Oblong.—When the design is to strike an object *direct*, however, the sphere is no longer the most advantageous form. For, by making the projectile elongated and pointed, the resistance of the air is very much diminished; and additional weight can be added without increasing the cross-section of the projectile; thus increasing its power of overcoming the resistance, and the penetration of the projectile when it strikes.

Taking the most approved form for elongated projectiles, the resistance to it is found to be about $\frac{1}{3}$ of that offered to a spherical ball of the same diameter. The resistance to the spherical ball is $\frac{1}{2}$ of what one of its great circles would experience. So that the resistance to a projectile moving point first is just $\frac{1}{6}$ of what it would be were it moving with the base to the front. The resistance increases as the surface against which it acts becomes more nearly perpendicular to the direction of this resistance.

Hence, if the projectile becomes flattened by the rammer in loading, the resistance is very much increased.

STONE balls were first used, but were found too brittle to resist the force of powder, and not sufficiently dense to produce the proper effect when striking.

LEADEN balls, although more dense and less brittle than stone, cannot be used in large guns on account of the softness of the metal, and the ease with which their form is changed when striking against objects offering even but slight resistance. Consequently, lead can be used advantageously only in small arms, and against animated beings.

IRON.—The use of *iron* in the manufacture of balls, dates from the XIV. century. The density and resistance of this metal allow the use of large charges. Consequently the effect of the ball is much greater, and it has an effect on stone walls vastly superior to stone balls.

SHELLS.—Soon after the adoption of iron balls, attempts were made to throw explosive globes, designed to act against an enemy behind his works. The first mention made of them is at the siege of St. Boniface, Corsica, in 1421. They were formed of two hollow hemispheres of stone, or bronze, joined by means of a hinge, a circle of iron, and keys. The *fuze* to this rude shell consisted of a sheet-iron tube, inclosing the priming, and riveted to one of the hemispheres. These were succeeded by shells cast in a single piece, either of bell-metal or iron, possessing much more solidity than the others, which often burst before leaving the piece.

CANISTER.—In firing against masses of troops at short distances, the advantages of a divided projectile, such as to strike a number of points at the same time, was early seen. The first used consisted of a box filled with old scrap-iron, which soon gave place to small iron balls, which of course carried further, and had many other advantages. Gibbon mentions the use of such *canisters* at the defense of Constantinople in 1453; and the grape-shot, canister, and spherical case or shrapnell, of the present day, are all modifications of them.

Projectiles are divided into two general classes, viz. *solid shot* and hollow shot or *shells*.

SOLID SHOT are divided into *balls*, or those used in heavy guns, and *bullets*, which are used with small ones. Solid shot being more dense than shells, are much more accurate in their fire, especially at great distances. They have greater power of overcoming the resistance of the air, and consequently greater velocity and penetration when they strike. They are made of cast iron, and used principally in *guns*. Their fire increases in accuracy and range as the size or caliber increases.

The *resistance of the air* is the principal cause of the decreased velocity and accuracy of balls. This resistance is proportional to the surface. A ball twice the size of another, meets with much greater resistance; but its weight is 8 times as great, which enables it to overcome that resistance with greater ease.

Two projectiles moving with the same velocity, the retarding force will be proportional to their surfaces, or to the squares of their diameters. But the velocity which will produce this retarding force is equal to the force divided by the mass of the projectile ($V=\frac{F}{M}$, since $F=MV$), which is itself proportional to the cube of the diameter into the density. Hence the losses of velocity caused by the resistance of the air in the two projectiles, are proportional to the squares of the diameters divided by the cubes of these diameters into the densities, or inversely proportional to the diameters into the densities.

With the same density, but different diameters, the loss is inversely proportional to the diameter; and the largest ball loses the least. Consequently, for great ranges, large balls must be used.

With the same diameter, but different densities, the most dense loses the least, so that dense projectiles have the greatest range.

And finally, in order that two balls, moving with the same velocity, shall meet with the same resistance, the respective products of their diameters by their densities, must be equal to each other. Thus, in order that a cast-iron ball shall experience the same resistance as an ordinary musket bullet of 0.65 inch in diameter, both having the same velocity, we must have the following relation:

$$x \times 7.207 = 0.65 \times 11.352,$$

in which x is the diameter of the iron ball; 7.207 is the density of iron, and 11.352 is the density of lead. Deducing the value of

x, we have $x=\frac{.65\times 11.352}{7.207}=1.02$, or a little over one inch. The weight of such a ball would be $=\frac{\pi}{6}(1.02)^3.2607=.145$ lb. or something over two ounces, instead of about one ounce, which is the weight of the musket bullet. Hence, cast-iron bullets for muskets would be inferior to leaden ones, as their loss of velocity is greater and the deviations more considerable; wherever, therefore, balls are used approaching the size of the musket bullet, lead is the best material.

Spherical projectiles, to be serviceable, should offer sufficient resistance to the action of the powder, in order that the initial velocity to be given them, may be great enough to produce the necessary results, such as penetration, &c. They should be as near spherical as possible; homogeneous in their structure; have their centers of gravity and figure as near together as possible; be as dense as possible, present no roughness on the surface, which would be liable to injure the piece; and if hollow, should have capacity to hold sufficient powder to fulfill the object for which they are fired.

The resistance offered by the air to shells, decreases as the interior is decreased or the thickness of metal increased. If we wish to find the diameter of a solid shot corresponding to a shell of a given density, the density of the shell being $\frac{2}{3}$ that of the shot, we will have, calling x the diameter of the shot, and D that of the shell, $x=\frac{2}{3}D$.

To find what should be the density of a 12-pound shell, in order that it experience the same ressistance from the air as a 6-pound shot, we will obtain, by calling d the required density, $4.52d=3.58\times 7.207$, and consequently, $d=5.7$, that is the density of the shell should be the $\frac{5.7}{7.207}=0.79$ part of that of the corresponding shot, and as that weighs 12.3 lbs., the weight for the shell will be $12.3\times 0.79=8.5$ lbs.

From what precedes it will be seen that lead, if hard enough, would be the best metal to use in projectiles, forged iron the next, and then cast iron, which is much cheaper than forged.

As early as the time of Louis XIV., hollow *elongated* projectiles were made use of, and must have been in great favor, since they appear to have been made of all sizes. The pointed form was found to aid the projectile passing through the air. The interior was divided into two compartments, the front one filled with powder and balls, the rear one with powder only, thus throwing the center of gravity well forward. Had these shells been provided with grooves on the rear part, they would have fulfilled all the conditions laid down on page 127, and have probably given better results.

As the length of the ball increases, its mass and weight increase, whilst the resistance remains the same. Doubling the weight of the ball thus renders the resistance of the air relatively one half less. If, for example, the oblong bullet weighs 730 grains, and the spherical one of the same caliber, 412 grains, the resistance to the first will be only the $\frac{412}{730}$ part of what would be offered to an oblong one of the same weight as the spherical ball. But this last experiences a resistance three times that of the elongated one of the same weight, so that the oblong bullet of 730 grains experiences $\frac{1}{3} \times \frac{412}{730}$ of the resistance (or about one fifth) offered to the spherical bullet of the same caliber (old).

The resistance is, however, still very great to pointed projectiles, it being estimated that it reduced the range of one experimented on in France, to one half of what it would be in vacuo.

Solid Shot.—In projectiles which are designed to act against animated beings, the diameter may be decreased considerably, and the length increased without impairing their efficiency. But for those which are to act against fortifications, masses of earth, &c., a certain caliber is necessary. Were the projectile of too small a diameter, and pointed in form, it would go through the obstacle without breaking or splintering it much, whilst spherical projectiles crush and break into pieces the objects struck, and they also have the advantage of deviating less from ricochets, which are very important in war, either as marking the point of fall, and allowing a rectification of the aim, or as a means of reaching an enemy behind his works.

Caliber.—The caliber of *balls* is expressed by the round number of pounds contained in them. Those used in our service are the following. They are made of cast iron.

128, 68, 42, 32, 24, 18, 12, and 6.

Shot to be used at sea, and on the seacoast, should be somewhat smaller than those for service inland, on account of their liability to oxidize from the dampness of the atmosphere. (For dimensions, see Appendix, page 26.)

Bar or *chain* shot consists of two hemispheres or balls connected by a bar of iron or chain. They take a rotary motion when fired, and are used with great effect against the masts and rigging of vessels, but are very inaccurate in their fire.

A *stand* of *grape* consists of nine shot of a size appropriate to the caliber used, which are held together by two rings, and a plate at each end of the stand connected by a rod or bolt. (For dimensions, see Appendix, page 27.)

Quilted grape consists of an iron plate and an upright spindle around which balls are placed and held in their positions by a canvas bag which is tied to the plate and then quilted on to the balls by means of strong twine, which is finally tied around the mouth of the bag. This kind is no longer used in our service.

Canister Shot is a tin cylinder with iron heads, filled with balls packed in with saw-dust. The heads are movable, and the edges of the tin are turned down over them to hold them in their places. The balls are made of such a size that seven of them can lie in a bed, one in the middle, and six around, making the diameter of the balls about one third that of the bore. These balls are all made of *cast iron*, except for the mountain howitzer, the canisters for which are filled with musket-bullets, which, as has been shown, meet with less resistance from the air, and retain their velocity longer than cast-iron balls of a much larger size. It would be better to make them of *wrought* iron for field-guns, as is done in France; as, besides being more dense, they would be less likely to break and injure the bore of the gun than cast-iron ones. (For dimensions, see Appendix, p. 28.)

It has been shown (p. 158) that a cast-iron ball 1.02 inch diameter experiences as much resistance from the air as the ordinary musket-bullet, where balls smaller than this are required, therefore, the musket bullet should be used.

Bullets are divided into spherical and elongated. They are made of lead, and used in small-arms. The caliber of spherical bullets is determined by the number which a pound of lead will make. Thus, our common musket is said to carry seventeen to the pound. But this is almost entirely superseded in use by the elongated bullet, the caliber of which is determined by its diameter or its weight in grains, usually by the former. There are but two sizes now in use in our service (see pp. 130–1); and in time it is the intention to reduce these to one, which may be used for either rifle, rifle-musket, or pistol. But the bullets, although precisely the same on the exterior, differ in the size of the conical opening, and in weight, the lightest being fifty grains less than the other, the weight of which is five hundred grains. (See Nos. 1 and 2, p. 131.) This difference was found necessary in order that the small charge of the pistol should have force sufficient to press out the sides of the bullet and rifle it properly, which it was found sometimes not to do with the thicker one. The longer these bullets are made, the less resistance is offered by the air, or, more correctly speaking, the greater the power of the bullet to overcome the resistance offered; and were it not for the want of consistency and the softness of the lead, this lengthening would be theoretically unlimited. But they soon become inconveniently long, and liable to be injured in shape, and it is found that, beyond a certain length and a certain number (three) of grooves, the bullet does not carry well. The length of those in our service has been fixed by experiment at 1.1 inch for the large, and 1.05 inch for the small size.

Rifles and Artillery.—Since the recent improvements in projectiles and long-range rifles, it has been customary to underrate the importance of the artillery arm on a field of battle, and the assertion is frequently heard, that the use of the rifle will supersede entirely the use of field-pieces in war, since their fire has a greater range and more accuracy than the field-pieces now in use. This, I am convinced, is a mistaken view. In the first place, it is true that long-range rifles are destined, in the hands of skillful marksmen, to play a very important part in battle, by picking off the cannoneers of the artillery from points beyond the range of this last, *provided* they can once get their sights properly arranged for that distance; but they have first to get their range. To do this, as very few men are at all accurate in estimating distances, trials have to be made; and the bullet makes so little dust in striking, and what it does make is scarcely visible at 1000 yds., that it affords the marksman but little opportunity to correct his aim. In the mean time the gunner is getting his range, which he is enabled to correct from the striking of the ball, which can be seen as far as it goes, and when he once gets it, and that not accurately and precisely, as the rifleman must, but approximately, he is enabled to let loose among his opponents a charge of from thirty to eighty musket-bullets at a time, or send a solid shot through them with sufficient force to disable, perhaps kill, half a dozen, and disorganize as many more by its moral effect.

When the rifleman gets his sight adjusted to the proper range, it is an easy matter for the artilleryman to increase or decrease his distance, rendering new adjustments of the sights necessary, and all this in the heat and confusion of a combat.

These facts, to say nothing of the great physical, as well as moral effect of a rapid and well-directed fire of half a dozen guns upon a body of infantry, seem to demonstrate that the importance of artillery upon the field of battle is rather increased than diminished, and should rather urge to improvement in its range and efficiency, than to its abandonment and under-rating.

The uncertainty of properly adjusting the sight of a rifle in the heat of battle, must be evident to every one; and it is stated that the present emperor of France, fully alive to the fact, has discarded the back elevating sights altogether on his rifles, or very much reduced their graduation, preferring to leave the aiming to the judgment of the men rather than possibly lead them into error by the presence of the sight.*

HOLLOW SHOT.

It has been stated that the fire of hollow shot increases in accuracy as it becomes heavier, or the interior space decreases. This also increases the penetration. On the other hand, the interior space should be large enough to contain sufficient powder or incendiary composition to produce a proper effect; whether this be to produce a great number of splinters with a certain velocity, to destroy by the explosions field-works, or to set fire to shelters occupied by the enemy. The last objects will be attained

---

* Napoleon's recent address to his troops, in Italy, warning them that long-range rifles are formidable only at a distance, and that great dependence must still be placed upon the skillful use of the *bayonet*, is a hint which they seem to have followed.

best by shells having a large interior space. Experiment shows that hollow projectiles fulfill the necessary conditions best when their mean density, or real weight, is equal to $\frac{2}{3}$ that of the solid shot of the same diameter.

DIVIDED.—Hollow shot are divided into *shells*, *spherical case* or *schrapnell*, *carcases*, and *grenades;* all of which are made of cast iron. Their caliber is determined either by the number of pounds contained in a solid shot of the same size, or by the number of inches in the diameter of the shell itself.

SHELLS are hollow shot the interior space being formed of a sphere concentric with the outer surface, thus making the sides of equal thickness *throughout*. They have a conical opening or *eye*, used to load the shell, and in which is inserted the fuze to communicate fire to the charge. Its axis is always coincident with a radius of the sphere.

Shells have sometimes been reinforced with a *culot*, or increased thickness of metal opposite the fuze hole, for the purpose of strengthening that part most exposed to the shock of the powder, especially in pieces with long, narrow chambers, which were formerly employed, and with an idea that it would cause the shell to fall with the fuze up, and prevent the failing of the fuze, from becoming stopped up with dirt when it fell.

The objections to the use of the *culot* are: that it separates the center of figure from that of gravity, thus diminishing the accuracy of fire and the velocity, by the increased resistance of the air, due to the irregular movement of the eccentric projectile; that the culot being thicker, presents more resistance to the powder than the other portions of the shell, and consequently a less number of pieces are formed. Experiments go to show that an exploding shell cracks generally through the fuze hole. The culot is not used at all in our service.

The resistance offered by a shell to the force of the powder increases with the thickness of its sides. The number of pieces produced when it explodes is the greater, all else being equal, as the metal is more brittle, and the eccentricity of the shell is less.

The *eye* should decrease in size with the interior space. Too large an opening might allow the escape of the gas without bursting the shell. It should not be too small, as this would prevent the use of a proper thickness in the fuze.

The French sometimes make their shells for sea-coast service with an additional eye, at an angle of 45° with the other, called a *charging hole*, the object being to have the fuze already fitted in, ready for use, and allow the charge to be poured in just before the shell is wanted. It is found that the powder, when left in the shell for any length of time, rapidly deteriorates, from the dampness of the sea air. This arrangement, however, has the disadvantage of requiring the fuzes to be cut beforehand and without knowing at what distances they are to be used.

The French make use of what they call a *shell bullet* (Balles-obus), made of lead, hollow, and oblong, with the pointed end formed like the nipple of a gun, and pierced to communicate with the interior space, which is filled with powder. A common percussion cap is placed on the end, and the bullet is set home with a concave rammer head. Fired from a rifle, the bullet goes point foremost, and when it strikes explodes the cap and the powder in the bullet, blowing up any powder it may come in contact with, or setting fire to inflammable material. Caissons or ammunition wagons of any kind may in this way be destroyed.

The following are the shells used in the United States service:

For mortars, 13 in., 10 in., and 8 in., Fig. 94, Pl. 8; the last used also in the 8 in. siege howitzer.
" Columbiads, 10 in. and 8 in., Fig. 95, Pl. 8.
" Guns and Howitzers, 42, 32, 24, 18, and 12 pdrs., Fig. 95, Pl. 8.

All but the mortar shells are reinforced on the inside of the fuze hole, to give a greater bearing surface to the fuze, and prevent its being driven in by the force of the heavy charges used. The sides of these shells are also thicker than the corresponding sizes for mortars, as larger charges are used with them.

The mortar and columbiad shells are handled by means of two ears placed one on each side of the eye, which serve for attaching a pair of shell-hooks. The other shells have no ears, but rope handles are fixed to the tin straps which fasten them to their sabots. (For dimensions, see Appendix, p. 27.)

SPHERICAL CASE, or *Schrapnell* Shot, as they are called, after the English general who brought them to perfection, are thin-sided shells, in which, besides the bursting charge, are placed a number of musket balls. Their sides are much thinner than those of the ordinary shell, in order that they may contain a greater number of bullets; and these acting as a support to the sides of the shell prevent it from being broken by the force of the discharge. The weight of the empty case is about ½ that of the solid shot of the same diameter. (For dimensions, see Appendix, p. 27.)

The calibers in use in our service are:

The 8 in.—42, 32, 24, 18, 12, and 6 pdrs., Fig. 96, Pl. 8.

They are all reinforced at the eye, to give a greater bearing for the fuze. Lead being much more dense than iron, the schrapnell is, when loaded, nearly as heavy as the solid shot of the same caliber; but on account of the less charge which it is necessary to use to prevent rupturing the case, their fire is neither so accurate nor the range so great as with the solid shot. But when the schrapnell bursts just in front of an object, the effect is terrific, being in fact pretty much the same as a discharge of grape from a piece at short range.

The range and effect are both much increased by the present method of loading. This is done by placing in the proper number of balls, and then pushing a stick, grooved on both sides, through the eye to the bottom of the case. Through the grooves melted sulphur or rosin is poured, and when cool the stick is withdrawn, leaving in the center of the mass a small *chamber*, in which the bursting charge is placed.* This arrangement places the powder entirely free from contact with the bullets, and it is consequently not liable to be ground up by them while being transported or when the shell is fired. The powder can be placed in this chamber and allowed to remain without fear of damage or danger, and be all ready for use when required. Being, besides, in a compact mass, instead of scattered among the bullets, its power is much greater and it acts more effectively in throwing the bullets outward from the center. This advantage is further increased by the adoption of the admirable Boarmann fuze,

* Schrapnell are now usually loaded by placing in the bullets and pouring the melted sulphur in until the case is full. After the sulphur has cooled, the space for the powder is bored out by a cutter, which removes both the sulphur and portions of the bullets from the space. This is a quicker method, and gives a more compact projectile.

which can be screwed into its place beforehand, and guaged on the field, in a moment.

The charge used in these shells is sufficient simply to rupture the case and release the bullets at the proper point in front of the object. Their execution depends, then, upon the velocity which the shell has at the moment it bursts.

The Boarmann fuze being adopted, the reinforce at the eye becomes useless, and may as well be dispensed with to allow more room for bullets. This would also make the firing more accurate, as the shell would be more concentric.

CARCASSES, Fig. 97, Pl. 8, are shells having, besides the usual eye, three others, which are placed at equal distances apart, and tangent to the great circle of the shell which is perpendicular to the axis of the first eye. They are filled with combustible composition, primed at the four holes with quick-match and mealed powder, and are used to set fire to an enemy's works, the additional holes being to allow a more rapid escape of the flame.

GRENADES, Fig. 98, Pl. 8, are of two kinds. The *hand-grenade* is a small shell thrown from the hand or in baskets from the stone mortar. *Rampart-grenades* are larger, and are used to roll down a breach in its defense, to throw over the ramparts, &c. Any kind of shell, unfit for firing either from being defective in form or solidity, may be used for the purpose. 6-pdr. spherical case. shot may be used as hand-grenades.

WAR-ROCKETS.—A *rocket* is a projectile which is set in motion by a power residing within itself. It therefore performs the part both of a piece and a projectile. The cases for *war-rockets* are made of sheet-iron and lined with paper or wood veneer to prevent the composition from touching the metal and rusting it, which would destroy the missile. They are filled with a composition of niter, sulphur, and charcoal, in the same way as described for signal rockets; but are generally filled solid by means of a ram or press, and the core then bored out. At the top end either a solid shot or shell is placed, and riveted to the case, a recess being cut out of the lower part of the projectile, which is there cylindrical, to fit into the case. When a shell is used, it is perforated through the diameter which coincides with the axis of the case, and a fuze driven into the opening next to the composition. When the composition burns out, fire is communicated to the fuze, which, in its turn, explodes the charge in the shell. The hole at the top of the shell enables the fuze to be regulated, by boring it out in part or altogether. The top hole is then closed with a wooden plug or a screw.

The dimensions of rockets are indicated either by the weight of the projectile or the diameter of the case, in inches.

Two kinds of rockets have been used in this country, the *Congreve*, and *Hale's*.

CONGREVE'S, Fig. 99, Pl. 8, has, like the ordinary sky-rocket, a directing-stick, but instead of being tied to the outside of the case, it is inserted in a socket placed directly in rear of the case, the flame escaping through holes around this. This modification was introduced by Sir William Congreve, who was also the first one in modern times to make use of metal cases; but he is not the inventor of the rocket, which has been known from time immemorial in China and India, in both of which it had been used as a war missile.

These rockets have been made of immense size, the largest weighing as much as three hundred pounds, but have never been adopted to any very great extent; for, although very formidable and destructive, especially when used against cavalry, their fire is very inaccurate. The motion of the rocket is due to the pressure on the case produced by the reaction of the gas escaping through the vents, and depends upon the mechanical principle of the equality between action and reaction.

Rockets are fired from troughs or tubes mounted on adjustible tripods, so that the necessary angle of elevation can be given to them. They may also be simply laid upon the ground, having the necessary slope, and fired singly or in volleys. In the latter case, they are connected by a piece of quick-match communicating with the priming, by lighting which the rockets go off in rapid succession.

Special troops have, in some of the European powers, been formed and armed with these projectiles, which are carried in wagons. Each man carries into action several rockets ready for use, and tied to his saddle. In this country, however, no such organization has been made; although, rockets were used to a limited extent in the war with Mexico.

In the Austrian service they appear to have been adopted to

a much greater extent than anywhere else. But they are there made with a shell very much larger than the diameter of the case, to which it is fastened with tin-plate straps. They are fired from tubes into which they are placed from the front, with the shell, which is too large to enter, projecting from the mouth. Large angles of elevation are used, and the rocket, after accomplishing a short range, drops its shell with its fuze (previously regulated) burning.

Hale's (Fig. 100, Pl. 8) differs from any other rocket in having no guide-stick. Direction is given to the rocket by imparting the *rifle* motion to it. This is effected by placing in the rear part a number of holes oblique to the axis. The gas, escaping through these, acts upon the air and gives the rotary motion to the case, whilst it propelled forward by the action, or rather *reaction*, of the gas escaping at the main hole at the extreme end. Within a year past, these oblique holes have been changed from their position in rear of the rocket, reduced to two in number, opposite the center of gravity, and fire instead of being communicated at the end is applied at one of these holes, and rapidly spreads in both directions in the interior.

Hale's rocket, by dispensing with the long and unwieldy guide-stick, is a great improvement; but it is not the only one which has been made. The great difficulty in rocket-firing, is to get them to start in the right direction. Long before it has attained anything like its maximum velocity, it commences to move, and the moment it loses the support of the tube or trough, it begins to fall or "dip," and before the constantly increasing velocity is great enough to overcome this disposition, the rocket will probably ricochet on the ground; and this, especially with Hale's, is apt to throw it very much out of its course, and add to its otherwise inaccurate firing. Mr. Hale has striven to overcome this difficulty by placing his rocket behind a strong spring, which holds it until it has acquired force enough both to overcome its own inertia and the strength of the spring, when it is released with a much greater velocity, and but little of its former disposition to "dip."

A great objection to these rockets made with metal cases, is that from the expansion and contraction of the metal, cracks and flaws are formed, after a time, which give passage to the flame, increase very rapidly the rate of combustion, and sometimes cause the rocket to explode like a charge of gunpowder. Rockets, when kept any length of time, are particularly liable to this accident. It is stated that in New Mexico, where the climate is very dry, the common sky-rocket cannot be kept for any length of time without being subject to the same defect; and they are sometimes restored to their former condition by soaking them for a short time in water, and then drying them.

MANUFACTURE OF PROJECTILES.

In the manufacture of projectiles, iron moulds have sometimes been used, but are found to make an inferior, brittle article, liable to be easily broken, principally from the more rapid cooling of the metal. The moulds are now made of sand, similar to that used in casting guns; though a less refractory sand is needed, as the mass of metal is less, and possesses, consequently, a less amount of heat. It is, as before, mixed with clay-water, to give it form and consistency.

MOULDING.—The model consists of two polished hemispheres of copper, which, fitted together by means of a groove in one and projecting edge in the other, form a perfect sphere. One of these hemispheres is placed on a board or other plane surface. Over this is placed one half of the *flask*, a sheet-iron box in two parts (Fig. 101, Pl. 8) made to fit each other, for the purpose of containing the mould. Each half has a movable bottom, taken off when the sand is placed in. Each one of the copper hemispheres has in the bottom a hole and thread of a screw, *c*, into which a handle can be placed to lift the model out of the mould, and on the outside at *d*, a corresponding hole and thread into which the handle *b* is now screwed. A round stick, *a*, is held in a suitable position against the board on which the flask rests, and the moulding is driven compactly in until the flask is full to the line *ef*, when it is accurately leveled off, the handle *b* unscrewed, and with the stick *a* removed. The bottom is placed on, secured in its place, and the whole turned over; the board, *g h*, taken off and the other half-model and flask adjusted on top of the first ones, *dry sand* being sprinkled on top of the half-mould formed, to prevent the other from sticking to it. Fig. 102.

After screwing the handle of the other half-model in its place, this flask is filled, the handle removed, and bottom put on in the same way as at first.

The top half is then taken off and turned over, and both half-models are taken out by screwing in the handle at *a*, and lifting them up carefully so as not to break the mould; a passage is cut at *c*, across from the channel *b*, and if casting solid shot, the hole left by the handle at *d* is closed with sand. Any parts which have been broken away are now repaired by hand, and the whole interior is covered with coke-wash; the mould is placed in an oven to be thoroughly dried,* after which the two parts are fastened together, with the two apertures *b* and *e* uppermost.

Casting.—The metal, in a proper state of fluidity, is brought from the furnace in a bucket or ladle, formed as shown in Fig. 103, Pl. 8, of iron, coated with clay, having wrought-iron or wooden handles, and poured into the mould at *f*, entering at the side to prevent injury to the form. As it rises, the air escapes at *e*, which also serves as a dead-head to collect the scoria, if any enters, and furnishes metal to supply the shrinkage caused by cooling.

Core.—In casting shells, the mould is made in the same way, but a *core* is needed in addition. This is a sphere of the proper size, made by compressing the moulding composition on a *stem b*, Fig. 105, Pl. 8, by means of two cups, Fig. 104, Pl. 8, the requisite compression being given by screws placed at *a*, *b*, and *c*. This core is, by means of a guage, placed exactly in the center of the mould, and supported in that position by the stem placed in the hole, which, in casting solid shot was closed. The core being subjected to greater heat than the other portion of the mould, should be made of a more refractory sand. The stem, besides supporting the core, forms the fuze-hole for the shell. It is formed of a thick wire covered with the moulding composition.† After the casting has become cool, the core is broken up and removed; and the projecting portions at the gate, *c*, Fig. 102,

---

* Until recently, the moulds were not dried before casting, but casting after the mould is dried is found to produce a much smoother, better shot than when cast wet or "*green.*"

† The stem is sometimes made hollow, as at *a*, *a*, *a*, Fig. 105, Pl. 8, to allow the escape of any gases which may form from the effect of the heated metal.

Pl. 8, and around the base where the two halves join, are taken off with a hammer and chisel.

Polishing.—A number of the balls are now placed in a large revolving iron cylinder, which, by friction, polishes and makes the surface more uniform; after which, and before any lacker or grease is placed on them, they are inspected.

All projectiles for large ordnance are made of cast iron, though other metals have been used, and, until recently, the Mexicans used copper to a great extent, and, perhaps, still do so.

Bullets.—Projectiles for small arms are made of lead, and the method of *compression* is now used in preference to casting, and furnishes a more uniform and homogeneous bullet.

Materials.—All projectiles for our service, for cannon are manufactured in private foundries, and inspected by officers of the ordnance, before they are received into service. After which they are covered with a coating of lacker, and placed in piles of different sizes until they are wanted. The *compressed* elongated bullets are now made in our arsenals, by a very ingenious piece of machinery which turns them out very rapidly and very perfect.

The iron used in the manufacture of projectiles should be what is commonly termed *gray* or *mottled*, and should be of good quality, especially for spherical case-shot, which requires more care on account of the thinness of the sides.

Inspecting.—The manner of inspecting shot and shell, and the instruments used, will now be described. To ascertain if they are of the proper weight, several parcels, of from twenty to fifty, are weighed, being taken from the pile indiscriminately. If any are found smaller than the rest, they are weighed separately, and rejected if they fall short of the proper weight by a small fraction, which has been successively reduced as the improvements in the art of casting enabled a higher standard to be reached. They generally exceed the required weight.

To find the weight of a cast-iron shot of any diameter, multiply the cube of its diameter in inches by 0.134. The result will be the weight in pounds. If the weight of a shell is required, use in this rule the difference between the exterior and interior diameters in place of the diameter. $\frac{1}{6}\pi D^3$ being the solid contents of any sphere, and 0.2607 = the weight of a cubic inch of cast iron, the weight of a cast-iron sphere will be $=\frac{1}{6}\pi D^3 \times 0.2607 = \frac{3.1416}{6} \times 0.2607 D^3 = 0.134 D^3$.

The multiple in the case of lead balls is 0.2142.

To find the diameter of a cast-iron shot of a given weight, reverse the rule: divide the weight by 0.134, and the cube root of the quotient will be the diameter in inches.

The shot is inspected while perfectly clean, and before becoming rusty, so that the eye can detect any flaws or imperfections in the metal. If any attempts have been made to fill these with iron, cement, &c., the shot is at once rejected without further examination. Such holes as are found are probed with a *steel punch*, or struck with the pointed end of the inspecting *hammer.*

This hammer weighs about half a pound, and is flat at one end for sounding shot and shell, and conical at the other. Cavities over 0.2 in. deep, cause rejection.

THE RING-GUAGE, Fig. 106, Pl. 8, is a ring of iron with a wooden handle, used to determine the diameter of the shot. Two sizes are used. The largest is 0.02 or 0.03 in. greater than the *true* diameter of the shot, and the smallest 0.02 or 0.03 less than the true diameter. The shot must pass in any direction through the large guage, and not at all through the small one.

(For dimensions see Appendix, p. 28).

The size of grape and canister shot is determined by using a large and small guage attached at the opposite ends of the same handle, Fig. 107, Pl. 8. The surface should be smooth and free from seams.

THE CYLINDER-GUAGE, Fig. 108, Pl. 8, is a cast-iron cylinder with reinforce bands on the exterior, and an interior diameter equal to the diameter of the large ring-guage. This is placed on blocks of wood, with one end about 2 inches higher than the other, in such a position as to be easily turned so that it will not be worn in furrows by the shot rolling through it. The shot is then rolled through. They should pass through without *sticking* or *sliding.* In this last case, it shows that some one diameter is too large. In case they stick, they are pushed out from the lower end with a rammer.

(For dimensions see Appendix, p. 28.)

The soundness or strength of shot is proved by dropping them from a height of 20 feet, on an iron block, or rolling them down an inclined plain of that height against a shot at the bottom.

Shells and hollow shot are inspected in the same way, but require in addition the following instruments:—

CALLIPERS, *Fig.* 109, *Pl.* 8, *for measuring the thickness of the metal at the sides*, which consist of two bent arms movable on a common pivot, and showing on a graduated arc the thickness of the metal; or, Fig. 110, Pl. 8, of one strait arm which is placed tangent to the outside of the shell, and one bent, which is inserted in the shell, the thickness being shown on a graduated limb which joins the two.

Callipers, Fig. 111, *for measuring the thickness of the shell at the bottom*, which consist of two straight arms connected by a circular piece. One of these arms is inserted in the shell, and the other, being movable, shows on a graduated side the thickness of the metal.

GUAGES, *Fig.* 112, *for the dimensions of the fuse hole and thickness of metal at that point.*—These are pieces of plate metal having inclined sides to fit the fuze hole, with the proper dimensions marked on them for each caliber.

A pair of hand bellows, and a wooden plug to fit the fuze hole, and bored through to receive the nose of the bellows.

The shell is sounded with the hammer, to see if it is free from cracks. The position and dimensions of the ears are verified; the thickness of metal is measured at several points on the great circle perpendicular to the axis of the fuze hole, at the bottom and at the fuze hole. The diameter of the fuze hole, which should be accurately reamed out, is measured with the guage; and the soundness of the metal about the inside of the hole is ascertained by inserting the finger.

The shell is now placed in a tub with water, deep enough to cover it nearly to the fuze hole; the bellows and plug are inserted in the fuze hole, and air forced in. If there are any holes in the shell, bubbles of air will rise through the water. Should there be any cavities in the metal, those portions will dry more slowly than the others.

Shot and shells rejected in the inspection are marked with a cold-chisel with an X, the shells near the fuze hole, the shot near the *gate*, or point where the metal entered the mould.

The shot and shell, as soon as received, are covered with a coating of lacker, which should be renewed from time to time as required.

Preservation.—Shot and shell are preserved in piles according to kind and caliber, under shelter if practicable, where there is a free circulation of air. The width of the bottom tier may be from 12 to 14 balls, according to the caliber.

The ground should be prepared, by raising it a little above the surrounding surface to throw off the water, leveling it, ramming it, and covering it with a layer of clean sand. Bury a tier of unserviceable balls about $\frac{2}{3}$ of their diameter in the sand. Place the fuze holes of shells down in the intervals, and not resting on those below. Each pile is marked with the number of balls it contains.

The base may be made of brick, concrete, stone, or with braces and borders of iron.

Grape and canister shot should be oiled or lackered, put in piles or in strong boxes, on the ground floor, or in dry cellars, each box marked with its kind, caliber, and number.

Each pile should contain only one kind and caliber of shot, and the bed may be covered with coal-ashes, or any thing else which will not promote vegetation.

There are three kinds of piles used, with oblong, square, and triangular bases; and it often becomes necessary to calculate rapidly the number of balls contained in them. To do this, the following formulas are used:—

In an oblong pile which has a rectangular base, let $n=$ the number of balls in the width of the base. In the triangular end the number of balls in the different horizontal layers will increase in arithmetical progression from 1, at the top to n, which is the bottom row. The sum to the nth term, or the number in the end, will then be $=\frac{n\,(n+1)}{2}$. Representing by m the number of balls contained in the *upper* edge of the pile, that in each of the lower edges parallel to it will be represented by $m+n-1$. Considering the pile as a triangular prism whose bases are oblique to each other, its contents will be equal to the number in one of the triangular bases (the end of the pile), multiplied by the mean of the three parallel edges, representing the altitude of the prism—that is, by $\frac{3m+2n-2}{3}$. Hence, the number of balls is $=\frac{n(n+1)}{2}\frac{(3m+2n-2)}{3}$ and the rule,—*Multiply the number of balls in a triangular face by one third the sum of the three parallel edges.*

If the base of the pile is square, m becomes $=1$, and the formula reduces to $n\frac{(n+1)}{2}\frac{(2n+1)}{3}$; and if it is triangular, m is $=1$ as before, and another of the parallel edges reduces to 1, while the third is $=n$, and the formula becomes $n\frac{(n+1)}{2}\frac{(n+2)}{3}$.

If we have given the number of balls to be piled and the width of an oblong pile, m the length of the top row, and the two sides of the base can be at once deduced from the formula.

If a pile consists of two piles joined at a right angle, to find the contents of it calculate the number contained in one as a common oblong pile, and the other as a pile of which the three parallel edges are equal.

CHAPTER VI.

ARTILLERY MATERIAL.

Under the term *artillery material*, are included all carriages and machines used for the service of the pieces, for the transportation of the munitions of war, and for the establishment of the movable bridges necessary on the marches of an army. In the United States' service, the term is more restricted, and does not include the apparatus for bridges, which is placed under charge of the engineers; a special company having been formed during the Mexican war, and kept up ever since.

A *gun carriage* is the machine on which a piece is mounted for maneuvering and firing.

Those first in use consisted of blocks of wood or frame structures, upon which the ancient bombardes were fixed, by means of iron work. These machines, made in imitation of the mountings of small arms, bore the names "fust," "fustage," "house," the first of which seems to be derived from the latin *fustis* (club, baton), from which the French have got their term affût (gun carriage).

Bombarde.

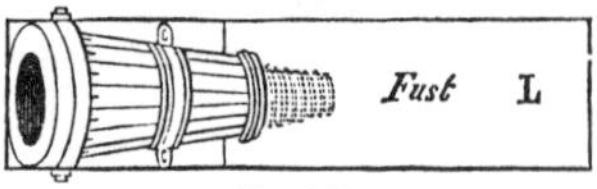

Fig. 113.

These bombardes were sometimes mounted on their carriages by means of a screw formed behind the breech, L, Fig. 113; and later, a kind of carriage, O, Fig. 114, was made use of. It consisted of a framework which could be raised or lowered by the use of iron pins placed in the uprights of a firm structure.

Bombarde.

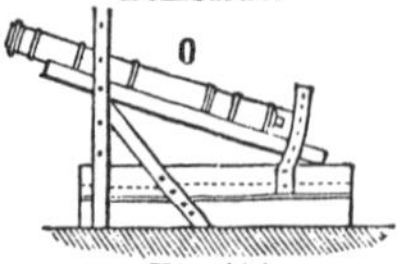

Fig. 114.

In the next form, Fig. 115, the framework was mounted upon wheels, or rather rollers, by means of which the piece was transported and fired horizontally.

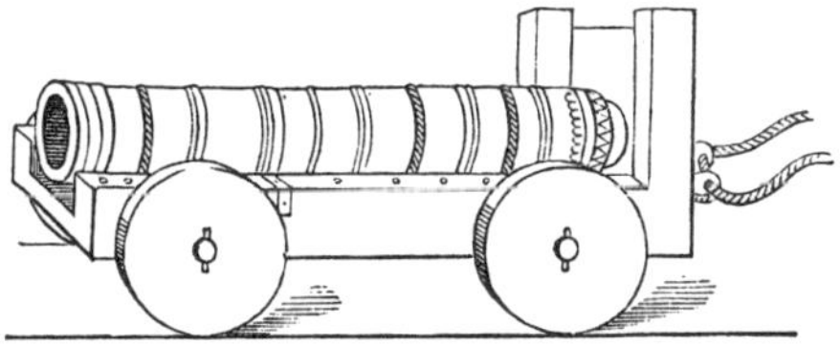

Fig. 115.

When the arm was of small caliber, it was mounted upon a light wooden horse, or placed on a small wheel-barrow, which served to transport and fire it.

When trunnions were first used on guns, the carriages had

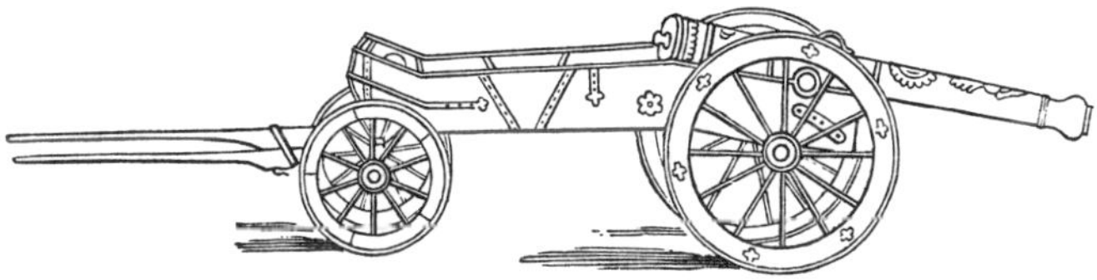

Fig. 116.

assumed pretty much the form they now have. And from the time of Louis XI. and Charles VIII. (1461), a limber was used with the carriage, and taken away, as now, when the piece was to be fired. The horses were harnessed to it in single file, as in heavy drays of the present day.

The carriages of pieces of small caliber had no limber, and those of mortars and bombardes were mounted upon four small wheels.

Fig. 117.

At this period, the other part of the material consisted of carts carrying powder and projectiles.

The largest-sized pieces were carried on wagons.

From the time of Henry II. (1547) boats were transported on carriages in the rear of armies, and used for the construction of bridges.

Valliere's System.—The structure of gun carriages in France, has undergone but little modification from the earliest times; and even Valliere's system did not change it. His system regulated simply the caliber of the pieces, and did not extend to the carriages, the forms of which were very variable. Gun and howitzer carriages preserved the old form. Those for mortars were not so high, and had no wheels. As formerly, the supplies were transported in carts, the horses, being still harnessed in single file. Copper pontoons took the place of the boats.

In battle, a pile of balls, one of wads, and a barrel of powder, were placed near each piece. The charge was taken from the barrel and placed in the piece with a kind of long-handled ladle (Fig. 118).

Fig. 118.

This kind of artillery, devoid of all mobility, was more particularly suited for the attack and defense of fortified places.

Gribeauval's System succeeded that of Valliere in 1765, and was the first in which the different parts were perfectly regulated and made uniform, so that the parts of any carriage could be used on any other of the same kind.

A tongue was used and the horses harnessed double, which enabled more rapid movements to be made, and shortened the columns of troops very much. From this period, we find established a distinction between siege, garrison, sea-coast, and field artillery; and in the last-named the charges were carried in boxes called caissons, all ready for use.

Desirous of giving to his system the greatest possible lightness and mobility, Gribeauval calculated the minimum dimensions which the different parts of his carriage should have; but, from this resulted the grave inconvenience of a great multiplicity of kinds, which rendered their replacement in war very difficult. He replaced the pontoons, which were not suitable for navigation, by large boats, better fitted for making bridges.

The mode of harnessing the horses in this system, was especially calculated to preserve them. In the year XI. attempts were made to modify the system, but were not successful except in the bridge equipage, which was made much lighter.

Each piece had a particular gun carriage, and its own sized wheels. The greater part of the wagons also had distinct wheels. The limbers were simplified, and some could be used with several different carriages. Eighteen different kinds of wheels were used. In 1814, there were in France seven kinds of field gun-carriages, six of siege, four of garrison, four of sea-coast, and four of mortars. These are now reduced to two of the 1st and 2d; four of the garrison and sea-coast, one mountain and four mortar carts.

U. S. Carriages.—The most essential properties of the present artillery material both of France and this country are: its mobility, and the reduction of the different kinds to the smallest possible number. All field carriages in the U. S. service have the same limber, and the wheels are all of the same form and height, and fit on the same axle-tree. They differ only in strength and weight. There are two numbers: No. 1, the lightest, is used for the 6-pound gun carriage, the caisson, forge, battery-wagon, and the limbers of all field carriages; No. 2, the heaviest, for the 24-pound howitzer and 12-pound gun carriages.

For siege service, all the wheels are the same, and used indiscriminately on all gun-carriages, limbers, and mortar-wagons.

The same limber is used for all siege-carriages.

Gribeauval's carriages were provided with swingle-trees and movable splinter-bars, like a stage, with a view to give the horses all possible liberty of motion. This arrangement made the two parts of the carriage dependent upon each other; allowed but little flexibility; the connection between the two often broke; the smallest obstacle would overturn, or separate the two parts; and the caisson could not turn with sufficient ease. It was difficult to limber and unlimber, the trail being very heavy, and it being necessary previously to lift off from its position on the trail the heavy box which contained the ammunition. Manœuvring with the prolonge in presence of the enemy became necessary, which rendered the draught extremely difficult, and caused many accidents.

The swingle-tree, being liable to get out of order and break, is now entirely dispensed with. In field-carriages, the end of the tongue is supported by the horses; and the two parts being united at a single point, are perfectly flexible, and can pass over the roughest ground without accident. The separation of the two parts is effected without any difficulty, and the use of the prolonge in the presence of the enemy is now entirely abandoned.

The siege-carriages serve not only to fire the pieces from, but also to transport them. Garrison and sea-coast carriages can be used only for the former; though some of them are so made as to be able to transport the piece for very short distances, as from one face of a fortification to another.

The French use bronze for their nave-boxes entirely, as the less friction of the iron axle-tree against it is favorable to the draught. Bronze is used in our siege-carriage wheels, but cast iron in the field-carriages.

The English have a field-carriage similar to our own ; but the fore-wheels are smaller than the hind ones, and instead of having one ammunition-box extending all the way across the limber, they have it divided in two, leaving an interval between to enable them to limber the piece, the pintle-hook being very long, and extending up above the bottom line of the boxes. Instead of a pin for the pintle-hook, a hinged hasp is used, which turns down over the top of the pintle-hook, and has to be raised and turned up in the interval between the two boxes before the piece can be unlimbered. No particular advantage is obtained from this arrangement; and it has the great disadvantages of lessening the capacity of the box for ammunition, and destroying, or at least impairing very much, its quality as a seat for cannoneers.

By means of the modifications made in the Gribeauval system, the mobility of the carriages is increased. Repairs, always so difficult in war, are made with greater ease, the parts being able to supply each other's places ; and the field artillery possesses all the necessary means to seize promptly the favorable occasions to act during a battle.

Gun-carriages may be divided into *movable* and *stationary* carriages. To the former class belong those used in the field and for siege purposes. The conditions which they should fulfill will now be examined.

Movable carriages are used for the transportation of the pieces, as well as for firing them ; and are, for these purposes, mounted on large wheels.

In firing, the carriage should yield to the recoil. Were it fixed immovably, it would soon be destroyed, no matter how great its solidity. Its weight should be proportional to that of the piece. If too heavy, it would soon be destroyed by the shocks of the piece. If too light, the recoil would be immoderate. The weight should always be less than that of the piece. Too heavy a piece upon too light a carriage will perform better service than the reverse arrangement, since the effort exerted by a piece is a function of its mass into the square of its velocity.

The carriage consists of two cheeks, connected together and

with the stock by assembling-bolts. The front part supports the piece, and rests upon an axle-tree furnished with wheels, the back end of the stock or trail resting on the ground, and forming the third point of support for the system.

The parts of the carriage between which the piece is placed are called the *cheeks*. In ancient carriages these cheeks extended all the way to the ground, joined to each other by transoms and bolts; but in the new, the cheeks are short, and fastened upon a stock formed of two pieces close together, and terminating in the trail.

The wheel is composed of a *nave*, into which the axle-tree enters; of a certain number of *spokes* fastened in the nave; and the circumference of the wheel, which is composed of a number of *felloes* equal to half the number of spokes.

The wheels are always made with a certain convexity called the *dish*. The obliquity of the spokes gives elasticity to the wheel, and protects it from shocks which, were the spokes in the same plane, would destroy it. The dish should increase as the ground to be passed over becomes more broken.

The object of the dish is, for the purpose of making the body of the carriage wider; to diminish the length of the axle-tree, thus increasing its strength; to throw the mud and water outside the wheels; and to tend, from the effect produced by the decomposition of forces, to keep the wheel close against the carriage, and prevent any tendency to run off the axle.

As the weight of the carriage should not be increased beyond a certain point, it is only by making use of friction, skillfully applied, that the recoil can be diminished. The friction of a wheel is proportional to the ratio between the radii of the wheel and axle. The recoil of the carriage may be diminished by decreasing the diameter of its wheels, or increasing that of its axle. For this reason, certain carriages (as, for example, mortar-beds) have been mounted upon rollers, and others upon wooden axles, with truck wheels.

The weight of the wheels, as it adds to the weight of the carriage, tends to diminish the recoil. If they are too heavy, they strain the axle-tree very much, and tend to bend it. The wheels slide along the ground at first, and commence to turn only after a certain length of time; but once they have acquired a certain velocity, they carry the carriage with them until this velocity is destroyed.

The position of the trunnions exercises some influence over the recoil. The higher they are placed above the axle-tree, the greater is the arm of the lever which tends to force the trail into the ground, the more energetic is this tendency, and the more is the recoil diminished.

For a given angle of fire below the horizon, or for a given length of carriage, the carriage tends to overturn to the rear, raising the wheels. The angle of incidence of the line along which the shock of the piece is transmitted to the ground, has a great influence on the recoil of the carriage.

For siege-guns and howitzers, the ancient carriages were very heavy, as were also the wheels. The axle-tree was of wood, the elasticity of which was necessary to prevent accidents. Though the carriage was very long, which diminished the angle of incidence of the force of recoil, the actual recoil was diminished by the great weight of the carriage and the friction of the axle in the nave.

In the more modern carriage, the axles are of iron, and the nave-boxes of iron or bronze, which decreases the friction. The trunnions are nearer the stock, and the carriage, though lighter, is shorter, and hence the pressure exerted by the trail on the ground is increased, and the recoil kept within proper limits.

Pieces, especially those which are light relative to their projectiles, strain their carriages a good deal, particularly when the firing is at angles of elevation. Thus, a howitzer carriage which will resist a horizontal fire perfectly, is liable to be broken when fired under a large angle. The only means to obviate this inconvenience is to favor the recoil at first.

Gribeauval's howitzers, which were very light in proportion to their projectiles, and which were mounted upon very heavy carriages with wooden axles, frequently caused accidents when they were fired at long ranges, or under large angles of elevation.

When the space in which a piece is fired is limited, its recoil may be reduced by using a rope, which, acting gradually, does not break the carriage. The rope is tied to the felloes near the highest point of both wheels, leaving a portion slack, which, catching on the stock when the piece recoils, forces the trail into the ground.

Mortars, which are still lighter than howitzers of the old model, and which are habitually fired under large angles, should never be mounted upon wheel-carriages; for their recoil being very violent and almost direct, would break and crush the axles. For this reason mortar carriages are without wheels.

In general, a gun carriage should be so made as to be easily placed in the direction of an object, and allow the piece different degrees of elevation or depression within the limits recognized as necessary for the kind of service in which it is to be used. It should also be such as to be easily manœuvred by the smallest possible number of men for any particular caliber, and its recoil should be restrained within proper limits.

Carriages with two wheels, or carts, enjoy some special advantages, but they have the disadvantage of requiring the first horse attached to them to bear a part of the load, which diminishes his draught powers, and the rate of traveling. This fact, recognized from the earliest times, led to the adoption of the fore wheels in the form of the limber, by which the cart becomes a four-wheeled vehicle, which runs easier, and is more serviceable.

It is apparent that for the proper manœuvering of the piece, it is of importance that the connection between the gun-carriage and its limber should be as easily and as promptly made as possible.

The gun-carriage and its limber united form a four-wheeled vehicle, which should satisfy the same conditions as other carriages.

In all kinds of carriages, the ease of draught is increased by enlarging the diameter of the wheels and decreasing those of the axles. There is, however, but little advantage in using wheels more than from 58 to 62 in. high. Those used in our field-service are 57 in., and those for siege-carriages, 60 in. high.

Formerly, the greater part of the axle-tree was made of wood, fitting into nave-boxes of iron, which caused a considerable amount of friction; but the elasticity of these axles was indispensable in preserving certain very heavy carriages, which did not run easy. Gribeauval adopted the use of iron axles for the greater part of field-carriages, with bronze for the nave-boxes.

This arrangement allows the axles to be made smaller, diminishes the friction, and favors the draft, and these last objects are farther attained by the use of grease.

Increasing the diameter of the wheel diminishes the draft in soft ground; as the large wheel bearing upon a greater number of points, makes a shallower rut. This is also the case with wide felloes.

The large wheel is also advantageous for surmounting obstacles. One wheel twice the height of another will surmount, other things being equal, an obstacle twice as high as will the small wheel.

On the other hand, as the dimensions of the wheel increase its weight and price become greater, so that beyond 58 inches, the advantages to be gained in a few particular instances are not sufficient to outweigh the objections to a further increase in diameter.

It may be remarked besides, that if the axletree be higher than the breast of the horse, there would be a decomposition in the tractive force, and the carriage would be difficult to manage in going down even the most gradual descents. In this view, then, there is a limit to the size of the wheel. As the horse acts from the shoulder, the traces ought to be inclined downwards from that point. For an unloaded horse, the most advantageous angle of traction appears to be about 12°; but when he carries a rider, as his shoulders are already loaded the most advantageous angle is reduced to 6°, which is about the angle made by the traces in the carriage now adopted. This limits the height of the wheel to about 58 inches. But as only one of the horses in each couple carries a rider, it results that the most advantageous angle of traction is really comprised between 6° and 12°.

When the carriages are designed to pass over very unequal broken ground, there is an advantage in the wheels of both parts being equal in height; for, were they unequal, the smallest—turning the quickest to get over the same distance—experience more resistance, and often slide in place of turning; besides which, a small obstacle, as a stone, would suffice to stop them.

When the ground passed over is very undulating, there is an advantage in the two parts of the carriage being connected by a single point, for then they might be on differently inclined ground without hampering or dragging each other. The force of traction of the limber is then applied to the axletree of the rear carriage, and its action is as direct as possible.

In this kind of carriage, the center of gravity should be as low as possible, in order that the carriage be less liable to turn over.

The distance between the axletrees of the limber and rear carriage is of great importance. As the carriage should be able to pass over the sharp crest of a hill, it should not be too long; for it might happen that the body would catch on the summit and the wheels be suspended on each side. Taking 30° as the maximum slope accessible to artillery, and we have for the distance between the two axletrees:

For field artillery, 99 inches. It really is 96 inches for the light pieces, and 101.7 inches for the heavy, in field artillery.

In field-artillery carriage, as the horses bear all the weight of the tongue, it is made as light as possible in order not to be too fatiguing for the rear team.

In siege artillery it can be made much heavier and stronger as it is held up by the trail bearing upon its rear end.

To facilitate maneuvering, the carriages should have the greatest turning capacity possible. Two-wheeled carriages are the only ones which can turn on their own ground. Four-wheeled ones always have to describe an arc of a circle. In carriages which have all the wheels equal, the necessary turning capacity has been given by reducing as much as possible the width of the middle part of the carriage, employing for that purpose a narrow stock. By this means, the modern carriage has at least as great a turning capacity as Gribeauval's, the fore wheels of which were the smallest.

In four-wheeled carriages, the two parts should be loaded in proportion to the diameter of the wheels. The front wheels, breaking the road and forming the rut, meet with more resistance than the hind wheels, and should not be too heavily loaded. Other things being equal the weight which they carry should be to that on the hind wheels as 2 : 3.

The length of the axletree and the *dish* of the wheel, determine the width of the *track*, which is the distance between the impression of the wheels on the ground measured from center to center. A knowledge of the track of a carriage and the length of its axletree, is indispensable in making reconnoissances of defiles and routes to be passed over by columns of artillery. The track of all our carriages is 60 inches, and the greatest length of axletree is 81.8 inches (siege); the field axletree being 78.84.

Field carriages should be able to carry a sufficient quantity of ammunition to prevent the possibility of the pieces being without it, and to seat the cannoneers when it becomes necessary to move at rapid gaits.

All possible liberty should be allowed the horses when harnessed, in order that the action of one may not shackle the others, that accidents may be as rare as possible, and that killed and wounded horses may be replaced easily and promptly.

A draught-horse can draw 1,800 lbs. 23 miles a day, weight of carriage included, on a good smooth road, and 3,000 lbs. on a paved road. At a trot, and on good roads, the weight is reduced to 840 lbs. In the French field artillery, each horse has to draw 720 lbs. In our service it varies from 530 to 760 lbs. for the field service, and from 900 to 1,000 lbs. for the siege.

As, upon ordinary roads, a horse can draw about seven times as much as he can carry, and the mule about the same, it follows that military stores and machines should be *packed* only when they cannot be drawn.

A good pack-horse or mule can carry from 250 to 300 lbs. 20 miles a day.

The quality and degree of mobility of the different kinds of material, should be in keeping with their destination. Hence results its division into different classes, according to the kind of service in which it is used. 1st. *Field* material. 2d. *Mountain*. 3d. *Siege* and *garrison*. 4th. *Sea-coast;* and 5th. *Bridge equipages*.

FIELD MATERIAL.

It is composed of six different carriages, all having the same kind of limber and the same sized wheel, so that any limber or any wheel may be used with any carriage, though if possible the heaviest wheel (No. 2), should be used on the carriages of the three heaviest pieces, 12-pd. gun and 24 and 32-pd. howitzers. The two parts of these carriages are connected by a stock, by means of a pintle hook and lunette. In consequence of this single point of connection, the tongue is not influenced by the movements of the rear part of the carriage, a matter of some importance in regard to fatiguing the horses.

The independence of the two parts requires the weight of the tongue to be supported by the horses, and it is consequently made as light as is consistent with its proper strength.

Thus constituted, these carriages can manœuvre easily on very rough ground, which they are often required to pass over. The disadvantage of the way in which the horses are harnessed to these carriages, results from the oscillations of the tongue, which is very fatiguing to them, although the improvements introduced in the harness have reduced this defect very much.

The six carriages are—

1. The carriage for the 6-pd. gun and 12-pd. howitzer.
2. " " " 24-pd. howitzer, and formerly for the 9-pd. gun also.
3. " " " 12-pd. gun and 32-pd. howitzer.
4. Caisson.
5. Battery-wagon; and
6. Traveling-forge.

1, 2, AND 3, THE GUN CARRIAGES, Fig. 119, Pl. 9, consist of two short cheeks of wood, bolted upon a stock and wooden axle-body, in a recess of which fits the iron axle on which the wheels are placed. The stock terminates in a *trail* and *trail plate*, which rests on the ground, and has on the end a strong ring called the *lunette*, which is placed on the pintle hook when the piece is limbered. In the stock is placed an elevating screw-box of bronze, in which the elevating screw fits.

The limber consists of a similar axle-body, axle, and two wheels; and on these rests a framework to receive the tongue. On top of the whole is an ammunition box, the top of which forms a seat for three cannoneers. In rear of the axle-tree is a *pintle hook* to receive the lunette of the trail. Connected with the framework in front is a fixed *splinter-bar* with four hooks, to which are attached the traces of the rear team of horses.

At the extremity of the tongue are placed two pole-chains, by which the pole is held up, and a pole-yoke with two movable branches, to prevent as much as possible the pole from oscillating and striking the horses. Each of these branches has on it a freely sliding ring, which is connected by a leather strap to the collar of the horse, and allows him greater freedom of motion. The pole-chain is hooked on to a ring on the breast-strap. The traces of each horse of the front teams are hooked on to those of the horse in his rear, just behind the collar, and are suspended by his sides in scabbards.

(For dimensions, weights, &c., see Appendix, p. 4.

4. THE CAISSON, Fig. 120, Pl. 9, is a carefully closed carriage used to transport ammunition. All caissons are the same on the exterior. They differ in their interior compartments, which vary according to the nature of the ammunition with which they are loaded.

On the axle-body of the rear part, and in a direction parallel to the stock, are placed three rails, upon which are fastened two ammunition boxes, one behind the other, and similar to the one on the limber; so that the caisson has three ammunition boxes, which will seat nine cannoneers. A wooden stock and lunette serve, as in the gun carriages, to connect the two parts together.

In rear of the last box is placed a spare wheel axle of iron, with a chain and toggle at the end of it. On the rear end of the middle rail is placed a carriage hook, similar to a pintle hook, on which the lunette of a gun-carriage whose limber has become disabled, may be placed, and the gun carried off the field. Under the hind axle a spare pole ring is placed, through which an extra pole is run, and sustained in its position under the stock by a key plate and key fixed beneath, near the lunette.

The caisson has the same turning capacity and mobility as the gun carriage, so that it can follow the piece in all its manœuvres if necessary.

The object of the caisson is to carry a supply of ammunition, the limber of the gun-carriage carrying the instruments for use with the piece, and such ammunition as may be necessary for its immediate use. The caisson has also slots left in the iron assembling-bars, between and in rear of the boxes, for the reception of an axe and a pick, and at the side and under the boxes, staples and other arrangements for securing a long-handled shovel and a spare handspike.

(For dimensions, weight, and equipment, see Appendix, pp. 6 and 15.)

5. The BATTERY WAGON, Fig. 121, Pl. 9, consists, besides the limber, of a long-bodied cart with a round top, which is connected with the limber in the same way as all other field-carriages. The lid opens on hinges placed at the side; and in rear is fixed a movable forage-rack, for carrying long forage.

One of these battery wagons accompanies each field-battery, for the purpose of transporting carriage-makers' and saddlers' tools, spare parts of carriages, harness, and equipments, and rough materials for replacing different parts.

Both this and the forge are made of equal mobility with the other field-carriages, in order to accompany them wherever they may be required to go.

The cover is made of light pine, tongued and grooved, and covered, after being painted, with canvas.

For dimensions and contents, see Appendix, p. 19 to 25.

6. THE FORGE, Fig. 122, Pl. 9, consists, besides the limber, of a frame-work of three rails and cross-ties, on which is fixed the bellows, fire-place, &c. The stock, held up by a prop, serves as a support for a vice. Behind the bellows is placed a coal-box, which has to be removed before the bellows can be put in position.

In the limber-box are placed the tools for use with the forge, horse-shoes, nails, and spare parts (iron) of carriages, harness, &c.

One of these forges accompanies each field-battery, and others are provided, equipped for general service with the army.

The forge should be as tight and movable as possible, in order to be able to repair promptly to the assistance of any carriages which may become disabled.

For dimensions, weight, and equipment, see Appendix, p. 6 to 25.

MOUNTAIN ARTILLERY.—The material for mountain artillery service consists of, a guncarriage for the 12-pd. mountain howitzer, without a limber; ammunition-boxes for the supplies and stores, and a portable forge, which is carried in boxes.

The Carriage, Fig. 123, Pl. 9, is formed like the field gun-carriage, but much smaller, the cheeks not being formed of pieces distinct from the stock, but all three made of two pieces bolted together. The axle-tree is of wood, which lessens the recoil, and gives an elasticity to the whole carriage better adapted to resist the shocks of firing. The wheels are but 38 inches high. Ordinarily, over rough ground, the carriage is transported on the backs of mules; but where it is possible to do so, a pair of shafts is attached to the trail in such a way as to keep it from the ground, and the piece is drawn on its carriage by harnessing one of the pack-mules to it.

The ammunition for the howitzer and the cartridges for small arms are carried in the ammunition-boxes on the backs of mules, two boxes to each mule.

The same kind of pack-saddle is used for carrying the piece, carriage, ammunition, and forge. The transom on each side of the saddle has a circular notch cut in it, to receive the trunnions of the piece, which is carried with the muzzle to the rear.

For the transportation of a single piece, its carriage, and ammunition, three mules are required.

On the 1st is placed the piece, and shafts of the carriage, weighing 251 lbs.

" 2d " carriage and implements, weighing 295 lbs.

" 3d " two ammunition-chests, weighing about 238 lbs.

To attach the shafts to the trail, the supporting bar is laid on the trail-plate, near the handspike-staple, the projecting knee in rear of the lunette resting on the cross-bar plate; these two pieces having holes through them, by means of which, with a pin, they are keyed together. For dimensions, weights, &c., see Appendix, p. 26.

The Portable Forge is designed for service in a mountainous country, where wheeled vehicles cannot travel, for the purpose of making repairs not only for the artillery, but for all other arms of service taken on such expeditions.

Its construction is as follows:

The *hearth* is of sheet iron, bent into a hollow form, and riveted to an iron *frame*. The *back* of the hearth is bent under the bottom and riveted to it. The *border* of the hearth is bent round the back, and is riveted to it and to the frame. The *back* of the fire-place is of sheet iron, connected with the back of the hearth by *two hinges*, which are riveted to each. The *edges* of both the back pieces are stiffened by strips of sheet iron riveted to them. An *air back* is formed by a piece of sheet iron, bent hot into a convex shape, and riveted to the inside of the plate of the fire-place. A *button*, turning on an axis which is riveted to the outside of the back plate, serves to fasten this plate to a stud in the front border of the hearth, when the back is turned down on its hinges.

The frame is supported by *three legs*, which are connected with it by bolts, so that they can be folded up close to the frame. The front leg is divided into two branches, which are bolted to *two eye-pieces* which are riveted to the sides of the frame. The two other legs are connected together by a cross-bar, with a nut at each end. This bar supports also the fork in which the bellows handle works. The legs of the frame have round tenons at the lower ends, with shoulders which rest on three socket plates attached to the side of the forge-chest, for the forge to stand on when set up for use.

The bellows handle is of iron, with a wooden head. It is attached to a *fork* which fits in a square hole in the cross-bar joining the rear legs of the frame. The lower end of the handle is hooked into a *connecting rod* attached to the rear end of the bellows; it is fastened to this rod by a sliding *catch* which is secured by a thumb screw. When the bellows is dismounted, this rod is hooked into an eye on the upper side of the bellows, to keep it closed.

THE BELLOWS.—The frame consists of an *upper*, a *lower*, and a *middle plank* (walnut), and *two ribs* (poplar), connected by a *cross head*, as in a common smith's bellows. There are valves in the middle and the lower plank.

A *bar* of iron, attached to the middle plank, terminates in two journals, which support the bellows, fitting in the joints of the rear legs of the frame of the forge.

The *nozzle*, of sheet iron, is inserted into the cross-head, above the middle plank. It enters into a *cast-iron pipe* which is attached to the rear of the forge back by means of a *bridle* bolted to the back plate of the hearth.

A *handle* is attached to a *plate* which is fastened on the upper bellows-plank. A *leaden weight* of one pound is attached to the inside of the lower plank, by the rivets which hold the eye-plate on the connecting rod.

The *bellows leather* (calfskin) is fastened to the planks by small bellows-nails.

A small anvil accompanies the forge, and is fitted into a block of wood, and fastened with an iron pin.

The forge is made to pack into two boxes similar to the ammunition boxes, and is like them carried on the back of a mule. Weight, 232 lbs., which includes the weight of the blacksmith's tools and materials necessary for the use of the forge.

The carriage maker's tools, and coal for the forge, are carried in two similar boxes on another mule. Weight, 115 lbs.

As all this material is made to be packed on the backs of mules, it can be carried over the roughest country.

The recoil of the carriage being considerable, it is limited as heretofore described (especially when the ground on which the piece is used is very narrow), by attaching a rope to the top parts of the wheels so as to press against the trail when the carriage recoils.

The Prairie Carriage.—The necessity for a small carriage for the mountain howitzer when used on our western prairies, has led to the adoption of a special carriage for that service, with a limber attached as in a field carriage. This gives a carriage less liable to overturn, and preferable in many ways for that service to the two-wheeled one. The limber is furnished with two ammunition boxes, placed over the axle-tree and parallel to it, and just wide enough for one row of shells and their cartridges.

The caisson for this new carriage is a two-wheeled cart with shafts, on which are fixed four boxes for ammunition, similar to those on the limber, but placed perpendicular to the axle-tree, and a fifth box for equipments, &c., in front of the other boxes, perpendicular to them and resting against their ends.

Siege Material.—The material for service in sieges consists of

1st. Gun carriages.

2d. Platforms for these carriages.

3d. Mortar-wagons.

4th. Mortar-beds.

5th. Platforms for these beds,—

And some other implements or machines which are used also in the field and in garrison, such as, the Gins, the (Siege and Field, the Garrison and the Casemate), Sling Cart, Hand Cart, Lifting Jack, Lever Jack, Truck, Manœuvring Blocks, &c.

1st. The *gun-carriages*, Fig. 124, Plate 10, are three in number, as follows:

One for the 12 pound gun.
" " " 18 " "
" " " 24 " " and 8 in. howitzer.

They are all constructed in the same manner, differing only in their dimensions. All the limbers and wheels are the same, so that they can be used in common.

The gun-carriage is similar in its construction to the field carriage, but is joined to the limber in a different way. Projecting upwards from the limber and in rear of the axletree, is placed a pintle, which enters a hole made in the trail from the under side, and a lashing-chain and hook keep the two parts together when once in position. The weight of the trail resting on the rear end of the tongue keeps this nearly horizontal, and relieves the horses of the weight of it, which, as it must be both long and heavy, is too much for the horses to carry.

The splinter bar is, as in field carriages, stationary, but the traces of the next team are attached to a movable bar which is connected with the end of the tongue. The tongue is furnished with pole-chains but no yoke, and the rest of the teams are harnessed as in field artillery.

It is not necessary for siege carriages to have the same degree of mobility and flexibility that field carriages have, as they are properly speaking transportation wagons for use on roads, and never intended for manœuvring with troops.

The axle-trees are of iron with axle-bodies of wood; which last, by its elasticity, renders the percussion of the piece less direct and violent.

On the upper surface of the cheeks, near the back ends, are placed two projecting bolts which, with the curve of the cheeks, form resting places for the trunnions, when the piece is in position for transportation. They are called the traveling trunnion-beds. When in this position, the breech of the piece rests upon the bolster, a curved block of wood bolted to the upper side of the stock. On each side of the trail, and perpendicular to it, a strong manœuvring bolt is placed to serve as places to apply the handspikes in manœuvring the carriages, or as handles in limbering, unlimbering, &c.

The traveling trunnion-beds are used for the purpose of distributing the load more equally over the carriage. Before the piece is put in the traveling position, the elevating screw is taken out and inserted from beneath in the screw-box, and the piece is run forward to its position on rollers, the carriage being limbered.

The limber consists of—

1 the fork, 2 the splinter bar, 3 the hounds, 4 the sweep bar, 5 the tongue, 6 the pintle, 7 the lashing-chain eye, 8 the axle tree (iron).

The sweep bar is of iron, and on it rests the trail which by its weight keeps up the tongue. (For dimensions, see Appendix, p. 5.)

The transportation of a 24-pdr. gun requires ten horses and five drivers; an 18-pdr. eight horses and four drivers; a battery wagon six horses and three drivers; and spare carriages—at the rate of one for every five pieces—require, each, six horses and three drivers.

When the 8-inch howitzer is mounted on the 24-pounder carriage, a *quoin* is used, instead of the elevating screw; the howitzer being too short to rest on the screw.

2d. Platforms. Fig. 125, Plate 10. As the siege pieces, to be of efficient service, must have their carriages on a solid foundation with the wheels on a level, a platform is provided for each. They are constructed at the arsenals, and should be as light as possible, compatible with sufficient strength to resist the shocks of the piece. All the pieces composing them are of the same dimensions, each weighing 50 lbs. They are made of yellow pine or oak, are 49 in number, 9 ft. long, 5 in. wide, and 3½ in. thick. One of these pieces is used as a *hurter* on the front part of the platform, to prevent the carriage running too far forward. The recoil is diminished and the water made to run off by giving the platform an inclination to the front. (For dimensions, see Appendix, p. 11.)

Laying.—To lay this platform, establish the center line of the embrasure, and stretch a cord on this line from the middle of the embrasure to the rear. This is the *directrix* of the platform.

Lay the two outside sleepers parallel to this directrix, their outside edges being 54 inches distant from it. The four other sleepers are laid parallel to these, the edge of each 15½ inches from the edge of the next. The upper surface of the front ends of these sleepers to be 50 inches, on a vertical line, below the sole of the embrasure.

They are laid with an elevation to the rear of 1½ inches to the yard, or 4½ inches to their whole length. This elevation may be determined by placing a block 4½ inches high on the front end of the sleeper, and laying a straight edge, with a gunner's level on it, from this block to the rear end, then so arrange the earth as to bring the level true in this position.

The next set of sleepers are laid against and inside of the first, overlapping them three feet, having the rear ends inclined outwards, so that the outer edges of the exterior ones shall each be 54 inches from the directrix, and the spaces between the rear edges of the others, the same as in the first set, viz., 15½ inches from the edge of one to the edge of the next, all having the elevation to the rear of 1½ inches to the yard, and perfectly level across. The earth is then rammed firmly around these sleepers, and made even with their upper surface. The first deck-plank, with a hole through each end for the eye-bolts, is laid in place, perpendicular to the directrix, its holes corresponding with those in the sleepers. The hurter is placed on it, and the bolts driven through the corresponding holes in these pieces.

The hurter should be so placed as to prevent the wheels from striking against the epaulment when the piece is in battery. If the interior slope has a base of 2-7ths of its height, the inner edge of the hurter should be 2½ inches from the foot of the slope. The other planks are then laid, each one forced against the preceding, the last plank having holes for the rear eye-bolts. By drawing out or driving in the outside sleepers, the holes through their rear ends are made to correspond with those in the last deck-plank, and the bolts are put in.

Drive stakes in rear of each sleeper, leaving their tops level with the upper surface of the platform. Raise, ram, and level the earth in rear of the platform, so as to have a plain hard surface to support the trail when the recoil is great.

The earth should be raised nearly as high as the platform at the sides, and well rammed, giving it a slight inclination outwards to allow the water to run off.

Another kind of platform, called the ricochet platform, is sometimes used with these carriages when the pieces are used for ricochet firing. The parts then made use of are—

1 Hurter, 8 feet long,	8 inches wide,	and 8 inches thick.	
3 Sleepers, 9 "	5½ "	5½	"
2 Planks, 10 ft. 8 in. long,	13 in. wide,	2¼	"
1 Plank, 7 ft. long,	13 in. wide,	2¼	"
1 " 2½ "	13 "	2¼	"

and some stakes, and the platform is constructed in the following manner:—

Place the hurter perpendicular to the line of fire, and secure it by four stakes, one at each end and two in front, 31½ inches from the middle towards each end; lay the three sleepers parallel to the hurter; the first, 16 inches from the rear edge of the hurter; the second, 43½ inches from the rear edge of the first; and the third, 43½ inches from the rear edge of the second. Lay the plank 31½ inches from the directrix of the platform to the center of the plank. Place the piece of plank 60 inches from the rear edge of the last sleeper, and bed it in the ground. Place on the last sleeper and this piece of plank *the* plank (84 inches long), its front end 106 inches from the rear edge of the hurter.

This platform will bear firing with charges as high as three pounds.

Platforms of this kind, of larger dimensions, may be used for guns and howitzers in firing at a fixed object with full charges.

3d. Mortar-Wagons,—Fig. 126, Pl. 10. This wagon is designed for the transportation of siege-mortars and their beds, or of guns, or large shot and shells.

The limber and the wheels are the same as those of the gun-carriage.

The body consists of a platform of rails and transoms resting on an axle-tree, the two middle rails being prolonged to form the stock, six stakes or standards are inserted in sockets on the sides of this platform, and used in securing the load.

The side-rails are prolonged to the rear, and furnish pivots for a roller placed immediately in rear of the platform. This roller has holes for the insertion of handspikes, and is used in loading the wagon; the guns, mortars, &c., being drawn up on the stock.

A muzzle bolster on the stock near the limber, and a breech hurter near the hind part of the wagon, are provided and used when long ordnance is transported on it. Mortars are usually carried mounted on their carriages.

(For dimensions and weights, see Appendix, p. 6.)

4th. Mortar-beds, Fig. 128, Pl. 10.—The carriage from which a mortar is fired, is called its *bed*. It has already been stated that the size of the angle of fire, and the violent reaction in a vertical direction which the mortar-bed experiences, prohibit its being mounted upon wheels. In consequence of this, also, the cheeks cannot be made of wood or bronze, the elasticity of which, being brought into play by the recoil, would cause the bed to bounce on the platform and throw the mortar from its place. Ancient carriages, of wood or bronze, were provided with cap-squares, to prevent this.

There are four siege mortar-beds:

The 8 in. mortar-bed,
The 10 in. "
The stone "
The Coehorn "

The first three are alike, differing only in dimensions, and made of cast iron, which has very little elasticity. The bed consists of two cheeks, joined by two transoms, all cast together in the same piece. The manœuvring bolts, placed on each side, one near each end of the cheeks, are made of wrought iron, and set in the mould when the bed is cast.

On the front transom is fastened a wooden *bolster*, grooved to receive the elevating quoin, which it is prescribed should be put in position in a direction perpendicular to the axis of the piece, but is usually for convenience placed obliquely.

Notches on the under side of the front and rear of the cheeks, give hold to the handspikes in throwing the piece to the right or left.

Cap-squares are used with these beds, but probably only for the purpose of preventing the piece from jumping from its place when fired at very small angles of elevation, as, for instance, in ricochet firing.

The Coehorn Mortar-Bed is made of a block of oak wood, in one piece, or two pieces joined together with bolts. A recess, for the trunnions and part of the breech, is made in the top of the bed; and the trunnions are kept in their places by plates of iron bolted down over them. Two iron handles are bolted to the bed on each side, by which four men can carry the bed with the mortar in its place.

The Eprouvette-Bed consists of a block of wood, on top of which is countersunk and bolted the bed-plate, which is a heavy circular plate of cast iron, having a rectangular recess, with sloping sides, so as to make it longest at the bottom. Into this recess the sole of the mortar slides. The wooden block is bolted to a stone block of the same size, which is firmly placed in the ground on a masonry foundation.

For dimensions and weights of the beds, see Appendix, p. 6.

Mortar-Platforms are made each of twenty-four pieces, similar to those used for siege-guns. Six of these are laid as sleepers perfectly horizontal, and parallel to the direction in which the piece is to be fired. The other eighteen are laid on top of these and perpendicular to them, the front and rear ones being bolted to each one of the sleepers. Between each pair of the latter (deck-planks), four *dowels* are placed, to prevent the plank from sliding out. These dowels are short, hard pieces of wood, which are fastened in the edge of each plank, and fit into corresponding holes made in the one next to it.

For dimensions, &c., see Appendix, p. 11.

The Rail-Platform, Fig. 127, is another kind sometimes used with mortars. It is simple, strong, and well-suited to positions where timber can be easily procured. It is composed of three sleepers and two rails, with stakes for keeping it in position. The rails are placed at such a distance apart that the cheeks of the bed can set on them. They and the sleepers are notched to fit, and driven together in the battery. The sleepers are imbed-

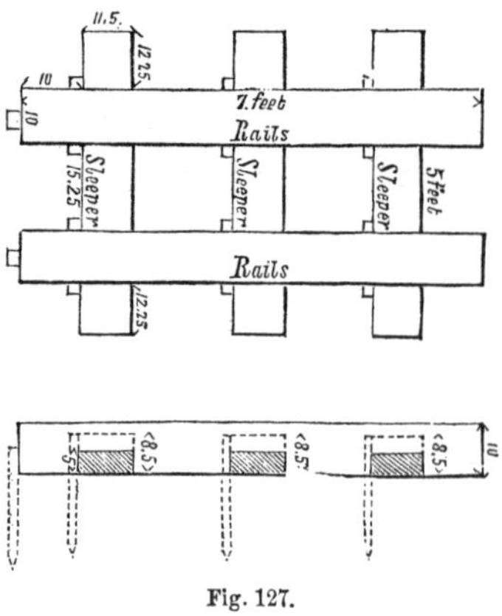

Fig. 127.

ded in the ground, and the earth well rammed around them, care being taken to place the rails horizontally. Stakes are driven in rear of each of the sleepers in the angles made with the rails, and at the rear ends of the rails. (For dimensions, &c., see Appendix, p. 11.)

THE HAND-CART.—A two-wheeled cart, being able to turn on its own ground, is more suitable than any four-wheeled vehicle for service in the trenches and parallels of a siege. The hand-cart consists of a light body with shafts, mounted on two wheels, which in the French service are the same as those used on their field carriages. The shafts are joined together at the ends, and supported just in front of the body by iron legs. The cart is used for the transportation of light stores in siege and garrison service.

THE HAND SLING-CART, Fig. 129, Pl. 10, is used in siege and garrison service, or, in fact, under any circumstances where light pieces or their carriages, or objects about that weight, have to be transported for short distances. It is made entirely of iron, except the pole, which is of oak. The axle-tree is arched, to make it stronger, and connected with the pole by strong wrought-iron straps and braces. In the rear of the axle a projection is welded to receive the end of a strong hook, the two being fastened together by a bolt. The end of the pole is terminated with an iron strap (through which is passed the handle), and an eye. The eye is for the purpose of attaching to the cart, when necessary, a limber or a horse, the handle, in the last case, being held up by hand.

This cart should not be used with heavier weights than about 4,000 lbs.; but in case of necessity a 24 or 32-pd. gun may be transported on it. For heavier guns or material the *large sling-cart*, drawn by horses or oxen, should be used. The French have a similar cart for carrying shot and shell.

(For dimensions, &c., see Appendix, p. 11.)

THE FIELD AND SIEGE GIN, Fig. 130, Pl. 10, is used for mounting or handling guns, or other heavy bodies in the field, or in the trenches of a siege. It consists, like all gins, of two legs and a pry-hole of spruce or ash as supporters, a windlass of oak, sheaves, pulleys, and a fall or rope. The two legs are mortised for the reception of three braces, on the ends of which tenons are cut.

These pass through the mortises, and are keyed with pins outside.

The top end of the pry-pole has on it an iron tongue, which fits in between the top of the legs, and all three are joined by a pin; just below which, and held between the legs by a bolt, two sheaves are placed to receive the fall, which is a strong four-inch rope. The legs are fastened together at top by a bolt.

Two and a half feet from the bottom of the pry-pole a round handle is run through it and fastened, and is used in handling or putting up the gin. The feet of the legs and pry-pole are provided with metal points, to prevent slipping when a weight is put upon the fall.

The windlass is provided with journals at the ends which turn in journal boxes, the plates of which are fixed to the legs by bolts. No ratchet wheel or pall is used, and the windlass is made fast by allowing a handspike, inserted in its place, to bear against the lower brace, or another handspike placed across the legs for the purpose. The ends of the windlass are left square, but the middle part is made round for the reception of the fall, which, in raising a weight, is coiled round it. The legs are about 14½ feet long, and the height of the gin, standing up, about 12 feet.

To put the gin together, the legs are laid on the ground, the outer sides up, the beveled ends together, and the windlass put in its place. The assembling bolt is put in, and the 1st, 2d, and 3d cross-bars inserted in succession and keyed.

To set up the gin, the bottoms of the legs are held in position by two men placing their feet against them, or holding handspikes against the lower cross-bar. The tongue of the pry-pole is inserted in its place and keyed. The gin is then raised by carrying up the foot of the pry-pole to its proper position, which is equally distant from the two legs, and 12 feet from the lower cross-bar. The fall may be attached by a man mounting on the third cross-bar; or, if he is too much exposed in that position, it may be done while the gin is lying down.

(For dimensions and weights, see Appendix, p. 11.)

The gin may be equipped with one or several pulleys, according to the weight to be raised,—a greater number being employed than is absolutely necessary, in order not to strain the rope. With one pulley a 12-pd. siege and garrison gun can be raised; with two, an 18-pd.; with three, a 24-pd., and with five or six, a 32 or 42-pd.

To Sling a Piece.—A small piece without handles may be slung by a piece of rope, the ends of which are tied together after passing it under the piece in rear of trunnions, then up over the trunnion, under the piece again in front and over the other trunnion, hooking both sides over the pulley-hook. Larger pieces are slung by making use of an endless rope or chain, when double somewhat greater in length than the piece. One end of this is looped round the neck of the cascable, and the other round a handspike or billet of wood inserted in the muzzle. Both strands are hooked in the pulley-hook just in rear of the trunnions, binding the two together, if necessary, with a cord or piece of small rope.

When the gin is put together and raised, that part included between the legs and pry-pole is called the inside; the outside being the part without the legs; the right corresponding to the right hand of a man standing at the middle and outside of the windlass, facing towards it; the left corresponding to his left hand.

Barbette Carriages, Fig. 131, Pl. 10.—The barbette carriage belongs to that class which has been denominated immovable. That is to say, it is used simply to fire the piece from, and not to transport it except for very short distances.

The gun carriages described under the head of siege material, may also be used in a fortification or garrison; and the mortar-beds heretofore described, are used either for siege or garrison service. The barbette carriage, however, is used only in a fixed position in garrison, and is a carriage on which a gun is mounted to fire *over* a parapet instead of through it, as siege or sea-coast guns usually fire. A *barbette-gun* is any gun mounted on a barbette carriage.

There are but two forms of the barbette carriage in the United States' service.

1. One for the *iron guns* and sea-coast howitzers (12, 18, 24, 32, 42 pdr. and 8 and 10 in.); and

2. One for the columbiads.

1. This carriage consists of a *gun-carriage* and a *chassis*.

The Gun-carriage is formed of two upright pieces of timber, nearly vertical, forming the cheeks. Behind these are placed two inclined braces, mortised into the uprights and designed to receive the force of the recoil. A horizontal piece (the transom and axle tie) runs from front to rear between these pieces, and the whole is firmly connected and braced by transoms and assembling bolts, thus forming a *triangular* framework, which is less liable than any other form, to become deformed from the shocks of the gun. The trunnion-bed is formed at the top of the upright, where it is joined to the brace; and the breech of the gun is supported on an elevating screw, working into a screw-box placed in the rear end of the transom and axle tie, near the front part of the rear transom.

The front transom is just under the gun, and is cut out to receive it. The middle transom is placed between the braces about their middle points, and is notched out to receive the upper side of the transom and axle tie. The rear transom is at the lower end of the braces, and under the transom and axle tie, into which it is notched, dowels being placed between the two to prevent sliding; the lower part of this transom is notched to receive the tongue of the chassis on which it slides. Between this transom and the transom and axle tie, the end of a lunette is placed projecting to the rear and fastened by a bolt, for the purpose of attaching a limber to the carriage.

The feet of the uprights and front end of the transom and axle tie are joined to an axle-body, in which an iron axle is placed. On the ends of the axle are fitted cast-iron rollers, which rest upon the rails of the chassis and support the front of the carriage. On the outside of the roller is placed an octagonal projection, on which the cast-iron nave of the wheel fits, secured by a washer and linch-pin. The spokes of the wheels are wood, inclosed within heavy iron tires.

Maneuvering bolts are inserted in front of the feet of the uprights, and in the carriages for pieces heavier than a 24-pounder in rear of these feet also. These bolts and the spokes of the wheels form the points of application for the handspikes, in maneuvering the piece. Maneuvering staples are placed in front of the feet of the braces, for the purpose of using handspikes to raise the rear of the carriage from the tongue of the chassis in running to and from battery.

The 32-pound gun and 8-inch howitzer, go on the same carriage. All other pieces have separate carriages.

Two sizes of rollers are used, one for the carriages of the 12, 18, and 24 pounders; the other, for the rest of the carriages. The roller part, however, is of the same diameter, and the octagonal projection accurately made, so that a wheel from any carriage will fit on any other. The wheels are all of the same diameter, 43½ inches.

No cap-squares are used with these carriages.

Besides serving to assist in maneuvering the gun-carriage, the wheels are also used in transporting the piece on its carriage for short distances; as, for instance, from one front of a fortification to another. To do this, the rear of the chassis is raised until the traverse wheels can be removed, when it is lowered to rest on the prop of the tongue and then on the traverse circle. Planks are then laid, inclining to the rear, to receive the wheels, which are backed on to them. A siege limber is backed up to the carriage, the lunette placed over the pintle and keyed, and the gun carriage drawn from the chassis.

The Chassis (the French for frame or framework) consists of two *rails* and a *tongue*, joined by three transoms. The tongue is in the middle and projects considerably beyond the rails, to the rear. At each end of the rails, on top, a hard piece of wood is notched in and bolted. They are called hurters and counter-hurters, and used to prevent the gun carriage from running from the chassis.

Rail-plates of iron to protect the wooden rails, are let in to the outside of the rails 0.2 of an inch below their upper surface, bolted at each end with an iron bolt, and fastened along the rails with wooden screws, which probably are less liable to break from the shock of the recoil.

At the rear end of the tongue a swinging prop is placed to support the end of the tongue when the piece is run back or the traverse wheels are removed. The lower side of the end of the tongue is notched out and a maneuvering loop fixed there, with a bolt and screws to assist in handling the chassis.

The same chassis serves for the 12 and 18 pdr. carriages. On the under side of each rail opposite the rear transom a mortise is formed, for the reception of a fork socket of iron which receives the *handle* of the traverse-wheel fork. Each of these forks receives a traverse-wheel, joined to it by an axle-bolt, and these support the rear end of the chassis. The front end is supported on a pintle-plate of iron; through which, and up into the middle of the front transom, passes a pintle or bolt, which serves as a pivot around which the whole system moves.

In permanent batteries, the pintle is fixed in a block of stone, and the traverse circle is an iron plate set also on stone. For temporary batteries, the pintle is attached to a wooden bolster, which is covered by a circular cast-iron plate, and attached by bolts to a wooden cross, picketed firmly into the ground.

A temporary traverse-circle is formed of plank pinned to sleepers and fastened to pickets, or secured to string pieces which connect the traverse-circle with the pintle-cross.

The traverse-wheels and their forks retain their places by the weight of the carriage and gun. The pivot-bolts of the traverse-wheels project to the rear, and are acted against by the hand-spikes in traversing the carriage.

The chassis for the heaviest carriages has in addition to the three transoms, four iron pipes placed in pairs between the tongue and rails. Through each pair a heavy bolt passes, and is fastened on the outside of the rails with nuts. The chassis slopes towards the front in order to diminish the recoil, and aid in running the piece into battery. (For dimensions and weights of these carriages, see Appendix, pp. 7 and 8.)

2. THE COLUMBIAD CARRIAGE, Fig. 132, Plate 11, consists of a *gun carriage* and *chassis*. Those for the 8 and 10 inch differ only in dimensions.

THE GUN CARRIAGE is a triangular frame work, consisting on each side of an upright, a horizontal rail or tie, and a brace, firmly mortised and bolted together, the two sides being joined by a transom and axle body at each end. These project below the lower surfaces of the ties, and fit in between the rails of the chassis, serving, like the flanges on the rollers in the other barbette carriages, to prevent the gun carriage from moving sideways off the chassis. The lower side of the ties is shod with a friction-plate.

Through the front axle-body and near the front ends of the ties, an iron axletree is placed, working in iron boxes fitting in the ties. On the projecting ends of this axletree (the axis of which is eccentric with the axis of the ends), the rollers or manœuvring wheels are fixed, the extreme ends of the axle being octagonal in shape to fit the wrench of the iron handspike.

These eccentrics are so arranged that when the centers of the wheels are at their lowest points, the surfaces of the wheels bear on the rails of the chassis and raise the gun-carriage tie from it; and when the centers are at their highest points, the surfaces of the wheels do not touch the chassis rails, and the carriage ties themselves are in contact with the rails.

A similar arrangement is made for the rear part of the carriage, except that the axle does not extend all the way through, but the wheel on each side has a projecting piece of axle which works into a box placed in the end of the tie.

The wheel is thrown into or out of gear, that is, made to bear on the rail of the chassis or relieved from it, by turning the axle with a wrench placed on the octagonal end.

In the direction of the radii of the wheels, but inclined outwards, mortises are placed for the reception of the end of the iron handspikes, by acting on which, while inserted, the wheels are turned and the carriage moved back and forth on the chassis. Ordinarily, when the wheels are thrown into gear, the carriage being back, it will run into battery of itself.

The elevating arrangement on this carriage is different from any other in the service. It consists of, first, an elevating screw which works into a screw bed below the second part, which slides in a vertical box, and carries on the top of it a movable *pawl* to fit into the notches cut in the breech of the gun. The pawl has a slit in it through which the elevating handspike is passed, and the gun raised by making use of the edge of the elevating box as a fulcrum. This arrangement is placed over the rear transom.

Chassis.—The chassis for this carriage does not differ much in its construction from those used with the other barbette carriages, heretofore described.

It consists, like them, of two rails connected by three transoms; but the tops of the rails are shod with iron plates, and the rear hurters are the large heads of heavy bolts which pass all the way through the rails. The front hurters are similar in form, but fixed to the front transom by a heavy plate and bolt.

Traverse wheels are placed under both front and rear transoms, and the chassis is pivoted on a pintle passing through the middle transom. Two of these wheels are placed under each end of the chassis, their axles being kept in place by straps bolted to the transoms. Recesses are cut in the under side of the transom for the wheels to turn in. The radius of the traverse circle being much less in this case than with the other barbette carriages, the wheels have to be placed in a much more oblique position in regard to the rail.

The Platform, Fig. 133, Plate 11.—The traverse-circle and pintle-plate may be set in masonry made of cut stone; or a wooden platform may be used. The last is a circular framework of heavy timber, notched and bolted together, and laid on a horizontal platform of thick plank. The space inside the traverse-circle is boarded over, and forms a platform on which the cannoneers stand when manœuvring the gun. The pintle-plate, which is of iron, must be raised above the level of this platform, in order to give support to the middle transom. This is done by placing it on a heavy circular block of wood, which is firmly bolted to the timbers of the main platform.

The arrangement of this carriage allows the piece to traverse the whole circle.

The Garrison Gin, Fig. 134, Pl. 11, is made heavier and stronger than the field and siege, as it is used for mounting heavier guns, and has not to be transported like the other with an army in the field. The legs are longer, and of course the gin higher, than the other.

The legs have but two braces, which are iron bars bolted and keyed to the legs. When the gin is dismounted these braces are turned on their bolts parallel to the legs, and bolted to them by the same bolts used in setting up the gin. The lower brace has a third eye made in it at the proper distance, and the upper one being just long enough to reach the second bolt hole in the leg to which it is fastened.

The pry-pole has no tongue, but fits between the tops of the legs, and is joined to them by a bolt which supports a clevis, the branches of which rest on each side of the pole. To the clevis a gin-block is hooked. The pry-pole has eleven cleats nailed to it, to serve as steps on which to mount to the top of the gin.

The ends of the windlass are provided with ratchets, which fit on hexagonal tenons cut in the wood. To work in these rachets, each leg is provided with a pawl joined to it by a bolt above the windlass journal-boxes.

The journal-boxes are of brass, and either fastened to the legs by bolts, or to blocks of wood which are fitted to the legs, and secured by heavy iron bands; or, if large enough timber is used for the legs, these beveled blocks can be formed out of the same sticks as the legs. (Appendix, p. 11.)

The Large Sling Cart, Fig. 135, Pl. 11, is used for carrying the heaviest pieces and carriages. It consists of a heavy, strong axletree, mounted on two wheels eight feet high, and surmounted by a bolster, between which and the top of the axletree the tongue and hounds are notched in and secured with bolts.

Two heavy hooks are placed in rear of the axle, and fastened to it by the stems passing through and being secured on the front side with washers and nuts. Two other smaller ones are fixed, inverted, to the front side of the bolster, and secured in the same way.

On top of the bolster an iron bed-plate is secured; and through this, the bolster, and the axle, passes the stem of a large double hook, on the upper part of which is cut a heavy square thread, to which is adapted a corresponding nut. Over the nut, with slots to fit corresponding projections on it, fits a large screw-handle, the projecting arms of which are parallel to the bolster, and turned up a foot and a half at the end. By turning this handle the hoisting-screw is raised or lowered. The part just above the double hook is made square in order that, passing through a square hole in the axle, the screw itself may not turn with the handle.

The end of the pole has a strap terminating in a lunette, for the purpose of attaching a limber to it. Either a field or siege limber may be used. Under the pole and near the front part of the wheels a long staple is placed, and secured by its branches passing through the tongue and being fastened on the upper side by washers and nuts. To this the cascable is lashed when the piece is slung on the hoisting-screw. The pole is provided with a strong pole-prop.

Guns are carried on this cart by using the trunnion-chain, the large rings being passed over the trunnions, and the middle link over the hoisting-screw hook. Other bodies are lashed to the axle and bolster hooks and the pole-staple, by means of the sling-chain and ropes, being first raised, if they are not high enough, with the hoisting-screw. In carrying heavy guns the strain should be taken off the screw as much as possible, by lashing them to the hooks and staple after they run up, and then turning the screw down a little. For dimensions, &c., see Appendix, p. 11.

THE LIFTING JACK, Fig. 136, Pl. 11, is a small but powerfully geared screw, with a horizontally projecting foot. It is principally used for raising great weights in confined places, or where the number of men is small. It is slow in its operation, but for a single lift is very convenient, and is found useful in extricating pieces from difficulties during the transportation; and, hence, one or more of them should accompany each siege train.

If the weight is near the ground, and is to be raised but a short distance, the foot is applied directly under it; but if the distance to be raised is greater, inasmuch as the foot cannot raise the weight high enough for it to rest on the head, it must be first propped up, and then a piece of scantling called a *lifting block* is placed on the foot, either flat, on edge, or upright, according to the distance through which the weight is to be raised. This block may be used also on the head of the jack, when it is laid down flat; or the same end may be attained by blocking up the bed of the jack.

The bed is made of a block of oak; and a notch is cut in the middle of it to allow the foot to come as near the ground as possible.

The stand is of cast iron, having an opening in one side to allow the passage of the foot. Four pivots project from the bottom of this stand in order to steady it on the wooden bed, to which it is joined by strong braces.

THE LEVER JACK consists of a *stand* and a *lever*. The stand is used as an adjustable fulcrum, for the lever made use of in siege and garrison service, for raising weights. The stand consists of two uprights of oak, connected at the bottom by a transom of oak and a heavy bolt; to these is mortised and pinned a heavy block of wood called the *bed*.

In the uprights, at convenient intervals, holes are bored for the reception of the fulcrum; which is a strong iron pin fastened to one of the uprights by a chain.

The lever consists of a heavy piece of timber, 15 feet long, with beveled ends. Near one end, and parallel to each other, two lever-plates are fixed with screws. These are made of wrought iron, and have circular notches cut in them of such a size as to fit the fulcrum-pin, and serve as resting places for the lever when in use.

THE MANŒUVERING-BLOCKS and implements used with them are—

THE BLOCK, Fig. 137, 8 in. square and 20 in. long, made of oak or other hard wood, as are the rest named under this head.

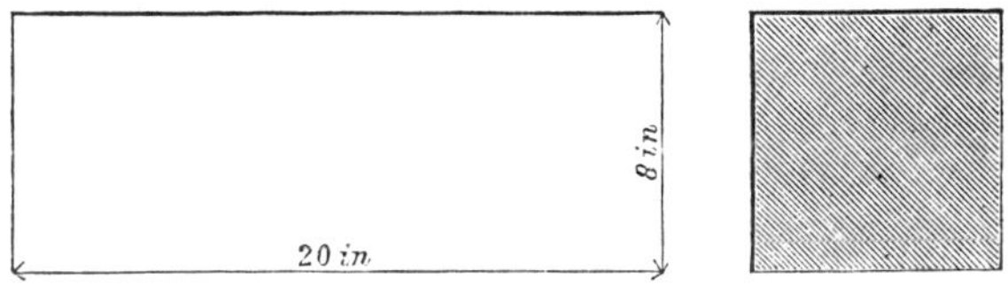

Fig. 137.

It is used for various purposes in mechanical manœuvers, such as serving as a rest for the piece when it is placed under it perpendicular to the axis; or in raising the piece to its carriage or lowering it from it, to form supporting walls on each side for the reception of the rollers or half-rollers on which the piece is resting. It is then placed parallel to the piece.

THE HALF-BLOCK, Fig. 138, is of the same dimensions as the block, except that its cross-section is 4 × 8 inches, in place of

Fig. 138.

8 inches square. It is used for the same purposes as the whole block, but when the distance through which the piece is to be raised is only half of what it is when the whole block is used. Quarter-blocks are also sometimes used, of the same length, 6 in. wide, and two thick.

THE SKID, Fig. 139, is a piece of timber the same size as the block, but 6 feet long. It is used, to support the rollers, half-rollers, or blocks on which a gun is lying, and is then placed

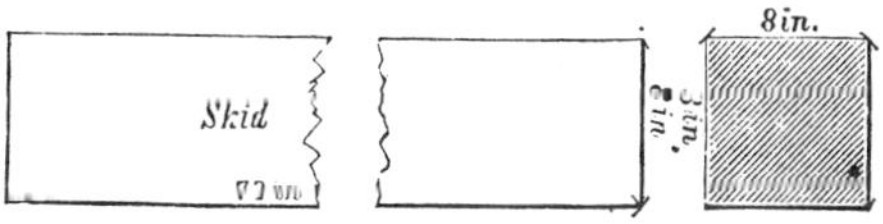

Fig. 139.

parallel to the axis of the gun; to roll the gun on, its height being sufficient for the trunnions of most guns to clear the ground; and as a foundation for the piles of blocks and half-blocks in raising or lowering the gun.

THE SHIFTING-PLANK, Fig. 140, is a piece of $2\frac{1}{4}$ in. plank, 1 ft. wide, 5 ft. 7 in. long, and beveled at both ends on opposite sides. It is used principally for shifting pieces from one carriage to another.

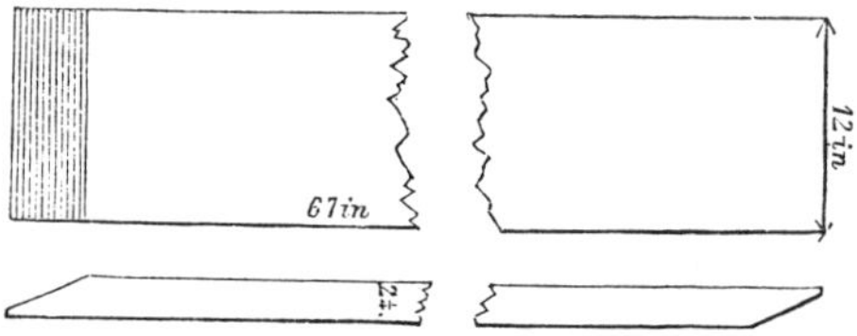

Fig. 140.

THE LONG ROLLER, Fig. 141, is a round piece of wood, 6 in. in diameter, and 3 ft. 6 in. long, having a groove cut round it in the middle, for the reception of the gun when placed upon it. It is used to move a gun in the direction of its axis, on skids, or a hard, smooth surface, such as a platform, &c. Two of them are used at once, and the gun moving upon them gains twice the distance passed over by the roller.

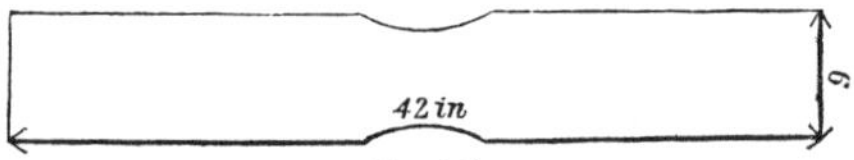

Fig. 141.

THE HALF-ROLLER, Fig. 142, has the same dimensions, but is round only on one side, square on the opposite side, and has no groove. It is used resting on the square side, when, in place of desiring to roll the gun, the object is to have a firm support on which the gun can have its ends alternately raised, as in mounting a gun on its carriage by means of the manœuvering blocks.

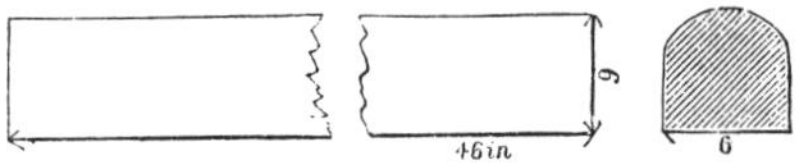

Fig. 142

THE SHORT-ROLLER, Fig. 143, is formed exactly like the long-roller, but is only 1 foot long. It is used in positions where the long-roller cannot act, as between the cheeks of a carriage in placing the gun in its trunnion-beds, or removing it from them, &c.

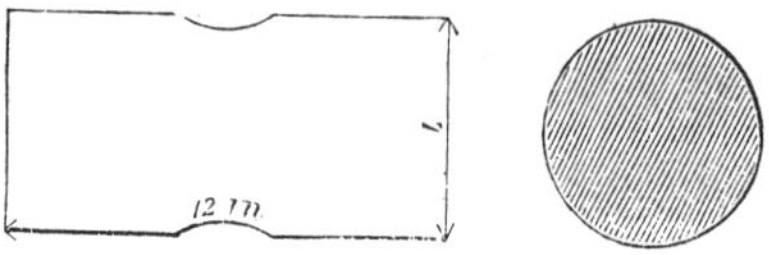

Fig. 143.

THE CHOCKS,—Figs. 144, 145, 146. In the different manœuvers with a gun and its carriage, chocks are necessary to steady the piece on the rollers, half-rollers, skids, or blocks; to arrest the progress of a roller, or prevent its turning in a wrong direction; and to perform the same offices for the wheels of a carriage. The different kinds used are here represented in the order in which they have been named:

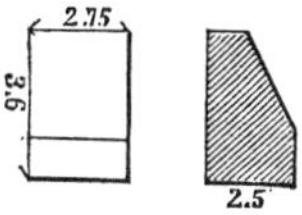

Fig. 144.

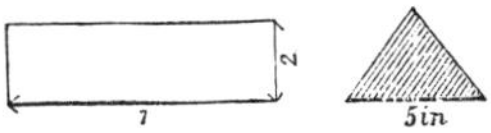

Fig. 145.

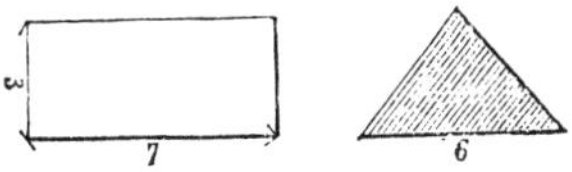

Fig. 146.

THE MANŒUVERING HANDSPIKE is beveled at the end on one side, for the purpose of allowing it to enter into places and under bodies where a square-pointed handspike could not go. It allows a gun, too, to be turned in its place, from this bevel presenting, when held against and under the gun, an inclined surface, on which the gun will slide.

THE TRUNNION-LOOP, Fig. 147, is a piece of rope about 18 in. long, having its two ends firmly spliced together, forming thus a ring which is placed over the trunnion, serving as a means of applying a handspike to slue the piece in different directions.

Fig. 147.

It is not deemed necessary to explain in detail the different manœuvers in which these implements are used, as they are all explicitly laid down in the Ordnance Manual and Heavy Artillery Tactics.

SEA-COAST MATERIAL.—Under this head will be described such carriages and implements as are especially applicable to our sea-coast fortifications, in which are also used the carriages and material described under the head of garrison material.

It consists of

1. Casemate carriages.
2. Sea-coast materials.
3. Casemate gin.
4. Casemate truck.

1. THE CARRIAGES for all pieces are alike, differing only in dimensions, with the exception of that used for mounting the 24-pdr. howitzer (iron) in the flanks of casemate batteries.

The carriage consists of a *gun-carriage* and *a chassis*. The different pieces mounted on these carriages are the 24, 32, and 42 pdr. guns, and the 8 in. columbiad.

The Gun-Carriage, Fig. 148, Pl. 12, consists of two cheeks joined together by as many transoms, and supported in front by an axletree on truck-wheels, in rear on a slide, fitted beneath the rear transom.

Each cheek is formed of two pieces, one on top of the other, and connected by dowels and bolts. On the under side, near the front, a notch is cut for the reception of the axletree, which is of wood (oak); and nearly over the axle, on the upper side of the cheek, the trunnion-bed is placed. The rear of the upper piece of the cheek is cut in notches, which give a better hold for the assembling-bolts than a uniform slope, and give purchases for the handspikes in elevating the piece.

On the inside of each cheek, just in rear of the axle, a vertical guide is fixed, to keep the carriage on the chassis. It is of wood, and bolted to the front transom and axle-tree. The top of the front transom is hollowed out, to admit the depression of the piece.

Under the rear transom a slide is placed, and if not made of the same piece, is bolted to the transom. It has on the under side a notch, to fit on the tongue of the chassis; and in the rear an eccentric roller, so arranged as to bear the weight of the rear part of the carriage or not, according as it is thrown in or out of gear.

Near the rear end of each cheek, and outside, a heavy trail-handle of iron is placed, and used in manœuvring the piece.

On the ends of the axle, truck-wheels are placed, with mortises made sloping outwards, in the direction of the radii, for the insertion of the handspikes in running from battery. These wheels are of one size for all the carriages, 20 inches in diameter.

The elevating apparatus consists of a cast-iron bed-plate, secured to the rear transom; an elevating screw and brass nut for it to work in; the nut being acted on by an oblique-toothed wheel, turned by a handle placed outside the right cheek. In the 8-in. columbiad carriage the cheeks are parallel to each other. In the others they converge towards the front. No casemate-carriage has been adopted for the 10-in. columbiad, as it is not designed for use in casemates.

For dimensions, &c., see Appendix, pp. 9 and 10.

THE CHASSIS consists of two rails and a tongue, joined by two transoms, and supported on traverse wheels both in front and rear, both of which require traverse circles to run on.

The outside upper half of each rail is cut out and shod with an iron rail, for the wheel to run on, the track being curved up at both ends, and provided with hurters, to prevent the carriage from running off the chassis.

A stationary prop is placed under the rear end of the tongue, to prevent the chassis upsetting backwards in firing very heavy charges, and may be used as a point of support in raising the chassis.

An iron fork is fastened by bolts to the under side of the front end of the tongue; and to this is bolted the iron tongue, through the eye in the other end of which the pintle passes. An opening in the masonry below the embrasure is left for this tongue, and it is secured in its place by dropping the pintle from the embrasure down through the eye of the tongue. (For dimensions, &c., see Appendix, pp. 9 and 10.)

THE 24PD. IRON HOWITZER CARRIAGE, Fig. 149, Plate 12, is specially adapted to the mounting of the howitzer in the flanks of casemate batteries, for the defense of the ditch; and both the gun carriage and chassis are narrower and lighter than any of the other casemate carriages.

The cheeks of the gun carriage are made of white oak, and connected by two iron transoms, the front one projecting below the cheeks, and resting on the chassis with a projection on the bottom of it, fitting in between the rails.

The bottom of the trail has the same slope as the upper surface of the chassis on which it rests, so that when its eccentric roller is out of gear, the rear parts of the cheeks fit upon the rails. The remaining portion of the bottom of the cheek makes an angle with the rail, and is furnished in front with a fork, and a roller which runs upon the rail of the chassis when the eccentric is in gear.

Each cheek has on the side near the rear, a trail-handle, and near the front a manœuvring ring.

In rear of the rear transom is placed an eccentric roller, having a projection on the middle of it just large enough to fit in between the rails of the chassis, and act as a guide to the trail of the carriage. When this roller is in gear, the weight of the trail rests upon it, while that of the front part of the carriage is thrown upon the front rollers, and the piece is then run easily in and out of battery; but the roller being out of gear, as when the piece is about to be fired, the weight all rests upon the rear part of the cheeks and the front transom, and friction is brought into play to diminish the recoil. Cap-squares are used with this carriage.

The elevating apparatus is similar to that used for the other casemate carriages. It has been suggested, that instead of having the handle of this apparatus four projecting branches, it should be made with a round tire similar to a wheel, when it could be worked easily by the gunner when engaged in sighting the piece.

THE CHASSIS consists simply of two rails, 3 in. apart, and joined by four transoms and assembling bolts.

Hurters on the rear ends of the rails only are used, as the bottom projection of the front transom prevents the carriage running too far into battery.

The front end of the chassis rests on the sole of the embrasure, the lower side being sloped off horizontally for the purpose. The end is provided with a pintle-plate and a strap of $\frac{1}{2}$-in. iron through which the pintle passes to the masonry beneath.

The rear of the chassis is held up by two traverse wheels arranged differently from those in the other carriages. A prop of iron with two branches at top, which are bent and bolted to the rails, is firmly braced with an iron bar to one of the transoms of the chassis. The lower end of this prop forms a socket for the reception of the stem of the traverse-wheel fork, which has two branches with a traverse wheel in each. The stem is bolted in the socket of the prop, and the arms are so bent as to make the wheels stand perpendicular to the radius of the traverse circle. The traverse wheels are slightly conical. (For dimensions, &c., see Appendix, pp. 9 and 10.

CURVED TRAVERSE CIRCLES.—It has been proposed to dispense with the pintle in these carriages, by adopting an oval-shaped traverse circle, and grooving the wheels to correspond. Experiment, so far, goes to show that the shock of firing is not sufficient, without the pintle, to throw the wheels from the track.

It might be more difficult to apply these traverse circles to the flank carriage; but it could be done by fixing a small casemate one in the sole of the embrasure, and fitting small grooved rollers to the under side of the front part of the chassis.

These traverse circles will require a man with a handspike at the front traverse wheel, as well as at the rear one, to prevent the wheels binding on the circles, by becoming oblique to them.

Iron Carriages.—For some years past, experiments have been going on in Europe, particularly in France, Russia and Prussia, in regard to the use of cast and wrought iron as a material for gun carriages. Cast iron is objectionable on account of the weight, and its great liability to splinter when struck by shot.

In the United States, similar experiments have been made, but more especially in regard to wrought iron, that being deemed the best material of the two. The results have been so satisfactory, that wrought iron carriage has been recommended for use altogether in garrisons, in place of the wooden, with the exception of the carriage for the 24 pd. howitzer in flank casemates.

The following are the carriages recommended for adoption.

1 Barbette carriage for 10 in. columbiad.
1 " " " 8 in. "
1 " " " 42 and 32 pd. guns.
1 " " " 24 pd. gun and smaller calibers.
1 Casemate " " 8 in. columbiad and 24 pd. gun.
1 " " 32 and 24 pd. guns.

These carriages are all made in a similar manner, differing only in weight and dimensions. When the same carriage is used for pieces of different sizes, movable trunnion plates to fit the trunnions of the smaller calibers are placed in the trunnion holes.

The Carriage, Fig. 150, Plate 12, consists of two cheeks of thick sheet iron, each one of which is strengthened by three flanged iron plates, which are bolted to the cheeks. Along the bottom of each cheek, an iron shoe is fixed with the end bent upwards.

In front, this bent end is bolted to the flange of the front strengthening plate. In rear the bent portion is longer, and terminated at top by another bend, which serves as a point of application for a wheeled lever, C, when running to and from battery.

The trunnion plates fit over the top ends of the strengthening plates which meet around the bed, and are fastened to the flanges of the latter with movable bolts and nuts.

The cheeks are joined together by transoms made of bar-iron, bent at the ends and bolted to the cheeks, instead of pipes and assembling bolts running through, which were used at first.

The cheeks are parallel to each other, and in order that the *base-ring* on the gun may not interfere with giving it the full elevation by striking against them, it is proposed to leave the base-ring off all sea-coast and garrison pieces, and to retain the preponderance by reducing the swell of the muzzle. These changes will, as has been stated in Chap. III., reduce the weight of the pieces, at the same time that, on Mr. Mallet's principle, they will increase their endurance.

The front of the carriage is mounted on an axletree, with truck-wheels similar to the present casemate carriages.

A carriage of this kind was submitted to every severe test to which it would be exposed in service, and sustained, satisfactorily, 4,000 rounds without requiring, for the last 2,000, any repairs whatever.

The Chassis is formed of two rails of wrought iron, the cross-section being in the form of a T, the flat surface on top being for the reception of the shoe-rail of the gun carriage. The rails are parallel to each other, and connected by iron transoms and braces which are bent at the ends and fastened to the rails with bolts. The chassis is supported as before, on traverse wheels, which work in forks fitting into sockets bolted on to the outside of the rails at the proper points.

It has since been proposed to have but four kinds of these carriages, viz.: a barbette carriage for each of the 8-inch and 10-inch columbiads, and one barbette and one casemate carriage for all other pieces; suitable trunnion-plates, transoms, and axles, being substituted, according to the size of the piece.

A prop is placed under the middle transom of the chassis, to provide against sagging.

The 10-inch carriage has braces bolted to both sides of each cheek-plate. All other carriages have braces on the inside only.

The thickness of the cheek-plates, the dimensions of the trough-beams for braces and transoms (rear), and the chassis rails, are the same for all carriages.

The axle boxes, axle-box washers, and truck wheels, are the same for all carriages.

The axles and transoms for all carriages differ only in their lengths.

The front * and middle transoms of the 8 and 10 inch columbiad barbette chassis, and the props for all the carriages, are wood.

The tongues for the wooden casemate carriages, and the traverse wheels, &c., for the wooden carriages, may be made to answer for the iron carriages.

The elevating screws, &c., are the same as are used for the wooden carriages respectively. A graduated arc may be attached by a hinge to the rear brace of the right cheek, and used in elevating the gun.

The chassis rails of all carriages have an inclination of three degrees (3°).

Lead or leather washers should be used under all the washers.

The traverse wheels for the barbette and casemate chassis may either be placed directly under the rails, or between the rails in sockets in the transoms. Both methods were tried, and answered equally well.

By placing the wheels directly under the rails, the distance between each pair of wheels is 5 inches greater than when they are placed in the transoms; and as the 24-pdr. barbette carriage is 12 inches higher and 7 inches narrower than the 8-inch casemate carriage, this method is considered to be the best.†

HEAVY SEA-COAST MORTAR-BEDS.—The bed for the heavy 10-inch mortar is the only one which has, up to this time, been adopted.

The cheeks are of cast iron, and somewhat similar in form to those in the beds of siege mortars, but in the front, the cheeks turn up to receive between them the front transom, which has, countersunk in and bolted to it, an elevating screw-bed through which works an inclined elevating screw which rises or falls by turning the nut fitted on it by means of a lever inserted into mortises cut in the direction of the radii of the circular nut.

---

* It is proposed to have the front transom of iron.

† As has already been stated, the socket plates are now bolted to the outer side of the rails.

Both the transoms are made of wood, connected with the cheeks by mortises and tenons, and secured by bolts running through and nuts on the outside. One of these bolts at each end, is longer than the others, and the projecting ends are made use of as maneuvering bolts.

Directly behind and underneath the position for the trunnions, a bronze bed-piece is placed to receive the shock of the piece. It consists of a large beam of bronze, with each end well let in to the face of the cheek.

Although no bed has as yet been adopted for the 13-inch mortar, it is presumed the bed for it would not differ essentially from that for the 10-inch.

The use of the elevating screw in place of the quoin, is rendered necessary by the great mass of metal to be raised or lowered in sighting the piece.

The Casemate Gin does not differ from the garrison gin, except in its height, which is less (about that of the field and siege gin), and the thickness and strength of the parts. These differences are necessary, on account of the contracted places in which the gin is used and the greater weights to be raised by it. (For dimensions, see Appendix, p. 11.)

The Casemate Truck, Fig. 151, Pl. 12, is used for transporting guns through casemate galleries and other contracted places in a fortification. It consists of a heavy framework platform made of two rails with three connecting transoms mortised into them; the tops of the transoms being hollowed out to receive the gun. The front and rear transoms have an iron transom-plate let into the underside and fastened; the ends of these plates terminate in eyes, into which rings are welded for hooking drag-ropes to the truck.

The platform stands on three wheels, supported in forks, the wheels and forks being the same as those used for the chassis in barbette carriages, except that the upper part of the stem for the front fork is made conical in shape, to allow the truck to change direction.

Two fork-plates are fastened under the rails, one on each, with their centers one foot in front of the rear transom. Into these, and running up into the rails, the stems of the rear wheel-forks fit.

The fork for the front wheel has attached to it by a bolt a long iron handle, with the end turned up and a wooden handle inserted in the eye. The conical top of the front-wheel fork is inserted in a socket made for it in the middle of the front transom and plate.

A Store Truck, such as may be seen in any commercial store, is used for moving boxes, &c., in storehouses, or in embarking and disembarking stores. To prevent the load from touching them, the wheels are protected on top by guard-plates of iron.

CHAPTER VII.

THEORY OF FIRE.

In spite of the great advantages derived from the bayonet in battle, the fact is now incontestible that the efficiency of a body of infantry resides essentially in its accuracy of fire; and this is made only the more apparent from the recent improvements in fire-arms. A *cool, well-directed* fire from a body of men armed with the new rifle or rifle musket, which is loaded with the same facility as the old musket, is sufficient to stop the advance of almost any kind of troops. But the very best disciplined men will, in time of battle, fire with precipitancy and at too great a distance; from which results a great loss of ammunition and of effect upon the enemy.

Target Practice. If instruction by target practice is of importance to well drilled and disciplined troops, when opposed to troops of the same kind, its importance cannot be estimated when the quality of the troops becomes worse, and the opponents the Indians of our western country, nearly all of whom are well versed in the use of the rifle or bow and arrow, and dependent on them for their daily subsistence.

In the recent encounter with Indians of Col. Steptoe's command in Washington Territory, many of the men were armed with the old *musketoon* (a most indifferent arm), and very soon expended all their ammunition in ineffectual fires against an enemy mounted on fleet horses, and armed partly with rifles, partly with bows and arrows, whose deadly shaft is shot by them with astonishing accuracy, and at a rate *exceeding* that at which an adept in its use can fire balls from a *revolver*. Charges of cavalry, with the saber, however gallantly conducted and successful for a time, could not succeed against such an enemy; and a hasty retreat in the night was the only resort, to avoid annihilation.

In the subsequent expedition, the men were armed with rifles and rifled-muskets, and after a *few weeks'* hasty drill attained such a confidence in their arms, that on the first encounter with the Indians the result proved very different.

Our service is perfectly devoid of any means of attaining proficiency in the use of arms of *any kind;* and the time will yet come, if some corrective is not applied, when this deficiency will result in a disaster vastly greater than the defeat of a handful of men in a distant territory.

Superiority in volunteer troops over regulars, when it does exist, is not found in those regiments recruited about our great cities, from men who never touched a gun before in their lives; but in those made up of young men from the woods, who, from daily use of the rifle from their boyhood, know its power; and this knowledge it is, which *in part*, supplies the place of a discipline they do not possess, and makes them *feel* that *with* discipline they are invincible. In confirmation of these remarks, such regiments as the South Carolina and Mississippi, in the Mexican war, are confidently referred to; both were commanded by men, soldiers from education and by profession.

To attain efficiency in the use of any arm, diligent and systematic practice is absolutely necessary; and this can only be obtained by establishing schools of instruction, like those at Vincennes, Saint Omer, Grenoble, and Toulouse, in France, from which, every year, officers and men well instructed in the *principles* of firing, may be sent out into the army at large, and impart the same system and efficiency to it. An attempt is now being made to establish such a school, for the artillery, at Fort Monroe; but the necessity for a school for the infantry arm, is just as great and more *immediately* pressing. The results of such schools will far more than compensate for the cost, and withdrawal from active military service of a small number of men, by the greatly increased efficiency of the rest.

In the successful and oft-repeated repulse of cavalry charges by squares of infantry, the main dependence is not in the use of the *bayonet*, but in the close, well-directed fire, delivered as the horsemen approach. This, breaking their formation and disorganizing their ranks, leaves them at the end of the charge with a wall of bayonets in their front, against which horses cannot be. forced unless at full speed and supported by numbers behind It is not pretended that a large body of compact horseman cannot *ride over* a *small* square of infantry. To hold such a doctrine would be to deny the fundamental principles which govern the relation between force and resistance. But with anything like parity of numbers, cavalry cannot break squares of infantry, simply from the injury inflicted *before* the shock takes place.

In order to become proficient in the use of fire-arms, a man must either learn the principles upon which his arm is constructed, as well as those governing the inflammation of powder and the passage of a projectile through the air; or he must, from long actual practice with the arm, understand the results of these principles. It is evident that a combination of the two will produce the best marksman in the shortest time.

The theory to be taught the marksman, will now be discussed.

Powder, when inflamed in a gun, developes a certain amount of elastic fluid, which escapes on that side where it meets with the least resistance, driving before it the projectile which is opposed to it.

VELOCITY is the space passed over by the projectile in a second of time. Its velocity, when it leaves the piece, is called the *initial velocity*.

The action of the powder, as has been before stated, is not instantaneous; and although the ball is less than $\frac{1}{200}$th part of a second in going out of the piece, it receives its velocity by degrees. The velocity of the gas developed is very great, and has been estimated as high as between 9,000 and 10,000 feet. This great velocity causes the first gas produced to spread rapidly through the interstices of the grains, and inflame the entire charge. The time, taken for the entire combustion of each grain, although very short, is greater than that required for the inflammation, and depends on the size and consistency of the grains.

Since the inflammation takes place through the interstices of the grains, it follows that hard ramming, by packing the grains and diminishing the spaces, should reduce the velocity of the ball. The ramming should be so regulated as to place the ball well against the charge, and to have it as near as possible under the same circumstances at each shot.

When the cartridges have been a good deal knocked about, the dust formed by the crushed grains obstructs the interstices between the grains, and, the charge fuzing instead of exploding, the velocity of the ball is diminished.

Since the smallest grains burn quickest, it follows that for any given length of barrel, there will be more powder burnt and a greater effect produced as the grains are decreased in size. For this reason, fine sporting-powder gives, with equal weights, greater velocity than musket-powder; and this difference becomes the more marked as the length of the gun and the charge are decreased.

Powder loses force as the amount of moisture contained in it increases. Thus, the effeet produced is less on a wet day than on a warm, dry one. Wet powder fuzes instead of exploding.

The initial velocity of the ball depends on—the charge; the quality of the powder; the length of the gun; the size and density of the projectile; on the amount of windage, and on the size of the vent, especially in flint-lock guns.

With a given length of gun, and particular projectile, there is a maximum charge beyond which no increased velocity is obtained. This charge must be determined by experiment; though the charges used are generally less than the maximum, the rule generally laid down being, that as the velocity increases very slowly from a third of the weight of the shot up to the maximum, it is not advisable to use a greater charge than one third, on account of the effect on the piece, the waste of powder, and the recoil. In this respect the modern improvements in small arms are very marked; for whilst with the old musket $\frac{1}{4}$th the weight of the bullet was used, with the rifled and altered musket we use but $\frac{1}{9}$ or $\frac{1}{10}$.

The longer a gun is, the greater length of time is the projectile submitted to the accelerating action of the gases; but past a certain length, the shocks and friction experienced by the ball overcome its increase of velocity. Experiment shows there is no advantage in giving to the rifle-musket barrel a greater length than forty inches, which is accordingly adopted for the new arm.

Powder developes a greater force as it meets with more resistance to its expansion, so that the heavier a projectile the greater becomes the quantity of motion it receives. Thus one projectile double the weight of another, receives from the same amount of powder a much greater velocity than one half of that given to the lighter one. This fact enables us, in mortars, where the weight of the projectile is almost directly opposed to the action of the powder, to use a smaller charge than would otherwise be necessary.

With a given charge, projectiles with the least density, and smallest diameter, receive the greatest velocity; but out of the gun the advantage soon disappears, for such projectiles meet with the greatest resistance from the air.

The charge necessary to produce a given velocity increases with the density of the projectile.

The new rifle-musket gives, with a charge of 60 and bullet of 510 grains, an initial velocity of over 950 feet; which, however, would probably be decreased when using cartridges which had been transported some distance, especially if in the hands of the men.

Recoil.—As the powder acts in all directions, the bottom of the gun is thrown back with a greater force as the charge increases. The action on the piece is greater proportionally than on the ball as the bottom of the bore presents a greater surface than the ball; and the gas continues to act on the gun after the ball has left it. The velocity impressed on the gun is called the *recoil.*

With the old musket, round bullet, and 120 grains of powder, the initial velocity of 1,426 ft. is obtained. The recoil of the gun, were it of the same weight as the bullet, would be greater, say 2,280 feet. That is, that the ball must have this velocity, in order that, striking the piece, it should produce the amount of recoil actually experienced by the gun. But velocities are inversely proportional to the masses moved; and the musket weighing about 166 times as much as the bullet, the retrograde velocity imparted will be only $\frac{2280}{166}=13.73$ ft., or about the same rate as a fast-trotting horse will travel. This velocity is so great that the marksman would be injured by the shock, if he did not hold the gun close against his shoulder, in such a way as to join the weight of his body to that of the piece. Supposing the resistance offered by him is ten times that offered by the gun, the recoil is reduced to 1.373 ft., which can be borne.

With the new rifle-musket, the ratio between the bullet and the piece is reduced to 138, the initial velocity to 963 ft., and the corresponding recoil to, say 1,500 ft.; so that the actual recoil of the piece is reduced to $\frac{1500}{138}=10.87$ ft. These numbers are not strictly accurate, but are near enough for our purpose.

The effect of the recoil on the marksman is farther diminished by the form given to the stock, which, being curved, decomposes the force of the blow. The same result is obtained in the pistol by the curve given to the stock, which changes the direct recoil into a movement of rotation around the hand.

The ball leaves the piece so rapidly, that the recoil has no appreciable effect upon its direction. This may be shown by placing a gun-barrel on each side of a square horizontal frame suspended by a single wire; the second barrel being for the purpose of balancing the first. One of these barrels being pointed and fired, the ball has no appreciable deviation, although the frame revolves, from the effect of the recoil, rapidly around the suspending wire. The recoil, then, probably begins to be felt only when the ball is nearly out of the piece; and as the recoil of a gun in a man's hands is sensibly in a right line, it follows that the recoil causes no appreciable deviations in firing from the shoulder.

In loading, the powder should be well shaken out of the paper, to prevent the formation of gas inside of this, which, forcing the paper against the sides of the bore, prevents it from leaving with the charge, and endangers the explosion of the next charge when loading, from the lighted paper. There is no danger of heating the piece by rapid firing so as to cause premature explosions, since long before it reaches 600°, the temperature at which gunpowder inflames, it is entirely too hot to handle.

Theory of Fire.—In the use of fire-arms, the three following lines, and their relation to each other must be known.

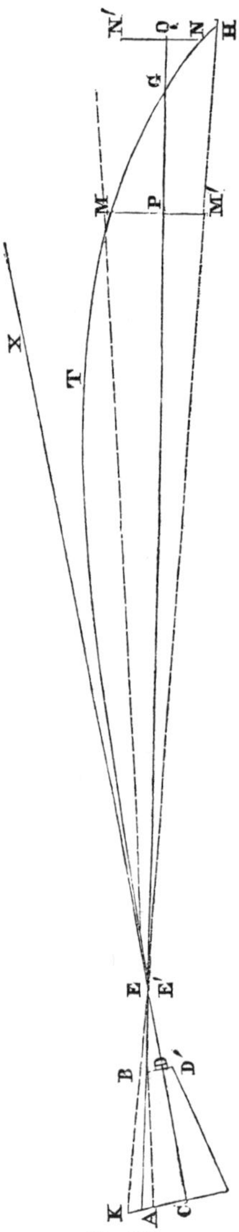

Fig. 152.

The *line of sight* is the visual ray A B E G, Fig. 152, passing through the highest points of the breech and muzzle of the piece, and directed on the object designed to be struck. The *line of fire*, or axis, is the axis of the piece C E X prolonged indefinitely; and the *trajectory* is the curve E T G H, described by the center of the projectile. All these lines lie, theoretically, in the same vertical plane, called the *vertical plane of fire.*

The ball starts along the axis, and but for its weight, would follow that line more or less closely; but as soon as it leaves the bore, its weight depresses it, and keeps it always below the line of fire C X. In vacuo, the trajectory would be a parabola; but in the air it is modified, and becomes less like that curve as the weight and velocity of the ball increase.

Ordinarily, the diameter of guns is greater at the breech than at the muzzle; and the *natural* line of sight, or *line of metal*, A B, makes an angle with the axis B E D, which is called the *angle of sight.*

The ball falling but little below the axis when it first leaves the piece, its trajectory cuts the line of metal at a point E′ (in small arms a very little below the axis), passes above it, and cuts it a second time at G, which is called the *point blank.* It is the point at which, in order to strike it, a gun is directly aimed. Its distance from the gun is called the *point-blank range.* The *range* is the distance from the piece at which the ball strikes.

The *angle of fire* is the angle which the axis makes with the horizontal. On horizontal ground the range increases with this angle up to a certain limit, depending upon the size and velocity of the projectile. At the limit, this angle is called the *angle of greatest range.* This angle in vacuo with a small velocity is 45°. In the air with small velocity and heavy projectiles it is a little less, but with long guns and great velocity, it is from 25° to 35°.

The point-blank range varies from several causes, the principal ones being: The velocity of the ball; its diameter and weight; the inclination of the line of sight; and the shape of the piece.

The velocity of the ball depends on the charge and the length of the piece.

The diameter and weight of the ball cause a certain amount of variation in the range. The larger and denser the ball, the more easily it overcomes the resistance of the air, preserves its velocity and accuracy of fire, compared with a smaller one of less density endowed with the same or even greater initial velocity. A leaden ball, for instance, will have a greater range and accuracy than an iron one.

The inclination of the line of sight with the horizontal produces very little change in the point blank range, unless this inclination is very great. When a gun is fired at a great angle of elevation, the action of gravity being almost directly opposed to the force of impulsion, the trajectory is shortened and the point-blank range diminished. The contrary is the case when the piece is fired under a great angle of depression, as gravity then acts *with* the impulsive force, and lengthens the trajectory.

Experiment shows that for angles within 15° above or below the horizontal, the variations in the point-blank ranges may be neglected, and the trajectory considered as constant. Hence, by fixing a gun in position, and varying the inclination of the line of sight, we get the same point-blanks which would be obtained by varying the position of the piece each time in order to make the line of sight horizontal.

Hence, it follows that to strike an object elevated or depressed 15° above or below the axis of the gun, the piece is aimed exactly as if on the same level with the object.

The greater the difference between the diameters of the breech and muzzle, the greater will be the angle of sight, B E D=X E G (Fig. 152), and the farther removed will the point blank be. Within a certain angle, when the points E′ and G are united, that is, the line of sight become tangent to the trajectory, there is no point blank, and none, of course, when this line becomes parallel to the axis (as is the case in our mortars), or when it diverges from the axis.

In the same kind of piece the form is almost identically the same; the angle of sight then remains the same; the charges and balls are constant; and the habitual angles of elevation of the line of sight are comprised between $\pm 15°$. The point blank and trajectory may, therefore, be considered as practically invariable. There is, then, a constant relation existing between the line of sight and trajectory, and the rules of firing will be known when the positions of the different points of the trajectory in reference to the line of sight, are known.

By the figure 152, it will be seen that to strike an object at point-blank distance (B G), the piece must be aimed directly at it. If the object is within point-blank range, as at P, the piece must be aimed below it, a distance $P M' = P M$, the height of the trajectory above the line of sight at that distance; otherwise, the shot would strike too high by that much. On the contrary, if the object is beyond point-blank range, as at Q, the piece must be aimed above it a distance $Q N' = Q N$, the distance of the trajectory below the line of sight at that distance; otherwise, the shot, after cutting the line of sight at G, would strike too low by the distance Q N.

The elevations and depressions of the trajectory are measured on the vertical lines passing through the points to be struck.

When the line of sight does not meet the trajectory, the piece must always be aimed above the point to be struck, a distance equal to the depression of the trajectory at that point.

When the angle of sight increases, the segment of the trajectory E′ T H above the new line of sight K H becomes greater; for the ball thrown under a greater angle must go farther, and the new point-blank, H, ought to be more distant than the first one, G.

As sighting above or below the mark is liable to many errors, and very inaccurate in practice, any method which will enable us to sight directly at the object, is preferable. This end is attained by the use of a *breech-sight*, or hausse (from hausser to increase), which being applied upright on the breech, increases its apparent diameter, and consequently the angle of sight, giving, as it were, a new point blank for every new length or division of the scale. On small arms, these scales, called *rear sights*, are connected with the piece by a hinge, and when not in use are folded down so as allow sighting along the *natural* line of sight. The line obtained by using a hausse is called an *artificial* line of sight.

MEAN TRAJECTORY.—A knowledge of the trajectory described by a ball is necessary in order to understand and apply the principles of fire. The curve may be calculated by means of an approximate equation; but it is better to employ this method in connection with the determination of points by practical experiments.

Points of the mean trajectory are obtained by firing from a stand, a great number of shot as nearly under the same circumstances as possible. With the musket, great variation is observed in successive shots, and it is the instability of the trajectory which renders musketry firing so uncertain.

These series of shots are fired at different distances, and at targets large enough to receive nearly all the balls. The piece is sometimes fired from the shoulder, using a rest, and sometimes from a light carriage, on which the axis of the piece is placed horizontal by means of a spirit level.

The piece being loaded and fired carefully a number of times, the balls, instead of following substantially the same course, deviate in every direction, forming thus a kind of cone with the apex at the piece, and having its cross-section increased as the distance from the piece increases. This renders it necessary to increase the size of the target as the firing distances become greater, and in proportion as the accuracy of the piece decreases.

As there is no reason why the ball should go to the right rather than the left, or too high rather than too low, all the shots are more or less liable to strike at the proper point; so that taking the general mean of the hits at each distance, one point of the mean trajectory will be obtained. Upon the knowledge of the *mean trajectory* of any arm, the rules with regard to its fire are founded.

The target for these experiments is made of very thin boards, or cotton cloth stretched over a frame. Through a point in the center, a horizontal and vertical line are drawn, as coordinates to which all the shots are referred; parallel lines to these are described at stated distances, to facilitate the reference.

The mean point obtained by this operation will of course be the more accurate as the number of shots fired is increased; and if the number is infinite, the result will be exact, as then all the chances of error will balance and mutually correct each other. The theory of chances, however, gives a sufficient approximation by using 200 shots at each distance. Those shots which strike after ricocheting, or do not strike the target at all, should be excluded.

The point of impact of a ball is the point where it strikes the target, and the mean of all the hits is called the *mean point of impact*, or the *center of impact*. It is a point of the mean trajectory.

To determine this point, the distances of all shots above the horizontal central line are added together, and also the distances of all shots falling below this line. The difference between these two sums, divided by the number of shots, gives one co-ordinate of the center of impact. The other is obtained by pursuing the same process in regard to the shots on the right and left of the vertical central line.

If the sum of the distances above the horizontal line is the greatest, the resultant point will correspond to a point of the trajectory within point-blank range. If the contrary is the case, the point will be one beyond the point blank.

To construct the curve from this data, an indefinite right line is drawn with perpendiculars erected at points corresponding to the different distances of the targets P P′, &c., Fig. 153; and on these are laid off the different elevations and depressions of the center of impact above and below the horizontal target-line. The curve described through these points will represent the trajectory. Where this curve crosses the right line will correspond to the point blank of the piece, and the distance of that point from the starting point will represent the point-blank range.

This supposes different elevations have been used in firing. If,

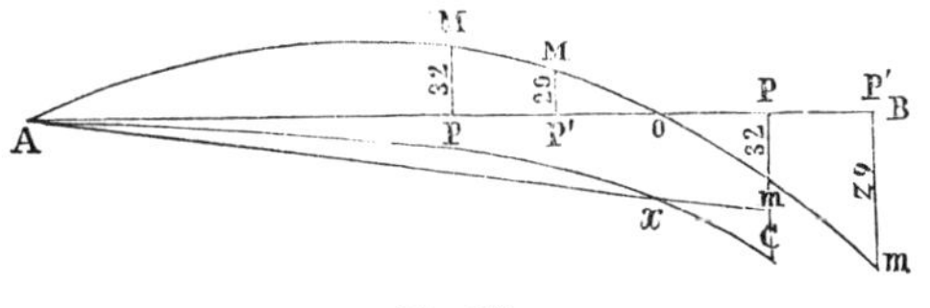

Fig. 153.

however, the axis of the piece has been kept horizontal, all the centers of impact will be below the horizontal co-ordinate; and n order to find the point blank, the line of sight $A\ c$ must be be drawn in its real position, and will give, by its intersection x with the curve, the point corresponding to the point blank.

The co-ordinates of the center of impact being known, the point itself is known, and its distance from the center of the target is called the absolute mean deviation.

It results from what precedes, that the mean trajectory occupies the center of the cone of despersion of the balls, and that the rules of fire, either for the direction or elevation of the shots, should be founded on the position of this curve.

For a small number of shots, with a given piece, or with an unskillful marksman, it might happen that the mean trajectory would not be comprised in the same plane; but when the number of shots is great, and proper precautions are taken, as there is no reason why the trajectory should leave the vertical plane of fire, it will, as a general thing, be found in it.

It is evident that for each marksman, and even for each piece, there is a particular cone of dispersion; and that the more skillful the marksman and more perfect the piece, the less will be the section of the cone for each distance.

It follows, therefore, that the best marksman is not necessarily he who makes the nearest shots, but the one whose shots are scattered over the smallest surface, or have the least absolute mean deviation. In firing, those marksmen should be excluded who do not hit the target every time, or who hit by ricochets.

In order to simplify, a point of the mean trajectory is taken as the common point of impact, and the best marksman is the one the sum of whose deviations, measured from this point, is the least.

The following are the points of the trajectories, as determined for the rifled-musket, the altered musket, and the rifle, from which the forms of the curves may be constructed:

POINTS OF TRAJECTORY OF NEW RIFLE MUSKET.

Distance, 200 yards; weight of ball, 500 grains; weight of powder, 60 grains.

Height (inches).....	14.5	17.7	19.3	19.6	16.	10.
Distance (yard).....	50	75	100	125	150	175

POINTS OF TRAJECTORY OF ALTERED MUSKET.

Distance, 200 yards; weight of ball, 730 grains; weight of powder, 60 grains.

Height (inches)	16.2	18.8	19.7	20.9	17.5	10.4
Distance (yards)....	50	75	100	125	150	175

POINTS OF TRAJECTORY OF HARPER'S FERRY RIFLE.

Distance, 500 yards; weights of ball, 400 grains; weight of powder, 50 grains.

Height (inches).....	85.	141.	145.	150.	97.	0.
Distance (yards)....	100	200	250	300	400	500

The cones of dispersion of the balls comprise all the causes of error in firing, whether resulting from the arm itself, from the projectile and the resistance of the air, or from the want of practice or skill in the marksman. The causes of irregularity in firing, although greater in the horizontal than in the vertical direction, are considerable in the latter. They raise or depress the balls, and change the ranges to an appreciable extent.

The general form of the cone of dispersion will present a curved surface, Fig. 154, which is concave outwards; for experiment goes to prove that when a variable cause acts an infinite number of times, the variations of this cause tend to neutralize each other, and we may then assimilate its effects to those of a constant accelerating force, acting in the same manner as gravity. In the same way that we see heavy bodies

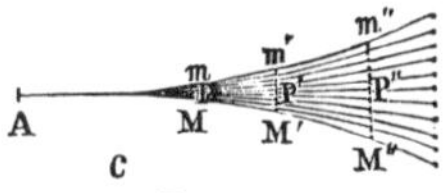

Fig. 154.

moving with an accelerated velocity, so the force with which the balls deviate goes on increasing, whilst the velocity of the ball diminishes. That is, the terms $P\,P'$, $P'\,P''$, &c., are constantly decreasing, whilst $P'\,M'$, $P''\,M''$, &c., are increasing, which will make the curve $A\,M'\,M''$, &c., concave.

The separation of the balls is not in proportion to the ranges; thus, if the distance is doubled, the separation will be more than doubled, and so on.

It is remarked that the cone of dispersion becomes longer, with equal deviations, as the velocity of the balls increases, or that the fire is more accurate as the velocity is increased; a principle known to ancient artillerists, who employed, for small arms, charges much larger than those now in use.

From what precedes, it will be seen that the dispersion of the balls results from causes, some of which are independent of the want of skill in the marksman, and others dependent upon it; that the accuracy of a marksman cannot be properly estimated from a single shot; and that it cannot be affirmed an observed deviation is the result of unskillfulness in the marksman, if the deviation is comprised within the cone of dispersion for that distance.

The objects of instruction in firing are, to analyze the causes which are independent of the marksman, to determine exactly their influence; and to rectify those which are dependent upon him, by imparting to the troops a thorough system of instruction; and for these objects, the influence of schools of practice upon the efficiency of troops, cannot be over-estimated.

The ordinary hausse or breech-sight will give the proper angle, provided the trunnions are horizontal, or the line along which the sighting is done, is in the same vertical plane with the axis of the piece. For, if this is not the case, the line of metal, or sight, marked on the piece, no longer passes through the highest points of the muzzle and breech. It, however, always meets the axis at the same point, and will be projected on the vertical plane of fire in $B'\,C'$, below $F'\,E'$, Fig. 155, which passes through the highest points in the new position of the piece, giving an angle less than the one intended.

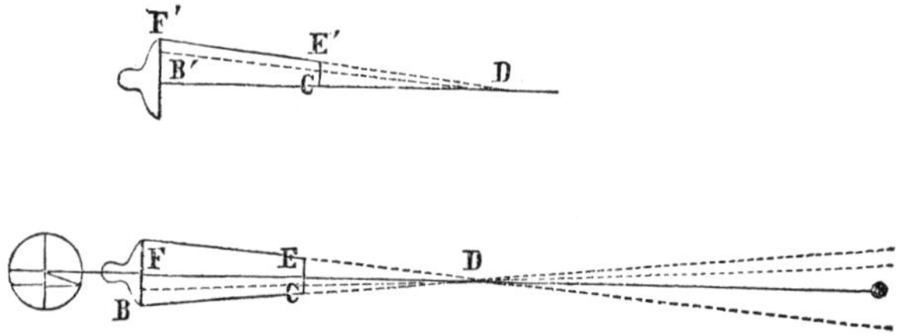

Fig. 155.

The line of sight marked on the gun, cutting the axis at *D*, passes to the opposite side of the plane of fire. Supposing the left wheel the highest, the marked line of sight prolonged, will pass to the left of the vertical plane of fire, and the piece being aimed by that, the shot will go too far to the right, or, on the side of the lower wheel. The amount of deviation will depend upon the difference of level between the two wheels, or in small arms, on the deviation of the sight from the vertical, and on the distance of the object. Therefore, to strike the object, the piece must be aimed high, and on the side of the higher wheel.

If time permit, the best remedy with cannon, for the case just mentioned, is to mark, by means of a spirit-level, the highest points of the breech and muzzle in the new position of the gun. The deviation in range is not so great or of so much importance as the lateral deviation, and the latter is completely obviated in large guns by the use of the Russian or pendulum hausse.

In mortars, where the line of metal is always parallel to the axis, these deviations do not occur, for if the axis of the trunnions is not horizontal, the vertical plane passing through the line of sight, will still be parallel to the vertical plane of fire, and may be taken for it. So, that it is not necessary to have the platform of mortars horizontal.

To determine the length of the hausse for any particular piece, aim at a point, *n*, Fig. 156, a short distance above the mark, and secure the piece in that position. Then mark off, on a line, *A H*, perpendicular to the axis of the piece, a point *H* in the same straight line with

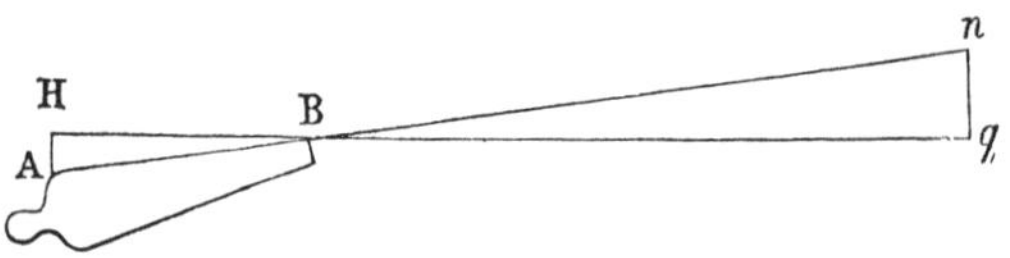

Fig. 156

the highest point of the muzzle, B, and the mark. $A\ H$ will be the height of the hausse for that distance. It is evident that the use of the hausse is equivalent to increasing the angle of fire; for the greater the length given to the hausse, the more must the breech be lowered and the muzzle elevated.

The lateral deviation for any particular piece, when the trunnions are not horizontal, can easily be calculated. Take a 6-pdr. firing at an elevation, by the hausse, of 3°, which will make the angle of fire 3°; call H the total height of the 3° point above the sighting point on the muzzle, that is=the height marked off on the hausse for 3°+the dispart. Let l be the distance between the two points, measured on a line parallel to the axis and passing through the highest point of the muzzle. l is, then, the base of a triangle of which H is the height. Calling φ, Fig. 157, the angle which the trunnions make with the horizontal, H sin. $\varphi=ab$, will be the horizontal distance between the sighting point, b, in its new position, and the vertical plane through l, which may be taken for the plane of fire. Calling D the distance of the mark and x the horizontal deviation at that distance, we will have the proportion

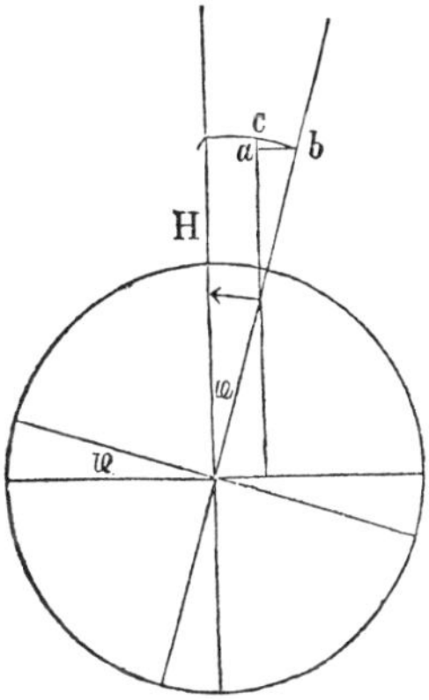

Fig. 157.

$$l : H \text{ sin. } \varphi :: D : x = \frac{H D \text{ sin. } \varphi}{l}.$$

Substituting in this the different values for a 6-pdr. firing at 3° elevation, range 1138 yards, and the trunnions inclined 20°, we have $x=\frac{3.076 \text{ in.} \times 1138 \text{ yds. sin. } 20°}{58.7 \text{ in.}}=73.4$ in., or about 6 feet, for the horizontal deviation at that distance due to an inclination of 20° in the trunnions.

To ascertain what difference of level between the wheels this will give, call it F, and using the *track* of the carriage=60 inches, as a radius, he have $F=R$ sin. $20°=20\frac{1}{2}$ in.

To get the depression of the rear point of the line of sight, ac, it will be equal to $H(1-\text{cos. } \varphi)=2\ H \text{ sin.}^2 \frac{1}{2}\varphi=0.18$ in., and the depression, y, at the distance of 1138 yards will be obtained by the formula $l : 2H \text{ sin.}^2 \frac{1}{2}\varphi :: D : y=\frac{2HD \text{ sin.}^2 \frac{1}{2}\varphi}{l}=13$ in.

WINDAGE. The diameter of the bore being greater than that of the ball, the latter does not travel along the axis of the gun, but ricochets along the bore; and when it leaves the piece, will do so in a direction oblique to the axis, and in a greater angle, as the last point of impact approaches the muzzle. This effect may take place in any direction, but is usually in the vertical plane of fire, increasing or decreasing the angle of fire, which is then called the *angle of departure*, sometimes 30′ greater than the angle of fire.

These ricochets produce a rotation in the ball, which modifies, more or less, the rotation which it naturally takes by rolling along the bottom of the bore. In striking, the point of contact is retarded in its movement by the shock and friction, whilst the opposite point being pushed forward by the gas, a movement of rotation from above to below is produced; and were the point of contact at the side instead of at the bottom, the rotation might take place around a vertical axis.

Any body projected through the air takes, naturally, a movement of rotation, around an axis passing through its center of gravity. The sphere having its center of gravity and figure nearly coincident, will be deviated less from its course by the resistance of the air, than a body of any other form.

Without the resistance of the air, the center of gravity would move the same as if the body did not rotate; but the movement will be different in the air, in consequence of the great resistance offered by it to a body moving with great velocity. This resistance, estimated at nearly 98 times the weight of the ball on first leaving the piece, acts with more force as the movement becomes more irregular, that is, as the mass of air struck and displaced, increases.

If a ball could be made perfectly spherical and homogeneous, the center of gravity would coincide with the center of figure, and the resistance of the air would be reduced simply to friction. As the friction is greatest on the front part, and least on the rear, where the air is rarified, the ball must deviate on the side where it meets with the least resistance.

As it is impossible to make balls perfectly spherical and homogeneous, the centers of gravity and figure must be separated. The impulsive force acts upon the mass, and the point of application may be taken as at the centre of gravity, G, Fig. 158, whilst the resistance of the air acts on the surface of the projectile, and the point of application of the force may be taken at C, the center of figure.

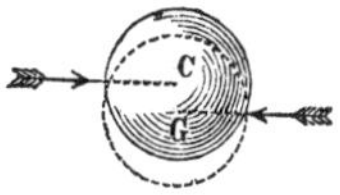

Fig. 158.

These two points not coinciding, the two forces will not be directly opposed, and a movement of rotation will be produced of C around G, or the lighter part of the ball around the heavier, which will be more or less energetic, according to the length of the lever arm, or the distance between the two centers.

This has been clearly demonstrated by using balls, having a portion of the metal bored out on one side, so as to make that consequently lighter than the other. By placing the ball in the gun, with the lightest side up, the rotation takes place from above to below, as is ordinarily the case. This increases the resistance of the air to the top hemisphere where the velocity of rotation is added to that of translation, and decreases it on the under side, where the velocity of rotation acts in opposition to that of translation. The resistance, then, on the upper side is greater than on the lower, the ball is forced towards the ground, and the range is decreased. The more metal taken out of the ball, the closer will it fall to the piece.

By reversing the position, the lightest half of the ball, which is below, moves first, and the rotation is from below to above. The resistance is then greatest below, the ball held up, and the range increased. By placing the light part to the right or left in the bore, the ball deviates very considerably to the left or right of the plane of fire.

A very ingenious instrument to show the unequal pressure on different sides of a rotating body, has been invented by Dr. Magnus, of Berlin, and noticed in "Delobel's Revue de Technologie Militaire." A small cylinder, Fig. 159, with a vertical axis, is placed in front of a fan-wheel, by turning which a current of air is forced against the cylinder. Two light vanes, *a* *a'*, are balanced on points on each side of the cylinder by weights on the opposite sides

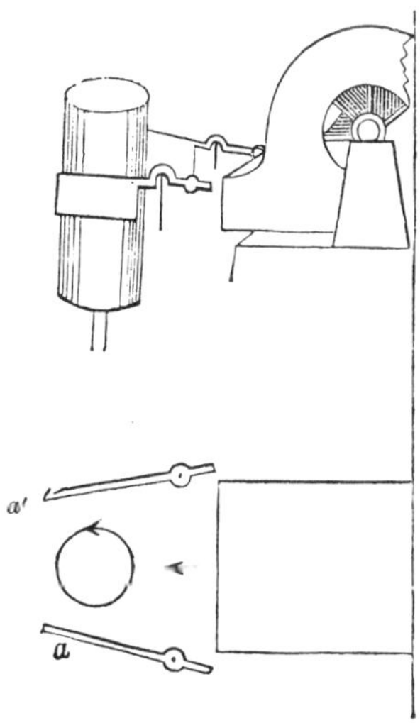

Fig. 159.

of the pivots. So long as the cylinder remains stationary, these little vanes are inclined at the same angle towards it, but the moment a rotary motion is given to the cylinder about its axis, the vanes are deflected at unequal angles, showing unequal pressure of the air, the greatest pressure being on that side where the motion of the cylinder and current are in opposite directions. If, for example, the cylinder be rotating from right to left towards *a'* as indicated by the arrow-head, the vane *a* will show the greatest deflection; and were the cylinder free to move like a ball, it would be pushed, out of its direction, to the left by the increased pressure on the right. This is exactly the effect which experiment shows is produced on a projectile, which has a motion of rotation like the cylinder, and in which the blast from the fan-wheel is replaced by a motion of translation of the body itself.

By another arrangement of the same apparatus, the effect is more distinctly shown by freely suspending a rotating cylinder in front of the fan-wheel, when it is found that it is driven round in a circle, the direction depending upon the direction in which it rotates. The effect is produced in a much greater degree when the cylinder is made to rotate eccentrically.

When the velocity of the current of air is very great in proportion to that of the revolving cylinder, the direction of the vane, Fig. 159, is but little different from what it was when the cylinder was stationary, whereas when the velocity of rotation is only a little inferior to the velocity of the current, one vane approaches very near to the cylinder, whilst the other recedes correspondingly. This corresponds with what is observed in practice; for the greater the velocity of rotation of a projectile in proportion to its velocity of translation, the greater will be the deviation.

This unequal pressure on the opposite sides of a revolving body is explained upon the following principle in regard to the movement of fluids.

"When a fluid jet, endowed with a certain velocity, flows into a fluid mass of the same kind, the pressure exerted perpendicular to the direction in which the jet is moving, is less than would be exerted in that direction were both fluids in a state of rest."

To demonstrate this practically, force a current of air through a tube near a lighted candle. When the velocity of the current reaches a certain point, the flame will be seen to approach the current, and as the velocity increases, it will assume a position nearly perpendicular to the direction of the current, thus showing a decrease of pressure, which is still further decreased by increasing the velocity of the current.

Forcing the current against an immovable cylinder, as explained above, the decrease of pressure on both sides is shown by the approach of the little vanes, but the moment rotation is given to the cylinder, a current of air is carried around with it, which, on one side, increases the velocity of the impinging current, diminishing the pressure, and on the opposite side does exactly the contrary. All the phenomena shown by this ingenious instrument, are known to occur in the practical use of spherical projectiles (not rifled). In rifled projectiles, the same cause of deviation does not exist as in them, the axis of rotation remaining always nearly tangent to the trajectory, the motion of rotation occurs in a direction nearly perpendicular to that of translation; hence, the pressure on all sides is nearly equal, and no deviation takes place. This, also, corresponds with practice, since the deviation or *drift* of rifled spherical balls is found to be almost inappreciable, although in elongated projectiles, to which this explanation of Prof. Magnus is not applied by the author, it is very marked.

DRIFT *in elongated projectiles.* Many explanations have been made with regard to what is call the *drift* (in French *dérivation*) in elongated projectiles, which is found by direct experiment always to take place in the same direction in which the barrel is rifled. That is, looking towards the muzzle, if the groove on the upper side of the bore curves towards the right, the projectile will *drift* towards the right of the observer. If the grooves curve toward the left, the projectile drifts to that side.

It has been proposed to rifle guns to the left instead of to the right, as is now generally done, in order to correct the deviation to the right, caused by the involuntary movement of the marksman's body when pulling the trigger, to which the drift was at first attributed; and it appears reasonable to suppose that this method of rifling would make a piece fire more accurately.

Were the axis of the projectile to remain always tangent to the trajectory, no deviation would occur; since the resistance of the air acting always in the direction of the tangent, the pressure would be equal on all sides, and the only effect would be to lessen the range. But the tendency of the axis is to remain parallel to itself as the ball moves through the air, and it really makes a small angle with the tangent, the point remaining above the trajectory, and the larger portion of the ball below it.

The forward motion of the ball causes a pressure of the air against it, which resists the rotary motion of the lower part of the ball more than does the more rarified air in contact with the upper part. The ball, yielding to the greater force, deviates to the right, in the case of a ball rotating from left to right. This is M. Panôt's theory with regard to drift; but the difference between the resistance above and below the ball is supposed not to be great enough to cause the deviations observed in practice.

Although Prof. Magnus does not apply the principle explained on page 238, to account for the drift observed in oblong rifle balls, and says expressly that it is not applicable to them, yet with some extension it would appear to account for all the phenomena observed.

The axis of the projectile making always a small angle with the tangent to the trajectory, the points on the surface of the right half have an oblique forward motion, which, adding to the motion of translation, the pressure on the right side is greater than on the left, where the motion of rotation is retrograde, and in a measure opposed to that of translation, thus increasing the velocity of the air along the surface, and decreasing the pressure. Under these circumstances, the projectile would yield to the greater pressure and go to the left, but for the fact that the center of gravity being nearest the front end, the rear presents the greatest surface to the action of the pressure, and that part is thrown to the left, whilst the point deviates to the right, causing the projectile to drift to the right. The longer the range, the greater will be the drift; for as soon as the axis points to the right, the resistance of the air on the left side of the point of the ball is increased, which will deviate the point farther and farther from the vertical plane of fire. This deviation is shown, by all the experiments in this country, to be greatest on projectiles which are without grooves; and although these grooves are in some countries, as England, rejected as useless, it is thought that their effect in diminishing the drift can be fully demonstrated.

They were originally placed on the projectile to act as the feathers of an arrow, offering a greater resistance to the action of the air than a smooth surface; and as the resistance of the air acts in the direction of the tangent to the trajectory, the moment the axis of the projectile, from its tendency to remain parallel to itself, ceases to coincide with this tangent, the air acts directly against these little surfaces on one side, whilst on the opposite side the resistance is correspondingly diminished, and the consequence is, the axis is forced back towards the tangent.

Now, according to the theory of Prof. Magnus, as we have applied it to oblong balls, there is no deviation unless the axis is oblique to the tangent; and after it becomes so, the deviation increases until these two lines become perpendicular to each other, when it is a maximium. The axis always makes a small angle with the tangent, and as the effect of the grooves is to lessen this angle, the result is a diminution in the drift. This effect has recently been shown on a large scale at the West Point Foundry, in experiments made with a 12-pd. iron field gun, with the Reed projectile. When made perfectly smooth, the deviation to the right at about 2200 yards was considerable, and was found to be decreased by turning upon the cylinder of the projectile two grooves similar to those used on bullets. The projectile weighed 15 lbs., and an elevation of 8° was used for the distance, with a charge of 1½ lb. of powder.

In firing at great distances, when considerable elevations are used, the angle made by the axis with the tangent is large, and the deviation, by the theory, should increase, as the position of the projectile then approaches more nearly the position in which the difference of pressure on the sides is a maximum. Such in fact is found to be the result in practice; the drift increasing rapidly with the range, and as the angle of fall increases.

If the axis of rotation of a ball be perpendicular to the vertical plane of fire, and both the centers of figure and gravity are in this plane, there will be no deviation from it; but the resistance of the air will have its maximum effect in producing variations in the range, which will become greater or less according to the direction in which the rotation takes place. But this condition of things is very unusual; and the rotation which results from the construction of the ball, combining with that which it receives in the piece, makes the position of the axis very variable, and produces the greatest variations in the firing.

The projectile may deviate in one direction, and then, the rotation in that direction being destroyed and replaced by another, may cut the plane of fire and deviate on the opposite side, describing thus, a curve of double curvature.

It may happen also that, on the descending branch of the trajectory, the axis of rotation may change its direction, and the ball be thrown back to the same side of the plane of fire on which it first deviated, as is often observed in practice.

As the resistance of the air is the cause of the deviations in shot, the more this resistance is increased in proportion to the weight of the shot, the less accurate will the firing be; for the deviating force, acting upon the surface, remains the same for all balls of the same diameter. The resistance offered to a common musket bullet, has been estimated at about 98 times its weight; and if we suppose a ball of the same size but only one half its weight, the resistance which *it* will experience will be $98 \times 2 = 196$ times its weight; and the deviations of course be much greater.

Balls will, then, experience more resistance on the part of the air, and deviate more, the smaller they become.

A bullet half the size of the common musket-bullet, will weigh only $\frac{1}{8}$ as much; and the resistance of the air due to its weight will be $98 \times 8 = 784$. But the surface of the small is only $\frac{1}{4}$ of that of the large bullet; and the initial effect of the resistance upon the small bullet will be $\frac{98 \times 8}{4} = 98 \times 2 = 196$.

All other things being equal, then, the largest, most dense, and most perfectly-formed balls will be the most accurate in their fire.

The movement of rotation from above to below, which the ball usually receives in the piece, tends evidently to make the axis of rotation perpendicular to the plane of fire; thus rendering the fire more accurate, at least as regards lateral deviations.

The velocity of rotation, being much less than that of translation, diminishes but slowly; while the resistance of the air, being at least proportional to the square of the velocity, diminishes rapidly. Hence, the deviating influence of the motion of rotation is greater near the end of the trajectory than at first.

That the velocity of rotation in projectiles is often very great, is shown by the fact that, after losing all their velocity of translation, they are sometimes seen to roll on the surface of the ground; and if any object is interposed to arrest the rotation, that motion is destroyed, wholly or in part; and all the force inherent in the ball is exerted to disengage it; and it will be thrown to a distance sometimes of 250 yards.

The motion of rotation explains how it is that when two similar projectiles are thrown with different velocities but under the same angle, 45° for instance, the one which has the greatest initial velocity will have only an equal and sometimes a less range than the other, if this last has a greater velocity of rotation. The velocity of rotation appears to increase in a greater proportion than the charge.

Each point of one of these balls, revolving around an axis of rotation, describes a kind of epicycloid; and this motion necessarily modifies the path described by the center of gravity of the ball.

If the axis of rotation coincides with the tangent to the trajectory at every point, each point on the surface of the ball describes a spiral; the deviations mutually counterbalance each other, and we have a *rifled* ball where the deviations are the least possible.

The eccentricity in balls and shells is very slight, never reaching in the former $\frac{1}{50}$ of the radius.

For a long time the deviations observed in shot were not attributed to this eccentricity existing in the bodies; and when it became known that this eccentricity was the cause of the resistance of the air producing such great deviations in the firing, every effort was made to make the projectiles as near concentric as possible, with a view to increasing the accuracy of fire.

It was not until quite recently that this very eccentricity was made use of to increase the accuracy and range of projectiles; and it was found that these could even be further increased by magnifying what had heretofore been regarded as the cause of the evils—the distance between the centers of gravity and figure.

Experiment demonstrates that:

Placing the center of gravity to the right in the bore of the piece throws the shot to the right;

Placing it to the left throws the shot to the left.

When placed below, the range is always shortened, and when above always increased.

A 12 pdr., with which experiments were made at Metz, in 1841, by Genl. Paixhans, gave, with its ordinary charge, shot, and elevation of 4°, a range of 1,500 yards. An eccentric shot, with a cavity on one side large enough to reduce the weight from 12 to 10 lbs., gave, with the center of gravity above, a range of 1,960 yards, being an increase of 460 yards; a matter of no small importance.

It was found that, whilst firing in the usual way, the mean deviation was $\frac{13}{100}$ of the range; it was reduced with the eccentric shot to $\frac{8}{100}$; and that the difference between the longest and shortest ranges with the solid shot was $\frac{276}{1000}$; whilst that between the extreme ranges of the eccentric shot was only $\frac{153}{1000}$ of the range. Thus showing a very decided improvement in the regularity of the firing, with a diminution of 2 lbs. in the weight of the shot. The lateral deviations are also decreased from $\frac{16}{1000}$ of the range to $\frac{2}{1000}$.

This discovery gives the means of increasing very much the accuracy of ordinary firing, either at long or short range. It would be very easy in the arsenals to give to the projectile, in fixed ammunition for the field service, the requisite position, and to indicate on the side of the sabot the positions corresponding to long and short range. So also with any projectile, a simple mark on the surface points out to the cannoneer, when loading, the proper position for his shot in order to give it the least deviation and longest or shortest range.

WIND.—The action of the wind on the projectile is another cause of inaccuracy in firing, and produces a greater effect as the projectile increases in size and decreases in density. Thus mortar shells deviate more than those from howitzers, and these more than solid shot.

With elongated projectiles the wind has a greater surface to act on, and produces a greater effect than on spherical balls, though sometimes in an opposite direction. Elongated projectiles are sometimes found to *work up* towards a wind instead of being driven off by it. This is due to the fact that the part of the ball behind, being the lightest, is most easily acted on, and being thrown away from the wind, the point is thrown in the opposite direction giving a deviation *towards* the wind.

The deviations arising from the action of the wind are very

variable, and no rules can be laid down for correcting them. Practice and close observation of the results of firing are the only correctives.

The effect produced on the ball is greater or less, according to the force and direction of the wind. If it acts uniformly and blows perpendicular to the plane of fire, the effect produced may be assimilated to that of a constant accelerating force, and the trajectory becomes a curve of double curvature, whose horizontal projection is convex towards the wind, and the deviations will increase in a greater proportion than the ranges. With small arms, the deviations are further increased by the action of the wind tending to throw aside the muzzle of the piece, especially when the trigger is pressed against.

Supposing the wind very strong (say 60 feet a second), and blowing perpendicular to the plane of fire, experiments show that firing at 160 yards the musket must be pointed on the side of the wind about 11 inches. The deviations increase very nearly as the square of the distance, so that the deviation at 320 yards is $11 \times 4 = 44$ inches, about.

The ball is not only thrown to one side by the action of the wind, but it may be raised or lowered according as the wind acts from beneath or on top.

When mortar shells are thrown to a very great height, they may meet currents of air which will carry them in an opposite direction to those near the surface of the earth, and the shells are seen to wind about in obedience to these currents.

OTHER CAUSES.—The inequalities of the ground between the object and the marksman deceive his view, and diminish the chances of hitting.

When the inclination of the line of sight is very great, and the object within the point-blank, it becomes necessary to sight below the object, in which case the line of sight will meet the ground in front of it. It is difficult to select the particular point to be aimed at, especially if the ground is broken. Experiment shows that firing from one point at an object above, is more exact and easier than in firing downwards.

When a moving object is fired at, the line of sight must be directed upon the point where it is presumed the object will be when the ball has passed over the distance between the piece and the object. For instance, a horseman moving at a gallop (6.66 yards per second), perpendicular to the plane of fire and at a distance of 160 yards, advances 3.33 yards during the half second that the bullet is passing over the 160 yards. The length of a horse being about 10 feet (3.3 yards), it is necessary, in order to be sure of striking him, to sight 1 or 1½ yards in front of his head.

THE MARKSMAN.—The greatest obstacle to precision in firing small arms, is the recoil of the piece; the soldier frequently, after having sighted well, loses the direction by the quick movement he imparts to the piece by pressing against the trigger. As has been already stated, the recoil does not sensibly effect the direction of the shot. The deviations therefore which result from this cause, must be attributed to the unskillfulness of the marksman. The principle object of instruction is, to habituate the soldier to being surprised by the explosion, by pressing gradually upon the trigger.

He is first taught to sight by placing his piece on a rest, as a wooden horse or bag of earth, which gives him sufficient steadiness to enable the instructor to verify and correct the pointing. The visual ray should just graze the rear sight; by placing the eye too high, the angle of sight is increased, and the shot goes too far.

The men should be taught the mechanism and object of the hausse, and be instructed to hold their arms squarely, with the hausse vertical. If the piece is held so that the hausse is inclined, say to the right, the line of sight is thrown on that side, and becomes so much the more faulty as the elevation at which the piece is fired increases. The shot will fall on that side towards which the piece is inclined, that is, in the case supposed, to the right. For the manner of calculating the effects of this error, see p. 235.

The men are also taught to take the easiest and most stable positions, either standing or kneeling; to sight and fire with blank cartridges, preserving immovable both the body and the piece.

To habituate the men to press gradually on the trigger, they are made to fire caps at a lighted candle, placed about 3 inches from the piece. If the piece is directed properly, the jet of gas produced by the cap will extinguish the candle. After this, they should be made to fire blank cartridges.

DISTANCES.—The estimation of distances is a very important part of the instruction for officers, as they will direct so much the better the fire of the men they command, when they are well exercised in measuring distances at a glance. This knowledge is of great advantage to them in manœuvring troops, as well as in time of war. Soldiers thrown out as skirmishers, will have a great advantage over an enemy if they know how to estimate distances with precision, for their fire will then be much more accurate and efficacious.

FRENCH SCHOOLS OF PRACTICE.—In France, each regiment of infantry has a school of practice established under the direction of the Lieut. Col. A captain is especially detailed as instructor of the firing, and is assisted by a subaltern from each battalion. Each company has a sergeant as the instructor of the firing, and an under-instructor.

Records of the firing are kept in each battalion, which are inspected by the Lt. Col., who also presides at a monthly conference at which the chiefs of battalions and captains are present. At these meetings, such portions of the instruction as are indicated by the colonel, are discussed.

The captain instructor gives theoretical instruction to the subalterns.

The subaltern instructors instruct the non-commissioned officers of their battalions, the non-commissioned instructors and their assistants, and have charge of the firing records, and under the superintendence of the captain instructor, and the chief of their battalion.

The non-commissioned instructors have charge of the training of the recruits.

The lieutenant of the armament has charge of the preservation and distribution of the ammunition, and keeping the materials in repair.

If schools of practice for our infantry regiments could be established on some such basis at one or two of the large western posts, now almost abandoned, the results to that arm of the service would be most beneficial.

CHAPTER VIII.

PRACTICE OF FIRE.

In the fire of any kind of arm, the probability of striking a mark is so much the greater as the caliber increases, as the trajectory is less inflected, and as the dimensions of the object fired at exceed the extent of the variations observed in several successive shots, fired as nearly as possible under the same circumstances. When the difference between these variations exceeds the dimensions of the object, the probability of striking it becomes very slight.

Two kinds of variations occur,—one laterally; the other in the direction of the range. The last is of the most importance, for the object generally is to strike within limits extending horizontally more than vertically. The limits in the last direction being 5 feet 5 inches, or 8 feet 4 inches, according as the fire is against foot troops or cavalry.

As the angle of fire increases, the variations in range become less and less up to the angle of greatest range, at which the variations are the least possible, for the kind of piece and charge used. Therefore, the angle of greatest range is the most advantageous in practice, when the object is to plant the greatest number of balls within a given space.

The angle of greatest range would be in vacuo 45°. Practically, it is a little less when firing mortars, but with larger pieces it is between 28° and 35°.

The fire of the ordinary musket is uncertain beyond 200 yards, the variations in height, from one shot to another, at that distance, often exceed one or one and a half yards. But when troops are in compact masses, the fire is still very effective beyond that distance. At 650 yards the musket ball is still deadly, and instances have been known where men have been killed or wounded at greater distances. In general, the infantry soldier, in the excitement of battle, does not make full use of the accuracy of fire of his arm, which has led some to think there is no great advantage in perfecting the accuracy of fire, of arms intended for the use of the mass of troops. But this is evidently a mistake, since out of a number of shots fired from any two pieces, that one which is the most accurate, will have the greatest number effective. So long as the fact remains of the great disparity between the number of balls expended and the number which proves effective, it is useless to give the men the means of firing a greater number of balls in a given time. It is, therefore, not advisable to arm our infantry with breech-loading arms, which, enabling them to expend the limited amount of ammunition carried on the person, in one third or one fourth the usual time, leaves them destitute perhaps at the most critical moment. The best position for such an arm is in the hands of superior marksmen, detached as light infantry, or in deliberate firing from behind obstacles of of some kind.

The effective range of the rifled spherical ball, is over 400 yards. The oblong rifle ball is effective at 1000 yards; but these arms exhibit their marked superiority when used by isolated marksmen.

Cavalry fire is generally of but little effect; though some examples of its efficiency have been exhibited, especially during the Peninsular war.

Solid Shot.—Beyond the limit of distinct vision, all fires become inexact; and infantry should not be fired at beyond 1000 yards, nor cavalry beyond 1200, unless the ground is suitable for ricochet firing, and the enemy's troops are in dense masses.

The precise effect of a single ball cannot be accurately stated. Cases are cited where thirty or forty men have been disabled by a single shot; but it is laid down as a principle, that a 6 or 12 pound ball will go through six men at 80 yards' distance.

As the distance decreases, the 12-pounder loses some of its advantage over the 6-pounder; but it is always superior in its fire, and experiments show that the larger caliber may compensate, by its greater accuracy, for the less number of its shots.

On hard, dry ground, the *rolling* is better than the *direct* fire, at very great distances, as, then, the shot bounds along the ground, striking objects at a number of points on its trajectory. On favorable ground, solid shot from field guns will range as high as 1600 or 1800 yds. The probability of striking an object is the greatest possible at 500 yards, at which distance $\frac{2}{3}$ or $\frac{1}{2}$ of the balls will carry. At 800 yards, the chance of striking an object is reduced to $\frac{1}{3}$ or $\frac{1}{4}$.

If a ball, moving with great velocity, goes through an object which offers little resistance, it does not displace it, and will make but a small hole; but if the ball has but little velocity, it splits and scatters the object into splinters.

A ball, passing through any of the fir-woods, makes a clean round hole, while oak is splintered and broken to a more or less extent. In this respect the former is, then, preferable to oak in positions exposed to the fire of artillery.

Solid shot is of very little effect upon earthen parapets, unless it passes through, in which case it soon knocks them down. At 600 yards, shot from field-guns will penetrate from 5 to $6\frac{1}{2}$ feet into newly thrown-up earth. Siege-guns will penetrate with their shot double this distance.

Shells, although less accurate than solid shot, produce a greater moral effect by their noise and force of explosion. They are, therefore, to be preferred in firing against cavalry, and are used also to reach troops masked by broken ground, or works, and to set fire to block-houses or villages, held by an enemy, and in general, against any wooden structure, which they break up and set fire to.

The large 8-in. and 10-in. howitzer-shells are very destructive either by destroying or setting fire to the shelters of an enemy, or by destroying the parapets and traverses of his batteries, the shells acting, in their explosion, like small mines.

When a shell is burst while stationary, the pieces are dispersed in almost every direction, with more or less force, according to the resistance offered by the sides. But if it is in motion, those parts of it in front will move forward with an increased, and those in rear with a decreased velocity; and in case the shell is moving at the time with very little velocity, this may be entirely overcome for the rear parts by the explosion, and they may drop at once to the ground, or even be thrown backwards.

In forming a good marksman with any arm, the first, and one of the most important steps is to instruct him how to estimate distances. Proficiency in this can be obtained only by constant practice.

The men should be first taught to measure distances by the pace, in which they should be practiced until they learn, in pacing off a distance, to take a uniform step of a yard.

In order to estimate distances by the eye, spaces are measured off, and men are placed at points along the line to show their different appearance at the different distances. For this purpose the ground should be level; persons are then stationed at different points, and the men made to estimate the distances to them. The distances are then paced, and, finally, measured. In these different operations the men are made to notice the different parts of the body and equipments, which become indistinct as the person is removed farther off. On a clear day, and with ordinary sight, at from 190 to 200 yards, every part of a man's body can be seen; and, although the details of dress and figure begin to grow indistinct, the grades of the officers can be recognized at these distances. From 400 to 480 yards, the face can no longer be distinguished, but the head, body, arms and movements, as well as the uniforms and muskets, can.

At 600 yards, the head and upper and lower parts of the body can be made out; and of the uniform, the accouterments and white pantaloons only can be seen.

From 750 to 800 yards, the body appears of an elongated form. Extended arms can be seen in profile, as also the legs of men in motion. The uniform can no longer be distinguished at 900 yards; but the files can still be seen, as well as the movement of troops, and the dust thrown up by a projectile ricocheting on dry ground. From 1100 to 1200 yards, the files can scarcely be distinguished, and the troops appear like solid masses, the movements of which can still be followed.

Stadia. A very simple aid in estimating distances, consists of a small stick, held vertically in the hand at arm's length, and bringing the top of a man's head in line with the top of the stick, noting where a line from the eye of the observer to the feet of the man cuts the stick, or *stadia*, as it is called in France.

To graduate the stadia, a man of the ordinary height of a foot-soldier, say 5 ft. 8 in., is placed at a known distance, say 50 yards; and the distance on the stick covered by him when it is held at arm's length is marked and divided into 8 equal parts. If the distance is now increased, until the man covers only one of these divisions, we know he is at a distance equal to 50 yds. × 8 = 400 yards. This instrument is not very accurate, except for short distances.

A much more accurate stadia is constructed by making use of a metal plate, having a slit in it in the form of an isosceles triangle, the base of which, held at a certain distance from the edge, subtends a man, (5 ft. 8 in.) say at the distance of 100 yards. A slider, *a b*, Fig. 160, moves along the triangle, being always parallel to the base, A B, and the length of it comprised between the two sides of the triangle, represents the height of men at different distances, which are marked in yards on the side of the triangle, above or below, according as the object looked at is a foot or horseman. In order to keep the stadia always at the same distance from the eye, a string is attached to the slider, the opposite end having a knot tied in it, which is held between the teeth while using the instrument, which is held in the right hand, the slider being moved with the left-hand finger. The string should always be kept stretched when the instrument is used, and the line AB in a vertical position.

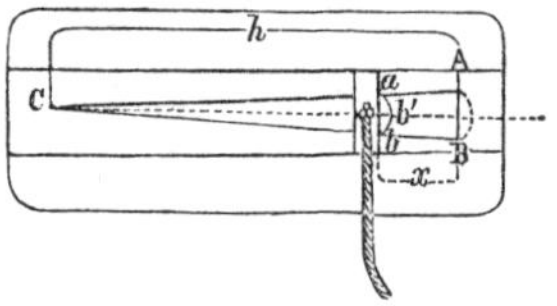

Fig. 160.

It may be graduated, either experimentally, by noting the positions in which the slider *a b* represents the apparent height of a man, at different known distances; or it may be graduated by calculation, as follows:

Calling the base of the triangle A B C, *b*, and the height *h*, the distance of the stadia from the eye *d*, and the distance to the object whose apparent height is *b*, D, H, the real height of the object, *x* the graduation corresponding to any other distance D′, and *b′* the base of the triangle *a* C *b* (Fig. 160), whose height is $h-x$, the similar triangles A O B, and F O G (Fig. 161), will give the proportion $d : b :: D : H$, and consequently $d = \frac{bD}{H}$.

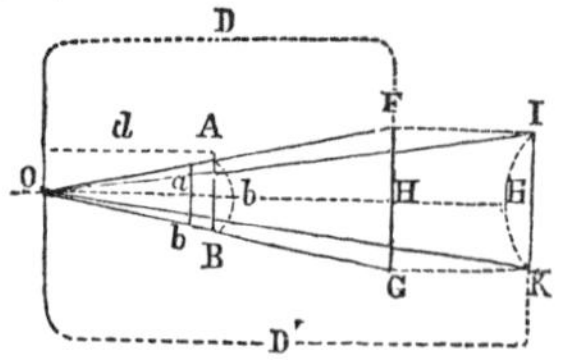

Fig. 161.

In the triangles $a\ O\ b$ and $K\ O\ I$, Figs. 161, we have

$$d : b' :: D' : H \text{ and } d = \frac{b'D'}{H}. \quad \text{Hence, } bD = b'D' \qquad (1)$$

The triangles $A\ C\ B$ and $a\ C\ b$, Fig. 160, give

$$h : h-x : b : b', \text{ and } b' = \frac{b\ (h-x)}{h} \qquad (2)$$

Combining equations 1 and 2, eliminating b', and deducing the value of x, we have $x = \frac{h\ (D'-D)}{D'}$

When $x=o$, we have $b=b'$, and consequently $D=D'$. From the equation $d = \frac{bD}{H}$ the value of b is found $= \frac{dH,}{D}$ which is the equation of the ordinary stadia. It gives $D = \frac{Hd}{b}$, d being the length of the cord, H the height of the object, and b that of the stadia scale.

TRAJECTORY.—Very little was known in early times of the form of the path described by a projectile moving through the air; it being supposed by some a compound curve, and by others a parabola with a vertical axis; and these errors continued until Sir Isaac Newton, towards 1723, demonstrated the modifications produced by the resistance of the air, which he estimated was proportional to the square of the velocity; and Benjamin Robins, the inventor of the ballistic pendulum, showed conclusively that the resistance of the air could not be neglected in computing the trajectory, by proving the range in vacuo many times greater than that in the air.

In a trajectory, the first or ascending branch $A\ B$, is of less and less curvature the nearer we approach the piece, and the greater the velocity. The point B, Fig. 162, where the projectile is at its greatest elevation, is called the *culminating point*. The descending branch $B\ C$ becomes constantly straighter, and approaches more nearly the vertical as the projectile is lighter and loses more of its velocity by the resistance of the air. On horizontal ground the descending branch is always the shortest.

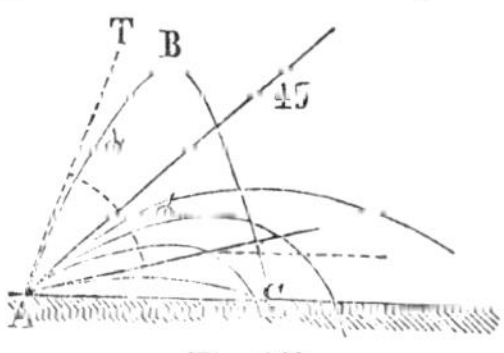

Fig. 162.

The *angle of fire* is the angle $T\ A\ C$, which the axis makes with the horizontal. When the gun and point of fall are on the same level, this angle is less than the *angle of fall*, which is the angle *B C A* made by the tangent to the trajectory with the ground at the point of fall.

As the weight of a projectile increases, and its velocity decreases, the resistance of the air decreases, the curve described becomes more like a parabola, and the *angle of fall* becomes more nearly equal to the *angle of fire*.

Ricochets.—When the angle of fall is small enough, the projectile rises and continues to move on, forming a series of bounds or ricochets, increasing in number as the angle of incidence decreases, the velocity increases, and the ground is more elastic and resisting. The projectile will ricochet under a greater angle on a hard resisting soil, than it will on a soft one. A ricocheting ball makes a furrow in the ground, and each time the angle under which it leaves the ground is greater, than that under which it enters it; for, having lost a portion of its velocity in passing over the first part of the curve, it has no longer the same power to overcome resistance, and must pass out by a shorter path than the one it followed in entering, and consequently the angle is increased, which causes the more or less rapid extinction of the ricochets.

An incompressible surface, like water, is still more favorable for ricochets than ground, and is attended, besides, with less loss of velocity. If the angle of incidence is too great, the ball buries itself in the ground, inclining to the side on which it meets with least resistance.

The kind of fire varies with the ground and the object to be attained.

A *direct fire* is one in which the projectile strikes the mark without touching the intermediate ground. This fire is generally used with *guns*.

A *grazing fire* is where the ball, rising but slightly from the ground, and striking it under a small angle, makes, at the end of its flight, a large number of ricochets. In such a fire, the ball should not strike the ground under a greater angle than 10°.

A *plunging* fire is where a ball, striking under a large angle, buries itself in the ground, losing its velocity entirely, or making only a few ricochets. The least angle under which this takes place is about 10°.

A *horizontal*, *parallel*, or *rolling* fire, is one in which the axis of field pieces is placed horizontal or parallel to the ground. In this case, the line of sight strikes the ground about 75 yards in front of the piece, and the ball meeting it under a very small angle, describes a number of low bounds or ricochets. This kind of fire is more efficient in striking a remote line than the *direct* fire, as the ball in the latter, rising higher, can strike an enemy in only a very small segment of its path. For the horizontal fire, however, a very hard, firm, and level ground is required, such as is very seldom met with on a field of battle.

A *ricochet* fire is used to strike an enemy concealed behind an obstacle, and to break up the ground he occupies, by the successive ricochets of the shot. It is used, especially in sieges, in flank and reverse fires, to destroy the defenses of the besieged, and dismount his guns. The ball having grazed the interior crest of the parapet adjoining the face to be destroyed, falls upon the terre-plein and ricochets along it.

The *angle of arrival* is the angle which the tangent to the trajectory makes with the horizontal at the interior crest; while the *angle of fall* is the angle which the tangent makes at the point of fall with the terre-plein. The size of this angle will determine the kind of ricochet. It will be *grazing* if the angle is 4 or less and *plunging* when it is comprised between 6° and 10°.

Small projectiles seldom ricochet under angles so large as from 7° to 8°. Large shells ricochet under larger angles than large shot.

Designating, Fig. 163, by H the height of the interior crest above the terre-plein, we shall have the equation,—The distance which the ball strikes inside the interior crest

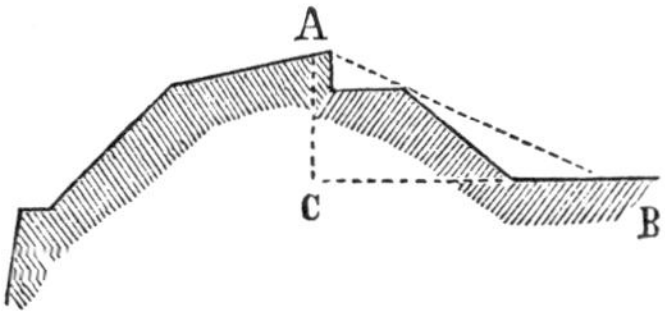

Fig. 163.

$$CB = H\frac{1}{\text{tang. } a} = H \text{ cotang } a;$$ a being the angle of fall. This distance being fixed, the kind of ricochet becomes known.

At great distances, either can be employed, but the grazing is generally used as more efficacious; though the trajectory must have a certain degree of curvature, otherwise many of the shot are lost by passing over the work.

As the projectile should strike the terre-plein as near as possible to the interior crest, it should pass over this on the descending branch of its trajectory; otherwise, continuing to rise, it would pass over the work entirely, or partly, and strike objects only on a small segment of its path, so that at very short distances a more inflected curve is necessary.

Upon terre-pleins the limit of ricochets seldom exceeds 10° unless the ground is very dry and hard. A ball will ricochet better in dry summer weather than in wet.

In *grazing* ricochets, the ball is given a great velocity, and the curves described are long and flat. In *plunging* ricochets the contrary is the case.

Solid shot, acting by their force of percussion alone, are not very effective when they have but little velocity. *Guns*, therefore, should not be employed under 200 yards, as it would be necessary to decrease the charges too much for ricochet firing. Howitzers can be used advantageously at the shortest distances; the size of the shells and their force of explosion supplying their want of velocity.

The ricochet firing with field pieces differs but little from a rolling fire; when this fire is used against works, it resembles that used in the attack of fortifications.

The *fire of greatest range*, is that obtained by giving the piece the greatest elevation it can take on its carriage, and employing the largest charge used for that caliber; but as this kind of fire is very injurious to the carriage, and often breaks it, it is employed only in testing.

Horizontal fires, are those in which the axis of the piece has small angles of elevation. *Vertical fires*, are those where the angle is very great, as in mortars.

The *rules of fire* for ordnance which fires horizontally, results from the knowledge of the *mean trajectory*, which, like the mean trajectory of the musket, comprises all the causes of error which can modify the fire; but the trajectory of artillery projectiles being more constant in its form, the results are more accurate, a consequence of the fact that as the caliber of the arm increases, each particular trajectory approaches more nearly the mean trajectory.

When canister is fired, the balls, having broken the case, escape in the form of a cone, and striking against each other, and against the sides of the piece, diverge more or less according to the distance; but the greater part of the balls are found to be near the center, and the rules of fire are still based upon the knowledge of the mean trajectory.

For *vertical* fires, the rules are based upon the knowledge of the mean range, and the relation which can be established between charges and the corresponding ranges.

The ranges of pieces of the same caliber, fired under the same angle and with the same charge, will be different from various causes, among which are,—

The different diameters of the bore and vent, from the widening of these parts.

The difference in the diameter and weight of the projectiles.

The difference in the quality of the powder, and especially in its hygrometric state, and the manner in which it inflames and communicates motion to the projectile.

The thermometric, barometric, and hygrometric state of the air; a variation in the temperature and pressure, increasing or diminishing sensibly the range.

The vibrations of the piece and the projectiles, especially when the latter are hollow, are not without their influence, particularly when the piece is formed of a thin and elastic tube; and it is from this last reason that pieces whose recoil has been almost entirely prevented, fire less accurately than those which are allowed to recoil freely. The vibrations are so much the more violent as the charge and length of piece are increased; and experiment proves that a musket fires less accurately when solidly fixed, than when fired from the shoulder.

Loading.—The manner of loading the different kinds of pieces being given in the tactics, it will be necessary here, simply to lay down some general principles.

In long pieces, the *vent* should always be kept carefully closed, while the loading is going on, especially when sponging, to prevent the current of air from passing out and collecting there pieces of thread, paper, &c., from the cartridge bag, which would retain fire in the gun, and cause premature explosion the next time the gun was loaded. This precaution is the more necessary, when the sponge fits the bore tight, and acts as a piston. The sponge should be well pressed down against the bottom of the bore, and turned, so as to leave no remnant of the cartridge bag. In mortars, where a sponge is seldom used, or when it does not fit tightly, the stopping of the vent is not necessary; but it should always be cleared out with the priming wire before the powder is placed in. Mortar-shells should be let down gently so as not to be forced into the chamber, or crush suddenly any powder they may meet.

The use of sabots is avoided when firing over the heads of our own men.

It may sometimes become necessary to fire a shell from a mortar too large for it; in which case it is wedged in on different sides with pieces of soft wood, and the space between it and the bore filled in with earth.

Large shells may be fired from guns or howitzers by placing the breech in a hole 2 or 3 feet deep, and resting the piece lengthways on two pieces of timber, inclined about 45°, the elevation being regulated by a quoin under the chase. The shell is then placed upon the mouth of the piece, and held there by a cord, which passes around it and is fastened to one around the neck. In this way the shell closes the mouth of the piece. The loading is as easily and as quickly performed as in mortars, and the firing is tolerably accurate. To insure the fuze taking fire, it is primed with a piece of quick-match.

For want of shells to set fire to works in the field, hot shot may be used, the forge serving as a furnace to heat them in. Canister shot may be replaced by filling pasteboard cylinders, having heads of elm, 1½ inch thick, with musket balls placed in beds, and sealed with good plaster; the weight of the whole not to exceed that of the shot, and to be fired with a charge of $\frac{1}{5}$ of that weight, and not at a greater range than 300 or 400 yards.

Charges.—The charges to be used depend upon the kind of fire, the purpose of it, the nature of the ground, object fired at, &c.

In mortars, the usual angle of elevation is 45°, generally a little less; and the charges vary according to the distances. The charges are rectified by knowing the increased range corresponding to a given increase in charge.

In early times two men were required in firing a mortar, one to fire the piece, the other to light the fuze, which could not take fire from the blast, earth being packed in to destroy the windage, and with an idea that it prevented the shell from being broken. This method of firing gave rise to many accidents. Should the piece miss fire, the shell burst in it, and if the fuze failed to take fire before the charge, it did not burst at all. As soon as it became known that the fuze could be lit by the blast of the charge, the packing of earth was discontinued, and the present method of firing was adopted.

Long pieces were at first loaded with loose powder, carried to its place at the bottom of the bore by means of a long-handled ladle; and although cartridge bags were sometimes used, it was not until Gribeauval substituted the use of them entirely that they were permanently adopted. To this celebrated Frenchman are we also indebted for the use of the hausse, or breech-sight, previously almost entirely unknown. Before its adoption, pieces were fired with but little accuracy, except at point-blank range; and it is singular that so many objections should have been urged against the use of an instrument now so indispensable to the proper sighting of artillery.

In *field guns*, when firing solid shot, the charge is usually about $\frac{1}{5}$ the weight of the shot. For spherical case and canister, the charge is less. These projectiles are always fixed to a block of wood, called a *sabot* (Fig. 164), to which the cartridge is also attached; forming what is called a round of *fixed ammunition*. Fig. 165.

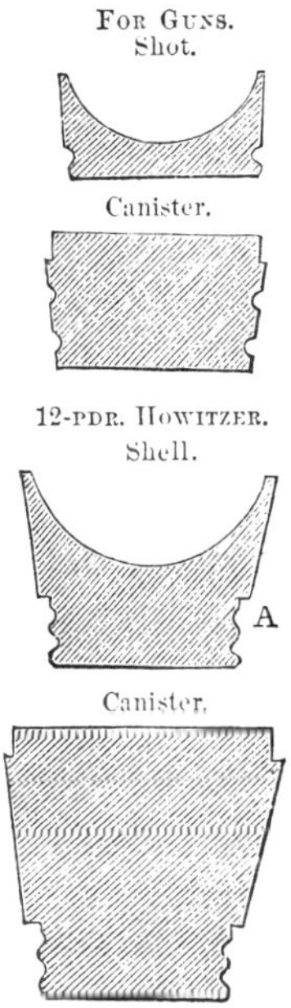

Fig. 164.

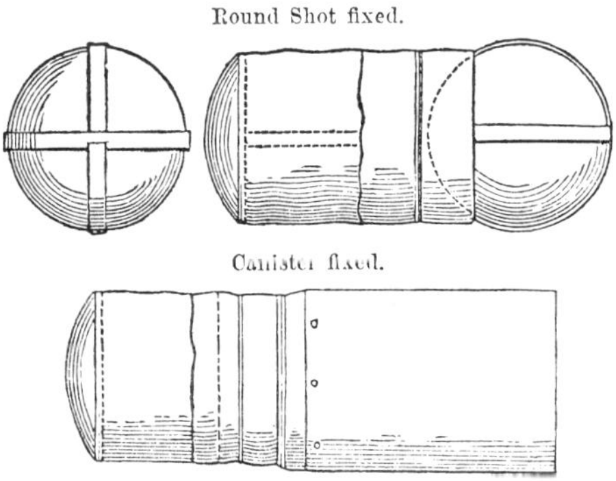

Fig. 165.

In the 12-pdr. field-howitzer, also, the ammunition used is

fixed, A, Fig. 164; but with the other howitzers the projectile and charge are separate, the latter being attached to a block of wood called a *cartridge-block*, Fig. 166, the object of which is to give a finish to the cartridge and fill the chamber, the dimensions of the block being so calculated for each different charge as to reach to the mouth of the chamber. The sabots used with these heavy howitzers are conical in shape to fit the connecting surface between the chamber and bore.

Cartridge Block.

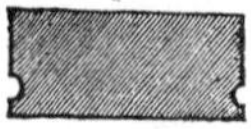

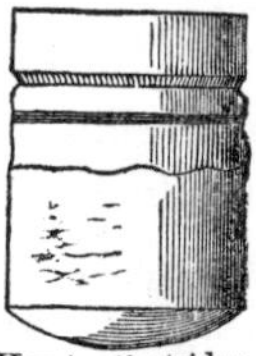

Howtz. Cartridge.
Fig. 166.

Care should be taken in loading to put the seam of the cartridge to the sides, so that it will not come under the vent.

In loading the 32 and 24-pdr. howitzer, the cartridge is first pushed carefully into the chamber without ramming, and the shell is then set home, also without ramming.

In loading the mountain-howitzer, it is best to keep the sponge moistened, especially when the piece is fired rapidly.

For the *siege and garrison* guns, the charge is $\frac{1}{4}$ the weight of the shot, which is increased in breeching to $\frac{1}{3}$, and sometimes to $\frac{1}{2}$. The French, whose guns of this size are made of bronze, use hay-wads between the charge and shot, varying the length of wad so as to prevent the formation of lodgments in the bore. Wads are not usually employed in our service. For firing hot shot, cartridge-bags are made double, by placing one within the other, care being taken to prevent the powder sifting out; and wads of wet hay or clay are used.

To Load with Hot Shot.—The piece should be sponged with great care, and the worm frequently passed into the bore. As a precaution, it is well to insert a wet sponge just before putting in the ball.

The muzzle being sufficiently elevated to allow the ball to roll down the bore, the cartridge is inserted, the mouth of the outer bag foremost, the fold down, and carefully pushed home without breaking it; a dry hay wad is placed upon it and rammed once; then a clay or wet hay wad and rammed twice; and finally, if firing at angles of depression, a wad of clay a half-caliber in length, or a wet hay wad, is put on the ball.

The charges for hot shot are from $\frac{1}{4}$ to $\frac{1}{6}$ the weight of the shot. With small velocities, the shot splits and splinters the wood, so as to render it favorable for burning. With great velocity, the ball sinks deep into the wood, is deprived of air by the closing of the hole, and chars instead of burning the surrounding wood. It should not penetrate deeper than 10 or 12 inches. Red-hot balls do not set fire to the wood until some time after their penetration. They retain sufficient heat to ignite wood after having made several ricochets upon water.

The wads are made of clay or hay. Clay wads should consist of pure clay, or fuller's earth, free from sand or gravel, well kneaded with just enough moisture to work well. They are cylindrical, and one caliber in length.

Hay wads should remain in the tub to soak, at least ten or fifteen minutes. Before being used, the water is pressed out of them.

When hay wads are used, vapor may be seen escaping from the vent on the insertion of the ball; but as this is only the effect of the heat of the ball on the water contained in the wad, no danger need be apprehended from it.

With proper precautions in loading, the ball may be permitted to cool in the gun without igniting the charge. The piece, however, should be fired with as little delay as possible, as the vapor would diminish the strength of the powder.

Furnaces for Heating Shot are erected at the forts on the sea-coast. These furnaces hold sixty or more shot. The shot being placed, and the furnace cold, it requires one hour and fifteen minutes to heat them to a red heat; but after the furnace is once heated, a 24-pdr. shot is brought to a red heat in twenty-five minutes; the 32-pdr. and 42-pdr. shot require a few minutes longer. Three men are required to attend the furnace; one takes out the hot shot, and places them on the stand to be scraped; another scrapes them and puts them in the ladle; and the third supplies cold shot and fuel.

Grates for Heating Shot.—In siege and other batteries, where there are no furnaces, a grate is used for heating shot. This grate consists of four bars 1.75 inch square, three feet long, placed four inches apart on three iron stands one foot in height. It is placed in an excavation one foot in depth, of the width of the grate, perpendicular at the back and side, open in front, the legs resting on bricks or stones rising about four or five inches from the bottom. A roof is made over it with hoops of flat iron, covered with sods and eighteen inches of earth, having in the back part a chimney six inches square.

The shot are placed on the back part of the grate, leaving one fourth of front part free; and under and on the front part the wood is put, cut in pieces about fourteen inches long and two inches thick. A thick sod is used as a register, to regulate the draught of the chimney, so that no flame can issue from the front. This grate, which will contain about fifteen 24-pdr. balls, heats them to a red heat in an hour, and will supply three guns. It requires the attendance of one man.

Heated shot do not return when cooled to their original dimensions, but retain a permanent enlargement.

Hot shot have been almost entirely given up since the adoption of the method of throwing large hollow shot from long pieces. These require but little preparation, can be used at once, and are more terrible in their effect, at the same time that they are about as accurate, since in order to prevent too great a penetration, the charge with hot shot has to be reduced.

Paixhan Guns.—In 1822, Lieut.-Col. Paixhan, of the French artillery, submitted, for the first time, his plan for throwing large heavy shells from long chambered guns (canons-*à*-bombes), in the same way that solid shot is thrown. Up to that time, shells fired from long pieces had been limited to the smaller calibers; and it remained for Paixhan to prove, after the greatest opposition on the part of others, that it was as practicable and almost as easy to throw shells to a great distance with slight elevations, as to throw shot.

His piece, which was provided with a chamber, seems to have been designed more especially for the navy, though its importance as a *defense* against shipping in sea-coast defense, was early perceived and taken advantage of; and his pieces, under the name of Paixhan guns, have been adopted in almost every service; and we are at this day witnessing the fulfillment of some of his predictions regarding the change likely to take place in the size and motive power of naval vessels from the adoption of this new arm.

The pieces are loaded like the guns, with the additional precaution of placing the fuze towards the muzzle, and either in the axis of the piece or above it, according to the manner in which the sabot is fixed. Sometimes, instead of using a sabot, the shell is prevented from turning in the bore by having attached to it a *ring* or *grommet* wad. Shells should not be rammed, but pushed gently home with the rammer.

The 8-*in. siege howitzer*, being loaded by hand, the fuze can be placed in its proper position without the assistance of a sabot, which, when the piece was fired in the trenches over the heads of the working parties, would be dangerous to them. The surface connecting the bore with the chamber is therefore made spherical, so that the shell will fit down over the mouth of the chamber.

In loading this piece, as well as with the mortars, care should be taken that no powder remains along the bore; and in firing under small angles, or at a depression, the shell should be wedged up, to prevent its rolling out.

In the *stone mortar*, the mouth of the chamber is covered with a thick oak plank having a number of holes bored through it, to allow the passage of the flame; and when shells are fired, they are laid in beds with their fuzes placed outwards, and cut to burn 15 seconds. The mortar is primed with a piece of quick-match long enough to allow the cannoneers to get out of the way of any fragments which might fall in the battery. The same precaution is not necessary when stones or balls are fired, though the mortar is then given an elevation of 60° instead of 30°. If the charge is too great, the projectiles are disseminated over too great a space. It should not exceed 1½ lbs. with a weight of projectile of 100 lbs.

In RICOCHET FIRING the charges are very variable, according to the distance of the mark and the kind of ricochet to be used. For grazing ricochets the charges are large, and the gun is fired at small elevations. For plunging ricochets, the reverse is the case. In general, the charges are very nearly inversely proportional to the *total elevation*, that is, to the height on the hausse + the dispart. The smallest charge for a gun with solid shot should not be less than $\frac{1}{40}$ of its weight.

POINTING. To *point* a piece is, to direct its axis on a mark and give it the elevation necessary for the ball to strike the point intended. The line of sight is usually marked upon the piece, so that the firing will be accurate only when this line is in the vertical plane of fire and the trajectory remains in this plane.

LONG GUNS. The *point-blank range* increases with the velocity, size, and density of the projectile.

The manner of pointing a piece to strike an object at, within, and beyond point-blank range, and the advantages of the hausse, have already been explained.

The line of sight and point blank obtained when the hausse is used are called *artificial*, in opposition to those obtained without it, which are called the *natural* point blank and line of sight.

The effect of the hausse, or breech-sight, is evident. The apparent increased diameter of the breech obliges us to elevate the piece, in order to have the new line of sight directed on a given object. This increases the angle of fire, and, consequently, the range.

For firing with point-blank range, the angle of sight may be diminished by using the hausse on the chase, or between the handles ; but this is seldom done.

With guns and howitzers, by means of the hausse, both the elevation and direction are given at the same time; but with mortars, which are fired at a great elevation, this cannot be done, and the piece is first given its elevation with the quadrant, either by inserting the arm in the bore, or placing it against the face of the piece, and afterwards the direction is given.

MORTARS are fired either over the parapet or through embrasures. In the former case, to establish the vertical plane of fire passing through the object, a short stake is driven in the interior crest, as near as possible opposite the middle of the platform. Sighting by this stake a long one is planted about one yard in front, and a cord being attached to the top of it, is stretched to the rear of the piece over the head of the first stake, and a third planted directly under the cord one yard in rear of the platform. The large stake is then removed, and the cord is attached to the small stake; and, when stretched over the rear one, gives the proper direction of the mark, and is not so liable to be disarranged by the enemy's shot as if attached to the large stake. When so stretched, a plumb-line attached to it will determine the vertical plane of fire passing through the object, and the center of the platform. The axis of the piece being placed in this plane, by means of the line of metal, will evidently have the proper direction; but should the shells fall habitually out of line, the direction of the piece is corrected by means of the *pointing-board*, a strip of wood with a notch on one side to fit on the rear stake, and graduated both ways from the middle, by means of which the mortar may be directed farther to the right or the left. Should the shell fall too far to the right, the cord is put to the right of the rear stake, and by varying the position, several times in succession, the proper direction will be finally reached.

If the mortar is to be fired through an embrasure, drive a stake in the middle of the embrasure, near the interior crest, and attach a cord to it, and a second in rear of the platform, as near as possible in line with the first, and the object; the last one being high, to clear the mortar with the cord, and allow the sighting. By means of the cord, and a pointing-board nailed to the rear stake, correct the shots as before. The cord is kept in its place by a little picket, which is attached to the end, and driven in rear of the rear stake.

Another method for directing a mortar, is for the gunner, with a plumb-line, to mount any object in rear of the platform, and cause a stake to be driven near the interior crest, and another in rear of the platform, in line with the first, and the object. Then, with the cord and pointing-board, rectify the firing as before.

The *angle of elevation* of a mark, is the angle which the line drawn from the object to the breech of the piece makes with the horizontal.

When the object fired at and the mortar are not on the same level, the angle of greatest range, instead of being 45° is $45° \pm \frac{1}{2}$ the angle of elevation or depression of the object. Thus to reach an object elevated 15° above a mortar, the angle of greatest range would be $45° + 7\frac{1}{2}° = 52\frac{1}{2}°$; while, if the object was depressed 15°, the angle would be $45° - 7\frac{1}{2} = 37\frac{1}{2}°$.

The advantages of the angle of greatest range are:—1. Economy of powder. 2. Diminution of recoil, and less strain on the piece and its platform. 3. More uniformity in the ranges. 4. Differences of a few degrees above or below cause only a slight difference in the ranges.

Should the distance to the object be very short, and the velocity of the shells not great enough on the descending branch, they must be fired under an angle of 60° ; when, of course, greater charges must be used for the same distance, the ranges obtained with 60° elevation being about 1-10th less than those with 45°. A shell fired at 60° elevation, as it goes higher and obtains a greater velocity in falling, will strike more directly and with greater violence upon the vaults of the magazines and shelters of the enemy, and will have a greater force of penetration.

For want of howitzers, mortars may be used for ricochet firing. The least angle of elevation for mortars on their beds being 30°, and the ricochet angle being but 15° at most, the rear of the bed has to be raised up.

Stone mortars are pointed in the same way as other mortars, but the angle of elevation, when firing stone, is 60°, in order that the stones may have a greater force of percussion. In firing grenades, the angle is about 33°, to prevent the shells from having too great a force in falling, by which they would be buried in the ground, and their explosion become less effective.

Ricochet and vertical fires being used to batter a surface, and not an isolated point, as is the case with direct fires, can be used at night as well as in the day-time.

Guns and howitzers are fixed for night-firing by measuring the distance of the handle of the elevating screw about the screw-box, or cutting a stick which will fit between the breech and trail. The direction is fixed by nailing down on the platform two strips of wood, one outside one of the wheels, the other along the trail, at such distances as to allow between them and the wheel and trail the interposition of movable strips, which are taken up when the piece is ready for firing, to prevent the stationary ones from being torn up by the recoil.

With mortars, the elevation is fixed by marking the quoin, or by tacking a strip upright on the bolster, so as to touch the chase. The direction of the bed is established by one movable and one fixed strip of wood along one of its cheeks.

Garrison and sea-coast carriages are the easiest to arrange for night-firing, the direction being fixed by simply chocking the wheels of the chassis.

Night-firing is evidently not so accurate as day-firing, but with due care is very efficient.

SOLID SHOT FROM GUNS.—The effect of solid shot increases with the caliber and velocity. As the shot acts simply by its force of striking, the quantity of motion imparted to it should be as great as possible.

For any given piece and charge, the probability of striking an object is the greatest possible at the point-blank range, and diminishes more and more beyond this. The largest pieces have the greatest point-blank ranges, and are, consequently, for any distance, the most effective, and may, from this fact, compensate for their numerical inferiority.

Beyond the limit of distinct vision the fire loses a great part of its efficiency, so that infantry should not be fired at much beyond 1,000 yards, nor cavalry at more that 1,200, unless the ground be very favorable for ricochet firing, or the enemy is in dense masses, and in favorable positions, as, for instance, in column.

When the ground is hard and dry, the rolling fire is more suitable than the direct, for great distances; as with the latter the shot strikes but at one point, and is too high to strike intermediate points.

When the ground is favorable for ricochets, the shot from field guns will range as high as 1700 and 1900 yards, after making a number of more or less extensive bounds. The probability of striking, however, is greatest at about 500 yards, at which distance two thirds or one half of the shots will range. At 800 yards the chance of striking is reduced to one third or one fourth.

A ball striking against a scarp, or wall, makes in it a funnel-shaped opening terminated by a cylindrical one, the mouth of which, *A*, Fig. 167, is from five to eight times the diameter of the ball. The shock, besides, splits and breaks the masonry in a circle varying from four to five feet in diameter for the largest calibers.

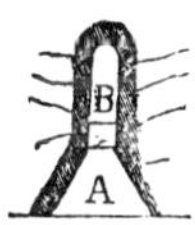

Fig. 167.

Shot from field guns will easily penetrate walls from one and a half feet to two feet thick; but good masonry four feet thick is safe against them, unless a regular breach is made, which is a long operation. Experiments show that breaches can be successfully effected with 12-pounders.

SHELLS, in consequence of irregularities attending their construction, are less exact in their fire than shot, especially at great distances, their effect being estimated at two thirds that of the corresponding shot. In consequence of their noise and explosion, however, their moral effect is greater than with shot, and they are preferred, therefore, in firing against cavalry. Their extreme range is sometimes as great as 2000 yards. The number of ricochets they make is very variable, and depends on the nature of the ground. The 24 and 32 pound shells break into eighteen or nineteen deadly pieces, which are sometimes thrown 600 yards.

Shells may be made use of to open breaches in intrenchments, as on penetrating and exploding, they act like small mines. It is estimated that for every two pounds of powder which they contain two cubic yards of earth are displaced. The 32-pdr. howitzer, on account of its greater accuracy, is more suitable for this pur pose than the 24. These, with their maximum charges, send their shells from four to five feet into earth newly thrown up.

Shells are very effective against block-houses and all other wooden structures, which they knock down, break to pieces, and set on fire.

The total range of the mountain howitzer sometimes reaches 1200 yards, after the shell has ricocheted three or four times. The 12-pound shell breaks into twelve or fifteen pieces, which are sometimes thrown 300 yards.

The shells of the 8-inch siege howitzer, in consequence of their great weight, and the quantity of powder contained in them, 2 lbs. 9 oz., are very destructive, either by destroying and setting fire to the enemy's shelters, or by destroying his parapets, traverses, or épaulments. In sea-coast defenses, these shells are formidable against ships over 3000 yards distant. The number of effective splinters is from twenty-eight to thirty, and when filled with musket balls (486) (schrapnell), they become still more destructive.

SCHRAPNELL shot have only enough powder placed in them to rupture the case and set the balls free. Consequently, their effect is greater the greater the velocity they have at the moment of bursting. 12-pound schrapnell are effective up to 800 yards.

GRAPE and CANISTER shot leave the piece diverging from each

other, in the form of a cone, the greater part of the balls being in the center, and the extreme ones separating about one tenth of the range. When fired at too short a distance, the balls occupy too small a space to produce the proper effect; and at too great a distance they diverge too much, and strike on too extended a surface. Good results can be obtained at from 300 to 600 yards, but the maximum effect is produced at from 400 to 450 yards.

When firing at very short distances over hard, dry ground, a suitable dispersion of the balls may be produced by firing very low, and allowing the balls to ricochet.

The range and effect of grape-shot become greater, at ordinary distances, as the size of the balls is increased. For this reason, the grape from 12-pdr. guns and 32-pdr. howitzers, is more effective than that from 6-pdr. guns and 12 or 24 pdr. howitzers, and the 6-pdr. grape is no longer used in the United States.

The fire of grape-shot from the mountain howitzer is not of much effect over 300 yards.

If the balls ricochet over hard, dry ground, grape-shot may be effective up to 800 yards, and the pointing is of but little importance; but with broken or soft ground, ricocheting cannot be depended on, and the angle of elevation should be regulated with care.

Mortar Shells.—These shells are designed to force the vaults or magazine roofs of an enemy, and to set fire to them, as also to destroy his blindages and shelters. They act by their explosion as well as by percussion. They break up the terre-pleins and ditches, destroy the "pas de souris" and other communications, and end with making communication between different parts of the work very difficult.

As these shells penetrate to a considerable depth, their splinters do not produce much execution if the shells bury themselves in the ground before bursting, as some of the pieces then remain in the ground, and the rest come out at so great an angle that they are not dangerous.

The advantages in mortar-firing are on the side of the besiegers. For the works of the besieged present a great mark, which it is not difficult to hit; whilst the batteries and works of the besiegers form but narrow bands around the work, which are difficult to strike, especially as the deviations in the direction of the range are much greater in mortars than the lateral deviations.

The greatest difficulty in firing these shells is to regulate the charges properly. Very great differences are found to exist between ranges obtained under the same circumstances, and these increase with the range, whilst the lateral deviations are much less. Firing at a bastion or demi-line at 600 yards, scarcely one fifth of the shells will fall into the work; at 400 yards, about one third.

The best method of firing these shells is to direct the mortars and place the batteries in such a way, that the shells will pass over the works fired at in the direction of their greatest extent.

The 8-in. shell being the same as that used for the siege howitzer, gives from the mortar the same number of fragments (30), some of which are thrown to 600 yards. The 10 and 13-in., in consequence of their greater thickness, do not form so many fragments. When firing against men, it is best to use small angles of elevation, and 8-in. shells are preferable to the larger calibers.

These shells will penetrate compact earth from 15 to 38 inches at 600 yards. The earth displaced is in the same proportion as for howitzer shells, two cubic yards for every two pounds of powder contained in the shell. The diameter of the top of the hole made is usually two or three times the depth. Sometimes the largest shells are broken by falling on a stone pavement. Good masonry arches, three feet thick, are sufficient to resist the largest shells falling with their maximum velocity.

Mortar-firing at sea is very uncertain, but when shells do strike a vessel, the effect is terrible. A 10 or 13-in. shell falling upon a three-decker, will sometimes pass entirely through and sink the vessel.

Light and fire balls, according to their size, are fired from mortars of corresponding calibers. With a charge of $\frac{1}{25}$ of its weight, they are thrown from 600 to 700 yards.

Time of Flight.—The time of flight for siege mortars, at an elevation of 45°, with ordinary charges, is nearly equal to the square root of the range in feet divided by four. The experimental length of the fuze may be given according to this rule.

To Ascertain the Distance by the Report of Fire-Arms.—Multiply the number of seconds which elapse between seeing the flash and hearing the report by 1100; the product will be nearly the distance in feet.

The smallest angle under which siege mortars can be fired without raising the rear part of the bed, is 30°.

The coehorn mortar is either fired from behind entrenchments, like other mortars, or it may accompany troops in effecting lodgements in towns and fortified places.

As the shell is without ears, it should be strapped with tin, having loops attached, through which a cord is passed for the purpose of lowering it into the bore. The chamber being cylindrical, a sponge is used.

In firing stone mortars, the stones or shells leave the piece, as in firing grape-shot, in the form of a cone which meets the ground in a more or less irregular section, from 30 to 50 yards wide by from 60 to 90 yards long. Some of the grenades fall near the battery, but the greater part are found within a circle of from 20 to 30 yards in diameter. Each grenade bursts into 12 or 15 fragments, which are effective within a radius of 10 to 20 yards, and some even as far as 300 yards.

Rapidity of Firing.—Siege mortars can be fired conveniently at the rate of twelve rounds an hour continuously; but they may, in case of need, be fired with greater rapidity. In general, the rate of firing depends on the caliber of the piece. A 24-pdr. takes about double the time to load and fire it that a 12-pdr. does.

For field guns, 30 or 40 seconds are required for the 6-pdr., and one minute for a 12-pdr. The mean rate is about one shot per minute, but when close pressed, and firing at objects not difficult to hit, two or three shots per minute can be fired.

The firing of howitzers is always slower than with guns, a field howitzer requiring 1½ minute to load and fire it.

The mean rate of siege guns is about 8 shots an hour, though this can be increased on an emergency to 12 shots; and in breaching, as many as 20 shots per hour have been fired. But this rapid firing is very injurious to pieces, especially bronze guns, which heat rapidly, soften, and lose their resistance.

Rockets can be fired more rapidly than any other kind of projectile; and if large quantities of them are on hand, they constitute the best means of setting fire to towns. They are fired from troughs or sheet-iron tubes, the former being the most rapid but the least accurate means. The elevations are given with a quadrant, and the direction with a plumb-line.

Recoil. The recoil of the carriages varies between very extended limits, according to the nature of the ground. Generally, howitzers recoil more than the corresponding guns, and the 6-pdr. recoils less than the 12-pdr. The 24 and 32 pdr. howitzers fired with large charges sometimes recoil 10 yards; whilst under some circumstances the 6-pdr. gun will recoil but 1½ or 2 yards. The mountain howitzer will recoil 11 or 12 yards, but may be limited to 4 by using a rope on the wheels.

When the piece is on a platform, as with siege pieces, the recoil is seldom more than 4 yards, unless the platform is wet, when it increases from the reduction of friction on the trail.

The recoil on garrison carriages is very slight, as is also the case with mortar-beds.

Penetration.—For the penetration of shot and shell, see the Ordnance Manual, p. 368 to 373.

It is remarked in firing leaden bullets into earth or water, that they are flattened the more as the velocity is increased. This affects their penetration, so that fired at a distance of 50 yards the penetration is greater than at a distance of 25 yards. The same effect is noticed, though not to the same extent, in firing into other materials. Beyond a certain limit, the flattening of the ball does not sensibly affect the penetration.

Tables of Fire.—Tables of fire should be considered simply as a means of limiting the number of trials to be made before finding out the range of a gun, and not as giving accurate results which will be the same under all circumstances. The first shots always produce the greatest effect on an enemy; and it often happens in battle that the smoke and dust prevent subsequent rectifications in the firing, unless a certain amount of information is possessed in regard to the effects of the piece. In giving this information, the tables of fire are of use.

If, in experimental firing, great differences are observed between two successive shots, fired with the same charge, same powder, from the same piece, under the same angle and at well-known distances,—we should not be surprised to find differences between the tables of fire and the results obtained in the presence of an enemy. These tables, then, give simply the first term of a series of trials, by the aid of which a skillful gunner arrives at accuracy in pointing. This first term is, however, often of the greatest importance, since, by knowing it, a great deal of time is gained.

For these tables for ordnance, see Appendix, pp. 40 to 48; and for the initial velocities and penetration of small-arms, pp. 49, 50, and 51.

BREACHING.—A *breach* is an opening made in the wall of a fortified place, to facilitate the entrance of a storming party.

In permanent works, the wall being formed of masonry, little if any part of which is exposed to the fire of artillery at any great distance, it becomes necessary to establish breaching batteries, for the purpose of knocking it down. These batteries are established in the covered way, or on the crest of the glacis, where the guns can be brought to bear upon the wall.

Before the use of iron as a material for balls, stone was used, which, not being so resisting, broke against the solid wall without producing much damage; and the wall had to be knocked down by commencing at the top, where the least resistance was offered, and continuing to fire down the face until sufficient debris was formed to make a ramp from the bottom of the ditch. This method was very laborious, and generally formed a very steep and often impracticable slope.

The effect of an iron ball upon masonry is noticed at page 267, Fig. 167. The ball, when it strikes, is acted upon by the elasticity developed by the shock; and is sometimes thrown back 150 yards, so as to be dangerous to those in the battery. Pieces of masonry are also sometimes thrown from the wall 50 or 60 yards. The opening made and penetration depend upon the kind of material of which the wall is made, and upon the caliber and velocity of the projectiles used.

If the axis of the piece is perpendicular, or nearly so, to the face of the wall, no difficulty is experienced in making the ball penetrate; but where the ball strikes the wall at a much smaller angle, it is not so easily done. Against solid masonry, a ball striking under a less angle than 33°, will glance off without producing much effect; and when no greater angle can be obtained, the same spot must be fired at until a hole is started. Then the cut is made with more ease, though with nothing like the rapidity with which it is effected when firing directly against the face of the wall.

The breaching batteries having been established, several pieces in each, the number depending upon their caliber and the size of the intended breach, the first thing to do is to make a horizontal cut in the wall. The determination of the proper position for this cut is a matter of importance, as on it depends in a great measure the practicability of the breach. The ramp formed of the falling masonry and earth diminishes as the height of the cut is increased; so that if the cut is made too high, a part of the scarp wall will remain standing above it, and if it cannot be ascended will require a long time to knock down.

If, on the contrary, the cut is made too low, the amount of material for forming the ramp is increased, and may mask the openings first made before the cut is completed.

Hence, the most suitable height is that which will furnish about sufficient material to reach the cut. This height is determined to be nearly equal to the thickness of the wall at the cut. This thickness, when not previously known, can be deduced from the dimensions necessary to be given to the wall to resist the pressure of the earth of the rampart and parapet.

A profile may be constructed, by ascertaining as near as possible the width of the ditch, that of the covered way, the height of the scarp, the thickness of the parapet, the height of the counter-scarp, and crest of the covered way. From this profile, the position of the cut is determined, so that the resultant ramp shall have a slope of 45°. The height of this cut should never be less than $\frac{1}{4}$ the height of the scarp. If the ditch is a wet one, the cut should be made at the water's edge.

The number of pieces in the battery, and the length of the breach to be made, will determine the field of fire of each piece, and the length of cut to be made by it.

The angle of depression for each piece should be determined as near as possible beforehand, and when once satisfactorily obtained, should be marked on the elevating screw.

Each piece commences to fire on the right or left of its portion of the cut, spacing its shot about 1½ yards for 24 pds., and 1 yard for 18 pds., the successive positions of the trail and wheels being marked on the platform. Returning, the piece is fired at the intervals between the shots in the first round, the positions of the piece being marked as before. This firing is continued back and forth, firing at the most prominent points in the cut, until it is finished, which is known to be the case when the earth falls through, throughout its length. Fig. 168.

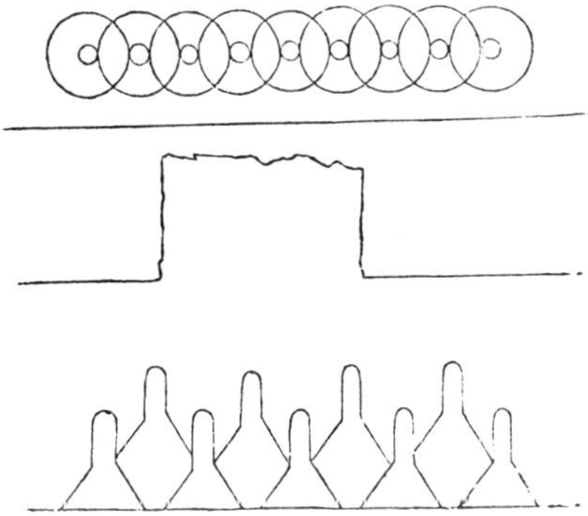

Fig. 168.

Frequent inspections are made to see that the cut progresses uniformly; and if some portions are formed more rapidly than others, adjacent pieces may be turned against the slowest parts.

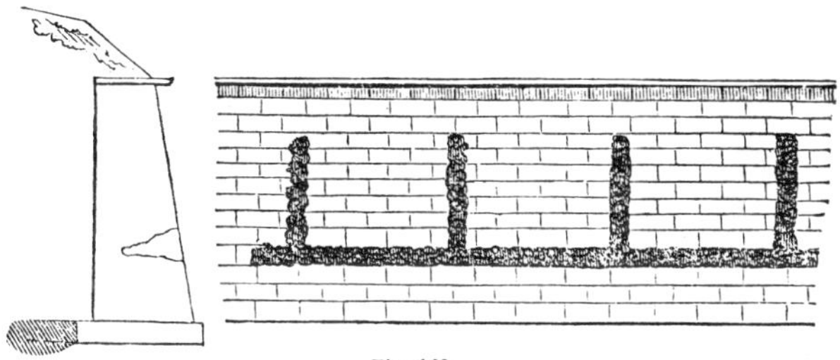

Fig· 169.

The vertical cuts, Fig. 169, are made between the horizontal cut and the top of the wall. There must be one at each end of the breach, and intermediate ones at a distance apart of not more than 10 yards, and closer if possible, so that no part of the wall may be sustained by more than one counterfort. The number of these cuts should not exceed the number of pieces in the battery, so that each one may be fired at by at least one piece. The firing is commenced just above the horizontal cut, and continued upwards, slowly at first, so as not to encumber the cuts with the falling material. The shots are spaced at first about half a yard apart, and after that the salient points fired at. The extreme cuts should progress at least as rapidly as the intermediate ones, so that the whole weight of the wall will come upon the ends, and assist in bringing it down. If necessary, the adjoining guns are directed upon the extreme cuts.

If the wall does not fall by the time the vertical cuts are finished, a few volleys fired against the spaces between the cuts will bring it down. The firing is then directed against the counterforts, which are knocked to pieces. If the earth is still sustained by its tenacity, loaded shells are fired into it. These, acting like small mines, soon bring down the earth on top of the masonry, and form a practicable breach. Fig. 170.

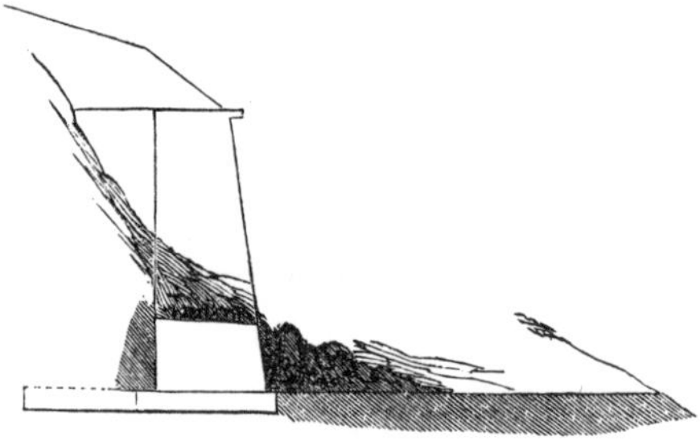

Fig. 170.

If time and other circumstances permit, these shells are fired from 8-inch howitzers, which replace the breaching guns; but if this cannot be done, shells are fired from the guns themselves.

The time necessary to make a practicable breach depends upon the length of it, the material of which the scarp is made, the number and caliber of the guns, &c. For a breach 20 or 30 yards long, at 40 yards from the battery, 1,500 shot of large caliber are required; but when the firing distance is greater, the estimate must be larger, on account of the less penetration and accuracy of each shot. Thus at 500 to 600 yards, 9000 or 10,000 projectiles, or even more, may be required.

CHAPTER IX.

FUZES.

A Fuze is the contrivance by which fire is communicated to the charge in a shell. It consists, essentially, of a highly inflammable composition, inclosed in a wood, metal, or paper case.

Wooden.—The oldest form has a wooden case, and is represented in the annexed drawing, Fig. 171. It consists of a conical plug of wood, of the proper size for the fuze-hole of the shell with which it is to be fired. The axis of this plug is bored out cylindrically, from the large down to within a short distance of the small end, which is left solid. At the large end a cup is hollowed out, and the outside of the plug is divided into inches and parts, generally tenths, commencing at the bottom of the *cup*. The cylindrical space is filled with composition, pounded hard, and as regularly as possible, and the cup filled with mealed powder, moistened with whisky or alcohol. The rate of burning is determined by experiment, and marked on a waterproof paper-cap, which is tied over the cup. Knowing the time any shell is to occupy in its flight, the fuze is cut off with a saw at the proper division, and firmly set in the fuze-hole with a fuze-set and mallet. Say the fuze burns 5″ to the inch. If a shell is 10″ in reaching the mark, two inches of fuze will burst it as it strikes. If it takes 8″, 1 in. and $\frac{6}{10}$ will be cut off; and so on.

Fig. 171.

The great disadvantage of this fuze is its irregularity, it being very difficult to pound the composition so that equal lengths will burn in equal times. So that the shell will either burst too soon, and a great part of its effect be lost; or it will burst after burying itself in the ground; when, from the high angle at which the pieces leave, they are rendered harmless to the enemy; or the shell will burst after it passes the proper point. This irregularity of burning, is common to all fuzes where the composition is driven in successive layers in a column which burns in the same direction.

Paper. The *Paper-Fuze*, Fig. 172, is a better form, and consists of a conical paper case, containing the composition, whose rate of burning is shown by the color of the case, as follows:

Black	Burns	2″ to the inch.
Red	"	3″ " "
Green	"	4″ " "
Yellow	"	5″ " "

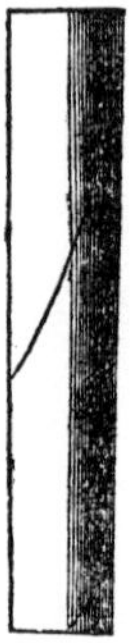

Fig. 172.

Each fuze is made two inches long, and the yellow burns, consequently, 10″. For any shorter time, the fuze is cut with a sharp knife. With this fuze, is used a fuze-plug, Fig. 173, having a conical opening, which is reamed out to fit the paper-case, when the shell is loaded, and the fuze is then pressed in with the thumb. The great advantage of this fuze is its simplicity; and the little trouble required to place it in the shell, which renders the numerous and complicated implements formerly used in field artillery, such as saws, fuze-setter and extractors, files, &c., unnecessary.

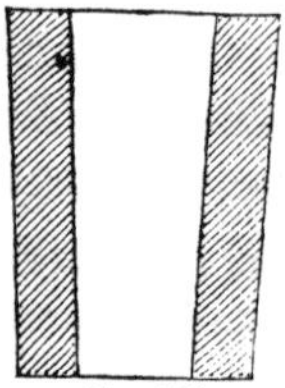

Fig. 173.

The wooden fuzes are still used for very long ranges, as in sea coast mortars. They are sometimes bored through at the proper position, instead of being sawed. When the fuze is so long as to render it likely that it will reach the bottom of the shell, it is cut obliquely, like a whistle; as by cutting it perpendicular to the axis, the whole base of the wood might be driven in contact with the bottom of the shell, and prevent the lighted composition from setting fire to the bursting-charge.

Boxer, Fig. 174. The English have modified the wooden fuze, in such a way as is said to render it more regular and certain in its effects.

The channel for the composition is placed slightly on one side of the axis, and filled with composition, as before, and two other smaller channels are made from the lower end, and filled with fine powder. Along the fuze a number of small holes are bored, communicating with these channels; the holes in one set being opposite to the intervals between those in the other set. These holes are numbered from the top, so as to designate the distance of each from the top of the column of composition. The main column communicates at the bottom with one of the small columns; and the two small ones communicate with each other at the same point. The holes are bored at such a distance apart, that planes passed through the axes of two consecutive ones, perpendicular to the axis of the fuze, will contain between them $\frac{1}{10}$ of an inch of composition, which takes about $\frac{6}{10}$ of a second to burn. The height of the column, exclusive of the priming, is one inch; so that the shell may be exploded at ten different points of its trajectory. The scale-holes are closed with pipe-clay, and numbered; those on the left, with even numbers, those on the right, with odd ones, commencing with 3; all painted red, and relieved, the first on a black ground, the second on a white. The paint preserves the wood, and the different colors prevent mistakes, by rendering the figures more conspicuous. The exact position of the holes is indicated by white or black points, according to the ground.

The cup of the fuze at the top is primed with mealed powder and quick-match, and the head of the fuze covered with a paper cap, to which is fixed a piece of tape to uncap it; and over this, is placed a tin cap.

This fuze, invented by Capt. Boxer, is designed specially for use with a kind of schrapnell, invented by him, in which a sheet-iron partition separates the powder from the balls, and a copper fuze-plug is screwed into the eye to receive the fuze, which is not put in until just before firing, like our paper-fuze. The size of these fuzes is the same for all calibers of schrapnells. Larger ones have been similarly made, and applied to mortar and other shell firing, where a greater time of flight is required.

To use the fuze, bore out with a gimlet the hole corresponding to the distance at which the shell is required to burst, boring, not only through the pipe-clay, but through the column of powder and the wood, which separates this from the fuze composition. Remove, also, in all cases, the pipe-clay from the hole marked 2.

The fuze is now placed in the plug, taking care to insert it in such a way, that the column marked with the even figures shall correspond with an opening left in the side of the fuze plug, by means of which fire passes in to the charge.

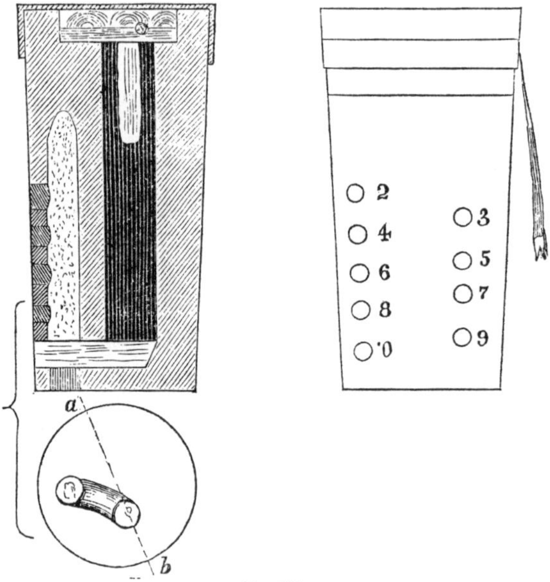

Fig. 174.

The tin cap is then removed, the paper one torn off, and the shell inserted in the piece.

When the fuze is fired, it burns down to the bored hole, through which the flame passes into the powder column, and communicates through the hole 2 with the charge in the shell, should it fail to communicate through the hole bored out.

These fuzes are said to have given very satisfactory results in experiments; but they appear entirely too complicated for ordinary use. The parts are numerous and complicated, requiring nice adjustment at the time of using; and no reason is seen why they should burn any more regularly than the paper fuze; whilst the certainty of exploding the shell appears to be much greater in this last, which also must be much the cheapest.

This fuze has been applied to the ordinary field shells, differing only in having a greater length, and a more extended graduation, and in not having the two columns of powder joined at the base, each of which transmits the flame directly to the charge.

The English use with their large mortar shells, a wooden fuze similar to ours, but differently graduated. It is divided into intervals, by grooves cut round the case, each one of which burns one second; and these are further divided into five parts, by holes bored through the wood and running in a spiral around the fuze. These holes are stopped with pipe-clay, which is removed as in the Boxer fuze, with a gimlet, just before inserting the fuze in the shell.

BRASS.—In consequence of the rapid deterioration of wooden fuzes at sea, the English navy has adopted metal ones altogether; but the French still hesitate, having found that the metal is raised, by the burning composition, to such a temperature as to explode the charge prematurely. The Russians early adopted the metal, and proved at Sinope, both the terrific power of the Paixhan shells, and the good quality of these fuzes.

The English fuzes are made of brass, Fig. 175, and are of three sizes, according to the calibers with which they are to be used. The fuze screws into the eye of the shell, the projecting head being covered with a cap in which is placed a leather cushion, pushed down by a brass-wire spring to close the mouth of the fuze, and prevent the fuze dust from sifting into the thread of the cap screw, causing accidental explosions when the cap is put on or taken off.

The inside of the fuze-case is cut into a number of small transverse grooves, which hold the composition firmly, and also prevent the flame from passing down along the sides of the composition.

Fig. 175 represents the smallest size of these fuzes. It is 1½ inch long, burns 2 seconds, and is used for distances within 600 yards.

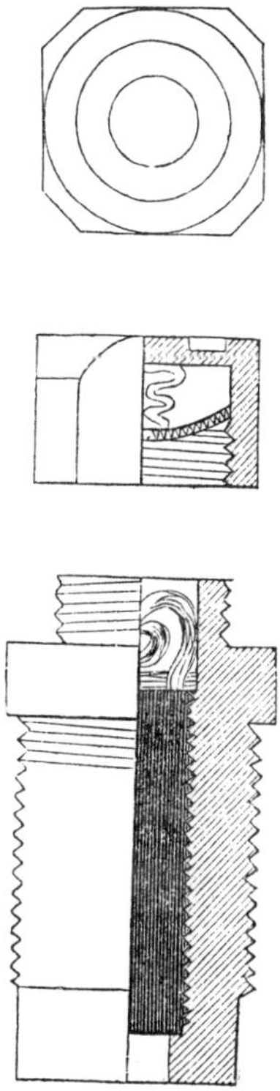

Fig. 175.

The other sizes are as follows,—

No. 1, 4 in. long, burns 20 seconds, used beyond 1900 yards.
" 2, 3 " " " 7½ " used between 600 and 700 yards.

The composition used in No. 1, is 13 nitre, 4 sulphur, and 11 mealed powder; in No. 2, mealed powder; and in No. 3, both of these are used.

On the outside of these fuzes are graduating lines; and the

column of composition is shortened to suit intermediate distances, by boring out from the lower end.

The cap is not taken off until the shell is placed in the piece; and we are warned not to unscrew the fuze itself at the same time, thus leaving the shell liable to explode in the bore; but this could be easily remedied by cutting the two screws in opposite directions.

SERRA FUZE, Fig. 176. This fuze consists of a cast-iron case, A (with a flat head), which is screwed into the eye of the shell to receive the fuze proper, B, which is made of bronze, and is screwed into the case when required, having for that purpose a square head. The end of the iron fuze-plug projects slightly into the shell, and is hollowed out and notched, in order to receive a priming to render the explosion of the charge more certain.

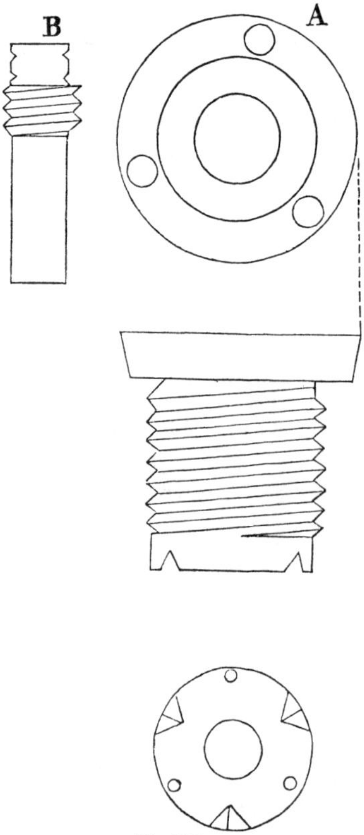

Fig. 176.

The fuze is filled with meal, powder, compactly and regularly driven. It is designed for use with schrapnell shot, and each caliber has three lengths, corresponding to the three most important distances, and differing in the time of burning, $\frac{4}{5}$ of a second in the 12-pounder, and one second in the 24-pd. (15^c), so that for the former piece, the fuzes would burn 2″, $2\frac{4}{5}$″, and $3\frac{3}{5}$″, and for the latter piece, $2\frac{2}{5}$″, $3\frac{2}{5}$″, and $4\frac{2}{5}$″.

As the fuze is not placed in its position until just before firing, the hole in the fuze-plug is kept closed with a tow wad, which has to be removed before inserting the fuze. This fuze is well spoken of, and has given satisfactory results.

It is recommended, as an improvement, that the plug or case be made of bronze, instead of cast iron, to avoid the danger of explosion when screwing the parts in place, to say nothing of the

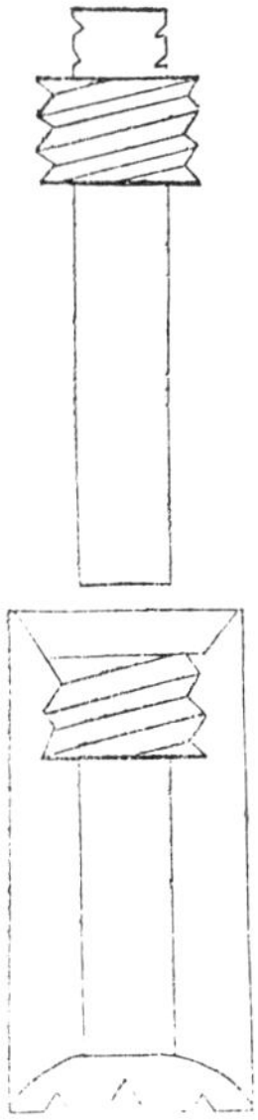

Fig. 177.

liability of the iron to rust, and render the unloading of the shell difficult.

This fuze has undergone some modifications, which are thought to be improvements over the original fuze of Major Serra. The plug, Fig. 177, is made smaller, with no screw on its exterior, which is instead made slightly conical, and is driven into the fuze-hole, like an ordinary wooden fuze. Instead of iron, it is made of a mixture of tin, lead, and zinc, one part of each. It can be applied to any shell without the trouble and expense of cutting a screw in the eye, consequently can be used with the shells already manufactured, without any change. It is less costly than that of Major Serra, and the same danger when inserting it does not exist.

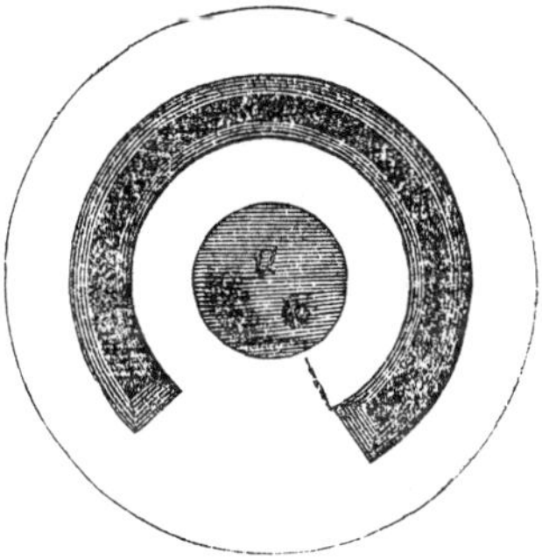

Fig. 178.

The Belgium or Boarmann Fuze, invented by Capt. Boarmann, of the Belgian army, is a great improvement on the old fuze, and by far the best and most regular of any now in use. The essential improvement consists in applying the composition pressure on the *side*, and burning it from the end, which does away with the irregularity resulting from driving and burning in layers.

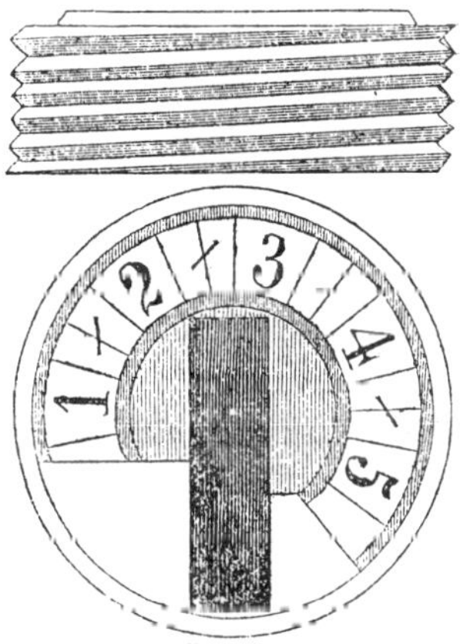

Fig. 179.

The fuze-case is made of metal (a composition of lead and tin), and consists, Fig. 178, first, of a short cylinder, having at one end a horse-shoe-shaped indentation, *one* end only of which communicates with the magazine of the fuze placed in the center. This horse-shoe indentation extends nearly to the other end of the cylinder, a thin layer of the metal only intervening. This is graduated on the outside into equal parts, representing seconds and quarter seconds, as represented in Fig. 179. In the bottom of this channel a smooth layer of the composition is placed, with a piece of wick or yarn underneath it. On this is placed the piece of metal represented in Fig. 180, the cross-section of it being wedge-shaped; and this is by machinery pressed down upon the composition, sealing it hermetically. The cylindrical opening represented at *a*, Fig. 178, is filled with fine powder, and covered with a sheet of tin, which is soldered in its place closing the magazine from the external air.

Before using the fuze, several holes are punched through this sheet of tin, to allow the flame to escape into the shell. On the side of the fuze the thread of a screw is cut which fits into one cut on the inside of the fuze-hole, and the fuze is screwed into the shell with a wrench, the projecting part of which fits into the indentation at *b*, Fig. 179.

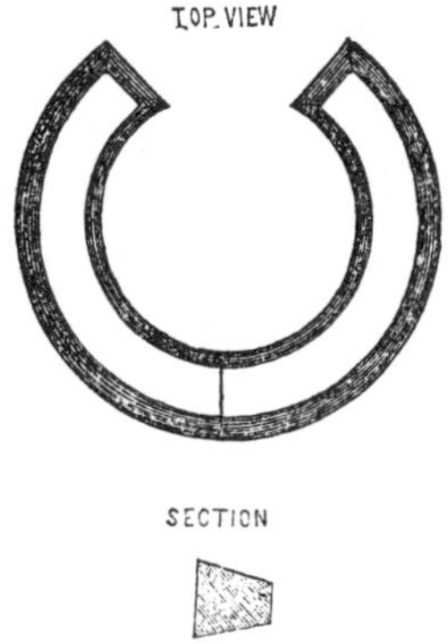

Fig. 180.

The thin layer of metal over the composition is cut away with a gouge or chisel of any kind, at the interval marked with the number of seconds which we wish the fuze to burn.

The metal of this fuze being soft, there is danger of its being driven into the shell by the explosive force of the charge. To prevent this, a circular piece of iron, of a less diameter than the fuze, with a hole through its center, and the thread of a screw on the outside, Fig. 181, is screwed into the fuze-hole before the fuze is placed in.

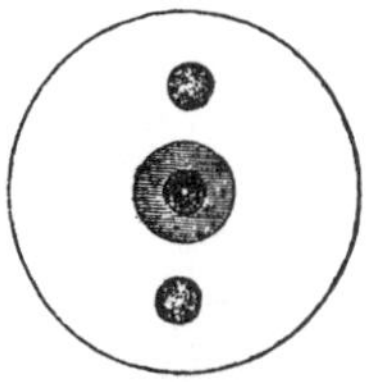
Fig. 181.

The regularity and certainty of this fuze are very great; and its use has, so far, been principally confined to light artillery in firing shells, and particularly schrapnell, in which these two requisites are so essential; but it has been applied to larger ordnance, with every promise of complete success.

One of the most important advantages of this fuze is, the fact that the shells can be loaded, all ready for use, and remain so any length of time, perfectly safe from explosion; as the fuze can be screwed into its place, and the composition never exposed to external fire until the metal is cut through. The only operation, then, to be performed under fire, is to gouge through the metal at the proper point, which may be done with any kind of a chisel, knife, or other instrument. It would be well, however, to examine some of the fuzes, to see that care has been taken to punch holes through the plate of tin which covers the magazine, as the powder contained in it might not be sufficient to blow this cover off.

This admirable fuze, which the Belgians have been experimenting with and improving for the last twenty years, is now acknowledged by all to be the most perfect of its kind extant. It has been subjected to all kinds of trials, and failed in none; and this can be said of no other fuze yet invented.

In 1852, they were submitted to a series of experiments in France, during which a number of schrapnells, with the fuzes not cut for bursting, were fired and afterwards recovered, and again fired with the fuzes properly cut. The results demonstrated that the fuzes resist completely all the shock which the projectile receives, either when in the bore or when ricocheting on the ground, without injury and without detaching itself from the shell. In order to demonstrate this fact more clearly, several shots were fired with a *rolling fire*, without cutting the fuzes, and the same fuzes were fired over again after duly regulating them, giving the most satisfactory results.

This fuze, however, did not hold undisputed the first place among fuzes in Europe; by far its most prominent rival being one invented by Capt. Splingard, of the Belgian army. Whether the former has lost anything by the comparison, after receiving "the baptism of the field of battle," I have not, at present, the means of ascertaining. In this country, the Boarmann fuze is the only one of the two in use, and the other almost entirely unknown.

Baden Fuze, Fig. 182. The Boarmann fuze has undergone several modifications, one of the best of which was made in Baden.

The metal, form of the parts, &c., are the same as in the Boarmann fuze, with the addition of a bronze stud and plate, used to confine a piece of quick-match in the recess shown at *b*, Fig. 179, and two pieces of common cartridge paper, cut in the proper shape to fit the horseshoe-shaped opening, one of which goes above, the other below, the charge of composition, to prevent it, when burning, from melting the metal above and below it.

The graduated scale is more extended than in Boarman's, the numbers expressing, instead of seconds, the bursting distances in hundreds of paces of 29½ inches in length ($0^m.75$).

The *charge-cover* is cast in a mould; the plate to cover the priming is cut out with scissors, and the paper is stamped or cut out with a stamp or punch.

The vent leading from the composition-channel to the magazine is closed, to prevent the entrance of composition when pressure is applied.

One of the pieces of paper is now laid in the bottom of the channel, and on it is placed evenly a layer of mealed powder, weighing exactly (3.29grm.) (51 grains Troy), on top of this, the second piece of paper, and then the charge-cover, which is pressed down to its place by machinery, having on its under surface the dies to impress the figures of the graduating scale, on the metal. The amount of pressure exerted is about 5000 lbs.; which forces the top of the charge-cover slightly below the surface of the fuze, forming thus a shallow annular hollow, in which is poured some of the melted metallic mixture, but at as low a temperature as possible, in order that the heat of this solder may not act on the sulphur in the fuze. When cool, the fuze is placed in a turning lathe, dressed off smooth, and any irregularities in the thread of the screw corrected. The vent is then unstopped, the magazine filled with fine powder, the cover placed on, fastened, and pierced.

Loaded.—Top view.

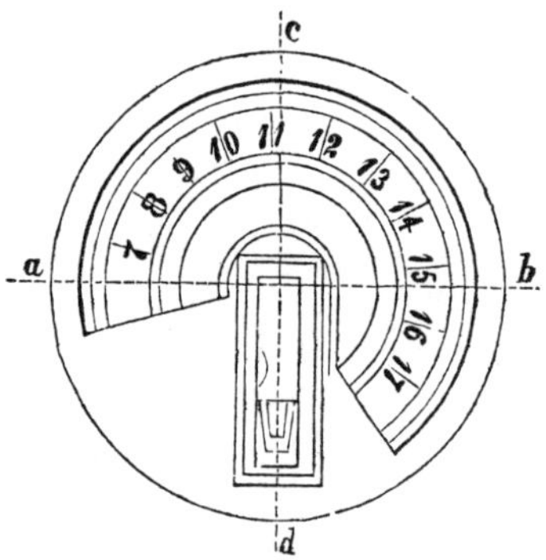

Sect. on *a b*.

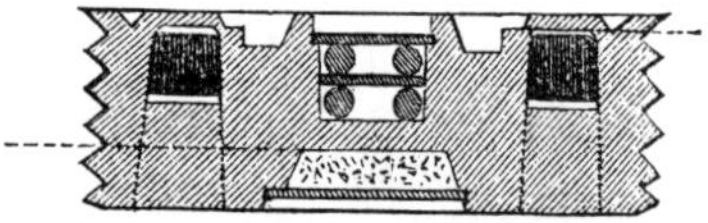

Sect. on *c d*.

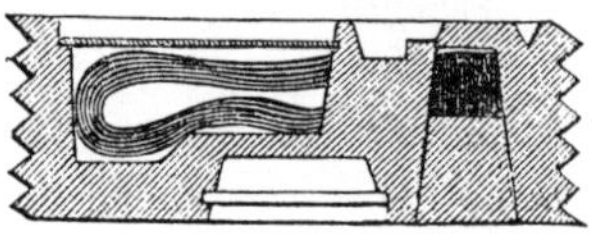

Fig. 182.

The fuze should burn 5½ seconds; and to test it, it is tried with a chronometer. The fuzes are the same for the three field-pieces,—6-pd., 12-pd., and 24-pd. howitzer (15^c); but the scale of graduation differs for each one, and is based upon the velocity of the projectile at different points of its trajectory; so that the maximum distance of fire for the schrapnells with which this fuze is designed to be used, is fixed at

1700 paces, (1393 yards), for the 6 pounder.
1800 " (1475 "), " " 12 "
1400 " (1147 "), " " 24 " howitzer (15c).

This fuze, like the original Boarmann, was designed for use with schrapnell shot, and is always used with what the French call an *obturateur* or stopper, which is made usually of wrought iron. The eye of the shell is divided into two parts, the largest on the exterior, and called the eye proper, and the contracted portion, next to the interior, called the *table*. Into this latter the stopper is screwed, and serves, with the offset between the two (which it would appear more appropriate to call the *table*), to prevent the soft metal of the fuze from being driven into the shell by the force of the charge. The stopper is perforated through the center, to allow the passage of the flame into the charge.

The stopper serves another purpose. In preparing the schrapnell the balls are placed in, and melted sulphur is usually poured in around a stick, to keep the balls in position. When it hardens, a chamber remains for the powder. After this operation, as it is not desirable to place the fuze in position until the shell is fastened to its sabot and all prepared for use, the stopper is screwed into its position, to protect the inside. To prevent its rusting it is covered with a coat of varnish, composed of one part asphaltum and one of oil of turpentine; and after this is dry the screw is oiled to make it work easy. Before screwing the fuze into its place, the hole in the stopper should be filled with powder.

The height of the fuze and depth of the large part of the eye are equal, so that the fuze, when in its place, does not project beyond the surface of the shell. The fuze being in position, a strand of quick-match is fastened in the priming-chamber (*b*, Fig. 179), bent in two under the stud, and folded back on itself on top of the stud. The chamber is then closed, by setting in its place, with a punch, the primer-cover.

The joint between the fuze and the shell is hermetically closed, with a mixture of six parts of powdered chalk and one of black-lead and linseed oil.

The fuze is further protected by a circular piece of paper, the fringed edge of which is pasted down; and on it is marked the date of fabrication of the fuze.

To regulate the fuze to any particular distance, the shell is rested against the nave of a wheel; and with a gouge, after tearing off the paper-cap, the priming-chamber is opened, and the metal cut through to the composition at the desired point on the scale. The gouge is now placed against the exposed composition, perpendicular to the face of the fuze, and a hole made, in which one end of the strand of quick-match is inserted.

When the piece is discharged, the quick-match takes fire, transmits the flames to the composition which burns in both directions from the cut. In one direction only, however, can it communicate with the magazine of the fuze. In the other direction it burns till the composition is consumed.

Objections are raised against the *stopper* used with this kind of fuze, as follows. It complicates the manufacture of the projectiles; the rusting of it interferes very much with the prompt loading of shells which have remained in store for any length of time; and the screwing and unscrewing of it when the shell is loaded, is very dangerous; it is stated that explosions have occurred even in screwing and unscrewing the bronze cap with the English naval fuze. The stopper, however, has the advantages of insuring the bursting of the shell with a small charge; of protecting the fuze against the shocks and friction of the balls; and of preventing the balls from knocking off the magazine cover and thereby preventing explosions.

It is deemed by some an objection to this fuze, that it cannot be primed until after it is fixed to the shell; but this objection is of but little importance, since it is found that the composition when laid bare properly, takes fire without the assistance of the quick-match.

The weak point of all these Boarmann fuzes is the difficulty of regulating them exactly and promptly on the field of battle, in a dense cloud of smoke, and especially at night; and although, when the tools and trouble required with the old fuze are remembered, this objection appears of little moment when compared with the many advantages of the fuze, it assumes a greater importance in view of the fact that a remedy can be, and has been, found for even this defect, in the ingenious invention of Captain Breithaupt, of the Austrian service. This fuze will first be described, and then the application of its principle to the Boarmann fuze.

Breithaupt Fuze, Fig. 183.—This fuze, called by its inventor, the *Field-artillery Fuze*, because he designed it for all kinds of shells used with field artillery, resembles in its general appearance the Boarmann fuze, and and is made of a mixture of tin and lead.

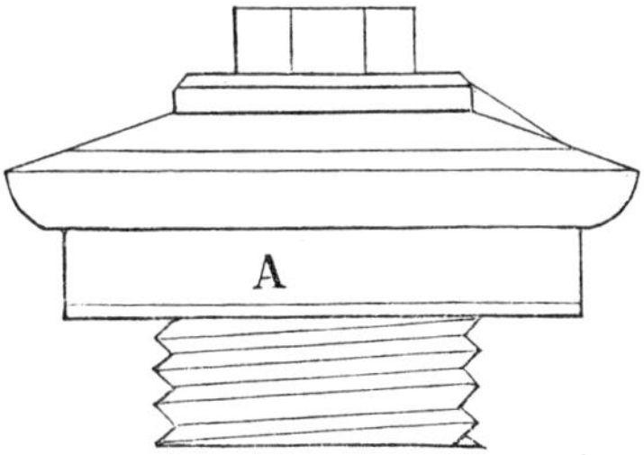

Fig. 183.

The body of it is formed like that of the Boarmann; but the outside has no screw, and it is placed in the eye of the shell by means of a projection from the bottom, which is threaded, and which screws into a corresponding aperture at the bottom of the eye.

The composition is laid in exactly the same kind of a channel, which communicates with the magazine in a somewhat different way from the Boarmann fuze, as is shown in the section, Fig. 184.

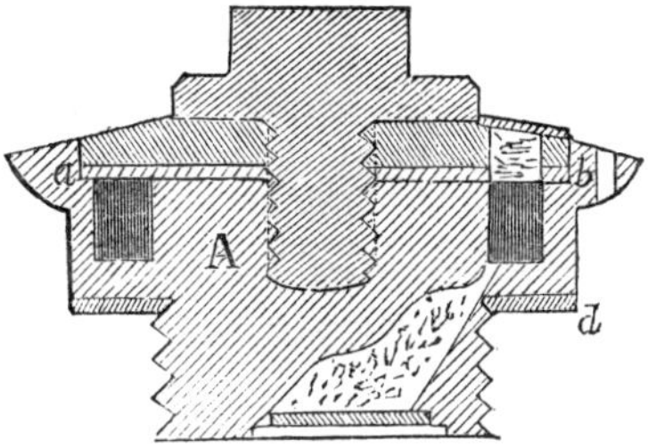

Fig. 184.

On the top of the part A, two mortises are placed, in which the prongs of the screw-driver fit when screwing or unscrewing the fuze from the shell. In the center of this part also is a socket, with the thread of a screw to receive the end of the *pressure screw*, which is represented in Fig. 185, and is made of iron.

Fig. 185.

The stopper and regulating disk, Fig. 186, is cast of the same kind of metal as the fuze, and has, through its center, a cylindrical hole for the passage of the pressure screw. A small priming chamber in the form of a hopper, is placed at the side, one of the lateral projections of which, marked with a red line, serves as an index to regulate it. A small projection on its surface gives a hold to the finger, to turn the disk on its axis, in order to bring the index opposite the required point on the scale of the fuze. A piece of pliable leather or skin, *a b*, is pasted on the under surface of this disk, in order to insure contact throughout, and preserve the composition from moisture. The composition is pressed into its position by machinery. It is meal-powder for schrapnell fuzes, but ordinary fuze composition for the fuzes of other shells. In the first case the fuze burns 7″, in the second 14″.

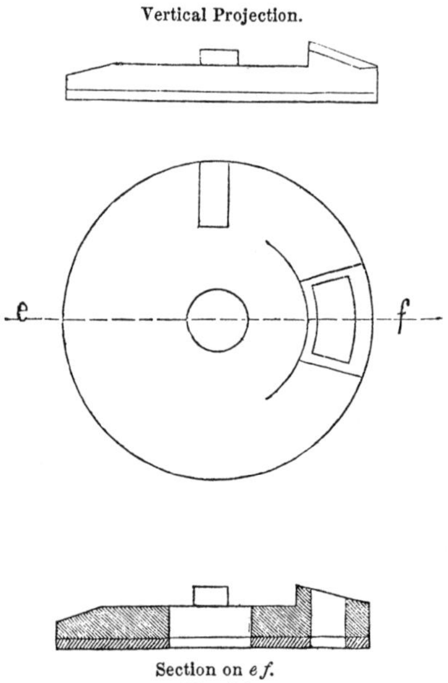

Fig. 186.

The regulating scale, Fig. 187, is a rondelle of white paper pasted on the outside upper edge of the fuze. It is made like a dial, with strokes, half strokes, and quarter strokes, radiating from the center, corresponding to intervals in the combustion equal to 1″, $\frac{1}{2}$″, $\frac{1}{4}$″, and $\frac{1}{8}$″. Upon larger fuzes, as those used for mortars and large howitzers, the scale can easily be carried to $\frac{1}{16}$″. The rondelles are cut with a stamp-iron or punch. Care must be taken in pasting the scale on to make the zero of it coincide with the edge of the solid part where the magazine vent debouches into the composition channel.

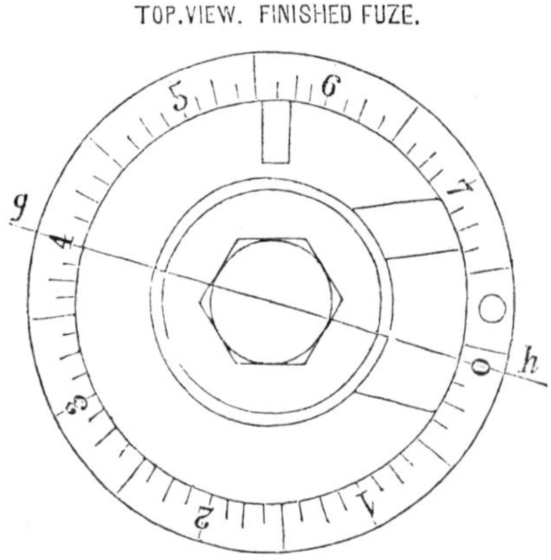

Fig. 187.

The priming consists of a small strand of quick-match, bent double, with the bend fixed in the priming chamber by means of a paste of meal-powder. The ends are free, so that when the priming is uncapped, the ends project outside the little, hopper-shaped chamber.

The priming-cap is a small, circular-sided trapezoid of gold-beaters' skin, which is pasted over the priming chamber by its four edges to protect the priming, as well as the fuze-composition, from dampness.

A rondelle of soft leather, *d*, Fig. 184, is cut out and fitted on the bottom of the fuze around the bottom stem, as a packing to the joint between the shell and fuze.

The magazine, and vent leading to the fuze composition, are filled with fine powder, and the bottom closed with a thin sheet of lead, kept in its place by four notches, a coat of varnish being given to the outside to keep out moisture.

Mounting the Fuze.—Unscrew the pressure-screw, and take off the disk. Screw the fuze carefully into its place until it will go no farther, when the upper edge of the fuze will be even with the edge of the fuze-hole. The disk and pressure-screw are now replaced, and the latter screwed home, and kept so, in order to protect perfectly the composition.

To Guage the Fuze for Firing.—The shell is placed upon the knees, the cannoneer squatting down for the purpose, and the proper division on the scale having been indicated, he takes the key of the pressure-screw, which fits on its head, in his right hand, and putting it on, turns the screw a quarter or half turn at most. Then, by means of the notch on the disk he turns this till the red line comes in front of the proper division of the scale, turns the pressure-screw down again as tight as possible, and tears off the priming cap.

Advantages.—The projectile is transported, and comes upon the field, ready for use. The guaging requires the exercise of no strength, and presents no more difficulty than that of the hausse, as in both cases the operation consists in simply fixing a slide, by means of a screw, at any desired point; and men can be taught in a few weeks to regulate the fuze in eight or ten seconds.

The only instrument needed to regulate and fix or unfix the fuze, is a key arranged like a tap-wrench. A priming-wire, nail, or any pointed piece of metal, could be made to supply the place of the key, by making a hole through the head of the pressure-screw.

In case of a change in the range being required, the fuze is not destroyed by being regulated once, and the guaging can be corrected, changed during the firing, or the cannoneers be exercised with the fuze without any additional cost; which is the case with the forms of the Boarman fuze only when the first cut made is for a greater range than is subsequently required.

In the majority of cases these schrapnells can be fired with quite as much rapidity as solid shot. And the ammunition boxes may be packed in such a way as to stow, in certain parts, a certain number of shells regulated in advance, and corresponding to the different distances at which this kind of projectile is usually fired. Chalk-marks, which can be effaced or changed at pleasure, will suffice to prevent all mistakes. This is the only fuze which possesses this great advantage; and the efficiency of all kinds of shells is very much increased thereby.

The ease with which the fuze is at all times fixed and unfixed, enables the supply of loaded hollow projectiles to be made only when they are required for service in the field; and the necessity will not exist to keep them for an indefinite period already loaded in the magazines. The charges of all kinds of shells can be removed or modified at any time, without danger and at short notice.

This fuze, like all others, is not devoid of defects. In guaging it, the pressure screw may not be screwed down so tight as to prevent the flame from running along the top surface of the composition; and although in experimental firing the index may be exactly placed opposite the proper point of the scale, it is by no means certain that the same accuracy will be attained on the field of battle, in the heat and excitement of a contest.

It may be that the leather packing will lose its pliability with time, and no longer protect the composition from moisture. This could be avoided by using vulcanized india-rubber.

It is thought that it would be an improvement to engrave the scale upon the metal itself instead of placing it on a strip of paper, which in service is liable to be worn or torn off, or the graduation made illegible.

Application of the Breithaupt Principle to the Baden Fuze.—The Breithaupt fuze having been invented after the Boarmann and Baden fuzes had been extensively adopted by different nations, a great objection against its adoption was the extensive changes required in the eyes of the shells already manufactured. To obviate this difficulty, it is proposed to transform the Baden into the Breithaupt fuze, by means of some slight changes, which will not affect the form or exterior dimensions, nor its scale or composition channel.

By comparing the fuze thus transformed, Fig. 188, with the Breithaupt fuze, it will be found that whereas in the last the guaging scale is fixed to the body of the fuze and the index to the movable disk, the regulating of the fuze being effected by turning the latter around until the index is brought opposite to the proper point on the scale, in the transformed fuze directly the reverse is the case; the index being on the body of the fuze, the scale on the disk, and the guaging done by bringing the desired point of the scale opposite the index. This index is formed by making a cut with a fine saw in the edge of the fuze metal, and filling it with a mixture of strong paste and red paint.

In making this transformed fuze, the body of it should be cast in such a way that the height of the composition channel should be at least double what it is when the fuze is finished, the excess of metal being turned off. The double height of the channel might be replaced by an auxiliary channel for charging, made in a steel disk, fixed over the other channel by means of a pin in the central hole made for the pressure screw.

The solid part of the fuze comprised between the two ends of the composition, should be of the same length as one of the divisions on the scale.

When the composition has been compressed in its place, its cross-section should be as near as possible square.

The vent between the composition and magazine should enter the former on the left of the solid part of the metal.

The index should correspond with the right edge of the solid part.

The scale should be numbered from left to right, like the face of a watch; and the right edge of the priming chamber should coincide with the left limit of the first division of the scale.

To avoid error, a small mark, in relief, like the figures of the scale, is placed diagonally in the right exterior angle of each division, to indicate the point which must be placed opposite the index.

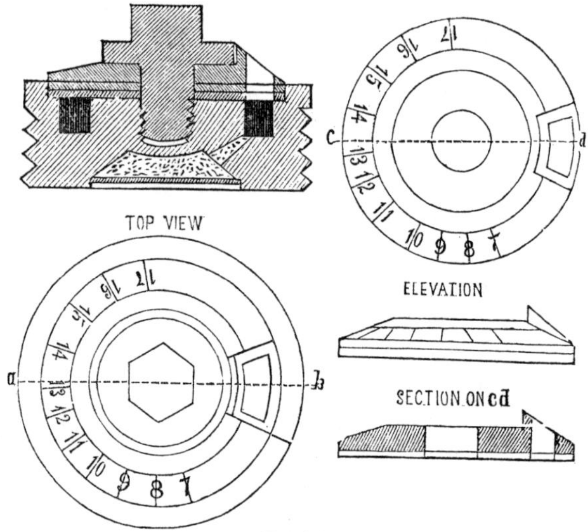

Fig. 198.

It is stated that this fuze will neither be dearer, nor more difficult to make, than the Baden. The seat of the graduating disk has to be cut in the lathe, and the disk itself stamped in a die.

To preserve the fuze better when attached to the shell, it is recommended to place over it a cap made of canvas, impervious to water, which is pasted around the edge of the eye, and has sewed to it a piece of cord, by which to tear it off before the fuze is used.

This, like the Breithaupt fuze, can be used with the schrapnell and all other shells in field service; and by making them of a larger diameter, or decreasing the rate of burning of the composition, they can be equally well used with shells of a larger caliber.

Such is the proposition made in Delobel's Review for 1857, making the old system of schrapnell conform to Capt. Breithaupt's system, without making any change in the eye of the shell.

For those which are still in the magazines convenient to the arsenals of constructions, it appears to me that the simplest way would be to make the form and dimensions of the eye conform to those of the new fuze: as, from appearance, a very little cutting would make the table of the eye large enough to receive the lower part of the Breithaupt fuze; and the upper part would not require much reaming out to make that of the proper size for the body of the fuze; and no stopper would be required.

Such is the Boarmann fuze, with its latest improvement, for which a universal adoption is predicted. It has added immeasurably to the importance and efficiency of the inventions of Generals Schrapnell and Paixhan; for without a good system of fuzes, schrapnell shot and paixhan guns may well be called "bodies without souls."

Of these several forms of the Boarmann fuze, the original is the only one which has, as yet, made its appearance in this country.

SPLINGARD FUZE.—Reference has been made to the fuze invented by Capt. Splingard, as a prominent rival, in excellence, of the Boarmann fuzes.

This fuze, invented in 1846, was declared in 1849, by the celebrated Capt. de Brettes, to fulfill better than any one then known the conditions required in a fuze for schrapnell. At that time, however, the Boarmann fuze, although it had been invented a number of years, was not known out of Belgium, being long kept as a state secret. The Boarmann, and especially the Breithaupt modification of it, stands now at the head of a long list of ingenious inventions, as the best fuze in existence.

It has long been recognized as a fact, that the fuze composition, driven in a wooden or metallic case, tightly fitting the eye of the shell, is liable to many injuries from shocks in and out of the bore, by which the composition is broken and cracked in such a way as to give passage to the flame and cause premature explosions.

Various were the means adopted to overcome this difficulty, and, among others, may be mentioned the cutting the inside of the case into grooves like a screw, and afterwards in the form of rings not communicating with each other, as the screw-shape was found sometimes, after the wood had shrunk, not to fulfill the object.

Such a defect was more especially noticed in schrapnell shot, the thin sides and short bearing surface for the fuze in them, causing the shocks to be more forcibly transmitted to the composition.

Capt. Splingard, by his invention, which has met with the most complete success, has succeeded in isolating the composition in such a way as to protect it completely from the injurious effects of these shocks.

The fuze, Fig. 189, consists of two parts. The *fuze*, properly so called, and the *fuze-plug*.

The first is a small cylindrical tube of hammered copper, the upper end of which swells out so as to form a kind of cup to hold the priming, and prevent the case from being driven into the shell when the piece is fired. This tube is filled with composition in the usual way, a conical opening one-tenth of an inch high being left in the bottom in order, when the flame reaches that point, that a larger surface may be ignited, thus rendering more certain the explosion of the shell. The cup is primed with mealed powder moistened in alcohol, and a piece of quick-match; and both that end and the other one are closed with rondelles of paper. The whole fuze is covered with a varnish into which glaring coloring matter has been mixed, in order that the fuzes may be easily distinguished apart, different compositions or lengths being used for the different distances.

Fig. 189.

Tubes of tin with the edges soldered together, may be used instead of copper.

FUZE-PLUG, Fig. 190.—The fuze-plug is made of wood, and fits the fuze-hole. The opening in the fuze-plug is in two parts; the upper, conical in shape, widening downwards. The other cylindrical, and only a little greater in diameter than the copper fuze, and much less than the upper part of the opening. The upper part is fitted with a cork having an opening just large enough to allow the entrance of the fuze.

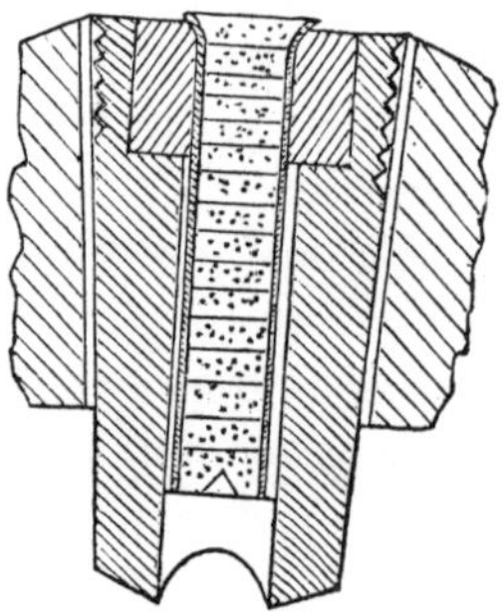

Fig. 190.

The fuze-plug is only a little longer than the fuze corresponding to the longest distance. On the exterior of the upper part, a number of grooves are cut around the fuze-plug to keep it in the eye of the shell, by placing around it a piece of leather, which, on its lower part, is pasted to the fuze-plug. To increase the elasticity of the upper part of the fuze-plug, four cuts longitudinal, are made with a saw nearly as far down as the bottom of the cork.

The lower end of the fuze-plug is crossed, perpendicular to its axis, by a semi-cylindrical groove, designed to prevent the balls in the shells from obstructing the outlet of the fuze. The opening in the plug is stopped with a wooden plug until just before firing, when it is replaced by the fuze.

Four or five fuzes of different lengths and of different colors, suffice for service with a 12-pdr. at all distances.

ADVANTAGES.—The advantages claimed for this system, are: its simplicity and certainty for field service; its allowing the use of a small fuze-plug, and consequently, fuze-hole, the size of which may be the same for all shells. The decrease in the size of the fuze-hole, makes the shell stronger, and by offering greater resistance to the force of the powder, increases the number of pieces when it bursts. The exterior surface of the projectiles is more uniform, as is also the mode of manufacturing them.

The fuze is small in volume, and, consequently, easily preserved; its manufacture is simple and cheap. Its use causes no danger in loading and unloading projectiles, nor any in transporting them, as the fuze is isolated.

The similarity between this system and the paper fuzes adopted in the United States' service, will be at once seen; though we have gone one step farther, and, making the fuze-tube of paper, shown that a conical form, fitting in a conical opening in the fuze-plug, renders superfluous the use of the cork heading to isolate the composition from the effect of the shocks of the projectile.

A great improvement in our paper fuzes would be made by discarding the use of colors to designate their rate of burning, having them of a uniform color, and marked with a contrasting color, with rings corresponding in number with the number of seconds which the fuze burns; or, what would be better, instead of rings, stripes, running the length of the fuze. Sometimes, when firing at short distances, the fuzes are cut in two or three parts, and used on subsequent occasions; so that were rings used, the designation of the fuze might be lost. Using different colors, arbitrarily, to designate the rate of burning, requires an effort of the memory to recollect the method, whilst the stripes or rings tell, at a glance, the rate of burning of the fuze.

Fuzes on exactly the same principle, and some of them copied from the Splingard, have been adopted in several of the European armies, among the rest those of Bavaria, Norway, Sweden, and the Netherlands.

In the Norwegian system, the fuze-case is made of paper, instead of copper.

United States Sea-Coast Fuze.—In the United States, a bronze fuze-plug has been adopted for use with heavy shells, instead of the wooden one. It, however, fits the eye in the same way, and is retained by friction.

It having been found that ricochets, more especially over water, were very apt to extinguish these fuzes, a safety-cap and primer combined, have been adopted in the navy, and found to succeed very well. Fig. 191. A recess in the top is filled with priming composition and covered, until the fuze is required for use, with a leaden disk which fits accurately the opening. A crooked passage filled with the priming conveys the fire to the fuze-composition beneath, and prevents water from being forced in, at least in sufficient quantity to extinguish the fuze.

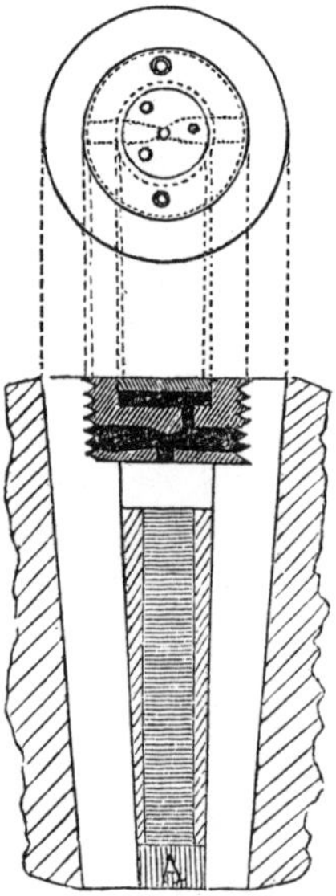

Fig. 191.

Another modification adds to the security of the fuze, and removes one great objection to the use and storage of loaded shells on shipboard. This consists of a small leaden plug, A, Fig. 191, which fits the interior end of the fuze-plug, and remains there until the shell is fired, when the shock of the explosion forces it out by blowing the less dense shell away from it, exactly as it is stated a recent powder-explosion in the streets of Wilmington, Del., blew the horses away from the shoes on their feet. No shock less than that from a charge of powder is sufficient to produce this result, so that even were the shell dropped from a considerable height on the end of the diameter opposite to the fuze-hole, the plug would not leave its position. Or if, by any accident, the leaden disk of the primer should be displaced, and the fuze take fire, the shell would not explode. In view of the increased danger on board ship, and the terrible accidents which have happened from shells, this improvement is a most important one. By means of it the shells can be loaded, the fuzes set, and nothing remain to be done except to remove the leaden disk of the primer when the shell is placed in the gun. The importance of the safety-primer is seen when it is stated that four out of five fuzes are extinguished by ricocheting on the water.

The best method of filling these fuzes is by machinery, as, no matter how skillful the workman, it is not to be expected that he can use exactly the same amount of force at each blow, and without this the composition is not uniformly compressed, and cannot burn equal lengths in equal spaces of time.

In the navy, a machine for the purpose has been in use since 1846. It consists of a driving shaft, moving vertically through a wrought-iron tube, which has on the exterior a strong square thread. On this works a nut by means of a disk attached to it of sufficient diameter to create the requisite power, and on the upper side of the disk a set of levers is placed.

The paper cases are placed in steel moulds, which fit so exactly the exterior of the cases as to support them against the pressure applied to the composition. Two or more of these moulds are placed on the edge of the circular plate on the lower part of the frame, which revolves so as to bring the moulds in succession under the driving shaft.

A ladle full of the composition is poured in the case, and the drift (which is of the same size as the mandril on which the cases are made) placed in position. The plate is then revolved around until the drift is brought under the driving shaft, the exact position being determined by a spring catch, which works into a notch in the edge of the plate.

The disk being given a quick whirl by the handles, the shaft descends and the pressure is continued until the sound of a bell gives notice that the lever has risen, and the action of the machine ceased. The motion of the disk is then reversed, raising the shaft sufficiently to allow the plate to be revolved, and another case, already filled with composition, brought into position under the shaft. This operation is continued until the column of compressed composition is rather longer than required, when the fuzes are removed from the mould, and placed in others of the exact length, the projecting portions being cut off with a sharp knife.

The power applied is usually 2200 lbs. The composition is doubled in density, and becomes very hard and firm.

The usual charge of composition is not more than sufficient to make, when compressed, a column equal in height to the diameter of the fuze channel, and more uniformity is obtained by using a less height than this.

French Schrapnell Fuze.—The French who, strange to say, appear to have paid less attention to the perfecting of schrapnell shot than many other nations, have not in consequence much of a variety of perfected fuzes. The good use made by the French troops in the affair of Traktir with this projectile, seems to have been the cause of more attention being directed to these important projectiles and their fuzes.

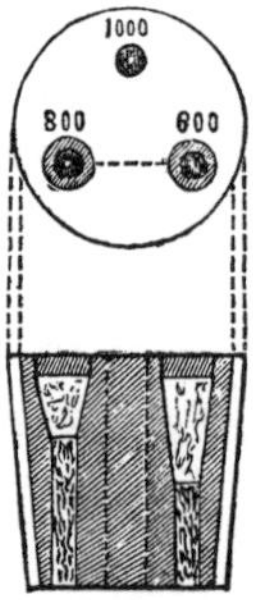

Fig. 192.

The fuze made use of is represented in Fig. 192, and is made of hard wood, having *three* channels parallel to its axis. These are filled to different heights with composition, corresponding thus to three different bursting distances. Each of these channels is provided with a tin tube in which the composition is placed.

The longest channel is always left open. The other two are closed with a covering of leather, over which is placed, for the shortest columns, a disk of rose-colored paper, for the other, one of blue. On these paper coverings are marked the distances at which the columns will cause explosion. These distances are also placed on the face of the fuze near the top of the channels.

The fuze is capped with a rondelle of fringed paper, over which is placed a plain rondelle of parchment with a piece of tape attached, by means of which the fuze is uncapped.

The proper channel is opened, and should a mistake be made, and the wrong one opened, it is only necessary to moisten the leather and replace it, opening the right one.

The composition in the three channels burn in $1\frac{1}{2}''$, $2\frac{1}{2}''$, and $3\frac{1}{2}''$.

It is proposed to modify this fuze by adding a fourth column, intended to burst the shell at only 250 yards' distance; but at this short distance canister shot will do quite as well, if not better, than schrapnell.

This fuze possesses one important advantage. It may be regulated very promptly by men who do not know how to read. It might even be done in the dark by replacing the different colored paper disks by knots or beads fixed to the priming cord.

Russian Fuze.—This fuze consists of a fuze-plug made of a mixture of lead and tin, conical in shape, the head of which projects beyond the external surface of the shell, and of a paper fuze, filled with fine powder rammed hard, which is introduced into the fuze-plug just before firing.

There are two sets of these fuzes, and three lengths in each, which burn—

In the first set, $2\frac{3}{4}''$, $4''$, and $5''$.
" second, $3\frac{1}{2}''$, $4\frac{1}{2}''$, and $5\frac{1}{2}''$.

These were the fuzes used by the Russians in the Crimea, since which a new system has been adopted, in which the fuze-plug is made of papier maché, and the fuze-tube of lead.

The number of fuzes is increased from three to five, and cut for intermediate distances, with a kind of scissors especially adapted to the purpose.

The plug is cast in moulds, and has at the top of the opening a cup in which the head of the fuze, similarly formed, fits. To fix the plug in the smooth eye of the shell, the upper part of it is wrapped with hemp soaked in liquid paste, and it is then driven into the eye. A bronze fuze-hammer is made use of to force the fuze to the bottom of the fuze-plug, the opening in which is stopped with tow, until the fuze is needed for use.

Objections.—The principal objection to these Russian fuzes is, the fact that on the field of battle it is necessary before using them to choose, regulate, cut, and fix them in the projectile, which objection is so much the more important in view of the great improvements in range, accuracy, and rapidity of firing recently made in small arms.

The preceding are the principal fuzes which, up to this time, have been experimented upon and adopted by different services. They are all called *time* fuzes, because the explosion of the shells with which they are used takes place after a certain lapse of time, depending on the length of the fusing column. The time which is required to consume a certain length of column is determined by experiment beforehand, and the fuze *timed* before being used, by appropriating a certain number of these lengths to a shell which requires a corresponding space of time to move from the piece to the point where it is required to burst.

Concussion Fuzes.—Many and various attempts have been made to construct fuzes which, from the shock of a shell when striking, will communicate fire to the charge and explode it.

A *concussion* fuze may be defined to be one which, taking fire like an ordinary fuze, when the gun is fired, continues to burn until the striking of the shell, when the shock, by producing some change in the condition of the fuze, transmits fire to the charge.

Such a fuze, in order to be serviceable, must not only produce explosion on striking, but it must not produce it from the shock of the explosion of the gun-charge, nor of that produced by the ricochets of the projectile in or out of the gun.

These fuzes have usually consisted of some combination of the highly explosive fulminates. But the extreme danger of using these, and the fearful accidents which they are liable to cause, have been great obstacles to their adoption.

The definition for a concussion fuze, as given above, is not without its objections; as undoubtedly the name is just as applicable to any other arrangement, not including a burning fuze, which sets fire to the charge on striking. The distinction is made for the sake of convenience; and only such as are described by the definition will be included under the head of *concussion* fuzes.

The attempts made to construct these fuzes date from a very early period; and probably many of these attempts, although partially successful, never became known, on account of the very general disposition to keep secret such inventions, in order that the authors of them might derive all the benefits resulting from their discoveries.

As early as the year 1637, mention is made of shells which took fire on striking the ground; and at various periods since that time such shells have been experimented upon,—many acting on the principle of having the part near the fuze-hole made heaviest, in order to strike the ground first. If it were possible to carry out this principle with any sort of certainty, the whole problem of concussion, or rather that of *percussion*, fuzes becomes solved. The moment, then, that the principles of the rifle are applied to large guns, so as to project elongated cannon-balls point foremost, as undoubtedly they will be sooner or later, the means of exploding shells on striking, or at a very short time after striking, become as simple as those used to fire off a musket.

It is not necessary to describe all the different attempts made to attain the desired object. Many of them proved successful, so far as the arrangement of the fuze was concerned. That is, the shells exploded when they happened to strike in a certain way; but the great difficulty still existed of compelling them to strike in that way.

In 1845 or 1846, Holland purchased, from the Duke of Normandy, the secret of a percussion projectile, which is said to have succeeded very well; but the construction of which has never transpired, having been entrusted under oath to only two or three officers of that country. It is, however, known to be of an extremely costly and complicated mechanism, that fulminating powder is used in its construction, and that it can be used only at very short distances. So that probably at this time, when the Splingard concussion-fuze is known, the secret can be of but little importance.

In this country, an ingenious contrivance has been suggested and experimented upon for some years, though it is believed with no very decided success. It consists of a bronze fuze-plug, solid at the outer end, and having in the body a square apartment, from which a vent leads into the interior of the shell. The sides of this little chamber are lined with a coating of percussion-powder, with the exception of the parts in the angles, and a small portion of each of the faces which are perpendicular to the axis of the fuze-plug. In the face farthest from the head of the plug, a small threaded hole is placed, for the purpose of holding in position a little metal ball, with a threaded stem attached. This stem is screwed into the hole, the inner end of the plug being movable, for the adjustment; the shell is attached to a sabot, with the fuze to the *rear*. When the shell is fired, the little ball breaks loose by the shock, strikes against the opposite face, where there is no fulminate, and drops into the lowest part of the chamber of the fuze, where it rolls about until the shell strikes, when the concussion between it and the fulminating powder produces the explosion. It is presumed the percussion ball is made of bronze, or some other metal more dense than that of the shell.

This fuze is open to the objections of all fuzes in which percussion powder is used. It requires great nicety of adjustment to insure the breaking loose of the ball from its stem; and if this last is too small, the ball may break from its position whilst the shell is being handled, and produce serious accidents.

Several of these fuzes succeeded admirably, but no definitely reliable results were obtained, and I have not heard that any subsequent trials have been more successful.

Both these last mentioned fuzes would be, under the definition given, not concussion, but percussion fuzes; but they are mentioned here merely in giving a history of the different inventions for making shells explode on striking.

Prussian Fuze.—From 1841 to 1847, numerous experiments were made in Prussia upon a concussion fuze invented in that country, on an entirely different principle from any yet mentioned; and although the success obtained with it has not been such as would warrant a very strong recommendation in its favor, a description of it may not be entirely unproductive of benefit.

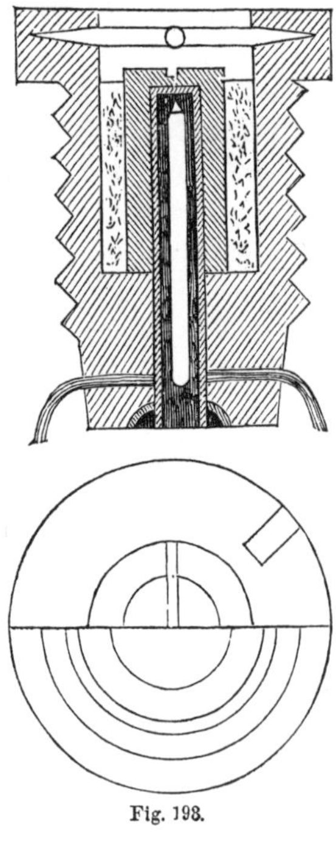

Fig. 193.

The fuze, Fig. 193, consists of three different parts.

1. The body of the fuze, or fuze-case, which holds the other parts, and is screwed into the eye of the shell, the top being flush with the outer surface. This part has been constructed of both metal (bronze and cast iron) and wood. The interior is divided into two parts, both cylindrical, and with the same axis. The one next the outside of the shell is much the largest in diameter. This case was made the same size for all calibers, but that part destined for the reception

of the fuze-composition, is longer for fuzes which are to be fired with small charges than for those with which large charges are to be used.

2. The percussion apparatus consists of a small glass tube, hermetically closed at both ends, partly filled with concentrated sulphuric acid, and wrapped with cotton thread soaked in a composition composed of—

70 parts, by weight, of chlorate of potassa,
10 " " flowers of sulphur,
20 " " white sugar, pulverized, sifted, and moistened with alcohol.

This covering is put on of such a thickness that the tube can just be inserted in a paper case which serves it as an envelope, and which fits partly into the smaller opening in the fuze-case, and partly into the thimble-shaped *breaker* of lead, which is inserted over it in the large part of the opening, and which is represented in Fig. 194.

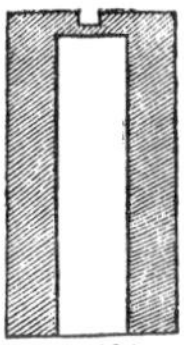
Fig. 194.

3d. The composition column. The explosive apparatus being in position, there remains between the thimble and the sides of the fuze-case a vacant space, which is filled with compressed meal-powder filled in by means of a hollow drift, the interior diameter of which is a little greater than the diameter of the thimble. When the composition reaches the top of the thimble, uncompressed mealed powder is filled in to the top of the case.

Should the firing take place under such short ranges as to run the risk of not consuming all the composition by the time the shell strikes, the time of combustion is shortened by piercing the composition with a small auger, in a direction parallel to the side of the thimble, and to the depth deemed necessary; or the rate of burning of the composition may be increased for short ranges, by mixing with it 1½ per cent. of pulverized charcoal.

On being fired, the thimble or breaker being supported by the composition around it, is not disturbed. But as this takes fire like an ordinary fuze, and burns down to the bottom of the breaker, it leaves this unsupported; and if the composition is all consumed when the shell strikes, the shock overthrows the breaker, rupturing the glass tube, setting free the sulphuric acid, and exploding the shell.

The same objections may be urged against this as against all those in which fulminating powder is used. It is of delicate construction and very dangerous, at least appears so to any one not experienced in its use, whilst the experiments made with it are far from demonstrating its success. Those made with a 24-pd. mortar, fired with a charge of 4 lbs. of powder, and an elevation of 15° over a sandy plain, resulted as follows,—

In 100 shots fired.	55	exploded at the 1st point of fall.
	7	" in the bore, or in leaving it.
	5	" at the 2d point of fall, or beyond that.
	26	did not burst, although the explosive apparatus acted.
	7	did not burst from the failure of the apparatus to act.

or about 60 per cent. succeeded, including the explosions at the 2d point of fall.

With a short 24-pd. gun, and a charge of $2\frac{1}{2}$ lbs. at 400 paces, fired against a target formed of skids 6″ square, the result was as follows,—

Of 100 shots fired,	14	exploded on striking the target.
	86	" in the bore, or in leaving it.

With a 11-inch shell gun, 15 lbs. of powder, and at an elevation of 1° and 2°, without a target, and using two variety of fuzes, numbered 1 and 2, the fuze-case in the last being made of elm, the result of 100 shots was that

21 succeeded, and
79 failed. } with No. 1 fuze.

and

70 succeeded, and
30 failed, } with No. 2 fuze.

With a 10-inch shell gun, 8 lbs. of powder, under 1° and 9° elevation, out of 100 shots fired with No. 2 fuze,

93 succeeded,
7 failed.

With large howitzers, charges of 10 lbs. and 5 lbs. under from $1\frac{1}{2}$° to 13° of elevation, and with fuze No. 1 made of beech and acacia wood : The beech case gave, out of 100 shots.

44 successful, and
56 failures.

The acacia case gave,

20 successful, and
80 failures.

These experiments go to show that the best material for the fuze case is beech-wood, and the worst, cast iron.

SCHONSTEDT FUZE. In 1852 Capt. Schonstedt, of Holland, invented a fuze very similar in its action to the Prussian, but had the advantage of acting both as an ordinary fuze and as an explosive one, and in having neither fulminating powder nor sulphuric acid in its construction. The principal points of difference between the two, will be readily seen by inspecting figure 195.

The case is made of a mixture of lead and tin, and the bottom part of it is made thick enough to allow the cutting of a side-channel, which enters the central one near its end.

The breaker is similar to the one in the Prussian fuze.

A tube of glass, open at both ends, and wrapped so as to fit, as in the Prussian fuze, takes the place of the closed tube.

The side-channel is filled with ordinary fuze-composition, and the space around the thimble with a composition which burns out in 2″.

The glass tube is filled with fine powder, and a strand of quick-match, the lower end of which last, is inserted in the mouth of the side-channel, where it enters the central one.

When the shell is fired the quick composition takes fire, and being consumed in 2″, sets fire to that in the side-channel, at the same time that it leaves the breaker unsupported. This upsets by the shock of striking, and the flame in the side-channel communicating with the powder and quick-match in the broken glass tube, explodes the shell. In case the explosive apparatus does not act, the shell acts like one with a time fuze, and explodes when the side-channel composition burns out.

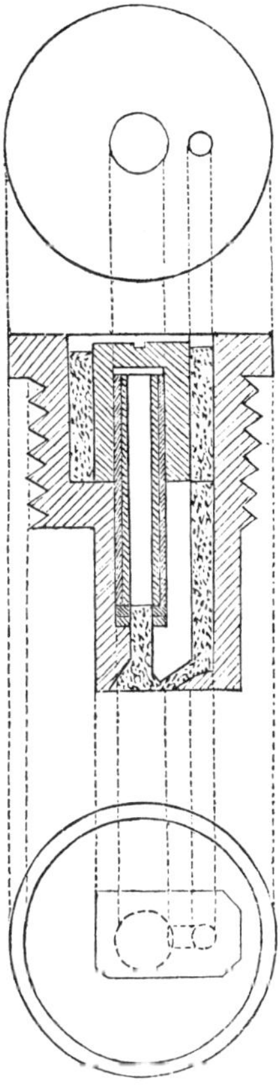

Fig. 195.

Trials made with this fuze in a long 24-pd howitzer, gave the following results, out of 31 shots, fired at a distance of 800 and 700 paces:

12 shells burst at the 1st point of fall.
5 " " " 2d " "
2 " " " 3d " "
1 " " " 4th " "

11 did not burst at all, the fuzes being extinguished by the sand over which they ricocheted.

Although this fuze has the advantage of dispensing with the dangerous contrivance in the Prussian fuze, the results obtained are not as satisfactory as those with the latter, and are not of such a nature as would recommend it as a reliable concussion fuze.

Snoeck Fuze, Fig. 196. We come now to a fuze of an entirely different construction from any yet described. It is the invention of Capt. Snoeck, of the Netherland Artillery, and was tried in Holland, in 1854.

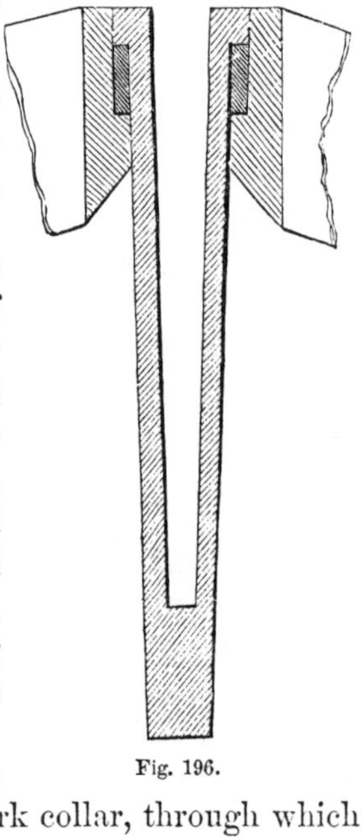

Fig. 196.

Its construction is based upon the property which cast zinc possesses, of being hard and tenacious at ordinary temperatures, but very brittle when heated to from 160° to 200° Reaumer, (417° to 482 F). Hence, a zinc fuze might resist, when cold, the shocks of the charge, and balloting in the bore; but when heated sufficiently by the burning composition, would break from the shock of the falling projectile, and communicate fire to the charge.

The fuze consists of a short wooden fuze-plug, fitted in the interior with a cork collar, through which the fuze passes; and the fuze proper, which consists of a zinc tube of a truncated conical form, having at the top a projecting band, which secures the tube in its position, and at the bottom a solid part, which, by its weight, assists in breaking the tube when the shell strikes. This tube is filled with ordinary fuze composition.

This fuze is constructed on a principle very similar to the Splingard, but its success has been much less satisfactory.

The trials were made with a 24-pd. bronze gun, with a charge of 2 lbs. of powder (1 kil.).

Some of the fuze-plugs were made of palm-wood, and others of beech. The cork collar, the object of which is to insure the complete closing of the fuze-plug, was about $\frac{1}{10}$ in. thick.

Three kinds of fuzes were used, differing only in the thickness of the sides and the form of the projecting ring at the top.

	No. 1.	No. 2.	No. 3.
	in. m.	in. m.	in. m.
Thickness,	0.07874=0.002	0.09842=0.0025	0.11811=0.003
Ring,	Cylindrical, 0.002 thick.	Conical.	Conical.

That part of the fuze touching the fuze-plug is made rough, to increase adhesion.

The lower half of the composition was firmly packed, as in ordinary fuzes; but the upper half, only sufficiently so to keep it from becoming detached by the discharge of the gun or the ricochets in the bore. The upper half, burning rapidly, should raise the temperature of the zinc to the required point in a second, whilst the lower half, burning slowly, should retain the fire to explode the shell when, by its striking, the tube is ruptured. The usual priming of mealed powder and quick-match was used.

Six unloaded shells were first fired, to see if the tubes resisted the charge of the gun and were not broken by the ricochets in the bore or on the ground. Of these—

2 were forced into the projectiles and broken to pieces.
3 succeeded completely, and
1 did not take fire.

Twelve shells were then fired loaded with small charges, and the fuze No. 1, to see if the tubes broke at the first point of fall. Of these—

1 was not recovered.
1 did not take fire.
8 exploded at the first point of fall.
1 " late after striking.
1 in which the effect was not noticed.

Twenty-four shells were next fired, 12 with the No. 2 fuze,

and 12 with No. 3, with increased charges in the shells. Of these—

3 No. 2 fuzes and 1 of No. 3 became detached on leaving the piece.

1 " " " " 2 " " " " before the 1st point of fall.

4 " " " " 1 " " exploded between 150 and 280 paces beyond the 1st point of fall.

4 " " " " 5 " " whose explosions were not noticed.

3 " 3 " exploded at the 1st point of fall.

The fuze No. 1 having succeeded the best, was the only one used in the rest of the experiments, after reducing the length of the solid part at the end.

Of 15 shots fired—

4 were exploded before the 1st point of fall.

3 " " from 100 to 960 paces beyond the 1st point of fall.

7 " " at the 1st point of fall.

1 the fuze did not take fire.

Considering only the fuzes No. 1 in the 2d and 4th trials, it will be seen that 60 per cent. succeeded, 16 per cent. exploded late, whilst an equal percentage exploded prematurely. For the first trial with a new fuze this is quite favorable enough to have warranted new experiments with a view of modifying and perfecting the invention. But these do not seem to have been undertaken, probably because the Splingard fuze becoming known, it has been deemed better suited to the exigencies of service.

The statement regarding the properties of zinc is not exactly true; for this metal is brittle at ordinary temperatures, and become malleable at a few degrees above 100°. It becomes laminable when heated to between 120° and 150°, and below or above these limits it becomes brittle. These facts may serve to explain some of the irregularities observed in the bursting of the shells.

It is possible, that firing at short distances, the burning composition does not heat the tube sufficiently for it to reach its maximum brittle point at its first point of fall; and on the contrary, in firing at great distances, the temperature of the tube may be raised so high as to inflame the powder in contact with it; that is, to from 265° to 270° R. Were this the fact, the tardy explosions are explained on the first supposition; and the spontaneous explosions, taking place between the first and second ricochets, by the second.

It would be well, therefore, to determine beforehand, by experiments with stationary tubes, to what degrees of temperature they are heated from the commencement to the end of the burning of the composition; using for this purpose thin bars of different fusible alloys, forming a thermometric series of four or five terms between 150° and 270°, stating, also, the relative degree of brittleness at different points during the combustion.

It may be, as was thought by the Holland Commission, to which this fuze was referred, that the premature explosions may occur from the tube becoming so brittle as to be broken by the movement of the charge in the shell, or by the inertia of that part of the tube between the solid end and the point of rupture. It might, then, be advantageous to surround the tube with an open sheath, with no bottom, of nearly the same diameter of the fuze-plug, and fastened to it in such a way as to protect the tube from the shock of the charge; at the same time, giving two points of support near its end to the end of the tube, to prevent the latter from breaking of its own weight. An open tin cylinder with the open spaces large enough to allow the broken part of the tube to fall out, would serve the purpose; though the experiments upon the temperature and brittleness of the fuze, as recommended in the preceding paragraph, might indicate that a progressive decrease of thickness of the tube would obviate every difficulty.

The explosions in the bore of the piece may arise from the passage of the flame between the sides of the fuze-plug and eye of the shell, or between the fuze and fuze-plug. This could be determined by firing some shells with no composition in the fuzes; and if it was found to be the fact, it would be well to make the fuze-plug of pasteboard and fix it in the eye by wrapping it in hemp soaked in liquid paste, as is done with the Russian fuze, and to increase the height of the cork collar, and even make it so large as to isolate the fuze entirely from the plug, as in the Splingard fuze. If this isolation should be found to interfere with the breaking of the tube at the point of fall, this could be remedied by increasing the weight of the solid part at the end.

To prevent, in casting the tubes, the formation of small holes or defects which might give passage to the flame, or, from the consequent thinness of the metal at those points, cause explosion of the charge by heating the metal there, it is proposed to make the fuzes from rolled tubes, which could be cut to the proper length, heated to 120°, and the solid end and reinforce band stamped on them.

With these proposed improvements, it is thought this fuze might be very much increased in value. It is simpler in its construction and cheaper than Splingard's, and being smaller in size, may be applied without difficulty to the eyes of existing shells of all calibers. Although constructed on a similar principle to that of the Splingard fuze, yet, as a distinct fuze, the merit of the invention certainly belong to Capt. Snoeck; as, after all, in inventions of this kind, he is entitled to the credit of making them, who first brings together into a perfect whole the scattered elements of a plan, and shows its successful operation. This seems to be the idea of the great artillerist (who, by his *invention* of the shell *system* for long guns, gave such an increased efficiency to the fire of artillery), when he says:

"We have invented nothing, innovated nothing, and almost changed nothing; we have simply united the scattered elements, to which it sufficed to give, with a little attention, the proper size and proportions to attain the important end we had proposed."

He is not the inventor who suggests the possibility of certain results following certain combinations, and allows his idea to lie dormant until practically tested by others, when failure leaves all the odium on the practical man, but success raises up many competitors who are ready to produce the evidence of their prior claims, to the exclusion of him who is the first to *demonstrate* that what has been predicted as possible, cannot be otherwise than as he has shown it. All great inventions have, therefore, their disputed claimants, and those in military science form no exception to the rule.

Splingard Concussion Fuze.—This fuze, invented by the same Belgian captain whose admirable system of time fuzes had already placed his name in the first rank of fuze inventors in 1846, first became generally known in 1850, although it had then been invented 20 years. During this period the knowledge of it was retained in Belgium as a state secret; and it would probably still have remained such but for the corruption of some agent not proof against the inducements offered to divulge it. It became known in England and Holland, when it was deemed advisable by the Belgian government to allow a description of it to be published; which was done, as much as anything else, for the purpose of forestalling the action of the speculators and counterfeiters, who sought to sell the secret to foreign governments.

The fuze is characterized by its simplicity, and easy manufacture, by its general application to all shells, and by the total absence of all those dangerous fulminating powders so generally used in concussion and percussion fuzes up to the invention of this one. In case of need, too, the present wooden one could without much difficulty be converted into a concussion fuze. It is perfectly safe in the hands of the artillery-man or in the magazine, and cheap enough to be generally used.

The fuze consists of two parts, the *fuze proper*, Fig. 197, and the fuze *plug*.

The fuze-case is made of cartridge paper, cylindrical in form, and filled with the ordinary fuze-composition, in the center of which is a hollow conical tube of plaster of Paris, open at the bottom; through which passes the flame when the cone, left unsupported by the burning away of the composition around it, breaks off from the shock of the falling shell.

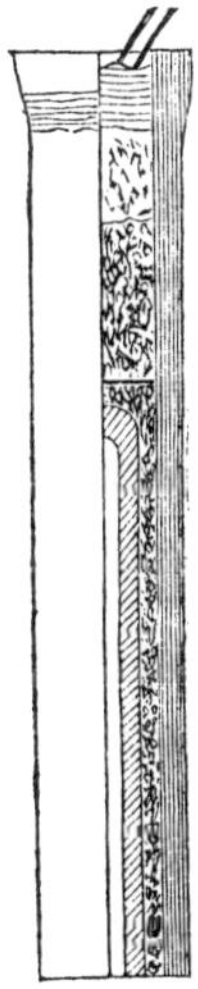

Fig. 197.

A strong paper is used, and is rendered incombustible by immersion in a solution of sulphate of ammonia.

The fuze is filled, like a rocket, on a spindle, using small charges, and taking care to pack the composition well; or it may be driven solid and bored out afterwards. In the bottom part of the fuze a slow composition is used, next to that a quicker one, and in the top part mealed powder.

The surface of the opening in the composition is covered with one or two coats of gum-lac varnish. When this is perfectly dry, plaster moistened with water is packed into the opening, so as to fill it completely; and while it is still soft, a small spindle is thrust in along the axis to such a depth as not to pierce the top of the cone. The proper direction is given to this spindle by a guide which is placed against the bottom of the fuze, and through which the handle of the spindle slides.

No fissures are formed between the plaster and composition, as the plaster enjoys the property of increasing its volume when combined with water.

The slow composition extends only a very little above the top of the plaster tube, in order to leave it unsupported very soon after fire is communicated to that part. In this way the same fuze may be employed, either at very small or very great distances.

The fuze-plug, Fig. 198, is of wood, and of the same form on the exterior as an ordinary wooden fuze. The interior is formed of three parts:

1st. The upper and largest part, in which is fitted a cylindrical collar of cork, through which the fuze is passed, and held in its position there by friction, isolated from contact with the rest of the fuze-plug. This arrangement, as in Splingard's time-fuze, protects the composition from the shocks of the discharge and ballotings in the bore.

2d. The middle part, which slightly exceeds in diameter the diameter of the fuze-case; and

3d. The lower part, which is very narrow, with the double object of allowing the passage of the flame into the shell, and forming an offset for the lower end of the fuze to rest upon.

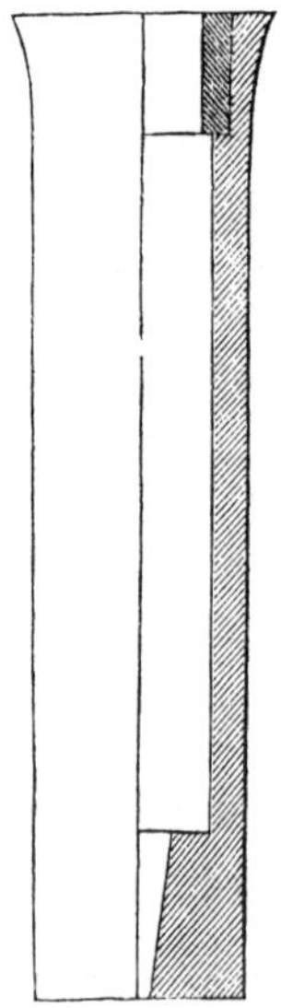

Fig. 198.

To prevent the powder in the shell from entering the fuze, a piece of light, open stuff, like gauze, is pasted over the bottom of the fuze-plug.

Employment.—The fuze-plug being fixed beforehand, the fuze is not introduced until just before firing. For the operation, no tool is required, it being pushed in simply with the hand.

In mortar-firing, the head of the fuze-plug may be allowed, as usual, to project beyond the surface of the shell, as there is no danger of its being crushed by the revolution of the shell in the bore; but with long pieces, and especially with those of large caliber, it must be allowed to project as little as possible. The shell, too, in such cases, should be attached to a sabot, as solid and stable as possible, in order to resist the effort to rotate,—a matter of considerable importance under like circumstances, no matter what kind of fuze is used.

Experiments with this fuze were made at Liege in 1850, and were so favorable as to cause their continuance the following year at Brasschaet, where a 11-in. (29[c]) mortar was used, firing at an angle of 60°, and a distance of 600 paces.

Out of 224 shell fired,

204 Burst at the moment of striking.
13 " after falling, and when all the fuze had burnt out.
3 " prematurely, near the end of their flight.
5 Did not burst, 3 of which in consequence of the fuze not taking fire.

More favorable results, as far as mortars are concerned, could scarcely be desired.

Some of the shells, loaded with the service-charge (4 lbs.) were fired against blindages, and showed that they had time to penetrate before the explosions took place.

A 24-pdr. howitzer was next tried, with 4 lbs. of powder, firing at a distance of 600 paces. Out of 31 shots fired,

10 Burst at the 1st point of fall.
17 " " 2d " "
4 " in the piece.

The eyes of these shells were so small that the heads of the fuzes projected more than an inch, which, in all probability, caused the 4 explosions in the piece.

The 17 tardy explosions are accounted for by the fact that the fuzes used were made for mortar-firing, in which the projectile moves comparatively slow, and the composition burnt too slowly to uncover sufficient of the plaster to allow rupture to take place. At the second point of fall, a greater distance was left unsupported, and the shells were exploded by the shock.

The obvious remedies in these two cases, are to quicken the burning of the composition, and make the fuze-plugs of such a size as not to allow any portion to project. These modifications have been made, and subsequent yearly experiments have fully demonstrated the complete success of the fuze in pieces of all calibers and kind.

The principle of the concussion-fuze being of more importance in firing against vessels at sea, or on the sea-board from seacoast defenses, than under any other circumstances, it is a matter of considerable importance to know whether the shell will burst while in the side of a vessel, or if the action of the fuze is not so quick as to burst the shell before it has penetrated sufficiently into the wood to produce its proper effect.

Experiments made in 1853, with long guns of large caliber and strong charges, fired against a target composed of a double thickness of strong beams, proved conclusively that the time which elapsed between the striking of the projectile, the breaking of the tube, the transmission of fire to the charge, and the explosion, is precisely what it ought to be to allow the shell to become properly imbedded in the wood.

The application of these fuzes may be universal; on board ship, in sea-coast defences, and in the attack of permanent works; whilst the power to burst at will field howitzer shells at their first point of fall, will very much increase the efficacy of field howitzers, simplify their rules of fire, and their supplies; and even in sieges, ricochet and enfilade fires are increased in power.

In addition to those advantages already mentioned, may be stated, the ease with which the fuze is prepared for use,—simply pushing it into the fuze-plug without any previous preparation.

The perfect freedom from danger in its use, and the little liability there is for its bursting anywhere within the first point of fall.

The ease with which it is preserved in magazines, requiring no more care than is used with portfires; an important quality in view of the difficulty of preserving fuzes driven in wood.

The reduction of the large supply of loaded shells now kept on board ship and in magazines, at imminent risk of destructive accidents; a much smaller supply being all that is now necessary, in consequence of the ease with which these fuzes are filled; and even where the shells are loaded, the fuze-plugs can be stopped with a good cork, and the fuze left out till the last moment.

The simplification which its adoption causes in the whole fuze system for hollow projectiles. The number of these fuzes for all calibers will be but two, whilst the same fuze-plug can be used for all (both time and concussion fuzes), the two differing only in the presence or absence of the plaster tube; the time-fuzes being regulated for different distances, as in the Norwegian service, by a longitudinal cut made in the paper of the fuze. The time and explosive fuzes could be designated by different colors painted on the cases, and a regulating scale would be all the addition necessary with the former.

It is possible experiment may show that this fuze will act if the principle of isolation is omitted; as there seems to be but little reason to suppose that the composition or plaster would be more liable to rupture than in the ordinary wooden fuze.

I was induced, soon after becoming acquainted with this fuze, to make some experiments, using metal as a fuze-plug, and glass tubes instead of plaster. The means at hand for making the fuzes were of a very inadequate kind, and other circumstances interfered to put an end to the trials. Some of the fuzes succeeded; but the unfitness of the glass as a material in place of plaster, was fully demonstrated when it was discovered that the burning composition, whether outside of the tube, or inside as in Snoeck's, completely fuzed the glass; and this was found in some cases to have been melted down so as to lose all appearance of the original form.

When the composition was not driven on the tube as a spindle, but placed in afterwards, it was found necessary still to make use of a thin coating of semi-fluid plaster, in order to completely close the joint between the composition and the glass, to prevent the flame from running down along it.

Percussion Fuzes.—A percussion fuze may be defined, one which, leaving the piece without being inflamed, communicates fire to the charge at the moment the shell strikes, by means of some change produced in its construction by the shock.

They have usually been constructed by making use of some of the dangerous fulminating powders; but even those which have given the greatest promise of success, have this great objection against them, and are of a complicated and delicate construction.

These fuzes are of more importance in naval warfare than any where else, and we find both England and France provided with percussion fuzes for the use of their navies. Both the systems are very complicated; and but for the importance attached to them in those countries, it would scarcely be deemed necessary to enter into a minute description of them.

French Fuze, Fig. 199 This fuze, the invention of Capt. Billette, consists of a wrought-iron fuze-plug to which is attached the explosive apparatus. The plug is screwed into the eye of the shell, stopping it completely, the rest of the apparatus being inside the shell.

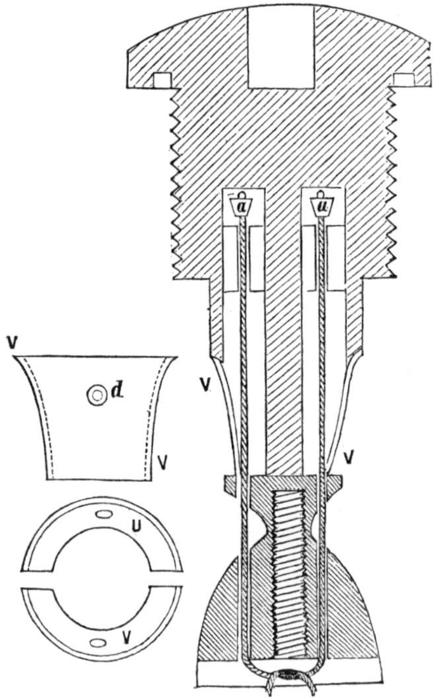

Fig. 199.

In two side channels are fixed by friction, by means of pieces of parchment, two small tubes of hard wood, one filled with chlorate of potassa, the other with sulphuret of antimony. Through these tubes pass two hard woolen cords, terminated at the upper ends by rubbers *a a* of copper. The other ends pass down through the cup and breaker of the fuze, and are tied together in a groove of the latter. In order to load the side channels, they are left open at the bottom as represented in the figure; that part of the apparatus being afterwards closed by means of the two halves of a hollow truncated cone *v v*, which are fixed to the body of the fuze by means of a screw at *d*. These half envelopes are made of bronze, and leave between them two openings through which the flame is transmitted to the charge. The breaker is fixed to the cup by means of a steel screw through its center.

The shells with which these fuzes are used are fixed to sabots, the fuze being placed in the hollow and exactly in the axis of the sabot.

When the shell strikes, the shock breaks off the steel screw connecting the cup and breaker at the point where they join; the weight of the breaker and strength of the screw being so regulated as to insure this effect. The weight and motion of the breaker act on the cords, producing the deflagration of the fulminates.

OBJECTIONS—If explosion is not produced in the bore of the piece by the shock of the discharge, it is natural to suppose it will not take place whenever the shell strikes, either on the fuze or on the opposite point of the same diameter, and very often when it strikes anywhere near the ends of this diameter.

The amount of temper given to the screw will exercise a certain influence upon the successful working of the fuze; for, if the temper is too little the rupture will not take place, and if it is too great, a ricochet on wa'er, or any other slight shock, may be sufficient to explode the shell.

It may happen that the powder and incendiary composition in the shell will not allow the breaker to act with that freedom which is necessary to produce explosion.

In spite of these objections, it must be admitted that as the French navy has adopted the fuze it must be better than it looks, and at least better than the percussion fuze heretofore used among them.

ENGLISH FUZE, Fig. 200.—This fuze, invented by Capt. Moorsom, has been but a short time in use in the British navy. It is very highly praised by Sir Howard Douglass, though he gives no description of it.

The body of the fuze is made of bronze, and is screwed into the eye of the shell by means of a key fitting into two mortises made in the head. The lower part is not threaded, and projects into the chamber of the shell.

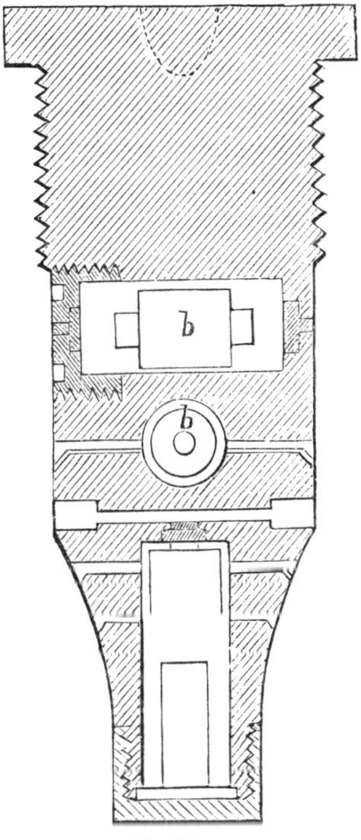

Fig. 200.

In the body of the fuze two cylindrical chambers are placed, one above the other, with their axes perpendicular to each other. In Fig. 200, which represents a section of the fuze through the axis, the upper chamber is shown by a section through its axis, the lower one by a section perpendicular to its. These chambers are both alike, with similar percussion apparatus; so that a description of one will do for both.

In the chamber is placed a solid cylinder of bronze, *b*, terminated at each end by a small projection, or piston.

One head of the chamber is movable, and when screwed into its place, its exterior is flush with the convex surface of the fuze. Holes are left on the exterior for the use of a key, and the head is screwed in, after the hammer is placed in the chamber and suspended.

In each end of the chamber is a small recess, a vent being bored through to it from the exterior of the fuze. These are both filled with fulminating powder.

A hole is drilled through the hammer at its middle point, and perpendicular to its axis, and is used to suspend the hammer, by means of a copper wire, in the center of the chamber. The wire passes through corresponding holes in the body of the fuze (seen in the section of the lower chamber), and are soldered at the ends in the curved portions of the holes near the surface of the fuze.

In the lower end of the fuze a third chamber is placed, with a percussion apparatus similar to the preceding, acting, however, in the direction of the axis of the fuze, and having but one end of the chamber provided with percussion-powder, the vent leading from which communicates with a cross-channel, having at each end a small chamber filled with powder.

The hammer, a cylinder of bronze, with a piston like the others on its upper end, is suspended in the same way, and has below it a copper-wire passing through holes in the fuze, and soldered like the rest.

At the bottom of this last chamber stands a cylinder of lead, fixed in its position by its base, which is pressed in a little offset, between the bottom end of the fuze and the cap which closes the chamber.

From the construction of this fuze it will be seen, that there are six points on the surface of the shell, the striking of which will produce the working of the apparatus; and that the more the point of impact is removed from these, the less will be the probability of the shell exploding, from the tendency the hammer will have to act in directions oblique to the axes of the chambers.

When the shell strikes, the suspension-wire of that hammer whose axis coincides with the diameter of the shell passing through the point of impact, or, is parallel to it, is torn loose, releasing the hammer, and allowing it to plunge forward and explode the fulminate, by striking it with the piston on its end.

It is doubtful, even when the shell strikes in the most favorable way, if the action of the hammer is sufficiently powerful to always produce explosion; and should it strike on the point opposite to the fuze, it would seem to be far from certain that explosion would follow, as the column of lead has then to tear itself from its position, and act on the third hammer to produce explosion in the fulminate in front of it.

Even, if the apparatus worked to perfection, when it did strike, from the very delicacy of its construction, numerous premature explosions in the bore of the piece are to be anticipated, to say nothing of those which may occur in handling the projectiles, or in ricochets.

The fuze, moreover, is so exceedingly complicated, and must be so costly,* that it can scarcely be considered as offering a good practical solution of the problem of percussion-fuzes; and as these fuzes are about as favorable as any of the kind in use, we may conclude that no practical percussion-fuze has yet been discovered for use with spherical projectiles. None that will at all compare for example, with the Splingard concussion-fuze in certainty and simplicity of action.

For projectiles which move point-foremost, and are always sure to strike at one place, the problem is an easy one. Several forms have been suggested; and although the problem regarding the projectile has not as yet been solved, it may be as well to describe one or two of these fuzes, in order to be prepared when the solution does come, which probably now will be at no distant day.

BOURBON FUZE, Fig. 201. This fuze was used with the percus-

* They cost about six shillings (English) a-piece.

sion projectile, to which reference is made on p. 303, invented by M. de Bourbon (Duke of Normandy), and which was guided by a kind of rudder, so as to compel it to move with the point to the front.

It consists of a bronze fuze-plug screwed into the eye of the shell, with a head larger in diameter than the other part and threaded on the exterior, by means of which a cap is screwed on, covering the fuze until just before it is used.

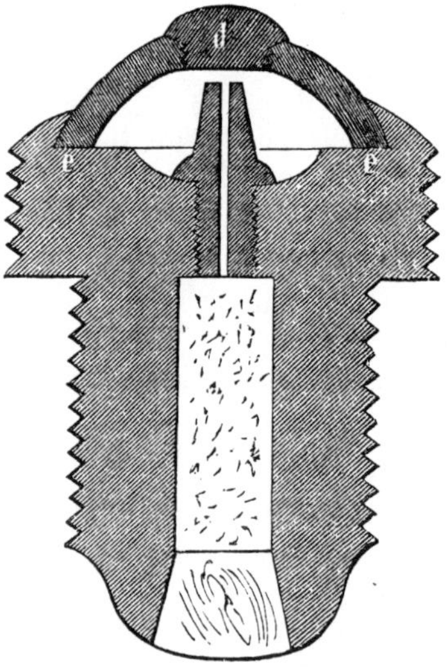

Fig. 201.

A cap of copper, *e e*, is fixed to the head of the fuze by means of several circular grooves in the fuze-plug; slots and projections being formed on the edge of the cap, and the rabbets of the circular grooves, which allow the cap to be inserted in its place. A threaded hole is placed at the highest point of this cap, in which the fulminating cap, *d*, is screwed just before the fuze is used. A steel nipple is screwed into the body of the fuze just under the cap, which, when the cap is exploded, conveys fire to the charge, communicating it first to the powder contained in the channel of the fuze. The bottom of this channel is closed with a cork stopper, which is blown out when the powder in the fuze takes fire.

The projectile being supposed to strike with the point first, the shock, in order to explode the cap, must be of sufficient force to flatten the cap, *e e;* and this cap has been made of such a thickness that nothing less than striking against the side of a ship, or other equally resisting body, is sufficient to cause explosion. No ricochet, therefore, on the water, will cause the shell to burst. Such a fuze would do very well for a rifled projectile of large size, did any good one of the kind exist, or for any projectile with an arrangement to keep a particular part always to the front.

It does not appear that the means used to produce this effect in the Bourbon projectile were at all successful. It is stated that the apparatus made use of, consisted of a stem of iron screwed into the bottom of the projectile, and having tied to the other end a line which uncoiled itself as the projectile left the piece. The stem acting besides as the axis of a kind of sabot, made of felt, oakum, &c.

In the experiments made with this fuze and projectile in Russia, it is stated that explosion took place always when the projectile struck point foremost, but that this condition was too seldom fulfilled for the mean result to be of sufficient efficacy.

It is predicted that the Bourbon fuze will be the type of all good fuzes for percussion projectiles fired in naval artillery from grooved guns.

Should it be found that shells armed with such fuzes burst too soon after striking, a small piece of fuze composition could easily take the place of the powder in the fuze, and delay the explosion of the shell any convenient length of time; and this modification may be applied to any percussion fuzes of this kind. It is now, however, pretty well known that the shell penetrates a sufficient distance to produce the proper effect, before the explosion takes place.

In the experiments made at West Point on elongated rifled projectiles, the exploding apparatus consisted simply of a nipple attached to a piece of metal, and having on it a common percussion cap. This was dropped into a small chamber left in the point of the projectile, and a head piece screwed on. When the shell struck, the nipple piece continued to move to the front until arrested by the head piece of the chamber, when the cap was exploded, and fire communicated to the charge. This gave a little more time for the projectile to penetrate than is allowed by the Bourbon fuze, and sometimes the shells appear to have passed through a solid target of beams three feet thick, bursting as they reached the opposite side. The destruction produced by a projectile bursting in this way after passing through a ship's side, would be very great.

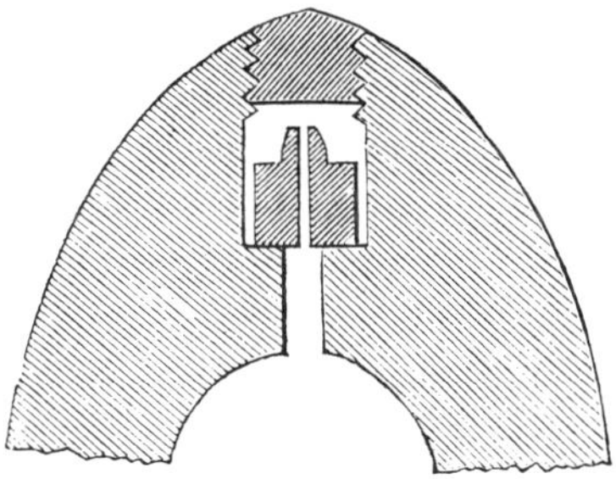

Fig. 202.

CHAPTER X.

ARTILLERY IMPLEMENTS.

For the purpose of describing these implements, they will be divided into classes with regard to the use made of them, as *Pointing* implements, *Loading* implements, *Priming* and *firing* implements; and these classes will be subdivided according to the pieces with which they are used.

POINTING IMPLEMENTS.

Quadrant, Fig. 203, Pl. 13. The first instrument used for giving the elevation to a piece of artillery was the common quadrant, made either of wood or metal. This is now used only with mortars, or in giving large degrees of elevation to long pieces, for which the ordinary breech-sight and tangent scale is not sufficiently developed. It consists of a quarter of a circle fixed to a long arm. The edge of the circle is divided into degrees, and the inclination of the arm to the horizon is determined by a plummet which is fastened at the center of the curve. This quadrant gives the elevation of a piece only to within a degree. For more accurate pointing a more complicated metal quadrant is used, having a movable limb fastened at the centre of the arc, and a spirit level attached to it. The end of this arm moves along the graduated arc, and has on it a vernier by means of which, parts of a degree can be read off. This quadrant is more especially intended for use with long pieces of large caliber, when greater accuracy is required than can be obtained with the breech-sight, or in firing at great elevations.

Breech-sight.—Fig. 204, Pl. 13. The breech-sight consists of an upright piece of sheet brass, supported on a foot piece, the lower side of which is curved to fit on the base-ring of the gun on which the sight is used. In the middle of the upright part a slit is made, and on one side of it is placed the scale. Along this slit a slider moves, and is retained in any particular position by a small screw which enters the slider from behind, and through the slit. The sighting is made through notches or small holes made in the slider. The middle of the sight is first placed over the highest point of the breech, a notch or mark being made on the foot for that purpose.

Formerly the highest point of the muzzle was marked either permanently or with chalk, and the line of sight passed through that point. The divisions on the scale were then determined as follows: Suppose a breech-sight or tangent scale for the 6 pd. gun is to be graduated. The natural angle of sight being one degree, the *dispart* will be equal to the tang. of 1° to a radius equal to the distance between the highest points of the swell of the muzzle and base ring, measured on a parallel to the axis of the piece. This distance in the 6-pd. gun is 58.7 in. and multiplied by the natural tangent of 1° (.0174551) gives the dispart 1.0246. The angles being very small, we may assume the tangents to increase in the same proportion with the arcs, and that therefore the tangent of 2° is equal to double the tangent of 1° and so on. Laying off then on the scale a distance equal to the dispart, we obtain the point through which the piece is sighted at an object on the same level, when its axis has an elevation of 2°. This being divided into 4 equal parts, gives us the points for each quarter of a degree; the different distances representing the tangents of each quarter of a degree to a radius of 58.7 in. It will be seen by this construction that if the natural tangents of 2°, 3°, &c., be multiplied by the radius 58.7 in. the distance obtained must first be decreased by the dispart before being laid off on the scale. The sight for any other piece is graduated in a similar manner. Muzzle-sights, equal in length to the dispart, are however now used, and the scale is laid off from the bottom in lengths, which are calculated as before, and represent the tangents of 1°, 2°, 3°, &c., to a radius equal to the distance between the two sighting points on the piece. These lengths are subdivided into smaller parts, which enables the piece to be sighted to within a small portion of a degree.

These sights are designed for use with the heavy sea-coast and siege, and garrison pieces, and the new method of graduating them has the advantage of rendering easier the application of the pendulum hausse to the pieces, as that instrument is always used with the muzzle-sight.

The Tangent Scale, Fig. 205, Pl. 13, is made of sheet-brass, with a flange along the lower edge, which latter is cut to fit the base-ring of the piece.

The upper edge is cut in steps, each *rise* representing the tangent of $\frac{1}{4}°$ to a radius equal to the distance between the highest points of the breech and muzzle, measured parallel to the axis.

In the middle of each step a notch is placed to assist in sighting. When the piece is sighted by these different notches, the axis is at an elevation indicated by the degrees and minutes under the notches.

This instrument has now fallen very much into disuse, the pendulum-hausse or breech-sight being preferred as more convenient in form, the first named correcting besides any error arising from the trunnions not being horizontal.

Pendulum-Hausse, Fig. 206, Pl. 13. The two preceding instruments give the proper elevation to the piece only when the trunnions being horizontal, the permanently marked highest points of the breech and muzzle are made use of, or when the trunnions not being horizontal, new points are designated by means of the gunner's-level. The pendulum-hausse serving as a corrective for a faulty position of the trunnions, does away with the necessity for the use of the gunner's-level or new sighting-points, which it would be very difficult, under many circumstances, to establish. This hausse is derived from the Russian service, and hence is called the Russian hausse.

It consists of an upright piece of sheet-brass, like the breech-sight, and has a movable slider and scale, similarly arranged. At the lower end is placed a bulb or disk, filled with lead. The scale passes through a slit in a piece of steel, and is connected with it by a brass screw, which serves as a point on which the scale vibrates laterally ; the slit is made long enough to allow the scale to assume a vertical position, in any ordinary cases of irregularity of the ground on which the gun-carriage may stand. The ends of the piece of steel are formed into journals or trunnions, by means of which the hausse is supported on the seat attached to the base of the breech, and is at liberty to vibrate in the direction of the axis of the piece.

Thus, in any ordinary variations, either in the level of the wheels or in the elevation of the gun, the scale is kept in the vertical position by the weight of the bulb at the bottom.

The *seat* is of iron, and fastened to the base of the breech by three screws, in such a way that the middle point of the steel bar, or pivot of the scale, shall be at a distance from the axis of the piece equal to the radius of the base-ring.

The *muzzle-sight* is of iron, and screwed into the highest point of the swell or muzzle-band. The height of this sight being equal to the dispart, a line from the top of it to the pivot of the scale, which is also its zero point, is parallel to the axis of the piece. This being the disposition, should the axis of the trunnions not be horizontal, the vertical plane passing through the central line of the scale, and top of the muzzle-sight, will be parallel to the vertical plane of fire, and to the axis of the piece which is in that plane, and may without sensible error be substituted for the latter. The scale will, therefore, always indicate correctly the proper angle of elevation of the piece.

Each piece must, of course, be provided with its own scale, seat and muzzle-sight, all of which are marked for the particular piece to which they belong.

The hausse, when not in use, is carried in a leather pouch suspended by a strap from the shoulder.

The manner of graduating the scale has already been described at p. 325. The divisions are the tangents of each quarter of a degree, to a radius equal to the distance between the top of the muzzle-sight and pivot of the scale, which exceeds, by an inch, the distance laid down in the tables as that between the semi-diameters of the base-ring and swell of the muzzle.

This instrument, the best of its kind now in use, has been adopted for our field artillery, and also applied to some of the heavier pieces; but with these its use is not of as much importance, since, the wheels of their carriages having generally a permanently level position, they are not liable to the same inaccuracies in firing as field pieces, which are used on rougher and more variable ground.

GUNNER'S LEVEL, Fig. 207, Pl. 13. This is an instrument for determining the highest points of the base-ring and muzzle, preparatory to the use of the breech-sight and tangent-scale. It consists of an upright of sheet-brass, supported on two feet, the line joining the bottom-points being perpendicular to the direction of a sliding pointer, which moves up and down in sockets on the upright. On a horizontal shelf, fixed to the upright by screws, is placed a spirit-level parallel to the line joining the feet. By means of the bubble, the feet are placed on a horizontal line on the base-ring, the slider pushed down until the point rests on the metal, and its position marked with chalk.

HANDSPIKE.—The handspike is used for giving the horizontal direction to the piece. That for the field-pieces, called a *trail* handspike, is round, and conical-shaped at the large end, to fit into the pointing-rings. Near this end is placed an iron stop, the stem of which passes through the handspike and is clinched on the opposite side. This stop slides through a slot in the large pointing-ring, and, when the handspike is turned, holds it in its position. At the small end a ring is placed for the purpose of suspending the handspike at the side of the carriage.

THE MANŒUVERING HANDSPIKE, Fig. 208, Pl. 13, is longer than the preceding, and is for service with the siege and garrison carriages, gins, &c. It differs somewhat in shape from the trail handspike, being octagonal in the middle; and, for the mechanical manœuvres, the large end is made square, and at the end tapers off on one side, so as to allow its entrance under a weight in getting a purchase.

This handspike is used with all heavy pieces for traversing the chassis of permanently-established pieces; for running the ordinary barbette pieces to and from battery, and also for the same purpose in siege-batteries, and for traversing the trails in these last.

SHOD HANDSPIKE, Fig. 209, Pl. 13.—For service with the garrison carriages and mortars, it is found advisable to place an iron shoe upon the large end of the handspike, to make it take a better hold upon the platform, especially when taking a very short purchase, in which case the ordinary handspike is apt to slip. The shoe passes entirely around the end of the handspike, to which it is fastened by three rivets, passing entirely through.

All of these handspikes are made of well-seasoned hickory or tough oak.

Pointing-stakes or wires are used for staking out the line of direction for mortar-firing. Any wooden stakes of proper size, may be made use of; one long one for placing in the exterior slope, a short one for the interior crest, and a medium-sized one for the rear of the platform. The *wires* are made twenty inches long, from round wire 0.2 in. in diameter.

Plummet.—Two of the pointing-wires being fixed in the parapet, as near as possible in the vertical plane passing through the object and middle of the platform, a plummet, which consists simply of a cord with a bullet or other weight attached to one end, is used to place the line of metal in the same plane.

Pointing-Cord.—This is any smooth small cord, one end of which is attached to the short stake in the interior crest, when pointing-stakes are used, and, being stretched back over the rear stake, serves to direct the mortar; the plummet, to facilitate the operation, being attached to it just in rear of the mortar, or held against it with the hand.

Pointing-Board.—This is a piece of board notched in the middle to fit on the rear stake, and graduated in both directions from the center, for the purpose of correcting the fire should the shells fall to the right or left of the object fired at. This is, however, of little importance if the pointing-stakes are accurately placed, as placing the pointing-cord to the right or left of the stake, instead of over the center, will be found usually a sufficient corrective.

The Quoin is simply a large wedge, used in place of an elevating screw under the chase of mortars and the breech of short howitzers, to keep them in the proper position when elevating. It has a handle on the large end, by which it is moved.

The Chock is a smaller wedge, with a handle in the side, used for chocking the wheels of the chassis of permanently established carriages, when the piece has received the proper direction.

LOADING IMPLEMENTS.

Rammers, Fig. 210, Pl. 13.—The *rammer-head* is cylindrical in shape, made of beech, elm, or other hard wood, and has the corners at the large end slightly rounded. It is bored to receive the tenon of the staff, to which it is fastened with a wooden pin, and around the neck is placed a copper band.

Rammer-heads for howitzers are countersunk to receive the head of the fuze, and prevent its being injured when the shell is pushed home.

Sponge-heads, Fig. 211, Pl. 13.—They are cylindrical, the corners at both ends slightly rounded, and are bored out to receive the tenon of the staff to which they are fastened with two wooden pins. They are made of poplar, elm, &c.

Sponges are made of woolen yarn, woven into a warp of hemp or flax-thread, making a tissue about half an inch thick, which is sewed up in the form of a bag, which fits the sponge-head, to which it is fastened with copper nails driven into the inner end of the head, near the staff, a strip of leather being put under the heads of the nails to prevent them from tearing out.

For the coehorn mortar the sponge is attached to a short staff, no rammer or sponge-heads being used.

For the columbiads, separate woolen sponges are provided for wiping the chamber and the bore. Stiff hair-brushes are also sometimes used with these pieces, in order to clean them more easily and thoroughly. The brushes are cylindrical, with hemispherical ends made to fit the bore or chamber. The sponge-heads for these pieces must also be made hemispherical at the bottom.

Staves, for sponges and rammers, are made of tough ash, and are turned in a lathe, with tenons and shoulders where they enter the head.

The rammer and sponge for a field gun or howitzer, or mortar, or the 8-inch siege howitzer, are attached to the same staff; for other pieces, they are on different staves, being thus lighter and more convenient to handle.

Sponge-covers, to protect and preserve the sponges, are made of strong linen or canvas, painted. The diameter of the bag is equal to that of the bore of the gun, and the length sufficient to allow the mouth to be drawn together, around the staff, by means of a cord inserted in the hem. A loop of canvas sewed to the bottom, serves as a handle by which to pull the cover off. The covers are marked, in white, with the caliber of the piece to which they belong.

LOADING-TONGS, Fig. 212, Pl. 13, for placing in the charge and shell of 8-inch siege howitzers, are formed of two arms so hinged together, that the bent ends of the short arms will enter the ears of the shell, and the grooved and widened ends of the long ones will clasp between them the cartridge. The implement is made of such a length, that the cartridge can be thrust into the chamber by reaching in one hand, holding the tongs with the cartridge in position.

SHELL-HOOKS, Fig. 213, Pl. 13.—For mortar and other heavy shells which cannot easily be handled, shell hooks are used. These consist of two bent iron arms connected by a pivot; one end of each is bent inwards to enter the ear of the shell, whilst the other ends are joined to a handspike ring, by two small rings. A handspike being run through the large ring, and the lower ends of the arms adjusted to the ears of the shell, these ends approach each other, taking hold of the shell as soon as the handspike is raised.

HAVRESACK.—To prevent accidents whilst the cartridges are being carried to the guns, they should be protected by some kind of a covering. For this purpose, havresacks are used with field pieces and mortars, and sometimes with larger guns. They are made of russet bag leather, the fronts and backs being connected by *gussets*, which form the ends and bottoms and allow the bags to be folded flat. The flap is of the same piece as the back, folds down, and is fastened to the front by a buckle and strap. A billet and a buckle strap, sewed to the back of the bag, form the shoulder-belt.

PASS-BOX.—This is a box used to carry the cartridges to guns in garrison or sieges. It is made of white pine, the sides and ends dovetailed together, and the bottom let in between the sides, and nailed to them and the ends. The top is fastened on with two butt hinges, and kept closed by a strong hook and staple. A wooden handle is fastened with screws diagonally on one end, by which the box is carried.

BUDGE-BARREL, Fig. 214, Pl. 13.—This is a strong barrel, made of oak, and bound with copper hoops. To protect the cartridges placed in it, and at the same time give easy access to them, it has a cover made of bag-leather, the lower edge of which is fastened under the top hoop by copper nails, and the rivets which hold the hoop. The mouth is drawn together by a double cord passing through holes in the leather near the upper edge. Each cord is 6′ long, and the ends pass through a conical hood of bag-leather, which covers the mouth when drawn together, and protects it from rain or sparks. Fig. 214 represents this arrangement for the mouth. The budge-barrel is used to convey cartridges from the magazine to the batteries, for the purpose of distributing them by means of the havresacks or pass-boxes to different pieces.

HOT-SHOT FORK, Fig. 215, Pl. 13.—Hot shot require particular implements for getting them from the furnace, carrying them to the pieces, and placing them in the muzzle. The fork is made of iron, fastened to a wooden handle, and is used to pull the shot out of the furnace. It has two prongs, which curve inwards and upwards, so as to retain the shot between them when once in position.

LADLES.—For carrying the shot to the pieces, there are two kinds of ladles. The first, Fig. 216, Pl. 13, consists of a ring and stem of iron, fastened to a wooden handle two feet long. The inner top edge of the ring is grooved out to receive the shot.

The other ladle, Fig. 217, Pl. 13, for carrying the largest shot, consists of a similar ring, to which stems are fixed for connecting one single and one double handle, so that two men can be employed to carry the shot, the double handle being to prevent the ladle from turning over.

TONGS, Fig. 218, Pl. 13.—To pick up any shot which may fall upon the ground, iron tongs, joined to wooden handles, are used. They consist of two arms joined by a pivot, each arm at its lower end being fastened to a semi-circular piece, which clasp firmly under the shot when the handles are closed.

LADLES, Fig. 219, Pl. 13.—For the purpose of removing the projectiles from guns, when it is not desired to fire them off, the common ladle is used. It consists of a ladle-head made of the same kind of wood and in the same way as a rammer-head, and the ladle proper, which is of sheet-brass, fastened to the head with copper nails. Ladle-staves, are made like those for rammers and sponges, and are fastened to the heads in the same way, with wooden pins.

Worms, Fig. 220, Pl. 13.—These are made of iron, twisted into the form represented in the figure, with a socket to receive the staff, to which it is fastened with an iron rivet. There are two sizes,—one for siege and garrison guns, the other for field artillery. They are used for withdrawing cartridges, rags, &c. from the guns.

Scraper, Fig. 221, Pl. 13.—This implement is used to scrape the residue of powder from the bore of mortars and howitzers, and remove it from the piece. It consists of a handle of iron, having a scraper at one end, and a spoon for collecting dirt at the other, both being made of steel.

Dredging-Box, Fig. 222, Pl. 13.—To render the fuzes of mortar shell more certain of taking fire, meal-powder is sprinkled over them, after the shell is placed in the mortar, from a dredging-box. It is made of sheet copper. The top fits over the box, and is pierced with holes for the escape of the powder.

IMPLEMENTS FOR LOADING SHELLS.

Powder-Measures are used for measuring, not only the charges for shells, but also those for the pieces themselves; but for the latter, when great accuracy is required, the charges should be weighed. They are made of sheet-copper, and cylindrical in shape. The bottom is made with a flange, turned downwards, and is soldered to the sides.

For dimensions, see Appendix, p. 31.

Funnel.—This is made of sheet-copper, and is used for pouring the charges from the powder-measure into the shell. It is of the ordinary shape, the upper edge being turned over, outwards, to stiffen it.

Fuze-Setter, Fig. 223, Pl. 13, for setting wooden fuzes in the fuze-hole. It is made of brass; the bottom is countersunk and cup-shaped, to prevent it from slipping off from the head of the fuze.

Fuze-Mallet, Fig. 224, Pl. 13, for setting the fuze home. It is cylindrical in shape, with a handle on one end, and is turned out of one piece of dogwood, oak, or other hard wood.

Fuze-Saw, for cutting wooden fuze to the proper length. It is a common 10″ tenon saw.

Gunner's Callipers, Fig. 225, Pl. 14, are made of sheet-brass, with steel points. The two branches are connected by a brass pivot, fastened on the upper side by a washer and screw. To prevent the screw from working loose, the upper end of the pivot and the hole in the washer are made square. Besides the graduations marked in the drawing, other useful data may be engraved on the reverse side.

The scale of inches and parts on the edge (see Fig.) is used for measuring lengths of fuzes, &c. The callipers are also used to measure the diameter of shot and the caliber of guns. For measuring shot, the points are placed at the opposite extremities of a diameter, when the size of the shot is shown by the figures placed on the small arc on the circular part of the arm near the joint, the inner edge of the other arm, or a mark on it, coming in succession opposite the different points which mark the sizes. To measure the caliber of a gun, the position of the points is reversed; they are pressed against the sides of the bore at the extremities of a diameter, and the caliber is read off from the line, on the scale marked "guns," with which the back of the other branch coincides. The graduations on the scale next below the one marked "guns," will give the diameters in inches. Those on the outside arc of the arm designate degrees and parts of degrees.

Fuze-Auger, Fig. 226, Pl. 14. After fuzes are set in position, it sometimes becomes necessary to reduce the length of the composition column, to do which the *fuze-auger* is employed. It consists of a steel *bit*, fastened in a wooden handle, at the lower end of which is a brass socket, with a bar, under which the graduated limb of the slider moves. The bit fits into the slider, and is, by a steel thumb-screw, fastened to it in any required position. The position of the slider, which determines the depth to which the auger bores, is regulated by a scale attached to it by a screw.

The Shell Plug-Screw, Fig. 227, Pl. 14, for extracting the corks or wooden plugs with which fuze-holes are stopped, is made of iron, and consists of a ring 2″ in diameter, to which is fastened a strong screw.

THE FUZE-EXTRACTOR, Fig. 228, Pl. 14, is used for extracting wooden fuzes from the fuze-hole when they have been too firmly driven to be withdrawn in any other way. It consists of an inner screw and stem of steel, riveted to an iron handle, and contained in a hollow steel screw, which works up and down by means of an iron nut with two handles. The hollow screw is prevented from turning by a slot and a feather in the frame, which is of brass. The nut is kept in place by four iron set screws, the points of which enter into a groove in the nut. To extract a fuze, the bottom of the frame is placed on the shell over the fuze head, and the inner screw screwed into the fuze, by means of the upper handle. The handles of the nut are then turned, which raises the hollow screw, and with it the inner screw and the fuze.

PORTFIRE CUTTER, for cutting portfires to place in the shells or for other purposes. It is simply a strong pair of steel scissors, with an indentation 1 in. wide and .4 in. deep made in one of the blades for the purpose of holding the portfire.

PRIMING AND FIRING IMPLEMENTS.

PRIMING-WIRE, Fig. 229, Pl. 14, for pricking the cartridge before priming when the quill or metal tubes are used. It is made of wire a little less in diameter than the vent, sloping to a point at one end, and at the other bent into a circle, which serves to hold it by, as well as prevent it from slipping through the loop on the tube pouch, where it is carried when not in use.

GUNNER'S GIMLET, E Fig. 229, Pl. 14, is of the same form and size, except that at the point a small screw is formed. It is used for boring out plugs which may have been inserted in the vent, or the stems of priming-tubes which may have become wedged in there.

VENT-PUNCH, Fig. 230, Pl. 14.—Sometimes these obstructions cannot be withdrawn by the gimlet, or an iron plug may have been inserted, in which case it is forced in with the vent-punch. This is a short wire of the same size as the preceding, cut off square at one end, and with the other brazed into an iron head. The head has a hole in it, through which a nail or piece of wire may be inserted, to aid in withdrawing the punch from the vent.

GUNNER'S PINCERS, Fig. 231, Pl. 14.—If the obstruction in the vent projects beyond the surface of the gun, or has a head, it may be withdrawn with the gunner's pincers, which are made of iron with steel jaws, and have on the end of one of the arms a claw for drawing nails, &c.

LANYARD AND HOOK, Fig. 232, Pl. 14.—The almost universal methods of firing cannons now is by means of friction-tubes, the only implement required being a lanyard, or cord of suitable length fixed at one end to a wooden handle on which it is wound when not in use, and having at the other end a small hook which enters the eye of the tube. When quick-match is used for priming, it is usually set fire to by using a piece of lighted slow-match held in a *linstock*, which is a short staff of oak or ash with a pointed ferrule to stick in the ground, and near the head a hole through which passes the end of the slow match.

TUBE-POUCH, Fig. 233, Pl. 14, for containing the friction-tubes, lanyard, and thumb-stall, and carrying the priming-wire and gunner's gimlet. The sides and ends are made of russet sole-leather. It has two covers, the inner one having end-pieces sewed to it which shut over the ends of the pouch. The outer one or *flap* is of the same piece as the back, and is fastened down by a strap to a brass button riveted to the bottom of the pouch. It is provided with a waist-belt, which passes through two loops sewed to the back of the pouch. Two small loops sewed to the inside of the flap-serve for carrying the priming-wire and gimlet.

THUMB-STALL, Fig. 234, Pl. 14, is made of buckskin, with the pad stuffed with hair, and a string at the end also of buckskin, to tie around the wrist. It is used to stop securely the vent whilst the piece is being loaded.

Besides the implements described in the foregoing classes, there are others which are used with the pieces in managing them, and which will now be described.

ROLLER HANDSPIKE.—This is for working the eccentric rollers of casemate carriages, and is made of round iron tapering to fit the mortise in the eccentric. It may be made single like a truck handspike, or with two branches to fit in both mortises of the roller at the same time.

Truck Handspike, for use with casemate carriages in running them from battery. It is made of round iron, tapered to fit the mortises in the periphery of the gun-carriage trucks.

Prolonge, Fig. 235, Pl. 14.—It is made of 3½" hemp rope. A *toggle* is fastened to one end by three *rings* and a *thimble*, which is worked into the rope; another thimble holds a *hook* at the other end of the rope; the splice at each end is sewed with marline. Two rings are lashed to the prolonge with ½" marline, for which purpose they have a straight side, which is made to fit the prolonge-rope. The prolonge is used with field pieces to attach the gun to the limber when firing in retreat, or advancing, instead of limbering up; for the same purpose in crossing ditches; for slinging a piece to a limber; for righting carriages when upset, and for various other purposes. It is 26′ 7″ long, and is carried wound around the prolonge-hooks on the trail of the piece.

Drag-Rope, Fig. 236, Pl. 14. This is a 4" hemp rope, with a thimble worked into each end, one of the thimbles carrying a hook. Six handles, made of oak or ash, are put in between the strands of the rope, and lashed with marline. It is used to assist in extricating carriages from different positions, by the men, for dragging pieces, &c. Length 28 feet.

Men's Harness, Fig. 237, Pl. 14, is made of 4" rope, with two thimbles and a hook like the drag-rope. Six loops, made of bag-leather, are attached to the rope in pairs, by means of knots worked in the rope, or by leather collars sewed to it with strong twine. As implied by its name, it is used for harnessing the men to the pieces, under circumstances where horses cannot be used, as in dragging them up very steep ascents, &c. Length 18 feet.

Vent Cover.—It is made of black bridle-leather, and used to strap over and protect the vent of a piece; a pin of copper or brass, fastened by two rivets, enters the vent, to prevent the cover from slipping. It is fastened to the piece by a buckle and strap, the latter passing around the breech, the length of it depending upon the size of the gun.

Tow-Hook, Fig. 238, Pl. 14.—It is made of round iron, with a hook at one end, and a small hammer welded to the other. It is used for unpacking the ammunition chests of field-carriages, the hammer being used for making any repairs on the strapped shot and shell and fixed ammunition, which may be found necessary, or for any other purpose.

SPONGE BUCKET, Fig. 239, Pl. 14, for field-gun carriages. It is made of sheet-iron. The top and bottom are turned over the sides and fastened with rivets; or the bottom may be fastened to the sides by a double fold, and stiffened with a hoop shrunk on above the seam.

The Float is of wood, fastened by two rivets to a cross-bar, and is put in before the top is fastened on. It serves as a cover to the bucket, by turning it after the ends of the bar are drawn through the openings left for them in the top. The handle of the float is fastened to it with two rivets, and is connected with the bail or handle of the bucket, by two rings and a chain. The bail is fastened to the bucket by two ears which are riveted to the sides. A toggle is fastened to the top of the bail by two links and a swivel, and is used to attach the bucket to the eye of the axle-strap on the gun-carriage. This bucket is used to contain water for dipping the sponge in when washing out the pieces.

TAR-BUCKET, Fig. 240, Pl. 14.—It is made of sheet-iron, and in the same manner as the sponge-bucket. The cover is fastened to the top by a rivet, on which it turns, and is kept closed by shutting over a stud riveted into the top of the bucket. The ears are, as in the other, riveted to the sides of the bucket, and a ring for suspending the bucket on its hook, is connected with the ears by two chains and hooks. The bucket is used for carrying tar along with the pieces, for use on marches.

WATER-BUCKET, Fig. 241, Pl. 14, for forge and battery wagons. It is made of a bottom and staves of oak, hooped with three iron hoops, riveted at the ends and to the bucket. The *ears* are let into the sides, and secured each with a rivet. The ends of the bail are bent and hooked into the ears, and at top is connected by a swivel to a single link.

These buckets are now made of leather, the seams sewed and riveted together, and the whole painted. This makes a much lighter, and quite as durable a bucket.

WATER-BUCKET, Fig. 242, Pl. 14, for garrison service. With some modifications in the form, this bucket is made in a similar manner to the preceding, but the handle has no link and swivel attached to it.

CHAPTER XI.

AMMUNITION.

THE term ammunition comprises the charges of powder and projectiles required either in the use of small arms or of cannon, prepared and ready to be placed in the piece; and in a more extended sense, is applied to all the stores, &c., used with the arms.

FIREWORKS.—By the term fireworks is meant the different pyrotechnical preparations used in warfare. They are divided into, such as are used to communicate fire to the charge in a gun; *fuzes*, which are used to communicate fire to the charge in hollow projectiles; war and signal *rockets; fire* and *light balls.*

MADE-UP.—In our service the ammunition and fireworks are made up in the arsenals, under charge of the ordnance department. Great care should be taken in their manufacture, especially in regard to some of the most inflammable.

The buildings should be as light as possible, in order to lessen the effect of an explosion; and the use of iron is avoided as much as possible in making them, and the tables, fixtures, and tools to be used. If iron nails are used in the floors, the heads should be driven below the surface and covered with putty; and as further precautions, use oil-cloths or carpets, and have them frequently swept. In rooms where powder is used, the men should be made to wear socks or, better still, slippers which are not worn out of the room. No more powder than is necessary is kept in the laboratory at a time, and the ammunition, &c., when finished, is taken to the magazine. Powder-barrels and ammunition-boxes must be carried, when moved, in hand barrows, and not rolled or dragged along the floor. Rockets, port-fires, fuzes, &c., should be driven in rooms by themselves, and where no powder or composition is present except that in use.

AMMUNITION FOR FIELD SERVICE.—This consists of the charge of powder contained in a cartridge-bag, either fixed to or separate from the projectile.

CARTRIDGE-BAGS.—The cartridge-bag should be made of wild-bore, merino, or bombazette, which should be composed entirely of wool, free from any mixture of thread or cotton, which would be apt to retain fire in the piece. The texture and sewing should be close enough to prevent the powder sifting through. Untwilled stuff is to be preferred. Flannel may be used when the other materials cannot be obtained.

The bag is formed of two pieces, a rectangle, which forms the cylinder, and a circular piece which forms the bottom. As the stuff does not stretch in the direction of its length, the long side of the rectangle should be taken in that direction, otherwise, the cartridge might become too large for convenient use with its piece.

The material is laid, sometimes several folds thick, on a table, and the rectangles and circles marked out on it with chalk, using for the purpose, patterns made of hard, well-seasoned wood, sheet-iron, or tin. (For the dimensions of these patterns, see table, p. 30, of the Appendix.) The pieces are then cut out with scissors.

The short sides of the rectangle are sewed together, allowing half an inch for seams, and the bottom sewed in with the same allowance. The sewing is done with woolen yarn, twelve stitches to the inch. The two edges of a seam are turned down on the same side and basted to prevent the powder from sifting through. The bottom edges are basted down on the sides. Bags to be used with fixed ammunition are not sewed all the way to the top. They should, when filled, pass through the small shot-guage of their caliber, and while empty, are verified by comparing them with a pattern, or having two marks on the table between which they should fit.

BLANK-CARTRIDGE BAGS, or those intended for immediate use, may be made of two rectangular pieces with semi-circular ends sewed together. The pieces are marked out with stamps made of one-inch board with a handle in the middle of one side, and on the other two projecting rims of copper or tin, parallel to each other, and half an inch apart. The outer one being at the edge of the board, marks where the stuff is to be cut, while the inner marks the sewing line. The bottoms of these rims are roughened to retain the chalk or paste used for marking. (See page 32, Appendix.)

Packing and Preserving.—The bags are preserved from moth by being packed with pounded camphor and black pepper, or dipped in water with arsenic dissolved in it, and then dried. Or they may be sealed up in packages of 50, in cases made of cartridge paper, with the seams carefully closed by pasting over them strips of thin paper. Each package is marked with the number and kind of bags.

They may be protected from moisture by being enveloped in water-proof paper, and inclosed in barrels or boxes, like cartridges for small arms.

Filling.—When special accuracy is required, charges are carefully weighed in delicate scales; but usually the bags are filled by measurement. For this purpose, 14 cylindrical, copper powder-measures are provided, each containing a known quantity of powder. (See Appendix, p.31 .) Each different kind is filled by itself. (See Appendix, p. 31, for charges.) Blank cartridges, and those for the 12-pd. gun, are simply tied firmly about the throat with twine. Those for *fixed* ammunition are left open until the projectiles are ready. Those for the 24-pd. and 32-pd. howitzers being used separate from the projectiles, are fixed to cartridge-blocks, to give them a better finish, to help fill the chamber, and to prevent them from turning in the bore when the piece is being loaded. These blocks, Fig. 166, are cylindrical in shape, with one groove, and the height depends upon the charge used. The grooved end is inserted in the mouth of the bag resting on the powder, and the bag being pulled up over it, a tie is made with a piece of twine in the groove. The mouth of the bag is then turned down, and another tie made over it, between the block and the powder, which prevents the powder from working up between the block and the bag. The superfluous part of the bag is then cut off, and the cartridge covered with its *cylinder* and *cap*, which will be described hereafter. The cartridge blocks are made of *poplar*, *linden*, or other soft wood. (For dimensions, see Appendix, p. 32.)

The Sabot, Fig. 164, for guns, is cylindrical or nearly so, in shape, and for the howitzers, conical. It is made of the same woods as the cartridge-block, and should be clear of knots and splits, and well seasoned.

24 or 32 pd. How. and Shell.

Canister.

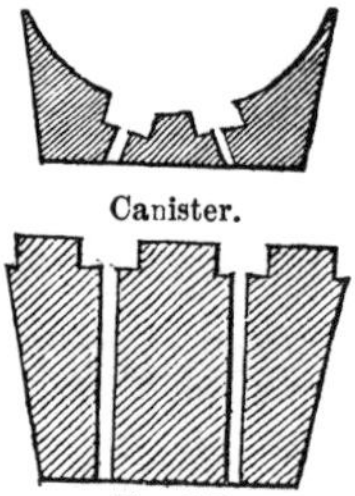

Fig. 243.

Sabots for shot and spherical case for guns have one groove for attaching the cartridge; those for gun canisters and for the 12-pd. howitzer shells, spherical case, and canister have two grooves. A, Fig. 164. Sabots for the 32-pd. and 24-pd. howitzers, Fig. 243, have no grooves, but are furnished with handles made of a piece of cord, passing through two holes, and fastened by knots countersunk on the inside. (For dimensions, see p. 34.)

The sabots are fastened to shot and shells with strips of sheet tin, which are cut with shears, and straightened with a wooden mallet on a block of lead.

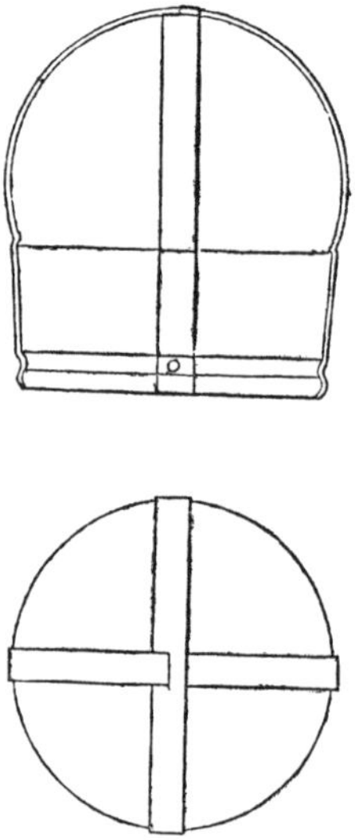

Fig. 244.

For shot, there are two straps crossing at right angles, Fig. 244, one passing through a slit in the middle of the other. For shells, there are four straps soldered to a ring of tin, or fastened to it by cutting four slits in the ring, into which the upper ends of the straps are hooked and turned down on the inside of the ring. Fig. 245.

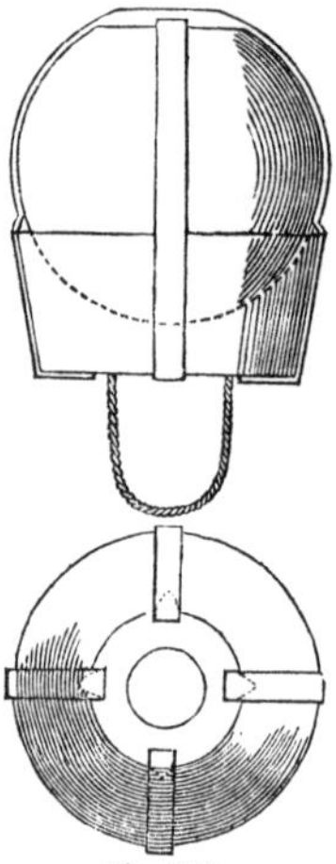

Fig. 245.

Strapped Shot and Shell.—To prepare these a helper knocks off the scales from the balls with a hammer, cleans and dries the interior of the shells if requisite, wipes the balls, and gauges them both before and after they are strapped. The workman inserts the roughest part of the shot in the cavity of the sabot, and strikes a few blows on the bottom of the sabot to make the shot enter; he can tell by the sound if the shot touches the bottom of the cavity; if it does not touch, he tries another sabot. With the edge of the hammer he bends one end of the strap which is not slit into the groove of the sabot, punches and nails it; he fastens the other end in the same manner, cutting off the superfluous length; he then nails the other strap, and with his hammer sets them both in, close to the ball, at the top of the sabot.

The sabots for 32 and 24-pounder field howitzers having no groove, Fig. 245, each strap is fastened by one nail on the side and two under the bottom of the sabot. Two men can strap, in ten hours, 130 shot, or 75 shells, cutting the tin from the sheet.

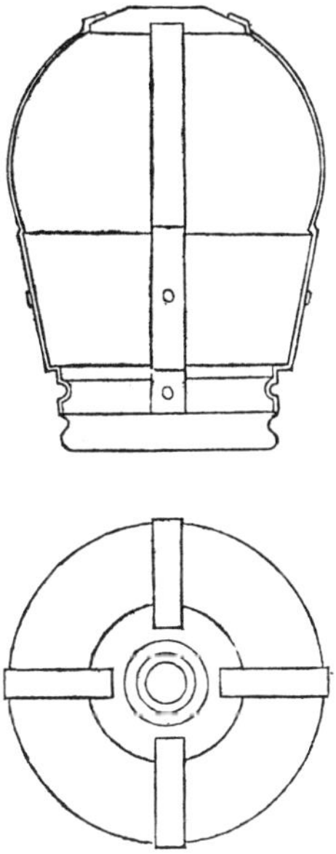

12-pd. How. Shell or Sph. Case.
Fig. 246.

If tin or sheet iron cannot be procured, straps may be made of *strong canvas*, one inch wide, sewed at the point of crossing. The part of the ball which is to be inserted into the sabot is dipped in glue; the straps are also glued to the ball; the ends are doubled into the groove, and secured by two nails in each end. Another method is to wrap round the ball a band of canvas, one inch wide, one half of which is glued to the ball, the other to the sabot.

When the shot is attached to the sabot by a single band of canvas, or when it is placed in the sabot without any strap, the cartridge-bag is drawn over it, and tied on top; for this purpose the bag should have an additional length of from $2\frac{1}{2}$ to 3 inches.

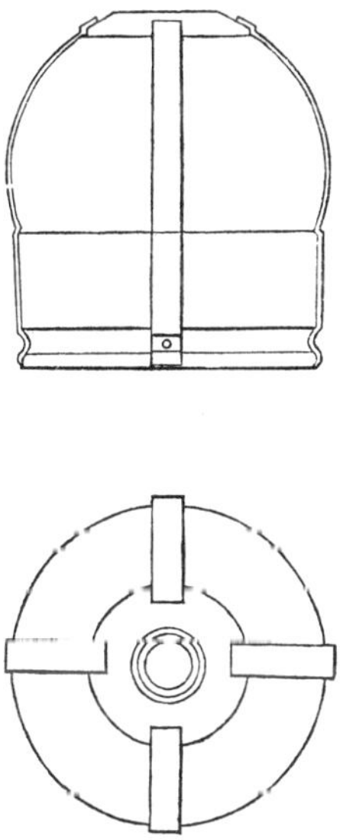

6 and 12 pd. Guns Sphl. Case.
Fig. 247.

When sabots cannot be obtained, place upon the powder a layer of tow, about 0.2 in. thick, forming a bed for the shot; tie the bag over the shot and around the tow; the bag requires to be one inch longer than for strapped shot.

The shells must be properly cleaned and dried before being attached to the sabots.

Fuze-Plugs, Fig. 248.—The fuzes for field shells and spherical case are inserted, at the moment of loading the gun, into wooden *fuze-plugs*, previously driven into the shells.

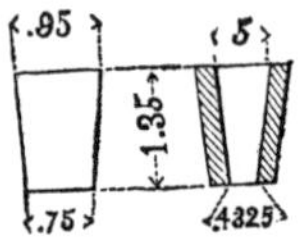

Fig. 248.

These plugs are made of *beech*, perfectly seasoned and dried, so that they may not shrink after they are driven.

(For dimensions, see Appendix, p. 34.)

Loading Shells.—They are set up on their sabots, the charges measured out in the proper powder-measure, and poured in through a copper funnel. The fuze-plugs are then driven in with a mallet, allowing the tops to project about 0.1 of an inch, care being taken not to split them. The holes in the plugs are then carefully reamed out, and stopped with tow wads, which are pressed in firmly with a round stick.

(For charges, see Appendix, p. 35.)

Loading Schrapnell, or Spherical-Case Shot.—The shot having been cleaned and strapped to the sabot, the balls are placed in. In the Baden service, before this is done, the balls and inside of the shot are given a coating of a kind of oil (colza), the object being to facilitate the loading, the formation of the powder-chamber, and the flowing of the sulphur. It prevents, also, the formation of cavities by the sulphur and the adhesion of it to the balls after the explosion takes place. This last is a very important point, since a serious defect in schrapnells loaded with sulphur, arises from the balls adhering together and not scattering sufficiently.

Formerly the powder was placed in loose, with the balls; but this arrangement was very liable to accidents, and if the powder remained in for any length of time before being used, it was ground up and became otherwise impaired. Loading with sulphur obviates all these difficulties, and is now almost universally adopted.

The balls being inserted, a stick a little less in diameter than the fuze-hole, and having a groove on each side of it, is inserted and pushed to the bottom of the chamber by working the balls aside. A better way is to make an opening through the balls with a stick, and then insert an iron spindle in three parts, the outside pieces being inserted first, and then the middle part, which is narrower than the others, being forced in between, like a wedge, leaving on each side of it an opening for the entrance of the sulphur.*

The shot is now placed in a sand-bath or oven, and brought to a proper temperature, which may be determined either by a thermometer or by making use of bars of fusible alloys, composed as follows:—

For the max. temperature.......	5 parts of tin.	For the min. temperature........	8 parts of tin.
	3 " " lead.		3 " " lead.
	5 " " bismuth.		5 " " bismuth.

The shot is at the proper temperature to receive the sulphur when the last alloy, being brought in contact with it, melts easily and the first melts with difficulty.

The melted sulphur is then poured in to fill up the interstices between the balls, the shot allowed to cool, and the sulphur to harden, when the stick or iron spindle is withdrawn, and the sulphur adhering to the sides of the eye and surface of the shot is removed. It is scarcely necessary to state, that case shot loaded in this way are not attached to the sabot until afterwards.

If a fuze-plug and paper fuze are to be used, the charge is placed in and the plug inserted exactly as for a shell; but if, as is most likely to be the case, the Boarmann fuze is to be used, the charge is inserted and the stopper and fuze screwed into their places, care being taken before placing the fuze in position, to puncture the covering of the magazine, so that the fire can communicate with the charge.

(For charges, see Appendix, p. 35.)

The use of grape-shot for field pieces has been discontinued for a number of years, it being considered that for the ranges of that kind of artillery, the shot of which canisters are made, are large enough, and the canister possesses the advantages of striking a great many more points at one discharge than grape. There is an advantage, too, in not having so many different kinds of ammunition for a piece.

Canister Shot, Fig. 249.—To form the cylinder for canister shot, the tin is lapped about 5 inches and soldered. It is then fast-

* The other and better method of loading schrapnell shot has been described elsewhere.

Gun-canister, fixed.

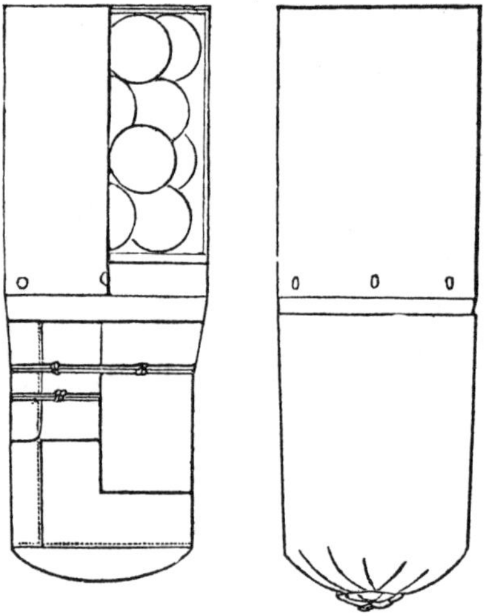

Fig. 249.

ened to the sabot with 6 or 8 nails, and a plate of rolled iron is placed at the bottom on the sabot.

The exterior diameter of each canister must be verified with the maximum shot gauge, and the interior with a cylinder of a diameter 0.02 in. less than that given in the table (see Appendix, p. 35), which should enter the canister.

Before filling, dip the cylinder in a lacker of bees-wax, dissolved in spirits of turpentine to prevent it from rusting. Give the balls and plates a coat of paint or coal-tar. To fill the canister, place it upright on its sabot; put in a tier of balls, filling the interstices with dry, sifted saw-dust, packing it with a pointed stick, so that the balls will hold by themselves when the case is turned over; and throw out the loose saw-dust. Place another tier of balls and proceed in the same way until the canister is filled; cover the top tier with a layer of saw-dust and put on the cover, which is a circular plate of sheet-iron, settling it well with a mallet in order to compress the saw-dust. The top of the cylinder is cut into slits about 5 in. long, which are turned down over the cover, thus securing it. After it is finished, the diameter of the canister should be verified with the maximum gauge of its caliber.

Fixed Ammunition.—This term is applied to the powder-charge and projectile when fixed together, forming thus one charge or round, which is inserted in the gun at the same time. All the ammunition for the field guns and 12-pd. howitzer is fixed. That used with the 24-pd. and 32-pd. howitzers is in two parts.

Shot Fixed.—The bags being filled, shaken, and gauged, are carried to the finishing-room. The mouths of the bag are twisted and pressed down on the powder so as to settle it; and the cartridges are then placed upright in a tub or box near the workmen, who are placed in pairs facing each other and seated on benches.

One of them opens the bag and levels the powder, and the other inserts the sabot of a strapped shot square upon the powder and draws up the mouth of the bag over the shot. The first now takes a stick with an eye bored through it, and passes about 4 feet of twine through the eye, making two turns and a double hitch at the top of the sabot, between that and the ball. He makes a knot in the end of the twine, inserts it in a slit made in another stick, and tightens the double hitch by rolling the twine on the sticks, and bearing on the sabot; he then takes out the end of the twine from the slit, ties it with a hard knot, which he tightens with the assistance of the sticks, and cuts the twine off near the knot. The second man turns down the mouth of the bag over the sabot, and the first makes a similar tie in the groove, and another below the sabot, the twine being lodged between it and the powder. The paper cylinder is then run over the cartridge and sabot, leaving about 2 inches of the end of the cartridge uncovered, and tied in the groove of the sabot with a tie like the others. The shot is then examined to see that the sabot is properly fixed and straight, on the same axis with the cartridge, and straightened if necessary. The seams of the bag should be between two straps and the knots, neither on the seams nor the straps. Fig. 165.

The cartridge is now passed through the large guage, which is 0.04 in. larger than the large shot-guage; and if found of the right size, the paper cap is put on, the cartridge laid on its side in a box, and afterwards carried to the magazine. Those cartridges which will not pass through the guage are returned to the men, who cut the strings, and put them up anew.

Fixed Canister.—Canisters are fixed in the same way as shot, except that with the canister, its sabot having two grooves, the first tie is made in the upper one. The paper cylinder is tied in the lower groove. The caps are made somewhat shorter than those for shot-cartridges. Figs. 249 and 165.

Fixed Shells, and Spherical Case, for 12-Pd. Howitzer.—The shells and spherical case are fixed in the same manner as the canisters. Fig. 246.

For the Mountain Howitzer.—The sabot having but one groove, the first tie is omitted, and the cartridge is covered with a cap only.

The great advantage in having fixed ammunition, is the facility and certainty in loading; and were it not for the fact that it would become entirely too unwieldy for prompt and convenient handling, it would be well to have all field ammunition fixed. (See Appendix, p. 36.)

Cylinders and Caps.—For the greater security of field ammunition, the cartridges are covered with paper cylinders and caps. They are both made together on the same former, which is a piece of board with slightly inclined sides, and rounded edges. Around this former the paper is placed, and lapped $\frac{3}{4}$ of an in. for pasting. The requisite length for the cylinder is cut from the smaller end, the rest forming the cap, which is choked at the end from which the cylinder is cut. For this purpose a cylindrical former of wood, with an hemispherical end is used, which should be bored through the axis to facilitate the drawing off of the cap.

The cylinder fits over the body of the cartridge and a part of the sabot, while the cap fits over the end; and when drawn off, which is always done when the cartridge is placed in the piece, leaves the lower end of the cartridge exposed, so that the priming wire or fire from the friction-tube can reach it without going through any paper. In firing shot, the cap when drawn off may be placed over the shot, to diminish windage. (For dimensions of formers, see Appendix, p. 32.)

Packing.—Field ammunition is packed in boxes of well seasoned stuff, $1\frac{1}{4}$ in. thick and dove-tailed at the corners. The cover is fastened with screws, and a rope handle attached to a bracket is fixed at each end. The boxes are painted olive color on the outside, and the kind of ammunition is marked on both ends in large white letters. The place and date of fabrication are marked on the inside of the cover.

For Guns.—Shot, spherical-case and canister fixed, are laid in two tiers across the box, the projectiles alternating with the cartridges at each side. The projectiles of the upper tier rest on those of the lower, and not on the cartridges.

For 32-pdr. and 24-pdr. Howitzers.—*Shells and Spherical-case Shot.*—Placed upright, the balls down, resting on strips of wood about 0.25 in. thick, placed lengthwise of the box and nailed to the bottom, so as to prevent the fuze-plugs from bearing on the bottom of the box. The balls are held down by small strips of wood tacked with sprigs to the sides of the box, over the sabots. The cartridges are laid on top of the sabots.

Canisters are packed in the same manner, omitting the strips of wood in the bottom of the box.

FOR 12-PDR. FIELD AND MOUNTAIN HOWITZERS.—*Shells and Spherical-case Shot, fixed.*—Placed upright, the balls down, resting on strips of wood, as for the other howitzers.

Canisters are packed in the same manner, resting on the bottom of the box.

In all the boxes, the small stores are placed in the vacant spaces on top of the ammunition.

The fuzes of each color are put up in a bundle, wrapped in water-proof paper of corresponding color, and marked with the time of burning. All the fuzes for a box are put up in one parcel, wrapped in water-proof paper, and marked: FUZES.

A layer of tow is placed in the bottom of each box, and the whole contents are well packed in tow, filling the box so as to be pressed down by the cover. About 3 lbs. of tow are required for a box.

SIEGE AND GARRISON AMMUNITION.—This consists of the charge of powder in a bag, and the projectile always separate from the cartridge.

The Cartridge-Bags are usually made of woolen stuff. They are made of two pieces, in the form of a rectangle with a semicircular end, which are marked out with stamps and sewed together as described for making blank-cartridge bags for the field service, and are filled, preserved, and packed in the same way.

Paper Bags.—Bags for heavy ordnance may be made entirely of paper. The bottom is circular, and one end of the cylindrical part is cut into slips about one inch long, which are pasted over the paper bottom on a cylindrical former. When a paper bag is filled, the open end is folded down about three-fourths of an inch wide, and this fold is rolled on itself down to the powder, and the part which projects beyond the cylinder is turned in on the top of it.

The bags are apt to leave paper burning in the gun, for which reason, those made of woolen stuff are preferable.

Bags are sometimes made of both paper and woolen stuff, by forming the cylindrical part of paper, and sewing to it a bottom of woolen stuff made of two pieces semi-circular in shape, and sewed together around the edge.

For Columbiads and Sea-coast Howitzers, the cartridge should always occupy the whole length of the chamber; for this purpose, in firing with reduced charges, *a cartridge-block* is placed in the bag, over the powder. The length of this block for any charge is easily deduced from the length occupied by 1 lb. of powder, as given in the table, Appendix, p. 33.

For Mortars, cartridge-bags may be made in the same manner as for guns, their dimensions corresponding to those of the chamber of the mortar. But as the charge is generally poured loose into the chamber, the bag being used only for carrying it to the mortar, a gun cartridge-bag of any convenient size may be used for mortar service.

For Firing Hot Shot, cartridge-bags are made double, by putting one bag within another, care being taken that they are free from holes.

For Ricochet Firing, or whenever very small charges are required, a bag for a smaller caliber may be used; or, after the powder is poured in, another bag filled with hay may be placed on top of it and pressed with the hands to reduce the diameter. After shaking this bag down and rolling and flattening the empty parts of the two bags, tie them with woolen yarn like a bundle of cartridges, placing the knot on top.

For Guns.

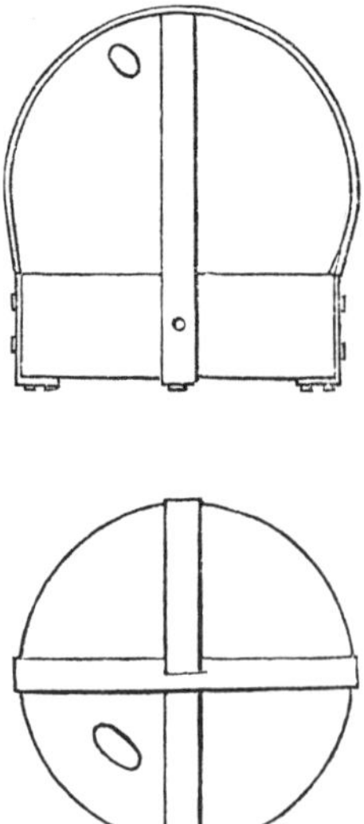

Fig. 250.

For Proving Ordnance, cartridge-bags are made of woolen stuff for small calibers, and of paper for large ones. They should be of the full diameter of the bore or chamber.

STRAPPED SHELLS.—The shot used with heavy guns is generally without a sabot, but the shells are strapped to sabots made of thick plank, with strips of tin as described for strapping shot for field service, except that the straps are not set in close to the shell above the sabot, as they are with the field ammunition. The shells are set in the sabot, and the straps put on in such a way that the fuze-hole may fall in one of the angles between two straps, its axis making an angle of 45° with that of the sabot. The straps are fastened with nails both on the side and bottom of the sabot. Figs. 250 and 251.

In loading, the fuze-hole is always placed in the upper part of the bore.

For Howitzers and Columbiads.
Fig. 251.

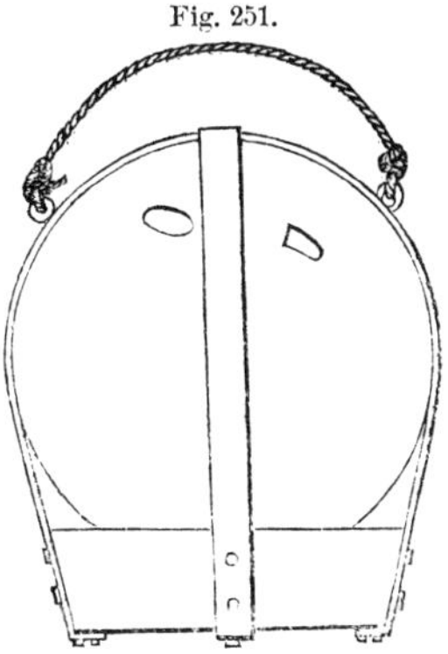

The columbiad and 8 and 10 inch seacoast howitzer shells being very heavy and inconvenient to handle, have a couple of rings or loops of tin soldered to one of the straps to receive the ends of a piece of rope, which is tied to them, and serves as a handle. Fig. 251. (For dimensions of sabots, see Appendix, p. 33.)

Canisters.—Canisters for the siege and seacoast guns, Fig. 252, are made by turning one end of the tin cylinder over the iron bottom.

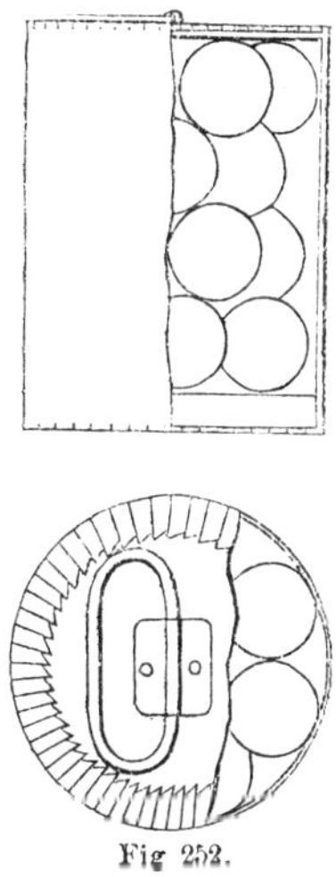

Fig 252.

They are filled and closed with the cover in the same way as field canisters, but have no sabot. A wire handle is fastened to the cover with rivets.

The shot is made of such a size that it will just fit in the cylinder in beds, the number in each bed depending upon the caliber of the piece.

(For dimensions, see Appendix, p. 29.)

In the 8-in. siege and seacoast howitzers, the canisters are attached to sabots in the same way as the field-howitzer canisters. The sabot for the siege howitzer has a hemispherical bottom, and that for the seacoast is conical, to suit the connecting surface between the bore and chamber in these pieces.

(For dimension of sabots, see Appendix, p. 33.)

These canisters have thick iron-wire handles on top, like the others, and are closed at the top in the same way. All canisters are formed of four tiers of shot.

(For their dimensions and weight, see Appendix, p. 27.)

Grape-Shot, Fig. 253, is formed by placing the bottom plate (see p. 160) upon a table, and placing in the indentations three shot of the proper caliber, and on top of these one of the rings, then three more shot, the other ring, the last three shot, all the shot breaking joints with each other, and finally the top plate, which being countersunk like the bottom one, is turned round till the top tier of shot fit in their places. The pin having been inserted from below through the lower plate before any of the shot were put in position, the top of it projects through the center hole in the top plate, and on it the nut is screwed tight, so as to make the whole firm and solid.

The pin in the stands of grape for guns is kept from turning, whilst the nut is being screwed on, by a projection on the head which fits into a slot in the countersunk for the head in the bottom plate. In the stands for the 8-in. siege howitzer, the only one provided with grape-shot, the sabot is attached by the pin passing through it first. The pin is prevented from turning by having a square head, which is countersunk in the sabot. The sabot may be fastened to the lower plate with screws, when a pin of the usual length may be made use of, or the sabot, which is the same size as the one used for canister in this piece, may be inserted separately.

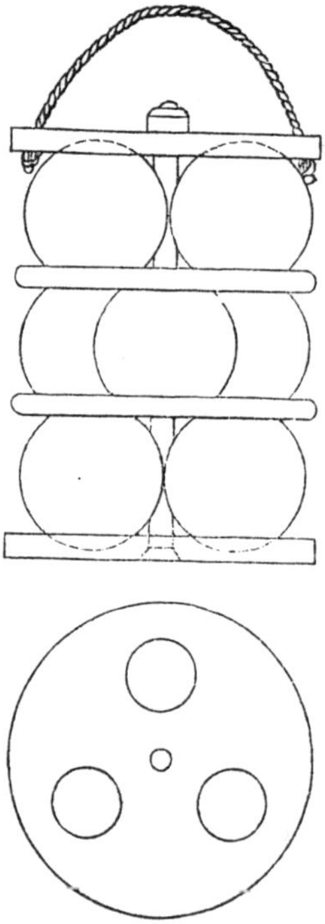

Fig. 253.

The same grape-shot which are used with the 8-in. howitzers are also used with the columbiad of that caliber, by adopting the sabot of the sea-coast howitzer, which serves for both pieces. The grape for these 8-in. pieces is made of 6-pd. shot.

The nut having been screwed on, the ends of a piece of rope are run through the holes in the upper plate, and tied beneath to serve as a handle. These holes are directly opposite to each other, $\frac{1}{4}$ inch in diameter, and $\frac{1}{2}$ inch from the edge of the plate.

(For the size of grape-shot and dimensions of the stands, see Appendix, pp. 29 and 30.)

Filling Mortar-Shells.—These shells are usually filled in the battery, and as they are required. Having been inspected, to see they are clean, dry, and in good order, place them on a block made for the purpose, or on rings of rope, or in indentations in the floor of the magazine, or on the ground, with the eyes up. The charge measured out in a powder-measure is poured in through a funnel, and any incendiary composition, such as pieces of port-fire, rock-fire, &c., inserted.

In the mean time the fuzes are being cut to the proper lengths by resting them in a groove in a block, or inserting them in holes made in the block, or in a post, and sawing them across with the fuze-saw. Or the fuze may be bored through with a gimlet at the proper point.

The fuze is then tried in the eye, and should enter $\frac{3}{4}$ of its length. If it does not, it may be reduced by rasping. The head of it is covered with tow to prevent the breaking of the composition, the fuze-setter placed on, and the fuze driven in until the head projects from 0.2 to 0.4 in. above the surface of the shell. (For the charges of these shells, see Appendix, p. 30.)

Shells for Columbiad and Heavy Guns, are loaded in the same way as mortar-shells, but as paper fuzes inserted in wooden or bronze fuze-plugs are used instead of wooden fuzes, the plug only is driven into its place, and stopped with tow after the bursting charge has been poured through it into the shell. The metal fuze-plug, with its safety cap, is described at p. 298. (For charges, see Appendix, p. 36.)

Wads.—For proving cannon, wads are made of junk, which is first picked, and then beaten in moulds until the compact mass assumes the requisite dimensions. The moulds for each caliber consist of two cast-iron cylinders of different diameters set in oak, or of two strong pieces of oak strapped with iron and joined by a hinge.

The junk is first placed in the smaller of the two moulds, and is beaten into a compact mass by using a cylindrical drift nearly the size of the mould, and a maul or hammer. The upper part of the mould is then raised, the wad taken out, and closely wrapped with rope-yarn in the direction of its axis, ending with a few turns around the middle of the wad, perpendicular to the axis. It is then placed in the large mould, and again beaten with the maul and drift, until its diameter is increased to that of the mould, when it is taken out as before, and its diameter verified by a wooden guage corresponding to the large shot-guage of the caliber.

WADS FOR FIRING HOT-SHOT, and other like purposes, may be made of hay wrapped with rope-yarn, and are made in the same way as junk wads; or they may be made entirely of hay, by twisting from the hay a rope of an inch or an inch and a half in diameter, and then commencing at one end and doubling it up about one caliber in length, twisting it all the time until it becomes nearly large enough, when the rope is wound around the wad perpendicular to its axis, and fastened with a hitch.

RING OR GROMMET WADS, consist of a ring of rope-yarn about 0.7 in. thick, with two pieces of strong twine tied across it at right angles to each other. The size of the ring is the full diameter of the bore, in order that it may fit tight and stop the windage. They increase the accuracy of fire, and are preferred to the other wads when the object is merely to keep the shot in its place, as in firing at a depression. They stop the windage best when placed behind the ball. They may be attached with twine to the straps, or to the balls themselves, or may be inserted, like other wads, separate from the balls.

MORTAR-FUZES are turned from beech, ash, or linden, well-seasoned, and free from knots. They are conical in shape, and made of different sizes to fit the different fuze-holes. They are then graduated on the exterior into inches and tenths, commencing at the bottom of the cup, by means of a steel guage applied to them while in the turning-lathes. The interior, cylindrical in shape, is bored out, to within about an inch of the bottom, and the cup hollowed out. (For dimensions, see Appendix, p. 37.)

The composition for filling them should be well pulverized, thoroughly mixed and sifted; and trial-fuzes should be made with the different compositions, and the time of burning tested. With the same composition, the time of burning will depend very much upon the manner in which the fuzes are driven.

The fuze is placed in a hole in a block of hard wood of the

same size as the fuze, and a quantity of composition sufficient to make, when driven, a column equal in height to one diameter of the fuze-column, put in; copper ladles of the proper capacity being used. A steel drift, shod with copper, is placed on this, and twenty-one blows, in volleys of three, are given with a wooden mallet weighing one lb., the drift being raised after each volley. This is repeated until the fuze is filled up to a certain point determined by a mark on the drift. Care must be taken that the charges are all equal, for which purpose a drift is passed over the top of the ladle to take off the surplus, and the blows should always be the same in number, and delivered with as near the same force as possible.

The pan containing the composition should not be placed on the driving-block, as the sulphur would separate from the rest of the composition, and collect together.

When the composition has reached to within about 0.2 in of the bottom of the cup, a charge of mealed powder is placed in, and driven in the same way as a ladleful of composition. The cup is filled with mealed powder moistened with spirits of wine or strong whisky; when dry, it is covered with a small piece of paper, over which is pasted a cap of strong, water-proof paper, marked with the number of seconds the fuze burns to the inch. For preservation and transportation, the fuze is capped with water-proof paper, linen, or serge, tied on and lackered. Fuzes are packed with tow, in boxes lined with water-proof paper.

Fuzes may be driven with *blind-fire composition*, which will not discover the flight of a shell in the night.

One ladleful of the common fuze composition, or of mealed powder, must be driven in the top of these fuzes.

Two kinds of compositions are used:

1. 6 parts of mealed powder and 4 of sifted wood-ashes; and
2. 10 " " " $9\frac{1}{2}$ " "

(For the ordinary composition, see Appendix, p. 37.)

Paper Fuzes *for Field Service. Making the Paper Cases.*—The case is made of a strip of smooth paper, rolled hard, on a mandril 0.35 inch diameter, and glued, after the first turn, with isinglass glue. The strip of paper is in the form of a rectangle joined to a trapezoid; it is rolled from the large end.

Whole length of the paper,		18.	inches.
Length of the rectangular part,		9.	"
Width of the rectangular part,		1.5	"
Width at small end,		0.4	"
Diameter of finished case before being cut,	At top,	0.52	"
	At bottom,	0.44	"

Log Paper (so called), or thin drawing paper, is suitable for making these cases. The dimensions of the strip of paper must be regulated by trial with the kind of paper used.

After the case is dry, it is smoothed by rubbing it with a fine file, and with sand-paper.

The material for filling these fuzes must be thoroughly incorporated, by being ground together with a muller. Trial fuzes should be made as before, and the rate at which the composition burns be determined. One pound of composition makes 100 fuzes.

Charging *the Fuze.*—A number of the cases are set upright in brass dies made in two halves, kept in place by pins and a wedge which fits in between one of the halves and a flange connected with the other. Or the two parts may be held together by a strong ring or socket inserted in a block of wood. The composition is then driven as has already been described for wooden fuzes, the mallet used weighing but ½ a lb., and the number of blows to each ladleful of composition being 15. Each ladleful makes a length of ¼ of an inch in the fuze.

The fuze is then placed in an iron guage, the bore of which has the same taper as the fuze, and cut off at both ends with a saw or sharp knife. It is then stained the proper color, according to its rate of burning. It has been before suggested that a better way to designate this would be by strips running along the length of the fuze.

For the compositions, see Appendix, p. 37.)

Fuzes for Heavy Guns, Columbiads, and Howitzers, are made in a similar manner, and differ from the preceding only in their dimensions and rate of burning.

(For dimensions and compositions, see Appendix, p. 38.)

The time of burning of these slow compositions is subject to considerable variations, according to the quality of the materials and the manner in which they are mixed. In making the fuzes, therefore, especial care should be taken to try the composition used, and vary the proportions to produce the required result. The compositions in the Appendix are given as approximate guides only.

The method of making fuzes of other kinds has been given in Chap. IX.

PRIMING CANNON.—The means first used to communicate fire to the charge in the gun, were of course of the most primitive kind. Loose powder filling the vent, and the application of a coal of fire, were probably the first employed.

For priming, the loose powder gave way to a strand of quick-match, and tubes of different substances filled with mealed powder, first moistened with alcohol or spirits of some kind, and afterwards dried; while the coal of fire was superseded by the slow-match, and afterwards by the lighted portfire. Finally the friction-tube took the place of both, simplifying and rendering much more certain the firing of the piece.

QUICK-MATCH is even now used to fire some kinds of ordnance, such as stone and heavy mortars, and sometimes in proving pieces. It is, however, extensively used in priming all kinds of fire-works, such as fire and light balls, carcasses, rockets, priming-tubes, &c. Inclosed in tubes, it burns much faster than in the open air, and more so in proportion as the tubes are smaller. It is so used in conveying fire very rapidly from one portion of a piece of fire-work to another, the paper tubes in which it is inclosed being called *leaders*.

It is made of cotton yarn, such as is used for candle-wick, of such a size that when doubled and twisted in the fingers it may be 0.07 in. in diameter. This is wound in a loose ball, of a convenient size, say weighing 1 lb., which will contain 1000 yards, and steeped in gummed brandy or whisky until thoroughly soaked.

A paste is made of mealed powder and gummed spirits, of about the consistency of flour paste, and the bottom of a wooden bowl or copper pan covered with a layer of it $\frac{1}{4}$ in. thick. On this is spread a coil of the yarn, by unrolling the ball, and distributing it equally over the surface of the paste, until there are five or six strands over one another. Over this is placed another layer of paste, and then one of yarn; and so until the vessel is full.

Care must be taken not to entangle the yarn; and the last layer is covered with a bed of the paste a little thicker than the others.

After the yarn has been soaking 3 or 4 hours in the paste, it is wound off on a reel, making it pass through a funnel filled with the paste, or through the hand with some paste in it, taking care that the different turns do not touch each other. Before it is dry it is dredged with meal-powder, allowed to dry slowly, cut from the reel in convenient lengths, and put up in bundles.

The gummed spirit is made by first dissolving the gum in the smallest possible quantity of hot water or vinegar, and afterwards mixing it with the spirits.

Match so prepared should be hard and stiff, and the composition should hold firmly on; 1 yard burns in the open air 13 seconds. The pound of yarn used requires 8 lbs. of mealed powder, $1\frac{1}{4}$ galls. of spirits, and $2\frac{1}{2}$ oz. of gum arabic; the match weighing when dried 9 lbs.

By using *vinegar*, a match is made which burns less rapidly in the proportion of 4 to 5, and with pure water in the ratio of 4 to 6. Alcohol makes a quicker match, but it cannot be gummed, and the composition falls off.

A slow kind of match is made by adding sulphur to the paste; with $\frac{1}{6}$ of sulphur, 1 yard of the match burns 22 seconds; with $\frac{1}{5}$, 33 seconds; with $\frac{1}{3}$, 53 seconds, and with $\frac{1}{2}$, 162 seconds.

When used for discharging cannon, quick match is set fire to by a slow match, portfire, or any other convenient material, the first being most frequently employed. When used to prime shells, carcasses, &c., it is set on fire by the flame from the piece.

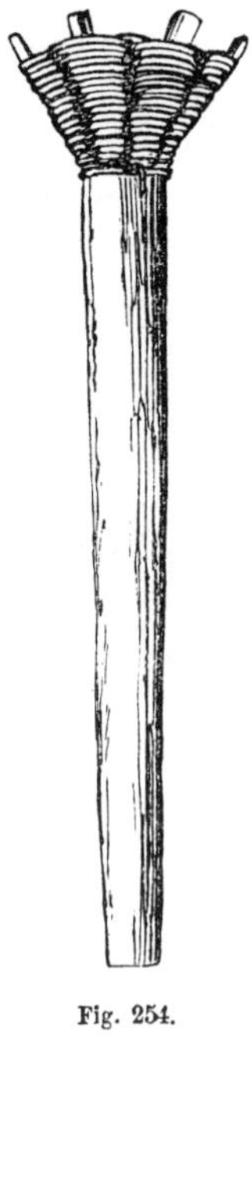

Fig. 254.

Quill Priming-Tubes, Fig. 254, are made from quills by cutting off the barrel at both ends, and splitting down the large end for about half an inch into 7 or any other odd number of parts; These are bent outwards, perpendicular to the body of the quill, and form the cup of the tube. Fine woolen yarn is then woven into these slits, like basket work, the end being brought down and tied on the stem; or a perforated disk of paper is pasted on them.

These tubes are filled by injecting into them, with a *tube-injector*, a liquid paste made of mealed powder and spirits of wine; but a better method is, not to make the paste so thin, and then press it in with the thumb. A strand of quick-match, 2 inches long, is now laid across the cup and pasted in there with the powder paste. A small wire is then run through the tube, and remains there until the paste is dry, and then withdrawn. This leaves an aperture, furnishing a quick communication for the fire along the tube.

A paper cap is placed over the cup, and twisted tightly around the tube under the cup.

Another way to fill these tubes is to stop the small end with sealing-wax, filling the barrel with rifle-powder, and then finishing the cup with paste and quick-match as before.

These tubes are preferable to metal tubes on ship-board or in casemates or block-houses, as the fragments blown out of the vent are not injurious, as is the case with the sharp, angular fragments of the metal tube, especially on board ship, where the men often go without their shoes.

When the tubes are dry, they are wrapped in paper in bundles of 10, and tied up.

METAL TUBES are made of an alloy of Banca tin, lead, and antimony, in the proportion of

50 lbs. of tin,		
50 " lead,		For 6,000 tubes.
1¼ " antimony,		

The metal is melted in an iron pan, and the tubes are cast in moulds which are kept at the same time at sufficiently high a temperature to scorch dry shavings. They should never be cooled by wetting.

The moulds have each a spindle to form the interior of the tubes, and are, during the casting, smoked occasionally with rosin or pine-knots, to prevent the tubes from sticking.

The tubes are drawn from the moulds by means of a lever fixed into the work-bench. If the spindle ever comes out, leaving the tube in the mould, a small screw is inserted into the tube, and being gently tapped with a hammer loosens the tube and withdraws it.

Another and better method of forming these tubes is to cut the metal into small disks, which are formed into tubes by being pressed through dies, which form the cup at the same time with the stem of the tube. This is the method always adopted, when these tubes are made; though, in consequence of the introduction of the friction-primer, the use of firing-tubes for our service is abandoned.

In the naval service, however, they still use percussion primers made of quills, prepared, as has been described, by pasting paper over the slits, on the top of which is placed a bed of percussion-powder, and covered with a coating of varnish, the piece being fired with a lock. This tube is found to work very well; and the quill being burnt to a crisp when the piece is fired, there is no danger of injury to the men.

These tubes are filled, primed, capped, and packed in the same way as quill-tubes.

1,000 of them require { 2½ lbs. of mealed powder.
2 qts. of whisky, or spirits of wine.

Slow-Match is used principally for the purpose of retaining fire in the shape of a hard pointed coal, to be used in firing ordnance, fireworks, &c. It was formerly used in field-batteries for lighting the portfires with which the pieces were discharged; but both are now entirely superseded by the friction-tube.

It is made of hemp, flax, or cotton rope, about 0.6 in. diameter, made with three strands, slightly twisted. Cotton rope, well twisted, forms a good match without any preparation.

To prepare hemp or flax rope,—boil it ten minutes in water holding in solution 1-20th of its weight of sugar of lead, or let it remain in the *cold* solution until it is thoroughly saturated; run it through the hands, to take the water from it; twist it hard by attaching one end to the hook of a twisting-winch, and putting a stick in a loop at a convenient distance for twisting; smooth it by rubbing it smartly with coarse mats, hair-cloth, or cuttings of buff-leather, commencing at the winch and rubbing always in the same direction, until the diameter of the rope is reduced 0.1 in., and until the tension and hardness are even; stretch it on poles or on a fence to dry, and put it up in neat coils of 25 yards each.

Match so prepared burns 4 inches to the hour. Plain cotton match burns 4½ inches to the hour.

In the absence of sugar of lead, the rope may be simply leached. For this purpose, it is put into a leach-tub, and steeped in pure water for 12 hours, when the water is drawn off. Ley prepared in a boiler with a quantity of ashes equal to half the weight of the rope, with 5 per cent. of quick-lime added, is then poured warm into the hopper. After it has run through the ashes, and remained sometime on the rope, it is drawn off, heated again and poured back into the hopper. This operation is repeated several times in the course of 24 hours, at the end of which the rope will be well leached. It is then taken out, twisted with sticks, and steeped for 5 minutes in hot water, stirring it at the same time, and finishing the match as before. This match burns 5 inches to the hour.

Bad match or old rope treated with sugar of lead, makes very good match.

Slow-match is packed in tight casks or boxes, which should be marked with the kind and quantity, place and date of fabrication. In burning, it forms a hard pointed coal, which readily communicates fire to any inflammable material with which it is brought in contact.

Portfires, which were formerly used to communicate fire to the priming tubes or priming of guns, are now entirely abandoned for that purpose, or so used only in special cases, as in proving guns; and are used principally as an incendiary material in loading shells, &c.

A portfire consists of a small paper case, filled with a highly inflammable, but slow-burning composition, the flame of which is very intense and penetrating, and cannot be extinguished with water.

The case is made on a polished steel former, 22 inches long, and ½ inch in diameter. The paper being cut to proper dimensions is laid upon a table, the end folded over, six inches parallel to the edge, and pasted. The former is now laid on the paper an inch from the double edge, and rolled up smoothly in it, making two revolutions after the pasted edge is passed. A thin coat of paste is now spread all over the remaining portion of the sheet, except an inch in width round the edge, and the sheet rolled on the former.

The *rolling-board* is made of a smooth piece of plank, having a strap tacked over the upper side near the end under which the hand is placed in using it. The case is laid upon the table and rolled with this board until quite firm and compact, the edge of the paper being prevented from rising and unrolling at first, by seizing it with the left hand every time the rolling board is raised. When the paper becomes hard the edge is pasted down, rolled smooth, the former withdrawn, and the case put aside. Length of case, 18 inches.

Portfire Composition is made of niter, sulphur, and mealed-powder in various proportions, antimony or steel filings being sometimes added. The following makes portfires which will burn ten minutes,

Nitre,	65 parts.
Sulphur,	22.5 "
Mealed-powder,	12.5 "

For others, see Appendix, p. 38.

The ingredients must be intimately mixed, by grinding with a muller on a mealing-table, rubbing them through the hands, and passing them through a sieve, regrinding the coarse parts.

Driving Portfires. Portfires are driven in a mould, Fig. 255, made of brass, and in two parts, held together by a socket at the foot, and four strong bands. The bore in the mould is the same length, (18 in.,) and the diameter (0.65 in.,) as the case. It having been put together, the case is put in position, and the bands driven firmly down.

Three drifts made of steel, tipped with brass at the lower end, are used for driving portfires. They are 22 in., 15 in., and 10 in. long, and in diameter 0.4 of an inch, or 0.1 less than the interior diameter of the case. Four small spiral grooves are cut on the surfaces of the drifts, to allow the composition to pass down. A hard wooden mallet, weighing one pound, is used for driving.

A small funnel is inserted in the top of the case, and the long drift passed through it to the bottom of the case. The funnel is then filled with the composition, and the driving commenced. Three blows are struck per second, raising the drift about half an inch with the left hand between the blows. The composition follows down the grooves, is driven compactly and with sufficient uniformity. The shorter drifts are used as the case is filled up.

Portfires should not be primed with mealed powder. Before the driving is commenced a piece of paper is inserted in the case, and driven like a plug at the bottom with the long drift; and when the case is full the top of it is turned in and beaten down, thus securing both ends.

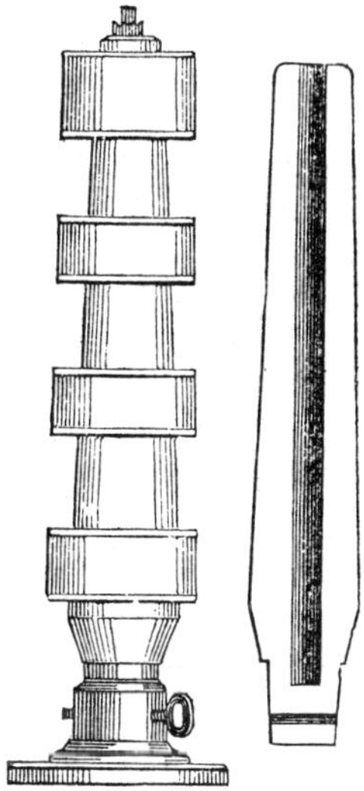

Fig. 255.

Portfires are packed in strong boxes, dove-tailed at the corners, lined with water-proof paper, painted olive color, and marked on one end with the contents, and the year of fabrication.

Portfires cut up into small pieces, and placed in a shell, form a very good incendiary material.

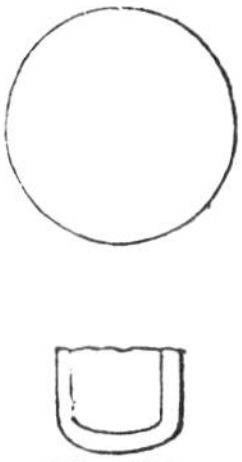

Fig. 256.

FRICTION TUBES, now almost exclusively used for firing guns of all kinds, are manufactured by machinery at the Frankford Arsenal, Pa.

Sheet brass, about $\frac{1}{24}$ of an inch thick, is cut into strips $\frac{3}{4}$ of an inch wide, and run through a punching machine, which cuts out disks of the size shown in Fig. 256. These are carried by a slider under another punch in the same machine, which forms them into cups, resembling the top of an ordinary percussion-cap.

Fig. 257.

These cups are annealed, and then passed through what is called the drawing-press, in which a punch pushes them through a die, increasing their length and decreasing their diameter. This is repeated six times, changing the punch and die each time, and annealing the metal after every *drawing*, to soften the metal and keep it from breaking or cracking. The circular disk has now assumed the form and dimension shown in Fig. 257.

These tubes are now taken to a third machine, in which a large cylinder with grooves, cut along its surface in the direction of its elements, receives a number of them with the closed ends all in one direction. As this feed-cylinder revolves on its axis, a circular saw cuts off in succession the broken ends at the proper distance; after which they are brought opposite a reamer, which smooths off the mouths of the tubes. In the same machine, a drill cuts a hole in the side of the tube, near the closed end, and drills a small hole directly opposite in the other side. While the reamer and drill are at work the large cylinder is at rest, and commences to revolve again as soon as the drill is withdrawn. Fig. 258.

Fig. 258.

The rough tube, Fig. 257, is passed twice more through the drawing press, which reduces its diameter two sizes, lengthening the tube considerably. A number of these are placed in a machine, and by means of circular saws, at proper distances apart, cut into lengths of about $\frac{5}{12}$ of an inch. The end of one of these small tubes is inserted in the hole in the long one, and soldered in that position under the blow-pipe. It is now ready for priming and wiring. Fig. 259.

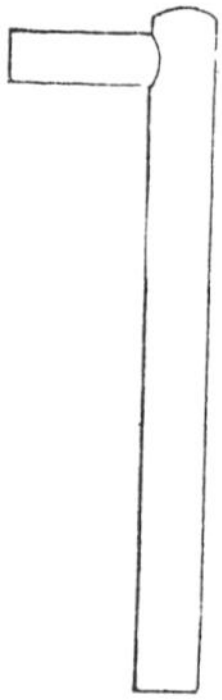

The wires are cut into lengths and flattened at one end. The flat end is then placed in a die under a press, and made rough and serrated for the purpose of acting on the friction-powder. Fig. 259. The extremity *a* is annealed in order to make it soft enough to bend without breaking.

The friction-powder used is made by mixing together two parts of chlorate of potassa and one of sulphuret of antimony moistened with gummed water. The mixing is done by grinding the wet materials in an ordinary cast-iron paint-mill,

Fig. 259.

keeping them at about the consistency of paste. The ground powder is placed in jars and laid by for use.

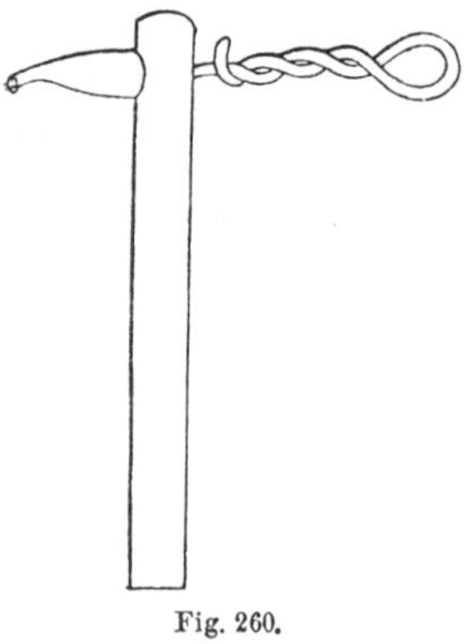

Fig. 260.

The short tube is now lined inside by stuffing it full of the moist powder and reaming it out afterwards, so that the serrated wire shall not be imbedded in the composition, but will have a lining of it on every side.

The wire is now passed through the short tube and the hole opposite to it in the side of the long one, the open end of the tube compressed with nippers, and the little tongue at the end of the serrated part doubled under to prevent any displacement. The other end is doubled and twisted by machinery, Fig. 260.

The head of the tube is then dipped in shellac varnish (made black by lamp-black), and when dry the long tube is filled with musket powder, its end closed with shoemaker's-wax, and retouched with varnish.

After being thoroughly dried, the tubes are bundled in packages of ten, the wire being bent down parallel to the stem for the convenience of packing, and ten of the packages placed in a tin box, in which is also placed a hook to be attached to the lanyard of the gun.

These tubes fulfill every requirement for a good, safe, and certain discharger of guns. A failure to ignite the charge is very unusual, and even after being soaked for a number of hours in water, the greater part of them will still act. If they become moist, they are restored by drying.

The first tubes used, instead of a wire had a slider of sheet brass incased in a sheet brass cup, which stood out perpendicular to the tube. The wire being bent down along the tube, takes up much less space in packing, and has besides another advantage. The vent of most pieces being inclined to the rear, the tendency of the pull made on the lanyard is to jerk the tube from the vent either completely or partially. In the last case, the side of the tube is drawn against the sharp edge of the mouth of the vent and bent over, so that when the slider does act and explode the friction-powder, all communication with the powder in the lower part of the tube is cut off, and the fire fails to communicate with the charge, Fig. 261. This may be obviated by holding the hand very low, so that the direction in which the pull acts is perpendicular to the axis of the vent; but the same object is better

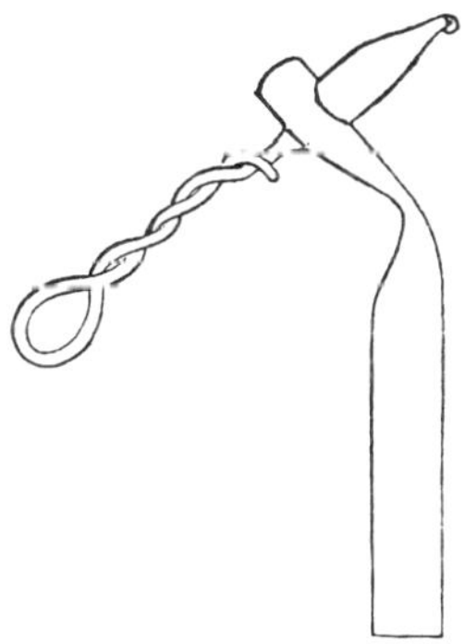

Fig. 261.

attained by not bending the wire to the full perpendicular. Then when the pull is made on the lanyard, the first tendency is to force the tube into the vent before the slider yields.

This fact has been taken into consideration in forming the new field gun and columbiads (or, rather *guns*, as they are now called), in which the axis of the vent is made perpendicular to that of the piece.

One great advantage of the friction-tube is, that it gives an enemy at night no clew to the position of your piece, as does the lighted portfire or slow-match.

French Friction-Tubes.—The French tube, of which ours is a modified and improved form, was adopted in the French army in 1847, and is somewhat differently constructed, as will appear by an inspection of Fig. 262. The large tube with a cross-head at top to prevent it entering the vent too far, is stopped at the top by a wooden plug held in place by an indented girdle. Below this and inside the large tube, is a smaller tube, the upper part of which is filled with a fulminating powder composed of $\frac{1}{3}$ chlorate of potassa and $\frac{2}{3}$ sulphuret of antimony mixed with gummed water, or alcohol. This is pierced through the axis to give passage to the friction-wire, the lower end of which is flattened and serrated, as shown in the figure. This serrated part remains in the empty part of the small tube, the lower end being clinched around the end of the tube. The small tube being prepared with the fulminate, the wire is introduced into the large tube, passing out through the hole in the axis of the wooden plug. The wire is then twisted on itself and bent down along the tube, as shown in the figure.

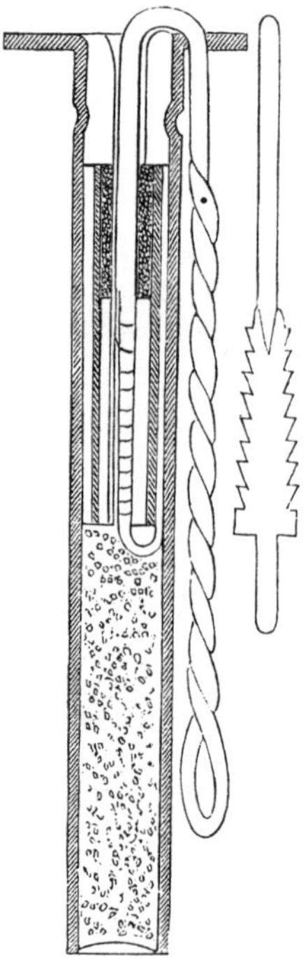

Fig. 262.

The tube is then finished by filling the lower ends of both tubes with musket-powder, and closing the end of the large tube with a mixture of wax and pitch. The opening around the wire where it passes through the wooden plug, is closed with a drop of wax which may be spread with the point of a knife, so as to close the opening perfectly.

Percussion Primers and Locks.—Before the introduction of friction tubes, percussion caps or wafers were used in connection with a lock, which was screwed to the body of field pieces, and to the lock-piece in heavier ordnance. Many different forms of these locks were invented, and now exist as curiosities in great numbers in military museums. They are no longer in use in our service, though still retained in the navy, as the best means of firing at the right moment a gun which is constantly in motion from the movements of the vessel.

The percussion-powder was placed in a thin layer between two circular pieces of cartridge-paper, united by glue, pressed firmly together, dried and varnished over with any water-proof varnish, forming the *wafer*.

The *caps* were made by forming the wafer at the bottom of a paper cap which fitted on the end of the lock hammer. This was probably the best form of the two, as the powder was then sure of being in the right place when the lanyard was pulled, whereas the wafer, being laid on the piece, might become displaced by the wind or otherwise, although it was partially secured by fastening to the bottom a little stem to enter the vent and keep it in place.

Ammunition for Small Arms.—To make the cylinders for blank cartridges, the paper is cut in the form of a trapezoid, with a pattern. The *former* is a cylinder of hard wood of the same diameter as the ball, concave at one end and convex at the other. The paper is laid on a table with the side perpendicular to the bases next the workman, the broad end to the left, the former laid on it with the concave end half an inch from the broad edge of the paper, and enveloped in it once. The right hand is then laid flat on the former, and all the paper rolled on it. The projecting end of the paper is now neatly folded down into the concavity of the former, pasted, and pressed on a ball imbedded in the table for the purpose.

Instead of being pasted, these cylinders may be closed by choking with a string tied to the table, and having at the other end a stick by which to hold it. The convex end of the former is placed to the left, and after the paper is rolled on, the former is taken in the left hand, and a turn made around it with the choking-string half an inch from the end of the paper. Whilst the string is drawn tight with the right hand the former is held in the left with the forefinger resting in the end of the cylinder, folding it neatly down upon the end of the former. The choke is then firmly tied with twine, by making two turns and a half-hitch around it.

For Ball-cartridges, the cylinders are made and choked as above, and the choke tied with two turns and a half-hitch, without cutting the twine. The former is then withdrawn, the ball inserted, and followed with the concave end of the former. Two half-hitches are then made just above the ball, and the twine cut off.

For Ball and Buckshot Cartridges.—Make the cylinder as before, insert three buckshot; fasten them with a half-hitch, and insert and secure the ball as before.

For Buckshot *Cartridges.*—Make the cylinder as before, insert four tiers of three buckshot each, as at first, making a half-hitch between the tiers, and ending with a double hitch. (For dimensions of paper for cylinders, see Appendix, p. 39.)

To Fill *the Cartridges.*—The cylinders are placed upright in a box, and the charge poured in each from a conical charger of the appropriate size; or they may be filled by a charger made by attaching to a large brass funnel two charging cylinders which communicate with one discharge-pipe at the lower end, and are alternately filled and emptied by the movement of a lever. The mouths of the cylinders are now folded down on the powder by two rectangular folds, and the cartridges *bundled* in packages of ten.

For this a folding-box is necessary. It is made with but two vertical sides, at a distance from each other equal to five diameters of the ball, and two diameters high. Two strips of wood nailed to the table will answer very well as a folding-box.

Bundling.—Put a wrapper in the folding-box, and place in it two tiers of five cartridges each, parallel to each other and to the short sides of the wrapper, the balls alternating; wrap the cartridges whilst in the folding-box, by folding the paper over them;

tie them, first in the direction of the length, then of the breadth, with a bit of twine fastened in a single flat knot.

A package of twelve *percussion-caps* is placed in each bundle of ten cartridges, at the end of the bundle.

The case for the caps is made like a cylinder for a rifle cartridge; it is choked at one end and tied; when the caps are inserted, it is folded like a cartridge.

The bundles are marked with the number and kind of cartridge.

Making Bullets.—Until a comparatively recent date, projectiles for small arms were made by casting; but they are now made altogether by pressure, which renders them more uniform in size, smoother, and more solid than the cast ones. Since the introduction of the elongated bullet, a machine has been adopted, which, receiving the lead in the form of round bars of about the diameter of the bullets, converts it into perfect projectiles with great rapidity.

Packing Bullets.—The bullets are packed in strong, hooped boxes for transportation. The cylindrical part of the elongated bullet is covered with a *melted* composition of one part beeswax and three parts tallow, care being taken to remove all the grease from the bottom of the ball, as, by coming in contact with the bottom of the case, it would penetrate the paper and injure the powder. This composition is sometimes put on before packing, but usually not until just before the cartridges are made up.

1,000 (about 75 lbs.) of the elongated, and 100 lbs. of the spherical bullets and buckshot, are packed in a box, which should be marked with the weight and kind, the place and date of fabrication.

Cartridges for Elongated Projectiles.—These cartridges differ so much from those used with the spherical bullet, and being now adopted in our service for all arms, a distinct description becomes necessary.

The *former* used is a cylindrical stick of the exact diameter of the bore for which the cartridges are to be made. This, making the exterior diameter of the case somewhat larger than the bullet will prevent the outer wrapper from binding around its base when the cartridge is broken.

In this cartridge, a *cylinder-case* and two *wrappers* are used.

The Cylinder Case, as Fig. 263, which contains the powder, is made of a rectangular piece of stiff rocket-paper, No. 4.

The Cylinder Wrapper, Fig. 263, in which this is inclosed, is a trapezoidal piece of paper (either of the kinds designated in the Ordnance Manual as Nos. 1, 2, or 3, may be used).

The Outer Wrapper, Fig. 264, is also in the form of a trapezoid, and should not be too strong, in order that the cartridge may be easily torn. That designated as No. 3 is recommended in preference to No. 1.

(For the dimensions of these, see Appendix, p. 39.)

To make the cartridge, place the cylinder-case on the cylinder wrapper, as shown by the broken lines on Fig. 263, and laying the former on the edge *f*, roll both on it, allowing a portion of the wrapper to project beyond both case and former. The end of the case is closed by folding this projecting part of the wrapper down against the end of the former (which is made slightly concave), and pasting the folds. It is then taken off the former and allowed to dry.

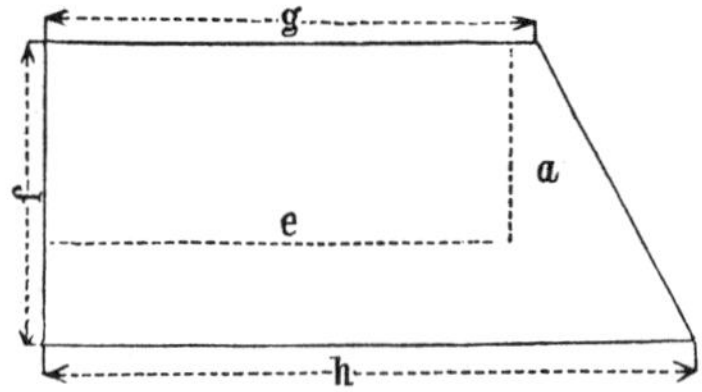

Fig. 263.

When dry, the former is again inserted in the case, laid along the edge *c* of the outer wrapper, and smoothly rolled up in it,

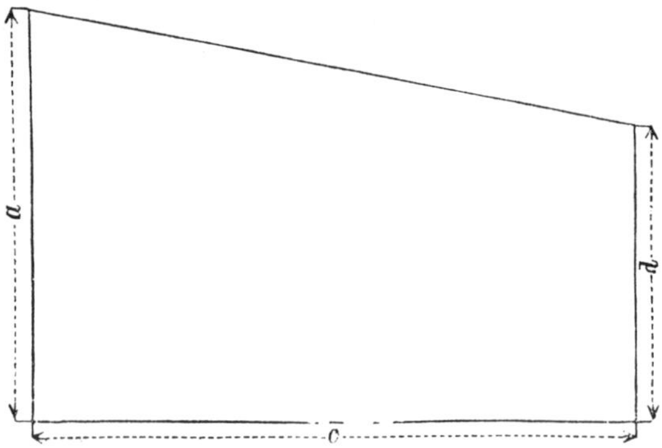

Fig. 264.

allowing sufficient of the end *a* to project beyond the bottom of the case to receive the ball and form the choke. The ball is then inserted in this end, the base resting on the bottom of the cylinder-case, the paper neatly choked around the point of the ball, and fastened with two half-hitches of cartridge-thread.

The former is then withdrawn, the powder poured into its place, and the mouth of the cartridge folded in the usual way, Fig. 265. This appears to be a better arrangement than is adopted in most of the European armies for the cartridge for the elongated ball. In most cases, the point is fixed next the powder, rendering it necessary to reverse the ball after the powder is poured into the gun.

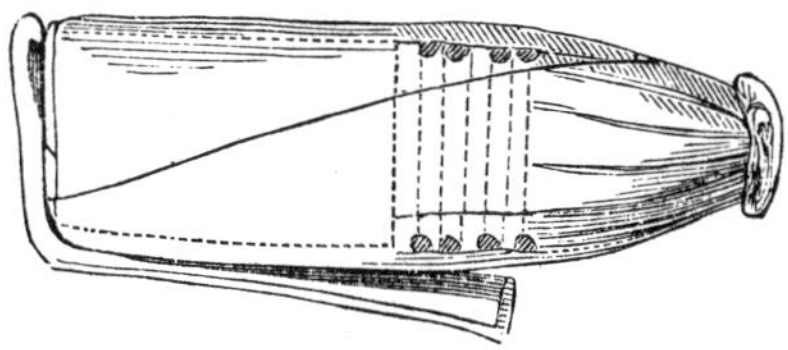

Fig. 265.

To use this cartridge, the fold is torn as usual, and the powder emptied into the muzzle. The ball is then seized between the thumb and forefinger, and the case struck sharply across the muzzle of the piece; this breaks away the paper, and exposes the bottom of the ball. By a slight pressure, the ball is forced into the muzzle clear of all cartridge-paper.

Percussion-Caps.—There are a number of different kinds of caps or priming-material used with small arms.

Until about the end of the last century, the almost universal method of firing small arms was by means of the flint-lock, which, itself, had succeeded as a great improvement, the use of the match, and wheel-locks, at the end of the preceding century, 1670. Towards the end of the last century, fulminating-powder was discovered, and as early as 1807, was used in sporting-arms as a more certain means of exploding the charge, and from this gradually extended, not only to the small arms used in the military service, but also to cannon; and it is now, with very few exceptions, universally adopted for small arms.

The powder first used was chlorate of potassa, which was found to oxidize and corrode iron and steel very rapidly, and was soon abandoned, being replaced by the fulminate of mercury, the action of which upon these metals is very slight. It was first encased in an envelope of wax, and afterwards placed at the bottom of a cap of thin copper, as now employed.

The difficulty of seizing and adjusting so small a body as the ordinary cap, under all circumstances, especially in the dark, or when the fingers are benumbed with cold, led to the adoption of a very large size, on the lower edge of which a projecting rim was placed to serve as a guide to the fingers.

It was soon found that these caps were so large and thick that the hammer sometimes failed to explode the powder, and this was obviated by a change in the nipple, the opening in which is made conical at top instead of cylindrical, so that the powder is forced by the action of the hammer upon a sharp edge, which presenting a much smaller number of points than the whole surface, never fails to explode it. The new system was applied to small arms in France in 1840–1, and in this country several years afterwards, flint-lock guns being changed to percussion.

The Common Percussion Cap is now made entirely by machinery. The machine used in the United States (Wright's), forms the cap from a sheet of copper, primes it, and turns it out completely finished all except the varnishing, which is done very rapidly in another machine.

A punch first cuts from the sheet of copper a piece in the form of a cross, which, sliding down an inclined plane, lands over a die, when a punch comes down upon it and presses it into the cap, the die and punch being so regulated that the slits in the cap do not extend to the bottom, which protects more completely the percussion-powder from moisture.

A horizontally revolving plate now brings a hook under the rim of the cap, and carries it around under the hopper containing the percussion-powder, which drops a charge of the *dry* powder in, and the plate continuing to revolve brings the cap under a punch, which not only presses the powder solidly into its place, but polishes the surface of it. This machine is capable of turning out 5,000 caps an hour, being furnished the copper in sheets.

Percussion-Powder.—The powder used in these caps consists of the fulminate of mercury mixed with half its weight of saltpetre. To prepare the mixture, the fulminate of mercury, which is kept in jars under water, is first drained and partially dried until it contains only 20 per cent. of water. It is then intimately mixed with 60 per cent. of its weight of refined pulverized nitre, by working with a spatula and wooden muller on a mealing-table. Whilst still moist, the mixture is passed through a common hair sieve, and is then dried with great care, passed a second time through the sieve, by rubbing it with the hand or a leather pad, until it is reduced to a fine-grained powder, when it is ready for use. The charge for each cap is half a grain.

VARNISHING.—The caps are varnished to protect the powder from the action of moisture. The varnish is made by dissolving the best gum shellac in rectified alcohol; and a single drop is placed in each cap, by means of a glass tube furnished with a sliding valve of iron wire, which allows the drop to escape when the tube is pressed against the bottom of the cap. The caps are placed in holes made in a board, 500 in each board. This is rapidly done by placing a number on the board and shaking it sideways, when the caps settle into the holes, and the defective or uncharged ones are easily discovered by inspection.

The varnish is also applied by a simple machine, which consists of a handle, to which is attached a number of little pistons or pins. A reservoir of varnish is placed on one side of this, and on the other a board containing the caps. The turn of the handle in one direction, loads each piston with a drop of the varnish, and a turn in the opposite direction deposits these in the caps.

MAYNARD'S PRIMERS are made by a machine in which the principal parts are three cast-iron plates, one on top of the other. Fig. 266. The bottom one has projecting from its surface, 500 small pins, which fit closely into as many holes bored entirely through the second plate. Springs are so arranged between these two plates, that the distance between them can be accurately regulated, governing thus the distance from the top of the pins to the surface of the second plate. These two plates are represented as completely separated in the figure, which is a section through the three plates perpendicular to their lengths, and through one row of holes.

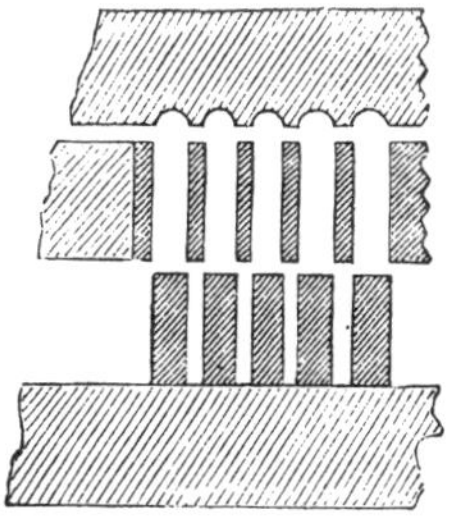

Fig. 266.

The third plate has countersunk in its lower surface, 500 little cups of the size of the primer to be made, which correspond when the plate is in its position to the 500 holes in the second plate.

Having adjusted the first and second plates, the percussion

powder, containing about twenty per cent. of moisture, is spread over the middle or charging plate with a steel pallette. All that portion of the holes above the little pins, is thereby filled, and the powder on the surface of the plate is removed. A sheet of strong, well-made paper, called parchment paper, is now slightly dampened and laid over the powder.

The third, or moulding-plate, is now placed on the paper, and the three plates transferred to a screw-press, where, pressure being applied, the paper and powder are forced into the little cups of the moulding-plate. The top plate is taken off, the sheet of paper adhering to it, and a sheet of thin, strong paper, well coated with paste, is smoothed down upon the other, and the plate and papers again subjected to pressure. The paste used to unite the two sheets of paper is made of

10 parts of gelatine and
1 " " sugar-house molasses.

After a few minutes' strong pressure, the plate is taken from the press, the sheet of primers removed from it and dried. After which, a coat of coloring mixed with shellac varnish, is applied, and the sheet dried again.

It is then cut by machine-sheers, into ten strips of fifty primers each, the strips given three coats of varnish, dried after each, and then coiled by inserting one end in a slit in the end of a little windlass and turning it. Each coil is then wrapped in paper, coated with linseed oil, and packed ten coils together in cylindrical tin boxes of the diameter of the coil.

FIRE-WORKS FOR INCENDIARY PURPOSES, SIGNALS AND LIGHTS.

INCENDIARY COMPOSITIONS for loading shells, &c. In shells designed to set fire to buildings or combustible materials of different kinds, something more than powder must be placed, to be scattered by the bursting of the shell, so as to make its effects felt to as great a distance as possible. The use of pieces of portfire for this purpose has already been mentioned. The others principally used are *fire-stone* and Valenciennes composition.

FIRE-STONE is a hard and highly combustible compostion, the flame of which penetrates and sets fire to most substances with which it comes in contact. There are several different compositions used, one of the best being,

Mutton Tallow, ...	1 part.
Turpentine, ...	1 "
Rosin, ...	3 "
Sulphur, ...	4 "
Nitre, ...	10 "
Regulus, (or pure) Antimony. . .	1 "

The last four substances are pulverized separately, and well mixed together by hand. In an iron kettle, the bottom of which only is heated in a furnace, or in a pan over a fire in the open air, the first two are melted together, and the pulverized ingredients thrown in, a small quantity at a time, the whole being kept constantly stirred with long wooden or copper spatulas. It is very liable to take fire from the heat of the vessel, if all the mixture is not kept constantly on the move. The operation is finished when the composition takes a bright metallic lustre, increasing in fluidity.

It is then poured out into wooden moulds and allowed to cool; after which it is broken up into pieces and introduced into shells, as required. It is also used in the manufacture of thundering barrels.

Another way of preparing it, is to cast it into cylindrical moulds, made of paper, and of two sizes, according to the shells in which they are to be used. The largest for 13 and 10 in. shells, and the other for 8 in. and 42-pd and 32-pd.

In the axis of the mould, a small paper tube is placed to receive the priming. The moulds are about 0.05 in. thick, and made by rolling rocket paper on a former, and fastening it with glue. The priming tubes are made with 4 turns of musket-cartridge paper.

For the convenience of filling, the moulds are placed upright in a frame of wood, a small spindle being fastened at the bottom of each to receive and support the priming tubes.

When the composition has become solid, the moulds are taken from the frame, and removed from around the fire-stone, and the priming-tubes filled with the composition used in the fuzes for mortar-shells, tightly driven. Both ends of the cylinder are then dipped in mealed powder, and the outside scraped clean of paper.

The cylinders are 3 ins. long, and can readily be introduced into the eyes of the shells, being from 1 to 1½ in. in diameter.

Valenciennes Composition, is made in pretty much the same way, being cast in cylindrical copper moulds, 6 in. long, and broken into pieces as large as the shell will admit without interfering with the fuze. It is made of

Nitre,	50	parts.
Sulphur,	28	"
Antimony,	18	"
Rosin,	6	"

Carcasses, are sometimes filled with fire-stone, and the holes for priming bored out before it becomes hard. The usual filling, however, is portfire composition mixed with a small quantity of finely chopped tow, and incorporated with equal parts of white turpentine and spirits of turpentine, until a proper consistency is attained. This composition is compactly pressed into the carcass with a drift, so as to fill it entirely. Plugs about ½ in. in diameter are then inserted in the holes, meeting at the center, forming when withdrawn, the priming holes of the projectile. In each of these holes, three strands of quick-match are inserted, long enough to project two or three inches outside, and packed in with portfire composition. The quick-match is coiled into the holes and fastened there by cotton patches, secured with glue or *kit*, until the carcass is wanted.

Common shells may be loaded as carcasses, by placing the bursting-charge in a flannel bag at the bottom and driving in carcass composition until the shell is nearly filled; 4 or 5 strands of quick-match are then inserted and secured, by driving more composition upon it. These shells, after burning as a carcass, explode.

Signal Rockets—These are usually made of two sizes, 1½″ and 2″, and are designated by the exterior diameter of the case. The case is made of paper. A sheet of No. 4 (19″ × 28″) makes two strips for the 2″ rocket, by cutting it in two, parallel to the short side; and when cut in the other direction, makes two strips for the 1½″.

The former, Fig. 267, consists of two parts, the longest one being square at one end, and having at the other, where it is slightly rounded, a conical opening, into which fits the spindle of

Fig. 267.

the second part, which is shorter than the first. At the point where the two parts join, both being rounded, is a depression which is to form the choke of the rocket. The diameter of the *former* is $\frac{3}{8}$ of the caliber of the rocket.

To make the case, the former is enveloped with a sheet of the paper, cut to the proper size and pasted after the first turn. It is then placed in the press and rolled tight, after which another piece of paper is rolled, pasted, and pressed on; and so on until the proper size is obtained. The press consists of a table having grooves in the top, of a proper size to receive the cases. On top of this is placed a heavy platform with corresponding grooves, and this is hinged to the table and raised by a lever to put the cases in. The cases are then rolled by slipping a handle on the square end of the former and turning it.

Choking.—To choke the case, it is wrapped at the joint of the former with a piece of strong paper, to prevent the choking cord from chafing it. A strong, smooth cord is attached to the top of a stationary post, passes over the end of a jointed limb which projects from it, and is secured to a treadle near the floor. A turn is taken around the case at the joints of the former with the cord, the foot pressed upon the treadle, and the case turned at the same time. As the paper yields to the pressure, the short part of the former is drawn out, until the case is sufficiently contracted, when the cord is taken off, the choke wrapped with strong twine, and *the former* removed. When the case is perfectly dry, it is trimmed to the proper length, so that the distance from the middle of the choke to the bottom shall be equal to that from the bottom of the spindle to the bottom of the mould, and the remaining portion equal in length to the distance between the bottom of the spindle and top of the mould.

Mixing the composition.—The composition should be well mixed, by passing it through fine sieves, and rubbing it in the hands. The charcoal, being the lightest ingredient, must be added after the nitre and sulphur have been mixed. Steel filings or antimony should be added after the charcoal; and whilst driving the rocket, the composition must be frequently stirred to prevent these heavy materials from settling to the bottom.

(For the proportion of the ingredients, see Appendix, p. 52.)

MOULDS for driving rockets are cast in a single piece and bored out to the proper caliber. The *spindle*, which is made of cast-steel, stands on and is connected with the base, of cast-iron, as is represented in Fig. 268. The mould being passed down over the spindle, is secured by a pin, which runs through both.

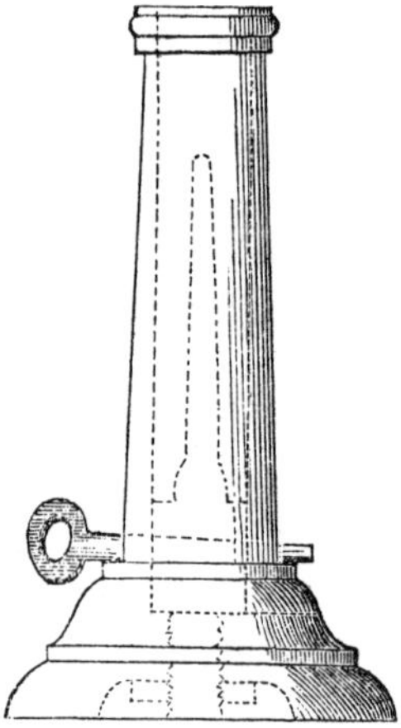

Fig. 268.

DRIVING.—The case is placed over the spindle, choke down, and settled with a mallet until it rests on the base of the spindle. The mould is then placed over it, and keyed to the base, which should rest on some solid foundation, as a large block of wood.

The composition is placed in from a ladle, which is made of such a size as to contain enough to form a column when driven equal in height to $\frac{1}{2}$ the interior of the case. In driving, four drifts (Fig. 269) are used, made of brass, or hard seasoned wood tipped with brass. The drifts have handles, strengthened at the top by copper bands. The first drift is pierced, to receive the whole length of the spindle, the second to receive $\frac{2}{3}$ of it, the third to receive $\frac{1}{3}$, and the fourth is solid.

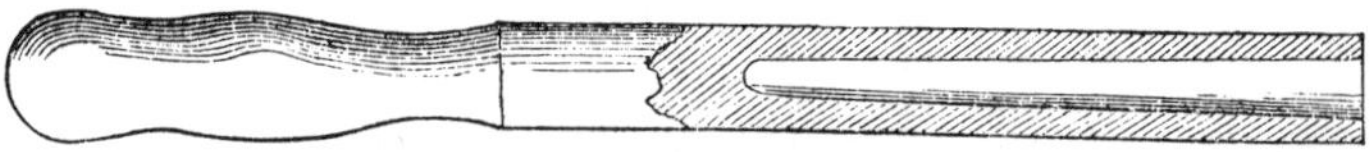

Fig. 269.

MALLETS, for driving $1\frac{1}{2}''$ and $2''$ rockets are turned out of hard, well-seasoned wood, and weigh about 2 and 3 lbs. respectively. The force to be employed in driving depends on the size of the rocket, the largest receives 30, and the smallest 25 blows, for each ladle full of composition.

The hollow drifts are first used, the shorter ones being taken as the case fills. When the composition reaches the top of the spindle, one more diameter is driven with the solid drift, and covered with a patch of stiff paper, cut to fit the case; and over this is driven a wad $\frac{1}{3}$ of a diameter high, of clay, or plaster of paris slightly moistened with water. This wad is afterwards pierced with a gimlet through to the composition, by means of which fire is communicated to the bursting-charge in the pot containing the ornaments.

Rockets are sometimes driven solid throughout, and afterwards bored out with a tap of the form of the spindle.

POTS are made of rocket-paper, by rolling two or three turns of it upon a former of the same diameter as the rocket-case, pasting it all well except the first turn on the former. The pot is two diameters long, and when attached to the rocket has an interior depth of one and a half diameters. In it are placed the ornaments of the rocket, and the charge of powder designed to blow them apart.

CONES are made of rocket paper, which is cut into circles of a diameter equal to twice the height of the cones to be made. Each of these circles, cut in half, makes two cones. They are rolled upon the former, Fig. 270, pasted and dried.

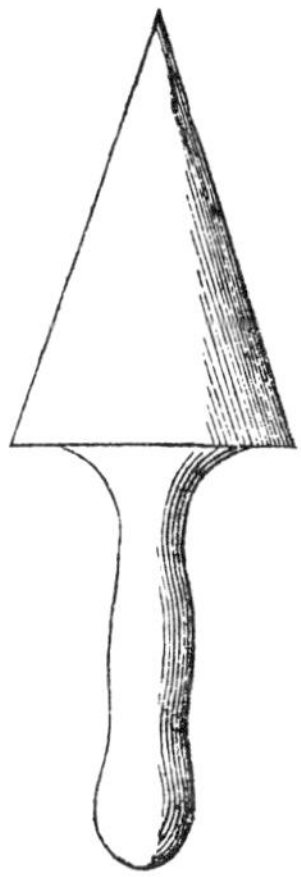

Fig. 270.

PRIMING.—A rocket is primed by coiling a piece of quick-match, about 2′ long, in the conical opening, and covering it with a cap of strong paper, pasted down or tied in the choke.

MAKING UP.—The pot is placed in position, by pasting the upper part of the case, and sliding it into the pot to the proper distance; or a ring of light wood may be used, which fitting inside the lower end of the pot, is placed over the upper end of the case, by taking off several folds of the paper down as far as is necessary. The plaster of paris covering having been pierced with a gimlet, the hole is filled with meal-powder, and the bursting-charge and ornaments placed in the pot with a slight covering of tow. The cone filled with tow (to assist in resisting the pressure of the air), and with its base cut to the same size as the pot, is placed on top of the latter, to which it is fastened by pasting over it a cone made of fine paper, the lower part being cut into slips, and pasted down over the pot. A slip of fine paper is then pasted around the joint, to give a finish to the rocket. The cone decreases the opposition offered by the air, and assists the rocket in penetrating it.

STICKS.—As a guide to the rocket, a stick made of dry pine, or other light wood, and nine times the length of the case, is attached to it with twine. The large end is fixed to the case, and is beveled off so as to decrease the resistance of the air. The side next the case is grooved out for a distance equal to $\frac{3}{8}$ the length of the case, which fits into it. Just below the bevel, and also opposite the choke, notches are cut to receive the twine by which the case is bound to the stick. (See Fig. 271.)

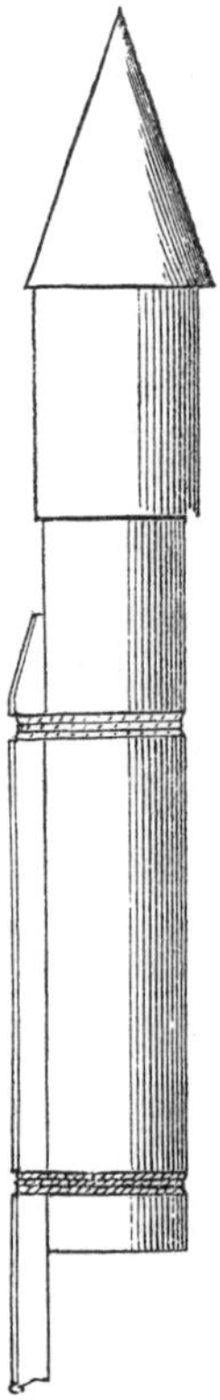

Fig. 271.

The other end of the stick is decreased in thickness to half that of the case end. The poise of rockets should be verified by balancing them on a knife-edge. Those under $1\frac{1}{4}''$ should balance at 3 diameters from the neck; between that and $2''$, at $2\frac{1}{2}$ diameters; and larger rockets at 2 diameters. If the stick is too light, the rocket will not rise vertically; and if too long and heavy it rises slowly, and will not arrive at its proper height.

DECORATIONS.—Various kinds are used, such as *Stars, Gold Rain, Serpents, Rain of Fire, Marrons, &c.*

STARS are the most beautiful decorations of rockets. They are made by driving the composition, moistened with alcohol and a small quantity of gum-arabic solution, in portfire moulds, without any paper case, and with a moderate number of blows; they are cut into lengths of about $\frac{3}{4}$ of an inch, and dredged with mealed powder. A more expeditious and better mode of making them is, to mould them in a brass cylinder of the diameter desired for the stars, and push them out with a rammer, cutting them into proper lengths as they are formed. Stars, after being dredged with mealed powder, must be dried in the shade. The gum arabic used in the star composition is intended to give such consistency to the stars that the explosion of the head of the rocket may not break them in pieces, and thereby destroy the effect. For composition, see Appendix, p. 52.

Gold Rain is made in the same manner as stars, observing to cut or mould the composition into pieces of equal size. The effect of this decoration is beautiful, and it is less troublesome than serpents. (For composition, see Appendix, p. 52.)

Serpents are driven in small cases made of rocket-paper or playing-cards, rolled over a former 0.4 inch in diameter and covered with two thicknesses of strong fine paper, the last turn of which is pasted. When dry, these cases are choked at one end, without being entirely closed, and are then charged about $\frac{2}{3}$ of their length with composition, by means of a small mallet, a drift, and a block of wood with a hole bored in it to receive nearly the the whole length of the case. The case is choked over the composition, and the remainder of it is nearly filled with mealed powder, upon which a small paper wad is placed; a clay head is then driven on it, and the end of the case turned down, to secure it; the other end is opened with a punch, and primed with priming paste, or a small strand of quick-match. Serpents are placed perpendicularly in the pot, with the primed end downwards. (For composition, see Appendix, p. 52.)

Rain of Fire is made with small cases 0.3 in. diameter, and 2 ins. long, two thicknesses of paper being sufficient. The end of the case is closed, and it is charged and primed like that for a serpent, except the powder for a cracker. Composition, 16 parts of mealed powder and 6 of charcoal. Another kind, which shows in sparks, is made of 16 parts camphor, 8 of niter, 8 of mealed powder, and 8 of tow. The composition is formed into a very liquid paste with gummed brandy, and after mixing in the tow chopped fine, is rolled into small balls about the size of buck-shot, covered with mealed powder, and dried.

Marrons are cubes filled with grained powder, and enveloped with two or three layers of strong twine or marline; to give them more consistency, they are dipped in kit; they are primed by punching a small hole in one corner and inserting quick-match. They are made of strong pasteboard, cut into the form of a parallelogram whose sides are in the proportion of 3 to 5, divided by 4 cuts from each side, extending one third of the width, and at equal distances apart, which prepare the paper for folding into the form of a cube of the size of one of the small squares thus marked out.

Packing Rockets.—The sticks are tied up in bundles; the rocket-case is wrapped with tow so as to be larger than the pot, the tow being confined with a piece of twine long enough to tie on the stick. The rockets are placed in a box on a bed of tow laid under the choke, and they are pressed closely together; tow is then carefully stuffed in between the heads of the rockets; each tier is also covered with tow.

War-Rockets.—The cases of war-rockets are made of sheet-iron, lined with paper or wood veneer. The head is of cast-iron, and may be either a solid shot or a shell, with a fuze communicating with the rocket composition. The case is usually charged solid, by means of a ram, or a press, and the core is then bored out.

The dimensions of war-rockets are indicated by the exterior diameters of the cases.

These rockets have been made of two kinds, viz.:

1. *The Congress Rocket*, which has a directing stick fastened to the tail-piece, in the axis of the rocket.

2. *Hale's Rocket*, which requires no stick, its direction being maintained by a peculiar arrangement of holes in the tail-piece, through which the flame issues.

Fire Balls are oval-shaped projectiles, formed of canvas sacks filled with combustible material. The sack is made of strong, close canvas, two or three thicknesses of which may be sewed together according to its strength. These are smeared on all sides with *kit* composition (see Appendix, p. 53) after the stuff is cut into three curved pieces, which are then sewed together to form the ball. The mouth is left open, and after the bag has been turned to bring the seams inside, the edges are turned outwards over an iron ring, and sewed to it. This ring must be large enough to admit the shell which is to go inside.

The shell is charged with powder and fitted with a slow-burning fuze. A tarred link, dipped in melted rosin, pitch, and tallow, is fastened with twine to the shell around the fuze, to insure the latter taking fire.

Charging.—The sack is now placed in a spherical mould formed of two heavy blocks fastened together with iron bands, and a layer of soft composition placed in the bottom (see Appendix. p. 53). On this is placed the shell with the fuze down, it being kept in place by twine passing through the sides of the sack, or by a piece of canvas sewed to the sides. The sack is filled with composition to the proper height above the shell, and the additional quantity of dry composition is placed in a heap in the center, finishing the filling with the dry composition, which is rammed in with a wooden drift and mallet. The sack is then taken out of the mould, the mouth-ring taken off, and the mouth closed by sewing or tying the pieces together.

To prevent the ball from being broken by the force of the charge, it is furnished with an iron bottom made of plate-iron, to which it is fastened with a cement which is poured hot into the bottom, the ball set in it and left to cool. The whole is then covered and strengthened with a net-work of spun-yarn or cord, which, commencing at the bottom, incloses the ball, and is formed at top into a strong loop which serves as a handle for carrying it.

PRIMING.—To prime one of these balls, four holes are made about three inches from the top, by driving in, two inches deep, greased wooden pins one inch in diameter. When the ball is to be primed, these pins are removed and the holes filled with fuzes, or fuze-composition, driven in with a couple of strands of quick-match. Room enough is left in the hole to coil up the quick-match, which is then covered with a piece of canvas, pasted down or fastened with nails. The priming should not be done until the balls are required for use.

Use.—These balls are used to light up an enemy's works, being thrown out from mortars for the purpose. As the inclosed shell is liable to explode at any moment, it is dangerous to approach them whilst burning.

LIGHT BALLS are made in the same manner as fire balls, but contain no shells, and are used to light up our own works.

TARRED LINKS are used for lighting up a rampart, or for incendiary purposes. They consist of coils of soft rope placed on top of each other and loosely tied together; the exterior diameter is six inches, the interior three inches. They may be made of pieces of slow match, about fifteen feet long; immerse them for ten minutes in a composition of 20 pitch and 1 tallow, and shape them under water; when dry, plunge them in a composition of equal parts of pitch and rosin, and roll them in tow or sawdust.

In making them, the hands of the workmen should be covered with linseed oil.

A *link* takes from 1 lb. to 1¼ lbs. of composition and ½ lb. of tow. Two of them are put into a rampart grate, separated by shavings. They burn one hour in calm weather, half an hour in a high wind, and are not extinguished by rain. The grates are placed about 250 feet apart.

Pitched Fascines.—Fagots of vine twigs, or other very combustible wood, about 20 inches long and 4 inches in diameter, tied in three places with iron wire, may be treated in the same manner as *Links*, and used for the same purpose; their inflammability is increased by dipping the ends in melted fire-stone.

CHAPTER XII.

FIELD ARTILLERY.

Origin.—Gustavus Adolphus seems to have originated the field or light artillery, using very small calibers, 1610 to 1622.

Frederick the Great improved the system, enlarged the caliber, and introduced horse-artillery into his armies, 1740 to 1786.

When first used in the French service, no system of manœuvers was adopted; the artillery was transported by contract; and, the teams being required only in time of war, no opportunity was afforded for the necessary instruction. The teamsters who were employed, often abandoned the pieces in times of danger, on account of which a large number of cannoneers were employed and provided with *bricoles*, to move the pieces on the field of battle in case of necessity, a means now seldom resorted to.

Napoleon, in 1801, having created his corps of drivers, field artillery became much more perfect, and was manœuvered with more ease and boldness, especially after the adoption of a regular system of instruction under Gen. Foy. Gribeauval introduced the battery formation.

The manœuvers first based upon the movements of infantry, were afterwards established on the principles of cavalry drill.

Use.—Light artillery has now become an indispensable arm in all armies; and it is brought to such perfection that no troops, however brave and well-disciplined, can do without it. Napoleon said, even before the many recent improvements were introduced that no body of infantry without the help of artillery could march for 1,000 or 1,200 yards against a battery of sixteen well-placed and well-served pieces, without being destroyed.

Artillery becomes of so much more importance when placed with raw and undisciplined troops; when the batteries, well posted and served, will atone, in some measure, for the deficiencies of the rest of the troops. In this case, the capacity of a general is exhibited by his knowing how to place, upon the point of attack, the few good troops at his disposal, and to protect the others with powerful batteries, which the enemy would in vain attempt to carry. But in order to succeed with such means, the genius of the chief must be capable of atoning for the moral inferiority of his troops, and his artillery possess a marked superiority over that of the enemy. Probably there never has been a battle fought in which this fact was rendered more apparent than it was at Buena Vista, where the batteries of Washington, Bragg, and Sherman performed such signal service, and atoned so well for the absence of a large well-disciplined force. Artillery may be also made to supply the want of a cavalry force, as well as to arrest a charge of cavalry, which can produce but little effect when made under the fire of well-posted and well-served batteries.

When serving with troops of the very best kind, artillery, although less important, is an arm the use of which cannot be dispensed with in battles of any size. Posted along the front of the line, it serves to maintain the combat; to spare the rest of the troops; to support their movements, and to make an opening for them to act. A portion of it, held in reserve, appearing at a decisive moment, produces by its rapid and terrible effect, the most important results.

In a flat, open country, with numerous roads, field artillery is of the most importance, as by its mobility and power of concentration, it neutralizes the importance and effect of the fixed positions of an enemy, and atones for the want of them on one's own side.

Its importance is diminished in a broken or mountainous country, where war is reduced to attacks on fixed positions; but even in this kind of war, where infantry plays the most important part, it is still of the greatest use in forcing difficult passages, setting fire to towns, and battering down houses or other obstacles.

The principal advantages of the fire of artillery over that of infantry, are—that the striking of the shot and shell gives a means of correcting the fire, not possessed by small arms. The dust thrown up by a shot striking may be seen at a distance of more than 1,000 yards. The moral effect due to the power of artillery is increased by the noise of the explosions, which can be heard in the midst of the most active firing of small arms.

Artillery is used against troops, either deployed or in squares, by forming openings preparatory to a charge of cavalry. It also repulses the enemy's columns of attack, throwing them into disorder.

The efficiency of the fire is, under similar circumstances, in proportion to the caliber; and for this reason a battery of 12-pdrs. produces greater effect than one of smaller pieces. The new 12-pdr. being calculated to throw shells as well as shot, and being light enough to pass over ordinarily rough ground, has many advantages over lighter pieces. Artillery fire is useful only within the limits of distinct vision. The greatest cannonading usually takes place at from 800 to 900 yards.

Artillery is of the greatest use in attacking posts and intrenchments; and a city which would resist troops without artillery, would quickly yield after receiving a few shot and shell.

Shells act by percussion and explosion both; but the firing with them is much less accurate than solid shot. The noise, however, made by them and the fear of seeing them burst, add much to their moral effect. They act effectively against cavalry in column or in several lines, by frightening the horses and demoralizing the men.

Whatever be the kind of piece used, it is apparent that success will depend upon the skill and address exhibited by the gunners in charge of it.

The quantity of artillery attached to an army is determined by the nature of the theatre of the war, the kind of troops, their number relative to that of the enemy, the character of the war, the size of the pieces, and their mobility. It varies usually from 1 to 3 pieces to 1,000 men. The latter number has been adopted in our service, 2 per 1,000, for service with the troops, and the third held in reserve.

The number of guns should be double that of the howitzers;

and for a defensive war, from $\frac{1}{4}$ to $\frac{1}{5}$ of the pieces should be 12-pdr. guns and 24 or 32-pdr. howitzers. For an offensive war, from $\frac{1}{7}$ to $\frac{1}{8}$ only should be of these calibers, otherwise the batteries would not have sufficient mobility.

As a general rule, the number of the artillery troops should be about one half that of the cavalry; though, as regards this, no fixed rule can be laid down, since artillery frequently supplies the place of cavalry.

Divided.—The following is the most recent division of the United-States artillery into kinds according to its duties. *Heavy* or *foot* artillery, is that portion which takes charge of and manœuvers the siege, sea-coast, and mountain artillery. *Light* or *field* artillery, is that portion which manœuvers field-pieces with troops in the field. It is divided into *horse-artillery* and *mounted-batteries.* In *horse-artillery*, the cannoneers, of which there are 11 to each piece, are mounted on horses, from which they have to dismount before attending on the piece, the 2 extra men holding the horses of the rest. In the mounted-batteries, formerly called foot-artillery, the cannoneers are on foot, and remain so during the manœuvers of the battery, except when it is desired to move at a very rapid rate, when they are mounted on the ammunition-boxes.

The horse-artillery was originally and is still designed for service with cavalry, receiving the lightest guns, which enables it to move at the same rate as the cavalry, and to keep it up for a considerable time.

In time of peace, or in the state of preparation for war, a battery of field-artillery is to be composed of 4 pieces. In time of war, of 6 or 8. In peace, each battery will have 76 men and 44 horses. In preparing for war, 100 men and 80 horses; and in war, 150 men and 110 horses. This supposes the battery to be formed of pieces of the smallest caliber, viz., 6-pdr. guns and 12-pdr. howitzers. For the new 12-pdr. or other heavy pieces, of course these numbers would have to be increased.

It is of the first importance that the fire of a battery be delivered at a good range with calmness and intelligence, those impetuous exhibitions of dashing bravery encouraged in other arms being more out of place in this. There are some circumstances, however, where several pieces, well harnessed, may advance to within 300 yards of an enemy, and overwhelm him with a storm of grape or canister shot; but these cases are very rare, and require much tact and resolution to know how to profit by them. One of the most brilliant feats of this kind probably ever performed on a field of battle, occurred at the battle of Palo Alto, with Col. Duncan, who, by a well-timed movement of two pieces of his battery, unlimbered in front of a large force of Mexicans attempting to turn the left flank of our line, and by a rapid and well-sustained fire, drove them back in confusion.

The principal object of artillery is, to sustain the troops in attack and defense; to facilitate their movements and oppose the enemy's; to destroy his forces as well as the obstacles which protect them; and to keep up the combat until an opportunity is offered for a decisive blow. Our mounted batteries have been so much perfected and increased in mobility, that they can move almost with as much celerity as horse artillery; and the latter has been practically abandoned in the United States. The men should be mounted on the boxes only when it is absolutely necessary, to avoid breaking down the horses. This rapid gait cannot, however, be kept up any length of time, as it can in horse artillery.

Horse artillery is, in France, considered indispensably necessary for service with cavalry, which having but little or no fire of its own, and acting simply by the shock of its charge, requires that the enemy should be kept at a distance, and first broken by the fire of artillery in order that the charges made may result in any practical good. Instances might be cited where the absence of its co-operation resulted in inflicting upon the enemy simply a few sabre cuts, when he ought to have been annihilated. This kind of artillery is, however, very costly, consuming a large number of horses, and should therefore be proportionally small in quantity. It should, however, be excellent in quality, bold, well manœuvred, even venturesome, appearing at and disappearing from different points, and multiplying, as it were, its action, which should be short and decisive. In France the horse artillery is nearly one-sixth of the artillery force.

In our service, but 2 companies in each of the 4 regiments of artillery, are authorized by law to be equipped as mounted batteries, and of these but 7 companies are now actually so equipped. The remaining 41 companies are equipped and serving as infantry, much to their detriment as an artillery force, and not very much to their improvement as infantry.

COMPOSITION OF BATTERIES.

Lightest.	*Heaviest.*
4 6-pd. guns.	4 12-pd. guns.
2 12 " howitzers.	2 24 or 32-pd. howitzers.
6 caissons.	6 caissons.
1 forge.	1 forge.
1 battery-wagon.	1 battery-wagon.
4 horses to each carriage.	6 horses to each carriage.

In batteries of only 4 pieces, it is usual to have one or two howitzers, and when there are two, it is well to have them together. But should the number of pieces be increased to 8, and there is a probability of the battery being divided, the howitzers should be distributed through it.

The light 12-pd. field gun of the new pattern, if adopted throughout the service, as it is probable it will be, will do away entirely with the use of howitzers in field batteries, the pieces being amply supplied with shells and spherical case.

To avoid accidents, only the objects absolutely necessary for use are brought into line during the battle, namely, the 6 pieces and the 6 caissons. These carriages constitute what is called the battery of manœuvre. Sometimes, but one caisson to every two pieces is placed in line; and if possible, the whole of them are kept out of fire by taking advantage of the inequalities of the ground, &c.

The carriages not employed, constitute the reserve battery, and are kept out of range during the battle.

The reserve furnishes supplies of men, horses, ammunition, and material.

In the French service, the two howitzers of a battery are both placed in the center section, in order that their fire, less accurate than that of guns, may not extend over a front greater than that embraced by the fire of the whole battery. In our service, it is customary to place one in each of the wing sections, in order to distribute the effect of the shells as much as possible, and have the benefit of shell firing in case it becomes necessary to detach one of the sections.

The composition of the field equipage for an army depends upon the number of pieces to be taken. The batteries are provided with all the accoutrements, small stores, and spare parts necessary for their service.

Besides the carriages composing the batteries, there are others which march in the rear, and form what are called the *parks*. There are two kinds of these; the *reserve* parks of divisions, or parts of an army, and *general* or *grand* parks. These parks contain the caissons for the transportation of the artillery and infantry stores, spare gun-carriages, battery wagons, park wagons, and forges. The caissons and spare carriages are drawn by only 4 horses.

The supplies of ammunition are fixed by allowing 400 rounds for each piece in round numbers for a campaign, distributed as follows: 200 with each piece in the batteries, 100 in the reserve parks of divisions, and 100 in the general park.

Additional supplies of all ordnance stores are placed in depôts convenient to the line of operations. These include pieces, projectiles, powder, gun-carriages, wagons, and all sorts of stores, and horses.

Batteries derive all their value from the courage and skill of the gunners; from their constancy and devotion on difficult marches; from the quickness and capacity of the officers; and especially from the good condition and vigor of the teams, without which nothing can be undertaken.

MARCHES.—As an officer has frequently charge of the convoys of artillery and other carriages, he must know how to conduct them properly, and keep every thing in good condition. He may be obliged, to provide, himself, from the means at hand, transportation for his supplies. To be able to do this, he should know the amount of work a horse is capable of performing, according to the condition of the roads, the distance, and the rapidity with which he has to travel.

A horse of medium strength can draw a load of 3,000 lbs. from 20 to 23 miles per day, over a paved road, and about 1,900 lbs. over a macadamized road. This includes the weight of the carriage. From 1,500 to 1,600 lbs., not including the weight of the carriage, is therefore a proper load for a horse over ordinary roads; and in case rough, broken ground is to be passed over, he should not be required to draw more than 1,100 lbs.

A horse moving at a trot can draw at the same rate as above, a weight of 800 lbs., not including the carriage, the weight of which would bring the load up to about 1,100 lbs. These numbers comprise the maximum load; and if the carriage is to move at a trot across fields, or in the open country, the weight, including everything, should not exceed 733 lbs.; or, if the carriage is not well constructed, 600 lbs.

A horse carrying a rider, loses his power of drawing in proportion to the increase of gait. This diminution, which is about $\frac{1}{2}$ when the horse is at a walk, becomes $\frac{2}{3}$ when he trots. Thus a team of 5 horses with a single driver seated on the carriage, will draw more than 6 horses, two of which are mounted with drivers.

In artillery carriages, half the horses carrying drivers, a team of six horses, moving at a trot, actually experiences a loss in traction of $3 \times \frac{2}{3}$, or two horses.

Admitting that the labor of the horse is inversely proportional to the road passed over, the labor of the off-horse moving at a trot at the rate of 20 miles a day, is reduced to 1,100 lbs., and that of a pair to 1,100 lbs.$+\frac{1100}{3}=$1,466 lbs., or 733 lbs. for each horse. Now, the weight for horse-artillery carriages not exceeding 660 lbs. for each horse, the horses are capable of moving at a trot for long distances, and in mounted batteries they can move very rapidly for short distances.

In proportion as the number of horses is increased, the relative force of each couple diminishes, in consequence of the difficulty of making them act together, and the results obtained are respectively :: 9 : 8 : 7 : 6, according as the teams are composed of 2, 4, 6 or 8 horses. More than 10 horses in a team are very difficult to manage. Hence, we conclude that teams are relatively stronger as their number is decreased.

For transporting short distances or over roads in good condition, the horses may be loaded more heavily. Wagoners make their teams draw more than 2200 lbs. per horse.

Before starting on the march, the inventory of the convoy should be verified. The carriages should be inspected, and the wheels especially examined to see that they are sound, and the tires not loose. When the spokes are loose and withdrawn from the nave, or the felloes separated on the exterior instead of the interior of the joints, the tires should be taken off, shortened, and replaced. The manner in which the loads are distributed, and teams harnessed, should be examined, and care taken that the traces are of the same length. The strongest horses should be placed in the wheel teams, the weakest in the middle, and a spirited horse should not be coupled with one of a contrary disposition. Those coupled together should be, as near as possible, of equal strength and of the same character and appetite, otherwise one is soon disabled from excessive fatigue, or want of nourishment.

In order to facilitate the inspection of the convoy it is divided into portions, which, under the command of officers or non-commissioned officers, are subdivided into parts, composed of a certain number of carriages, each inspected by a non-commissioned officer or trusty soldier. In this way the drivers are overlooked, and the carriages assisted in their march, either by manning the wheels or chocking them, or by the use of the shovels and picks.

The advance-guard marches with a certain number of sappers to repair the road; or, for the want of these, the laborers of the country may be called upon for the purpose. Marching through a friendly country, this work is done under charge of an officer sent in advance, and in an enemy's country, is confided to the commander of the advance-guard.

The carriages march usually in single file, at about 2½ miles an hour. The distance between the carriages on good roads, is one yard; on bad roads, from five to six yards. If the distances were not increased the carriages would run into each other, and cause accidents. The forges are placed at the end of the column, in order to facilitate the necessary repairs.

The most rapid rate at which a well-harnessed convoy can travel, marching at a slow trot, is 5 miles an hour. This gait can be kept up only for half the marching distances, and the teams should be brought to a walk an hour before reaching camp, and 15 or 20 minutes before each halt.

The carriages should be conducted in such a way as to allow of being turned in their positions in case of need; and for this purpose they should keep the right of the road, unless it should happen to be very wide. As soon as a carriage becomes disabled, it should be at once withdrawn from the column, repaired if possible, and if not, its load transferred to the neighboring carriages, leaving it to proceed empty; or even abandon it, if in the vicinity of an enemy. Care should be taken to lock the wheels going down hill, and chock them, at times, going up, to allow the horses time to breathe and recover their strength. Short halts of 10 minutes every hour must be made for resting, and allowing the rear carriages to come up.

The commander of the convoy, and those of the different divisions, stop frequently and allow their teams to pass them, to satisfy themselves that everything is in order. Forage is the only thing allowed to be placed on the carriages, and no smoking is permitted, especially in the vicinity of the powder-wagons.

On long and difficult marches, and during hot weather, a halt of an hour in the middle of the day is made, convenient to water, so as to allow the horses to drink and eat a little forage. In an enemy's country, but half of the teams are unharnessed at once, for watering, in order not to be taken by surprise.

With new material, ditches are passed in a direction perpendicular to their length; with baggage wagons and old carriages, they must be cut diagonally. It is better when practicable, to fill up ditches with fascines and straw, to avoid upsetting the carriages. Common agricultural carts, with the wheels removed, make very good bridges by which to cross ditches.

To ascend very steep hills, the column is halted, and the teams of the front half of the convoy doubled. These carriages are halted at the top of the hill, and the others brought up in the same way. The whole resume the march after a short halt.

Difficult fords are passed in the same way, except that the drivers remain on their horses instead of dismounting, as in the other case. If the wheels become embedded in the sand, they are disengaged by applying men to them. The teams are not allowed to stop to drink.

In very steep descents, covered with snow and ice, simply locking the wheels is frequently not sufficient. The horses should be rough-shod; and where the ice covers the ground in the most difficult parts, it should be cleared away by the workmen. It is sometimes necessary to unharness a part of the teams, and place them in rear of the carriages, to hold back and prevent them from running down too rapidly.

In night marches, it is necessary to watch the drivers attentively, and see that they do not fall asleep on their horses. At every halt, the harness is inspected to see that it is not tangled, which can be ascertained by stretching the traces.

The usual manner of marching, is in column of pieces, each piece followed by its caisson, and the other carriages at the end of the train. In the vicinity of the enemy, the latter are kept back some distance, and the battery is reduced to 12 carriages, 6 pieces and 6 caissons, which then occupy a length of about 190 yards for the heavy batteries harnessed with six horses, and for light ones harnessed with 4, about 132 yards.

Artillery should never march directly at the head or tail of a column, but should always have at least a battalion of infantry, or two squadrons of cavalry, in front and in rear of it, in order that in case of necessity it may have time to form *in battery*.

When a convoy composed of artillery carriages is obliged to enter a very narrow defile, the column should be followed by an extra limber, so that if obliged to retreat, each limber and rear train could be turned separately on its own ground, and the retrograde movement performed; which could not be done with ordinary wagons.

Horses should not be allowed to fall in harness, from fatigue, before they are replaced. The near horse becomes tired, usually, sooner than the other, and should be specially looked after and replaced when he begins to exhibit signs of fatigue.

Horses should not be worked immediately after eating, especially if they have to travel at a rapid gait. They should never be watered when heated, unless the march is to be immediately resumed. If they are watered from wells, the water should be drawn beforehand, and agitated so as to take the temperature of the air, or mixed with a little bran to render it less injurious.

A horse in harness, properly cared for and well fed, cannot travel more than 35 or 40 miles a day, and not this distance for any great length of time.

When regular forage has not been distributed, or cannot be obtained along the route, and the horses are obliged to subsist on green forage, the ration should be 80 lbs. per horse. Vegetables, such as turnips, carrots, and beets, are very nourishing, and do not debilitate the horses too much. If obliged to use the cereals, but few of the heads of wheat, and none of those of barley or rye should be given to the horses. Lucern, clover, and grass, are very heating, and should be given with judgment, otherwise, they ruin the horses by foundering them. With such food, the same amount of work cannot be expected from horses, as when they are properly fed, especially when in addition they are exposed to all kinds of weather in the bivouac.

The shoulders of the horses and all other parts in contact with the collars, should be frequently rubbed with wisps of straw, to restore the circulation of the blood. If the parts begin to swell, the horse's position in the team is changed. If time permits, the packing of the collar may be modified so as to change the points of contact. Without these precautions, the horses would commence to gall after seven or eight days of marching.

On arriving in camp, the teams are unhitched, the girths and breast-straps are loosened, and cruppers taken off. If the horses are very warm, straw should be placed under the blankets and the horses rubbed down. If they are covered with dust, sponge out their eyes, noses, mouths, fundaments and sheaths. In summer, when the roads are muddy, their legs should be washed and afterwards rubbed down with straw. When they are cool, unsaddle them, rub their backs dry, and examine them with care. Inspect the shoes closely every day. Horses sometimes refuse to eat, on account of thirst. They should be watered as soon as possible, which, in hot weather, is about an hour after reaching camp.

When the marches are long, shorten the time for grooming, or even dispense with it in the morning altogether, to allow the horses rest. If they sweat much, use wisps of straw principally in grooming them. In muddy weather, more time should be spent on the legs. (The tails should be tied up.) In dry and dusty weather, use the sponge and comb more freely. River bathing should be frequently used, when the temperature permits, unless the horse's hoofs are inclined to peel off, in which case they should be well greased.

Mountain Artillery.—With mountain artillery, each mule should be followed by a cannoneer, to observe the motion of the load, especially in ascending and descending slopes, and correct its position if necessary. The saddles should not be taken off until two hours after halting. Inspect the mules; apply remedies for any bruises; observe the defects of the saddles, and modify the padding if necessary. Dry the pack-saddles in the sun, and when dry, beat the padding lightly with a stick. The padding should be re-made every three or four months, and the upper part replaced when a saddle is changed from one mule to another. The saddle is carried by taking hold of the arcs of the tree, in order not to derange the padding. When the mules are heated and stopped in hot, wet weather, they should be sheltered as much as possible by covering them with cloths, sacks, &c.

If a mule falls, he is kept down by holding his head close to the ground, until he can be relieved of his load. In halts long enough to relieve the mules, the girths should be loosened, and tightened again before starting, care being taken to adjust the saddle-cloths. When a piece or carriage mule is wounded, or when in the vicinity of an enemy, the pieces should be drawn instead of carried. A cannoneer then attends each piece, holding by a rammer-staff which is run into the muzzle, which enables him to give assistance in passing over rough places, and prevents the carriage from upsetting. The staff should be withdrawn when crossing narrow ditches, to avoid breaking the handle. In case of a sudden attack, when the material is carried, a minute is sufficient, with well-drilled cannoneers, to unload, place in battery, and fire one round. The wheels should be greased at least once every two days, and the carriages should be drawn as little as possible during night marches.

Forage.—Common wagons should be made use of as much as possible to transport forage. A few horses can easily transport enough forage for a large number. With indifferent wagons and over bad roads, 4 horses can draw the rations of 40 for 24 hours. The ration is 14 lbs. of hay and 12 lbs. of oats, barley, or corn.

Ammunition.—At the halting places, the caissons should, if there is time, be opened, to see if the loads are in good condition. All necessary repairs to the carriages and harness should be made at the same time. The axletree should be greased every five days.

If powder is carried in any of the wagons, and especially if the barrels or caissons leak, every precaution should be taken to avoid accidents. For this purpose pavements should be avoided, passing through towns, and the carriages kept 40 or 50 paces apart. No fires should be allowed in the vicinity, and if the weather is hot and dry, the wheels will be occasionally wet with water.

Parking.—The columns of artillery or batteries are parked within reach of the troops, and at least 100 yards from any houses. The carriages are placed 4 yds. from axis to axis, the pieces in the first line, their caissons in the second. The reserve caissons, forges, and battery-wagons in the third and fourth. The distance between the lines is about 16 yards. If the forges are to be used, they should be placed at least 100 yards to the leeward of the battery, to avoid accident. This rule also applies to the camp-fires.

The number of sentinels to be posted over a park depends upon the number of carriages. A battery requires at least two, and during the night these are doubled. No one is allowed to enter the park without permission.

The position of the bivouacs will depend on the locality. The fires should be to the leeward of the battery; the ground easy of access and within reach of water and forage; a dry and sheltered position should be chosen for the horses. The camp being established, a portion of the horses should be sent on a foraging party, the rest unharnessed as heretofore directed, placed under shelter, when it is possible without compromising the safety of the park, and well groomed. Batteries, when forming a part of a line of battle, are encamped or bivouacked upon the same line with the rest of the troops.

Marching.—Before starting on the march, the commandant of a convoy should prescribe to all portions of his command what is to be done in case of an attack. He should procure all the information possible in regard to the country he has to pass through; the obstacles which may interrupt his march; the number and condition of the roads which lead to his destination, and the probability of meeting the enemy. If necessary, he takes, forcibly or otherwise, one or several guides, and from the information he collects, distributes his force in the advance-guard, the escort, and the rear-guard. The advance-guard, the head of which is formed of cavalry, precedes at a considerable distance the head of the column.

It reconnoitres the country, occupies the defiles, communicates with the commandant of the convoy by means of mounted men distributed along the route, repairs the roads, and searches the valleys and other positions. The escort marches on the flank, on the enemy's side. The rear-guard brings up the rear.

The carriages should march as close together as possible, and in two files if the road permits it. Pieces may, according to the importance of the convoy, be placed in the front, in rear, or on the flanks, or be given to the advance-guard, if it is large enough to derive any advantage from them.

In case of attack by light cavalry, the column is formed in two files, a convenient distance apart, and the front and rear closed, either with two pieces or with platoons of infantry. Thus formed, the march is continued, taking care to keep the carriages well closed up; and in case the enemy exposes himself, he is routed by a fire delivered at short range. But if he has infantry, and the attack becomes serious, the artillery should be placed in position for acting, the escort formed in line, and the enemy kept at a distance, whilst the convoy continues its march. If closely pressed, and there is hope of assistance, the carriages are placed in several ranks, and a square formed by facing the horses inwards, in order to protect them better, and prevent them from being stampeded. Recourse should be had to this means only in the last extremity, as it is much better to continue the march towards the point of destination than to halt, whereby time is lost and the command weakened. When the square has to be formed, care must be taken to avoid being surprised and overwhelmed by the enemy before the carriages get into position. An advantageous position should be chosen, where resistance would have to be offered only at a few points. The drivers all dismount. Pieces *in battery* are placed at the angles, the caissons and powder-wagons being in the interior of the square, and the escort delivers its fire from behind the outside carriages.

If the convoy contains nothing but powder, the escort must then cover it by forming a distance in front sufficient to protect itself from explosions. This requires that the escort of such a convoy be larger than that for an ordinary one.

If obliged to yield to superior forces, a vigorous resistance should be made, whilst preparing to abandon a part of the convoy, and carry off the remainder with as many horses as possible. The portion abandoned should be first as much damaged as circumstances will permit.

When the convoy is parked, a chain of posts is established around it in such a way as to protect it from surprise, and be a support to each other in case of attack. The points of attack can, if necessary, be strengthened by cutting ditches, and obstructing the roads and pathways leading to the park, with abattis, fallen trees, &c. The commandant with the escort takes a central position, or leaves a detachment there in case no serious attack is anticipated.

Tactics.—A battery going into line with other troops, is usually formed in column of sections, and deployed into line as the enemy is approached. Under ordinary circumstances the best formation is the column doubled on the center section, as the deploy is then toward both wings at the same time, and more promptly performed. Unless in extreme cases, the cannoneers should never be mounted on the boxes when the battery is within range of the enemy, as the explosion of a caisson might destroy nearly every cannoneer belonging to a piece.

When several batteries are united, they are formed by sections in one or several parallel columns, or in double columns on the centre, or still better, in two columns joined, and presenting a front of four pieces with the same intervals as in line. Sometimes they are formed in close column with a front of four or six pieces, and the batteries being spaced a distance apart equal to the interval between two pieces. When deployed, the distance between the batteries is double this.

When horse-artillery and mounted batteries are placed together, the former are placed on the wings, and the distances and intervals of the whole conform to those of horse-artillery; as in manœuvring, no regard is paid to inversions, it frequently happens that the batteries change their relative positions, and it is then necessary that each space should be large enough to contain a horse-artillery battery.

A close column of several batteries is deployed in the same manner as a column of cavalry; the leading battery moving off at an increased gait, and the others, obliquing to the right or left gain their intervals and form in line or battery to the front as usual.

The changes of front to fire to the right and left, are made on the wings in the same manner as with a single battery; but it is better to make these changes on the center battery. But four of these changes are practicable, viz., two to fire to the right by throwing the left wing to the front or rear, and two to fire to the left by throwing the right wing to the front or rear. In the other four changes of front, the pivot pieces would be masked by the rest of the carriages, and could not commence their fire soon enough. On this account the pivot carriages in these changes should be on the side towards which the fire is to be delivered.

In defensive battles, the contour of the ground is of the first importance, and if properly taken advantage of, may be made to double the force and importance of artillery.

Artillery cannot defend itself when hard pressed, and should always be sustained by either infantry or cavalry. The proposition made to arm the cannoneers with small arms, such as revolvers, short rifles, &c., is calculated to do more harm than good. They should be taught to look upon their pieces as their proper arm of defense, to be abandoned only at the very last moment. The fate of many a battle has turned upon the delivery of a few rounds of grape or canister at short range upon an advancing column; and if they have the means, how natural for men to resort to them for personal safety in time of extreme danger, forgetting for the moment that the fate of the whole army may be imperiled whilst they are defending themselves only! Let the rifles, therefore, be given to the infantry, and the sabres and revolvers to the cavalry; guard the artillery with these arms, and teach *them* that *their* salvation is in sticking to their pieces.

In a flat, open country, artillery is best sustained by cavalry, which is placed on the wings ready to charge the enemy in flank before he can get through the line of pieces. With very large batteries it is often necessary to place the supporting force in the rear, where they should take every advantage of the formation of ground, to protect themselves from the enemy's fire. In a broken country, artillery is more properly defended by infantry. It is formed in squares on the flanks and in rear of the centers of the batteries, so as to be able to move forward when necessary to meet the enemy. Infantry marching at a double quick, requires 2 minutes to pass over a distance of 200 yards, and cavalry at a gallop passes over the same space in $\frac{1}{2}$ minute. From these data the supporting troops are arranged so that they may reach the enemy before he arrives at the line of pieces, as the greatest disorder results should he succeed in forcing his way through the battery.

Batteries are usually placed at least 60 yards in front of the intervals between regiments and brigades, and upon their flanks, so as not to offer two marks for the fire of the enemy, or subject the troops placed in rear to a fire directed against the artillery. When the troops form in squares, the caissons are placed on the interior, and the pieces at the angles; and if the squares flank each other, the batteries may be placed on the dead ground on the prolongation of the diagonals.

The pieces may be fired in retreat, when the ground is sufficiently smooth, by using the prolonge, by which means the men may be mounted on the boxes immediately after firing, and the piece has simply to halt in order to open fire again, thus saving all the time necessary for limbering up, unlimbering, and coming into battery again; but, everything considered, it may be questioned whether this method of retreating (undoubtedly preferable to any other in maneuvering with the old heavy carriages, so difficult to limber and unlimber) is to be preferred to the ordinary one, now that the carriage and means of connecting it with the limber have been so improved. Retreating with the prolonge, should it be necessary to make a hasty retreat, the operation would be much delayed securing the prolonge and limbering up.

The commander of a battery should be acquainted with the movements which the troops are to perform, in order to be able to support them in case the orders of the general should not immediately reach him.

If the positions of the batteries are not definitely fixed, every circumstance capable of rendering their fire more effective and of protecting them from the enemy's, should be taken advantage of. For this purpose, the battery commanders, or several commissioned and non-commissioned officers, should reconnoitre with a view of giving each piece the most advantageous position.

Advantage is taken of the inequalities in the ground to protect the pieces and caissons, without regard to alignment or distances. A rise of from 2 to $2\frac{1}{2}$ feet in front of a piece, sometimes protects it entirely from the enemy's fire. The ground in front of a battery should be clear and destitute of bushes or depressions within rifle range, to shelter the enemy's skirmishers. The approaches to the position should be minutely inspected, in order to know how properly to conduct an advance or defend a retreat. An undulating surface is very advantageous, as the shot which strike the summit will generally ricochet over the battery; and as the dust raised by them cannot be seen by the enemy, he is apt to imagine he is firing too high, and thus loses many shots practicing for the range.

Attack.—Artillery should not be placed on rocky soil, in consequence of the splinters formed by the enemy's shot; nor, in the offensive, should it be placed on slippery or marshy ground, nor on ground intersected with hedges or ditches, which would interfere with its maneuvering. Several batteries should be so placed as to converge their fire upon the point of attack, whilst others check the troops opposed to them. Reserve batteries are placed on distant points (usually heights) to support the flanks of the attack.

Defense.—On the defensive, broken ground is sought for, to cover the front and flanks and keep the enemy for a longer time under the fire of the battery. The largest pieces and batteries of position are placed at the points where the enemy will probably make his chief attempts. Batteries, instead of being placed directly in front of the objects to be fired at, should be established obliquely so as to take the enemy's line or artillery slantingly, thereby producing greater damage.

A battery should not be established on a height the foot of which it cannot command, otherwise pieces have to be placed half-way down, in order to prevent the enemy from forming there in force for the purpose of storming the position. It is sometimes advantageous to withdraw to a greater distance, in order to avoid too plunging a fire. A well-practiced eye may then discover a position where the fire becomes grazing; but of course such selections can seldom be made, in consequence of other controlling considerations.

Firing from above to below is not very accurate at ordinary

distances. At great distances, the range is increased. The most advantageous command is $\frac{1}{100}$ of the distance, which gives a grazing fire and numerous ricochets. These positions should be sought after, and those avoided which give a command of more than $\frac{7}{100}$.

Firing.—The pieces should fire slowly at distances beyond 600 or 700 yards, in order to render their shots certain. They cease firing beyond 1,000 or 1,200 yards; otherwise the enemy, experiencing no damage from the shot, gains confidence, and advances with more boldness. Within 600 yards, the firing should be rapid, as it is then very sure, but it is only at the decisive moment that the rapidity is increased to its greatest limit. By wasting the ammunition, the supply destined for a whole campaign may be expended in a few hours' firing, preliminary to a battle.

Generally, the rate of firing should be much less than one shot per minute; for the whole supply of a six-pounder being about 400 shots, if the firing was at the rate of one per minute, the whole provision for a campaign would be consumed in about seven hours.

The firing should rather be too low than too high, in order to take advantage of the ricochets, and, besides, a ball which ricochets in front of a line produces a greater moral effect than one which passes directly through it. Volleys should not be fired, especially with small batteries, as the enemy is enabled to take advantage of the intervals between them, to charge. In changing position, advantage is taken of obstacles and favorable ground, to deceive the enemy as to the number of batteries he has opposed to him. An habitual slowness in firing renders this manœuver very easy.

In a combat with an enemy's artillery, the pieces should be separated, and given if possible a position more direct than usual, to diminish the chances of being struck. In general, the distance between the pieces should be a little greater than the maximum deviation of the fire, in order that the enemy, in firing upon one piece, may not strike the next one to it. The fire of two or three pieces ought to be directed on a single one of the opposite battery, to try to take it in flank, or obliquely, in order to dismount it. If possible, fire should be opened on the enemy's artillery just as he is bringing his pieces around to form into battery. This kind of a contest should, however, be avoided as much as possible, and the pieces fired against the troops, particularly if they are in deep columns, in order to retard their movements and throw them into disorder.

An oblique fire should be obtained on the line, and a converging fire directed to a single point, in order to make a sensible break, the moral effect produced by such a result being still more terrible than the physical. Cannonade incessantly the enemy whilst deploying his columns, no matter how terrible the fire of his guns. If his calibers are the largest, shorten the distance as much as possible, keeping beyond range of grape-shot. In this way, the enemy, as he fires within point-blank ranges, loses a part of his advantage.

Projectiles.—The kind of projectile to be used, will depend upon circumstances. Shot and shell should be fired against troops taken in flank or obliquely, against deep columns, and against artillery. The horizontal fire should be used against troops advancing in mass to force a bridge or defile, or marching over very smooth ground. Shot had better be used against infantry, and shells and schrapnell against cavalry, as this latter arm presents the highest mark, and enables the pieces of the bursting shells to do more execution. Moreover, the noise of the explosions frightens the horses and demoralizes the men. Against houses occupied by the enemy, or troops covered by the undulations of the grounds, shells are used also. A charge, when within short range, may be received by firing from each piece a solid shot on top of which is placed a round of canister or grape. The firing is then as rapid as possible, sponging may be dispensed with, and as the enemy approaches nearer, grape and canister alone are used, pointing very low at very short ranges, so that these projectiles may ricochet and scatter more. Grape or canister should not be fired at distances greater than 700 or 800 yards. Schrapnell should be used against troops deployed, or in column, by division or by squadron. Schrapnell and shells produce a greater moral effect, generally, than grape or canister.

The batteries should fire up to the last moment, if it is necessary for the protection of the troops. They ought never to expose their flank, unless to execute an essential movement; and then only when sure of performing it before being overwhelmed by the fire of the enemy.

On going into action, but one caisson at a time for two pieces should be opened, reserving to the last the ammunition in the limber-boxes of the pieces, in case they should be separated momentarily from their caissons. The pieces should never be without a supply. As soon as a caisson is emptied, it is sent to the reserve park.

If the pieces are dismounted, they are lashed under the limbers of the gun-carriages or caissons, and carried off. The enemy's pieces may be carried off in the same manner.

Woods, Attacking.—In attacking a wood, 12-pdr. guns should be used to destroy abattis, and counter-batter the enemy's artillery. The pieces should be covered, either by the undulations of the ground, or by parapets hastily formed of earth, fascines, or the bodies of trees. The enemy's artillery should, if possible, be battered obliquely or in flank.

Defending.—The position of a battery near a wood is not tenable, unless the wood is in our possession, and, within range, occupied by skirmishers. In defending it, the trees should be cut off at a distance of three feet from the ground, within a radius of at least 300 yards, and at most of 600. The position should be covered by abattis, and the pieces placed in barbette, enfilading the points of attack as completely as possible.

Defiles, Attacking.—A defile is attacked by placing the largest calibers in advantageous positions to counter-batter and dismount the enemy's artillery, whilst the lightest pieces endeavor to turn it, or batter it obliquely or in flank.

Defense.—Artillery should not be placed in front of a defile which it is to defend, as it might be stormed or cut off by the enemy. The best position is at 100 or 200 yards in rear, and should then be disposed in a curve, to obtain a converging fire on the defile. If a bridge at the end of a long causeway is to be defended, shot and shell should be fired horizontally. If the bridge has been cut, grape and canister will be fired at the enemy, whilst he is stopped by the cut. If the pieces are obliged to be placed in front of the defile, they should be strongly sustained on the flanks, to prevent being turned. If the defile is very long, an advantageous position for the artillery is chosen in the interior near the entrance.

Posts, Towns, &c., Attacking.—If a position is surrounded by a wall, it is breached with the largest pieces at hand by placing them at 400 or 500 yards, directing the fire so as to cut the wall near its foot. To set the place on fire, shells are used. In the absence of these, make use of shot heated in the traveling forges, or otherwise. If it is designed to occupy the place after it is taken, it should not be burnt.

Defense.—In defending a town or post, it is not necessary to place the artillery on the interior, unless it is fortified, or it is intended to hold it to the last extremity. In these cases, cut embrasures in the walls of entrenched buildings, propping up the floors if they are to support cannon. Defend the principal streets and all the approaches, by placing the largest pieces at the points where the greatest ranges are to be obtained, and under the protection of works susceptible of the longest resistance. Flank the approaches with small calibers, converging and crossing their fire on all the roads leading to the position to be defended. Place in advance only such light pieces as can be easily moved. Batter the neighboring valleys and thickets with shells, and unite the greatest amount of means upon the point of attack. In all cases set fire to the place if obliged to abandon it to the enemy.

Intrenchments, Attack.—A battery of 12-pd. guns and 24-pd. or 32-pd. howitzers, is established at from 400 to 500 yards, and covered, if there is time, by sinking a trench 20 in. deep, forming a parapet of the earth. Other batteries, composed of guns and howitzers, are established on the prolongation of the faces to enfilade and take them in reverse. Howitzers are placed to fire along the capitals, so as to burst the shells in the entrenchment. Under protection of these batteries an attempt is made to open a breach, for which purpose the parapet is knocked down with 12-pdrs., commencing at the top, and taking advantage of the embrasures, which serve as a commencement for the opening. Shells fired into the parapet crumble down the earth, and when the thickness is reduced sufficiently to allow solid shot to pass completely through, the breach soon becomes practicable.

The breach made, canister-shot and grape are fired upon the troops defending it; the columns of attack are supported by troops drawn up in échelon on their flanks, and arrangements made to prevent the enemy from retaking the offensive.

Defense.—When entrenchments have sufficient command over the surrounding ground, either from the strength of the profile or the position which they occupy, the salients should be armed with large guns and howitzers, firing in barbette in order to expose to fire all the surrounding points. The smallest calibers are reserved for the armament of the flanks, and are fired through embrasures. This disposition will prove advantageous if the different parts of the intrenchments project beyond each other much ; but if the works form almost a straight line, or are composed of parts projecting but little, placed on a horizontal site, and with a profile of small dimensions, the pieces had better be placed in the re-entering angles, and even in barbette, on the curtains.

In this position in fact the pieces command at almost as great a distance as when in the salients; they are less exposed to ricochet shots, and can fire up to the last moment, and at decisive periods; advantages which they would not possess if placed in the salients.

Field-batteries are established on the natural surface, are sunk or raised, and fire through embrasures or in barbette, according as time and the position occupied will allow.

The parapet of field-batteries is from 10 to 12 ft. thick. The height of the interior crest 8 ft.; that of the génouillère about $2\frac{1}{2}$ ft., which is also the height of the barbette parapet. The width of the embrasure-necks is about 20 in.; that of the exterior opening at the bottom, is half the length of the directrix. The checks are vertical on the interior, and have on the exterior a slope of $\frac{1}{3}$ their height. The bottom of the embrasure is so arranged that grape and canister can always be fired upon the columns of attack. It should never mask the line of sight, nor the fire of the piece. The distance between the pieces is 16 ft., and the width of the terre-plein 24 ft.

Ordinarily the terre-plein of the barbette-battery is raised to within $2\frac{1}{2}$ ft. of the interior-crest, which requires a large amount of filling in, for the terre-plein should not be less than 24 feet wide. Besides which, platforms have to be laid, or the wheels will soon form ruts in the newly thrown-up earth. It has, in consequence, been proposed to form barbettes on the natural surface, and cover them with a parapet $2\frac{1}{2}$ ft. high and 10 ft. thick. In this way no platforms are required, or nothing more than thick planks laid under the wheels; trenches may be dug on the right and left of the pieces for the cannoneers to stand in, to protect them from the enemy's fire.

Batteries sometimes consist simply of mounds hastily thrown up, when a few hours suffice for their construction.

The surface occupied by a piece or caisson drawn by six horses is 28 square yards. That occupied by a limber and its teams, 22 square yards. The front of all the carriages is 2 yards.

The artillery of intrenchments fires shot upon the enemy's batteries; grape, canister, and shells against the columns of attack; and at short distances canister is placed on top of the other shot.

RETREAT.—The retreat of troops is defended with a fire of grape and canister; and in such a case, horse artillery marching with the reserves, may be of great assistance, by furnishing its mounted cannoneers to charge upon the pursuing enemy.

Batteries execute a retreat with a fire by battery or by half batteries. They should retire slowly and without confusion. A specified number of pieces occupy rapidly positions selected in advance, and protect the movement of the troops by enabling them to rally or pass through narrow defiles. These pieces, whose fire should be rapid and well directed, should retain their position until the last extremity, and ought not to retire until a formal order to do so is received. Villages are obstinately defended, set fire to when abandoned, and the enemy kept out as long as possible, by throwing shells into them. If all hope of retaking the offensive is given up, the bridges on the route are blown up, and every other possible means taken to retard the advance of the enemy.

SELECTION OF HORSES FOR ARTILLERY SERVICE.

QUALITIES.—The horse for artillery service should be from 5 to 7 years old (the latter age to be preferred), and should be from 15 to 16 hands high.

The saddle horse should be free in his movements; have good sight; a full, firm chest; be sure footed; have a good disposition, with boldness and courage; more bottom than spirit, and not too showy.

The draft horse should stand erect on his legs, be strongly built, but free in his movements; his shoulders should be large enough to give support to the collar, but not too heavy; his body full, but not too long; the sides well rounded; the limbs solid, with rather strong shanks, and feet in good condition.

To these qualities he should unite, as much as possible, the qualities of the saddle horse; should trot and gallop easily; have even gaits, and not be skittish. The most suitable horse for the pack-saddle is the one most nearly approaching the mule in his formation. He should be very strong-backed, and from 14 to 15 hands high.

Horses with very long legs, or long pasterns, should be rejected, as well as those which are poor, lank, stubborn, or vicious.

The Mule, used with mountain artillery in carrying the pieces, carriages, &c., is preferable on many accounts to the horse, especially in a very rough country, where its surefootedness is an important quality. There are two kinds—the mule proper, or product of the jackass and mare, being considered preferable to that of the horse and ass. The former brays, the latter neighs.

The mule may be usefully employed from its fourth year to beyond its twenty-fifth. It is usually from 13½ to 15 hands high; is hardy, seldom sick, fears heat but little; is easy to keep; is very sure-footed, and is especially adapted for draught or packing; but is seldom used for drawing pieces, on account of its fear of fire-arms. What has been said in regard to the desirable qualities in the horse, is mostly applicable to the mule.

Selecting.—To choose horses, their attitudes and habits should be examined in the stable. Leaving the stable, they should be stopped at the door in order to examine their eyes, the pupils of which should contract when struck by the light. Out of the stable, they should neither be allowed to remain quiet, nor to be worried. Care should be taken against being deceived by the effects of the whip, cries, &c. The positions of a horse, his limbs, age, and height, should be examined at different times. He should be walked about with a long rein, observing the action of his rear extremities when he moves off, of his fore ones when approaching, and of both when moving with his flank towards you. The examination should be repeated at a trot, observing in what manner the horse gathers himself; whether he interferes, rocks in his motions, or traverses his shoulders or haunches. Rein him backwards; make one of the men get on him, and see if he is difficult to mount, and whether or not he bears too hard on the bit. Make him gallop a little, to judge of his wind, and see whether his flanks heave. Have his feet washed and examined carefully. Strike upon the shoe to determine whether he is easily shod or not.

Age.—The age of a horse is determined by the appearance of his teeth. When he is 5 years old, his mouth is nearly perfect with a full set (40) of teeth, 20 in each jaw; six of these are in front, and called *nippers*, or cutting teeth; a tush on each side of these, and on each side of the back part of the jaws six molars or grinding teeth.

At the birth of the colt, the 1st and 2d grinders have appeared, and in the course of seven or eight days after, the two central nippers force their way though the gums. In the course of the first month, the 3d grinder appears above and below, and shortly after another of the incisors on each side of the first two.

At the end of two months, the central nippers reach their full height, and before another month the second pair will overtake them. They then begin to wear away a little, and the outer edge, which was at first somewhat raised and sharp, is brought to a level with the inner one. So the mouth continues until some time between the 6th and 9th month, when two other nippers begin to appear, making 12 in all, and completing the colt's mouth. After this, the only observable difference, until between the 2d and 3d year, is the wear of these teeth.

These teeth are covered with a polished and very hard enamel, which spreads over that portion above the gum. From the constant habit of nipping grass, and gathering up the animal's food, a portion of the enamel is worn away, while in the center of the upper surface of the teeth, it sinks into the body of the tooth, forming a little pit. The inside and bottom of this pit being blackened by the food, constitutes the *mark* of the teeth, by the gradual disappearance of which from the wearing down of the edge, we are enabled, for several years, to judge of the age of the animal.

The teeth, at first presenting a cutting surface, with the outer edge rising in a slanting direction above the inner, soon begin to wear down until both surfaces are level; and the *mark*, originally long and narrow, becomes shorter, wider, and fainter. Fig. 272 represents the appearance of the animal's mouth at 12 months. The four middle teeth are almost level, and the corner ones becoming so. The mark in the two middle teeth is wide and faint; in the two next, darker, longer, and narrower; and in the extreme ones, it is darkest, longest, and narrowest. This appearance of the nippers, together with the coming of four new grinders, enables the age of the colt to be pretty nearly calculated.

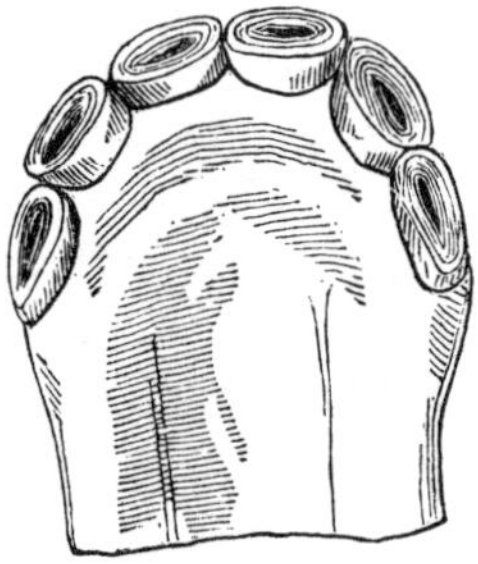

Fig. 272.

Six months after, the mark in the central nippers will be much shorter and fainter; that in the two other pairs will have undergone an evident change, and all the nippers will be flat.

At two years old, this change will be still more manifest, and the lower jaw of the colt will present the appearance represented in Fig. 273. About this period, too, a new grinder appears, making 20 in all, and a still more important change takes place. This consists in the formation of the permanent teeth which gradually come up from beneath, *absorb*, and take the place of the temporary, or milk teeth as they are called, and finally push the top parts of these latter out of their places. These permanent teeth are much larger and stronger than the first ones.

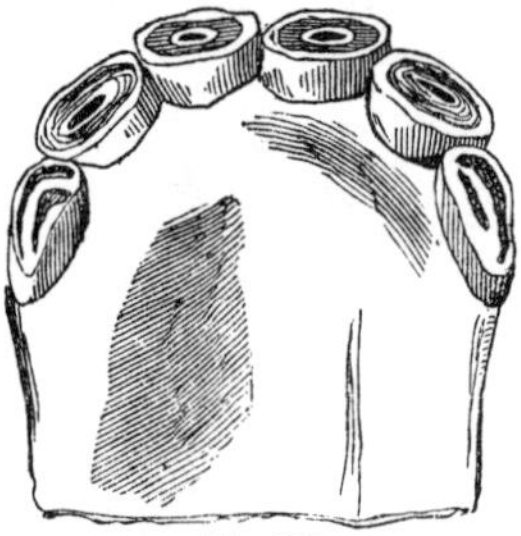

Fig. 273.

The teeth are replaced in the same order that they originally appeared, and consequently, at the end of the second year, the first grinders are replaced by permanent and larger ones; then the central nippers and so on. At the end of the third year, the colt's mouth will present the appearance shown in Fig. 274. The central teeth are larger than the others, with two grooves in the outer convex surface, and the mark is long, narrow, deep, and black. Not hav-

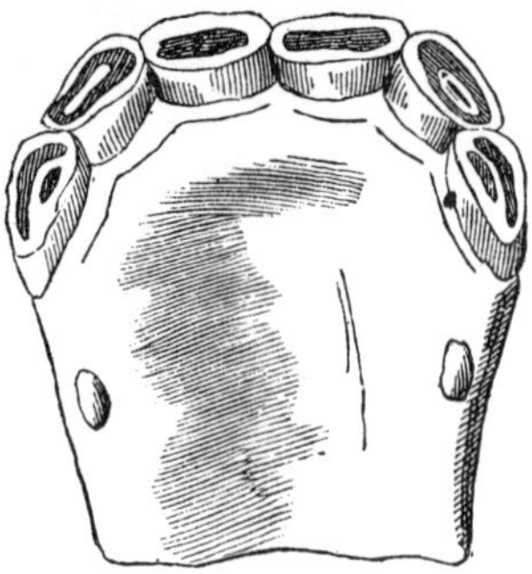

Fig. 274.

ing yet attained their full growth, they are rather lower than the others. The mark in the two next nippers is nearly worn out, and it is wearing away in the extreme ones.

A horse at three years old ought to have the central permanent nippers growing; the other two pairs wasting; six grinders in each jaw, above and below—the first and fifth level with the other, and the sixth protruding. The sharp edge of the new incisors will be very evident when compared with the neighboring teeth.

As the permanent nippers wear, and continue to grow, a narrower portion of the cone-shaped tooth is exposed to attrition, and they look as if they had been compressed. The mark, of course, gradually disappears as the pit is worn away.

At three years and a half, or between that and four, the next pair of nippers will be changed. The central nippers will have attained nearly their full growth. A vacuity will be left where the second stood, or they will begin to peep above the gum, and the corner ones will be diminished in breadth, worn down, and the mark becoming small and faint. At this period, too, the second pair of grinders will be shed.

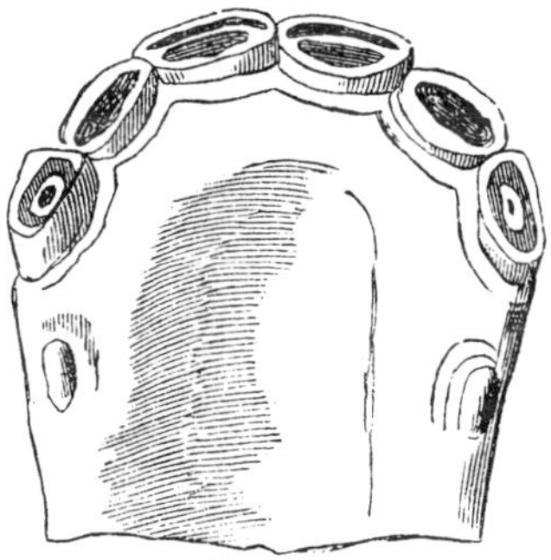

Fig. 275.

At four years, the central nippers will be fully developed; the sharp edge somewhat worn off, and the mark shorter, wider, and fainter. The next pair will be up, but they will be small, with the mark deep, and extending quite across them. The corner nippers will be larger than the inside ones, yet smaller than they were, flat, and the mark nearly effaced. The sixth grinder will have risen to a level with the others, and the tushes will begin to appear. See Fig. 275. The small size of the corner nippers, the want of wear in the others, the little growth of the tush, the smallness of the second grinder, the low fore-hand, the legginess of the colt, and the thickness and little depth of the mouth, will prevent the horse from being passed off as over four years old.

The tushes are much nearer the nippers than the grinders, but this distance increases with the age of the animal. The time of their appearance is uncertain, and it may vary from the fourth year to four years and six months.

At four years and a half the last important change takes place in the mouth. The corner nippers are shed, and the permanent ones begin to appear. The central nippers are considerably worn, and the next pair are commencing to show signs of usage. The tush has now protruded, and is generally a full half-inch in height. After the rising of the corner nippers the animal changes its *name*—the colt becomes a horse, and the filly a mare.

At five years the corner nippers are quite up, with the long deep mark irregular on the inside, and the other nippers bearing evidence of increased wear. The tush is much grown, the grooves have nearly disappeared, and the outer surface is regularly convex, though the inner is still concave, with the edge nearly as sharp as it was six months before. The sixth molar is quite up, and the third wanting, which last circumstance will be of great assistance in preventing deception. The three last grinders and the tushes are never shed. Fig. 276 represents the mouth of a 5-year old horse.

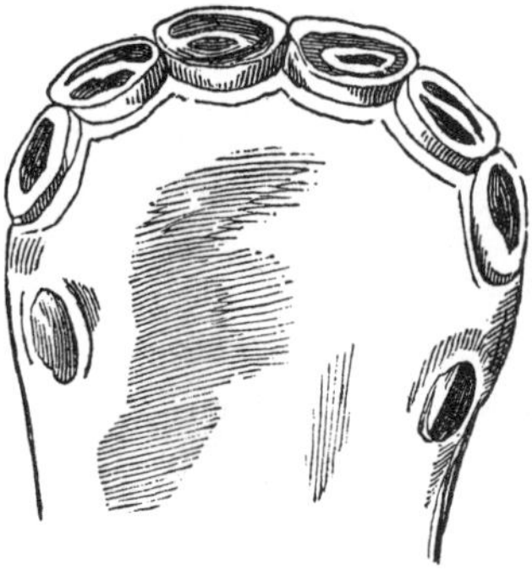
Fig. 276.

At six years the mark on the central nippers is worn out, though a difference of color still remains in the center of the tooth, and although a slight depression may exist, the deep hole with the blackened surface and elevated edge of enamel will have disappeared. In the next incisors the mark is shorter, broader, and fainter; and in the corner teeth the edges of the enamel are more regular, and the surface is evidently worn. The tush has attained its full growth of nearly an inch in length; convex outwards, concave within, tending to a point, and the extremity somewhat curved. The third grinder is fairly up, and all the grinders are level.

At seven years, the mark is worn out in the four central nippers, and fast wearing away in the corner ones. The tush is becoming rounded at the point and edges; still round outside, and beginning to get so inside. Fig. 277.

At eight years old, the tush is rounder in every way; the mark is gone from all the bottom nippers, and nothing remains in them that can afterwards clearly show the age of the horse.

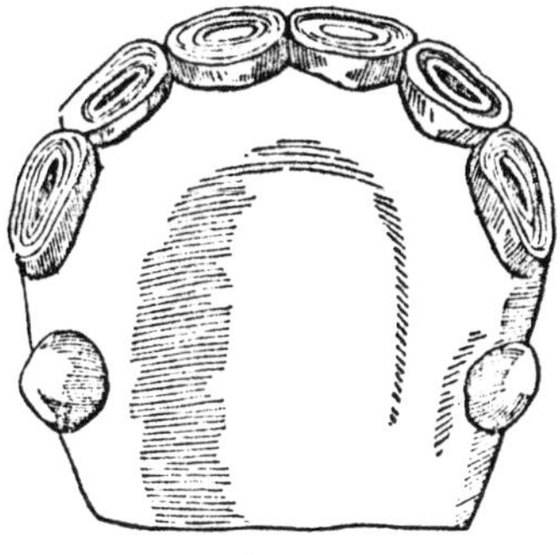

Fig. 277.

An operation is sometimes performed on the teeth of horses, to deceive purchasers in regard to age. This, called *bishoping*, after the inventor, consists in throwing a horse, 8 or 9 years old, and with an engraver's tool digging a hole in the almost plane surface of the corner teeth, of the same shape and depth of those seen in a 7-years old horse. The holes are then burned with a heated iron, leaving a permanent black stain. The next pair of nippers are also sometimes lightly touched. An inexperienced person might be deceived by the process; but a careful examination will disclose the irregular appearance of the cavity—the diffusion of the black stain around the tushes, the sharpened edges and concave inner surface of which can never be given again—and the marks on the upper nippers. After the horse is 8 years old, horsemen are accustomed to judge of his age from the nippers in the upper jaw, where the mark remains longer than in the lower jaw teeth; so that at 9 years of age it disappears from the central nippers; at 10 from the next pair, and from all the upper nippers at 11. During this time, too, the tushes are changing, becoming blunter, shorter, and rounder; but the means for determining accurately the age of a horse, after he has passed 8 years, are very uncertain.

The general indications of old age, independent of the teeth, are deepening of the hollows over the eyes, and about the muzzle; thinness and hanging down of the lips; sharpness of the withers; sinking of the back; lengthening of the quarters; and the disappearance of windgalls, spavins, and tumors of every kind.

The uprightness with which a horse habitually stands, has a great bearing upon his good qualities and endurance. Viewed in profile, his front legs should be comprised between two verticals, the one, A, Fig. 278, let fall from the point of his shoulder, and terminating at his toe; the other, B, from the top of the withers, and passing through the elbow. A line, C, passing through the fetlock joint, should divide the limb into two equal parts. The hind legs should be comprised between two verticals, A′ falling

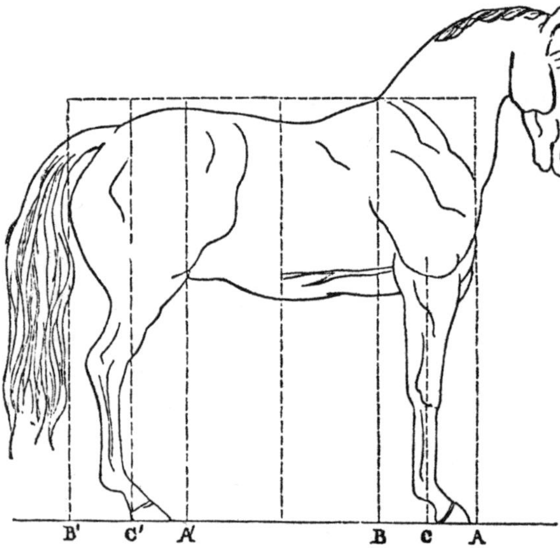

Fig. 278.

from the hip, and B′ falling from the point of the buttock; the foot at very nearly equal distances from these two lines. A line, C′, let fall from the hip-joint, should be equally distant from these two lines A′, B′.

Viewed in front, a vertical let fall from the point of the shoulder, should divide the leg along its central line. In rear, a vertical from the point of the buttock, should divide the leg equally throughout its entire length.

The height of the horse, measured from the top of the withers to the ground, should be equal to his length from the point of the shoulder to the point of the buttock. His chest, looking at him from the front, should be broad; and viewed from the rear, he should be broad, with good muscle, and strongly built.

TRANSPORTATION.

When artillery, horses, &c., are to be transported, whether by sea or land, an inventory of the whole, and a statement of the number of men or troops to accompany them should be made out, and the proper means of transportation applied for.

By Sea.—For transportation by sea, the inventory should state the number of articles, the weight of each, and the total weight of each kind, leaving a large column for remarks. In estimating the weight, increase the total by one half the weight of the small articles, such as accoutrements, tools, &c., which occupy considerable space in proportion to their weight, and apply for vessels sufficient for the transportation of the whole weight. *Decked* vessels are preferred, especially if ammunition is included among the stores.

The tonnage of vessels is estimated in tons of 2,240 lbs. The ballast should not count in the tonnage of a vessel, and if it is included, a deduction of one third must be made. The captains have usually a certificate of the tonnage of their vessels; but it may become necessary to make the estimate, in which case, multiply together the three principal dimensions of the vessel in yards, and divide the product by 41.9. The result will be the number of tons.

A statement of the distribution of the articles among the different vessels should then be made out, as well as an inventory of the number to be placed on each. The latter should be made out in duplicate, one copy being kept by the master of the vessel, the other, signed by him, by the person having the stores in charge, who also should see that the full capacity of the vessel is made use of, as captains are sometimes apt to say they are overloaded before they have received their full cargo.

The cargoes should be distributed in accordance with the object in view. For a single expedition, each vessel should be loaded with everything necessary for action at the moment of disembarking; so that in case of the delay of a part of the transports, it will not become necessary to wait for them. With each gun should be placed its equipments, ammunition, and the carriages necessary for their transportation, as well as the platforms, tools, instruments, and material for the construction of batteries. If a certain caliber is indispensable to any expedition, all the pieces of that caliber should not be placed upon one vessel.

The gun-carriages, wagons, and limbers should be dismounted by removing the wheels and boxes, placing in the latter the irons and tools necessary for putting the carriages together again.

Made-up ammunition should be well packed in boxes weighing from 100 to 150 lbs., closed with wooden screws or with bands, and provided with rope handles. Cartridges, fuzes, fire-works, and materials for them, in casks. Powder in barrels, of not more than 100 lbs. each. Sponges and rammers, worms and ladles, are united in bundles, according to caliber, for one or two pieces, and held together by two or three circular plates, with notches; the havresacks and pouches are placed inside of these bundles. Pioneers' tools, levels, rules, &c., each kind by itself, are placed in packages or in boxes. Hay-wads are placed in sacks, 100 or 150 in each. Each box, package, &c., must be marked with the kind and number of its contents.

The heaviest objects are placed beneath, commencing with the projectiles (including empty shells), then the pieces, platforms, gun-carriages, wagons, limbers, ammunition-boxes, &c. Boxes of arms and ammunition in positions the least exposed to moisture. Nothing should be placed in the lower part of the hold, where water is usually found. If it is necessary to occupy that part of the vessel, the objects least injured by moisture should be placed there.

If the disembarkation is to be made in the presence of an enemy, the vessels should be so loaded that some of the field-pieces can be landed at once, with their ammunition and equipments, with some chevaux-de-frise, and the tools necessary to throw up intrenchments, or facilitate the landing. These pieces may be placed upon the deck in positions where they will not interfere with the manœuvering of the vessel.

Horses.—In transporting horses by sea, great care should be taken, both of their food, and to prevent their being injured. With regard to the arrangements made for their reception and the manner of regulating them on board ship, the method pursued on board the English horse-transport steamer Himalaya, as described by Capt. McClellan, will be inserted, as the whole system is represented as very perfect, and well worthy of imitation.

Two rows of stalls, with the rear ends 2′ at least from the vessel's sides, are arranged on each deck. These stalls, Fig. 279, are each furnished with movable side-boards, a movable breast-board, and a fixed tail-board, all padded; the side-boards on both sides, the tail-board next to the horse and nearly to the bottom of the stall, and the breast-board on top and on the side next the horse. The padding used consists of felt, or raw hide (the latter objectionable

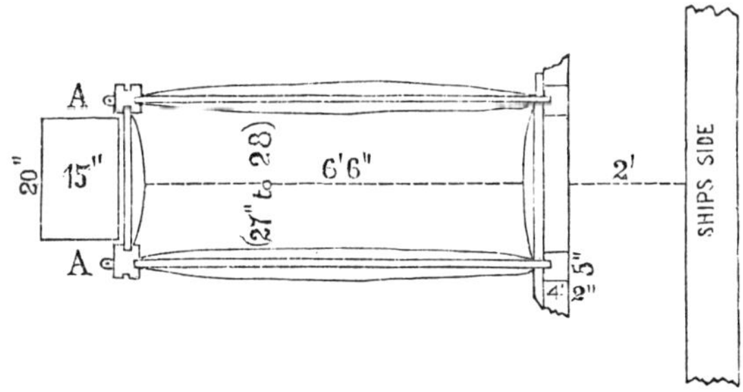

Fig. 279.

on account of the odor), stuffed with cow's hair wherever the animal can gnaw it, with straw in other parts. It is from 2″ to 3″ thick. The feed-troughs are of wood, bound on the edges with sheet-iron or zinc, and attached to the breast-boards with two hooks. The breast and side boards ship in grooves. Fig. 279 represents the horizontal projection of one stall. In front of each head-post a halter-ring A is placed, and over this near the top of the post is a hook, to which the sea-halter is hung when not in use. The feed-troughs, head-boards, and stalls are white-washed and numbered.

Fig. 280 represents a section of one of these stalls through the axis. The flooring is raised above the deck on battens, and is

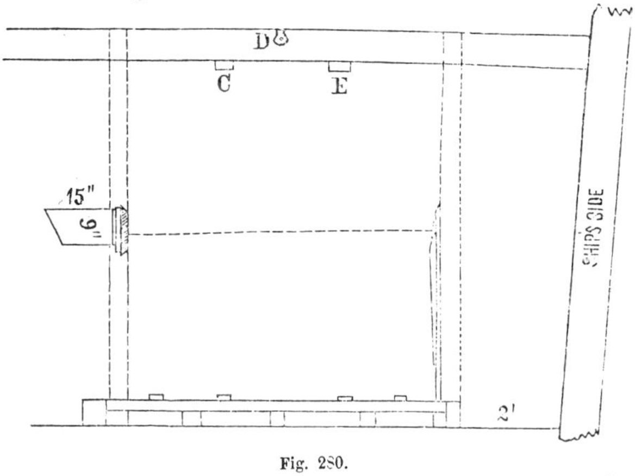

Fig. 280.

divided into separate platforms for every two stalls, so that it can easily be raised to clean the deck beneath; 4 strong battens are nailed across to give the animals a foot-hold.

Fig. 281 is a section through the side boards of a stall, and shows the dimensions of the timbers and height of side-boards, as well as the manner of inserting them in their grooves. *B* is

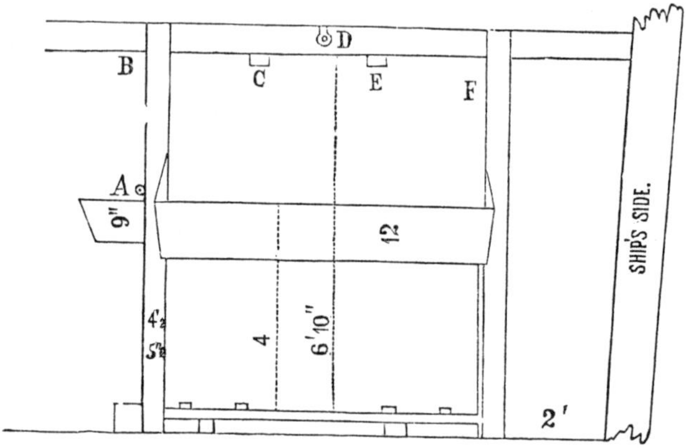

Fig. 281.

the hook for hanging up the sea-halter. This halter is made of double canvas, 2″ wide, and has two ropes, which, being fastened one to each post, keep the animal's head still, and prevent him from interfering with his neighbor. *C* and *E* are battens for securing the ropes of the sling, shown in Fig. 282. *D*, bolts, for the same purpose, when the sling is of the form represented in Fig. 283.

On the spar deck, the stalls are under sheds, every 8 stalls forming a separate set, so that they can readily be moved about when the decks are to be cleaned. Water-proof curtains are provided for the front and rear; a passage way of at least 2′ is left betwen the sheds and the bulwarks.

When practicable, a staging is erected alongside, that the horses may be walked on and off the vessel; when this cannot be done, they are hoisted on board in the sling, a small donkey engine being used for the purpose. In this way, horses may be shipped or unloaded at the rate of one per minute.

The slings are of canvas, of the shape and dimensions represented in Figs. 282 and 283. For hoisting in and out the horses, the sling is provided with a breast strap and breeching. On the main and orlop decks the sling ropes are

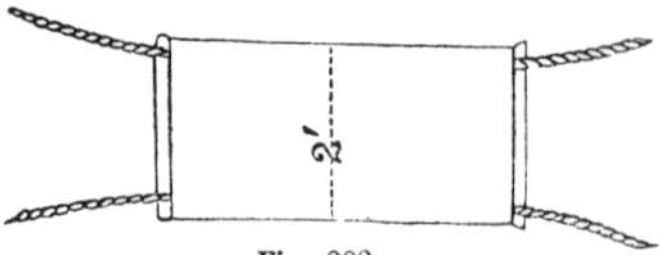

Fig. 282.

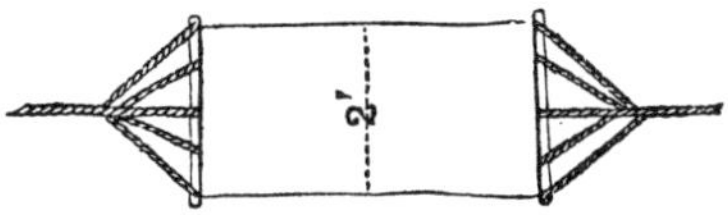

Fig. 283.

attached to the bolts; on the spar deck to battens. It was intended to adopt the sling represented in Fig. 282, as diminishing vibration.

At sea, the sling is used only when the animals show signs of weakness in bad weather, in which case about 1″ play is given to the slings, as it is only intended to prevent the horses from falling.

To place the horses in the stalls, all the side boards are removed except the one at the end of the row; a horse is then walked along to the last stall, and the other side board put in, and so on with all the rest. They should be placed in in the same order that they are accustomed to stand in the stable or at the picket rope. If it becomes necessary to remove a horse from his stall during the voyage, the breast board is taken away, and he is walked out.

All wooden parts are washed with some disinfecting compound, or simply white-washed. Chloride of zinc is freely used.

The decks are washed every day, and the stalls cleaned after every feed, especially at 7 P. M. From the spar and main decks, the stale passes off through the scuppers; from the orlop deck it passes to the hold, and is pumped out by the engine. On the Himalaya not the slightest disagreeable odor could be detected.

The feed-troughs and horses' nostrils are washed every morning and evening with vinegar. A scraper, brush, and shovel are allowed to every eight stalls.

A guard always remains over the horses, and in cases of necessity a farrier or non-commissioned officer is sent for.

Great attention is paid to ventilation. The orlop-deck, although hotter than the others, appears to be the most favorable one for the horses.

So long as cleanliness is preserved, the commander of the vessel does not interfere as to the hours for feeding, which are usually at 6 and 11 A. M. and 5½ P. M. If any horse refuses his food, the fact is at once reported. A supply of forage is always carried on board the ship. The horses drink condensed steam. The ration at sea was established at 10 lbs. of hay, 6 lbs. of oats, half peck of bran, and 6 galls. of water, as a maximum; but it is generally considered this is too great, and that ⅔ the allowance, except the water, would be ample, as it is found there is great danger from over-feeding at sea. No grain is given the day the horses come on board, but simply a mash of bran, which is considered the best habitual food at sea.

For the men, bunks and hammocks are generally used. Standing bunks are found to be very objectionable, on account of the difficulty of keeping them clean. Hammocks are regarded as preferable for men in good health, while many officers consider it best to provide neither hammocks nor bunks, but to allow the men to lie down on the fore-decks, with their blankets and overcoats.

When the transports are numerous, each one should have on the starboard and larboard, and on a broad pendant at the top of the main-mast, an easily distinguished number. By means of these numbers, which are marked on the bills of lading, the disposable resources of the expedition are known at any time. Vessels carrying some particular flag should be specially appropriated for the transportation of powder, fire-works, and ammunition, which may be separated from the pieces.

Disembarking.—The pieces required to act against the enemy at once are first disembarked, and the remainder of the cargo is taken out in the inverse order in which it was placed upon the vessels. The wagons which are to serve as transportation, are put together as quickly as possible. The different articles are placed together by kind, at a sufficient distance from the beach, in order not to embarrass the landing. All necessary precaution should be taken with the powder and ammunition which is not to be transported immediately, to prevent confusion, damage and accidents.

If it becomes necessary to tranship, or leave any articles upon the vessels, the facts should be carefully noted on the manifests.

The ships' crews are used in embarking and disembarking, using for these purposes the yard-arms and tackles. It is ordinarily sufficient to furnish them with rollers and skids, in order to place the articles convenient to the tackle. Under some circumstances, it becomes necessary to establish bridge abutments, shears, gins, &c. For the want of the ordinary means, a temporary crane may be established. To do this a long mortise is cut in a beam about $\frac{1}{3}$ of the distance from its end, and upon the ground is fixed a frame-work, furnished with a strong vertical pin. The beam is laid on this frame with the pin in the mortise, like an ordinary pintle, but in such a way that the ends of the beam can be raised and lowered. The shortest part of the beam is then turned towards the load, and the different weights being slung to it, are raised by lowering the opposite end, previously raised to make the lashings shorter. The beam is then turned around on its pintle until the weight is in the proper position, when it is lowered gently and unlashed. If a tree or beam fit for the purpose cannot be obtained, several small pieces may be lashed and pinned together.

Railroad Transportation.—In railroad transportation, when several trains are required, they should be in proportion to the power of the engines employed, and full loads should be placed on them regardless of the divisions made in the command, such as batteries, half-batteries, sections, &c.

The men are provided, before starting, with provisions to last during the trip, which should be cooked and carried in the haversacks. The canteens are filled with water; the French, in warm weather, mixing brandy with it. As the horses can eat in the wagons, even whilst the train is in motion, hay (pressed if possible) should be distributed at the rate of about 8, 14, or 24 lbs. per horse, according as the trip is to last less than 12, between 12 and 24, or more than 24 hours. A feed of oats (half a ration, 6 lbs.) is carried in bags and placed in the baggage wagons. It should not be given to the horses on the road, but after they have arrived at the terminus.

The horses are carried in cattle-cars, or if possible, in box-cars, which are covered. They are provided with bars at the doors to prevent the horses from backing out when the doors are opened. By taking care to keep the horses quiet, however, these bars may be dispensed with. The saddles, &c., the valises of the drivers, and the bags of oats, are placed in the baggage cars, which should be provided with brakes.

The "material" is carried on trucks or common platform cars.

The command should be at the station at least two hours before starting. The horses should have finished feeding about two hours previous to their arrival at the station, as they are then more docile. The baggage should arrive half an hour before the command, under charge of an officer, and be loaded under the direction of the employés of the road.

The cars should be arranged as near as possible in the following order: 1st, a baggage wagon; 2d, a truck carrying the beams, platforms, &c., necessary in disembarking; 3d, the horse-cars; 4th, the cars for the men, one, at least, of which, should be provided with a brake; 5th, trucks loaded with material; 6th, baggage cars (with brakes) loaded with saddles, &c. Cars with brakes should always be placed at the head and tail of the train.

Guards should be detailed and so stationed on the train as to preserve order both when in motion and during stoppages. The commanding officer should pay especial regard to the wishes of those having the train in charge, and enforce an observance of the road regulations in his command.

On arriving at the station, the commander at once divides his command and material into the portions to occupy the different cars.

Horses.—An officer is detailed to superintend the embarkation of the horses. He furnishes each car with two bundles of litter, and places forage along the long side of the car opposite to the door. A non-commissioned officer is charged with loading the saddles, &c.

The men are, under an officer, formed into detachments proportional to the importance of the material to be embarked.

In a battery, the front and middle teams are unhitched and assembled, under charge of a non-commissioned officer, with the saddle horses, in a convenient position near the station. The carriages are drawn up near the cars by the rear teams, which are then unhitched and placed near the others. The horses are divided, according to the capacity of the cars, into groups, so that those of any carriage may be, as much as possible, placed together. The several groups are arranged in front of the cars they are to occupy.

The saddle horses and near horses of the teams are unsaddled, but not unbridled. In bad weather the blankets may be placed on and secured with the surcingles. The harness is left upon the draft horses, and the traces, breeching, &c., lashed to the collar in such a way that the whole will be fixed as compactly as possible in rear of the horses' breasts. The crupper, breast-strap, girth, and blanket, are placed together on the shabrac and secured with the surcingle. The stirrups are tied together. The saddles, so arranged, are carried and laid on the ground at a designated point near the baggage-wagon, as are also the valises, &c., of the drivers.

As soon as the horses of the first carriage are ready, the officer in charge of the horses directs the loading to commence. The driver of the leading team, assisted by another man, leads successively his two horses in, lowering their heads to clear the top of the door, and placing them at the right-hand end of the car, with their heads on the side opposite to the door. The driver of the second team, in like manner, places his horses at the opposite end of the car. The driver of the third team places in his horses and the saddle horse belonging to the carriage. The drivers remain in the car. If there are four teams in place of three, and the car not large enough to hold them, the remaining team should be placed in the next car. Should any horse resist entering, the next one should be taken and the refractory animal led in immediately behind him. The docile horses should as much as possible be placed in first, and mild means are employed in preference to any other in inducing them to enter.

When the last horse enters, the cannoneers on the outside place in the bar, remove the ramp, and close the door, after which the bar is removed and passed to the men inside the car. The horses are tied to the head bar with the halter strap, which enables them to be unbridled, if the order should be given. In this case, three or four bridles are united, tied together near the head-stalls by the reins of one, and attached to one of the bars of the car near the corner, by the same reins. The awnings with which the cars should be provided, should be left up, unless the state of the weather requires them to be let down on either side. When the last car is not completely filled with horses, they must be kept in place in one or two groups, by bars placed across the car in the same manner as the door-bars. Two movable seats should be provided, for seating two of the men left in each car.

Each car of men or horses is placed in charge of a non-commissioned officer, or the oldest soldier in it. His duty as chief of the car is to maintain good order, and to see to the execution of all directions that are given.

SADDLES, &c.—The non-commissioned officer in charge of the saddle car, has the saddles loaded by placing one man in the car, who packs them away as they are handed to him, packed and tied up as directed, by another man outside. The first saddle is placed upon a bundle of straw with the valise resting against the side of the car; the saddles of the other teams belonging to the same carriage, are piled on top of this in their proper order. The saddles of the other teams are piled successively in the same manner, so as to form separate piles for each carriage. The officers' saddles are placed on top of the piles. The chief of the car, with two assistants, inside the car, takes a memorandum of the arrangement adopted.

CARRIAGES.—Trucks are the most convenient vehicles for the transportation of the carriages of a battery. Two field carriages, one and a half, or a single one, can be placed on one truck, according to its size.

The carriages are unlimbered and placed, just as they are, upon the trucks. The following are the principles which should govern the manner of loading. 1st. The weight should be distributed equally over the surface of the truck, occupying the least space possible. 2d. Arrange the carriages, in such a way that the ends of the tongues and the extra wheels shall not project beyond the edge of the truck; and 3d. Consolidate, by chocking and lashing with great care every part of the load in such a way that the whole is rendered perfectly solid and firm in position.

With trucks 14′ long, two are required for every three carriages. They are loaded at the end by placing on, first, the rear train with the stock to the rear, and running it forward until the wheels strike the front end of the truck, when the stock is rested on the floor. A limber is then placed on with the tongue to the front and raised, and run forward until its wheel touch those of the rear train already in place; a second rear train is now put on with the stock to the front resting on the floor, taking care not to allow its box, if it has one, to touch that of the limber in front of it. In like manner, a rear train and limber are placed upon the second truck. The third limber is placed in the rear with its tongue to the front, and resting on the floor, under the carriages already on the truck.

With trucks 17 feet long, two carriages can be placed on a single one. In this case, the rear train is placed on as before, but the limber is backed towards it till the wheels touch, or if possible overlap those of the rear train, and the tongue is rested on the floor; the second limber is placed on in the same way, and the second rear train with its stock to the front, resting on the floor.

When these trucks are so situated in the depót that they cannot be loaded from the end, the carriages are first placed upon the second truck, from which they are crossed to the first on a bridge communicating between the two, and arranged on it as already described. The second truck is loaded like the first, by making use of a third, and so on. The last truck is loaded as hereafter described, for those which can be loaded only at the side.

Trucks which can be loaded only at the side, should be at least 15 feet long to carry a carriage and a half, and 19 feet for two carriages. To load the former, place two rear trains, or a limber and a rear train successively, in position at the ends of the truck, as before described, with the stock or tongue to the rear, and resting on the floor. Then introduce between these two a limber, raising it by hand, with the tongue up so as to pass one of its wheels over the tongue and stocks of the trains already in position.

For the trucks 19 feet long, the first carriage is established at one end of the truck, as described with the truck 14 feet long. The rear train of the second carriage is placed in the same way at the other end, and its limber is passed in by hand with the tongue up, passing one of its wheels over the stocks of the rear trains to get it into its position.

In trucks less than fifteen feet long, which can be loaded only at the side, the two parts of a carriage are placed, one at each end, the stock and tongue to the rear, and then placed closer together by making the wheels on one side overlap; the stock being on the floor, the tongue elevated. In some cases the spare wheel is taken off, and either laid flat on the floor or leaned against one of the carriages.

During the operation of loading, the tongues should be held

by a sling fixed at the end. They are afterwards strongly lashed to one of the parts of the rear train already on the truck. The lunettes of the caisson stocks, which are provided with extra tongues, rest upon blocks high enough to prevent the weight bearing upon the ends of the tongues, and consequently on the keys which hold them in position in their stirrups.

As soon as a truck has received its load, the wheels of the different trains are locked together with cord from .5 to .6 inch in diameter, chocks are placed under the wheels and nailed to the floor, and the stability of the whole secured by tying the carriages to the rings of the truck. Straw ropes, or other means, are made use of to prevent friction between the parts.

The Men.—The men, with their knapsacks and side-arms, are divided, under the superintendence of an officer, into portions corresponding to the capacity of the cars. Each division is conducted promptly to the car it is to occupy, the men entering first going to the end farthest from the door, and so on. They seat themselves, holding their arms between their legs, the stock or scabbard resting on the floor. Fire-arms should never be laid on the seats or stood in corners, except when leaving the cars at the principal stopping places and stations.

Numbering.—The commanding officer should cause each car to be marked in chalk, with the number of the piece to which the men, horses, and baggage occupying it, belong.

Inspecting.—Immediately before starting, the commanding officer and conductor of the train inspect the cars to ascertain that every thing is in order. They should see that the couplings of the cars containing the "material" are short enough to insure the contact of the buffers.

The officers then enter the car assigned them.

Regulations.—The command being all aboard, the men are strictly prohibited putting their heads or arms out of the car while it is in motion; passing from one car to another; uttering loud cries of any kind; and from leaving the cars at the station before the signal for doing so is given.

The men with the horses, keep them from putting their heads outside the car. They feed them with hay from the hand, until they get used to the motion, hold them by the bridle or halter, and quiet their fears whilst the locomotive is whistling. In case of any accident, they make a signal outside the car, by waving a handkerchief.

If at any station the commander deems it necessary for the men to leave the cars, after the time indicated by the conductor, he informs the officers of the length of the halt. The officers remain in the vicinity of the cars containing their men, in order to direct and govern their movements. The guard posts sentinels wherever it is necessary, especially at the doors, to prevent the men from gathering near or opening them.

At a given signal on the bugle, the men leave the cars in order, and without side-arms. The men in the horse-cars get out over the side. If it becomes necessary to open the doors of these cars, the door-bars are first placed in position.

About the middle of the trip, as near as possible, the police-guard and men with the horses, are relieved. At each halt of more than ten minutes, the commander, or some other officer, and the conductor, inspect the cars, and especially those which carry the ammunition wagons.

Five minutes before starting, a bugle-call gives the signal for entering the cars. At the station immediately preceding the terminus, the horses are bridled, and the forage is collected and formed into one bundle for each car.

During feeding time there should be at least one man to every two horse-cars. In general, oats should be distributed only after the horses leave the cars. Hay is fed by hand by the drivers whilst the train is in motion. In ordinary weather, the horses are watered only when the trip exceeds twelve hours; and even in this case they need but little, and a single ordinary-sized pailful suffices for two horses.

UNLOADING.—On arriving at the terminus, the officers leave the cars first, and a bugle-call gives the signal for the men. The officers assemble the cannoneers, with their knapsacks and side-arms on, and form them into detachments according to the number and position of the discharging points.

The officer who superintended the embarkation of the horses, assembles the drivers, and if necessary, some of the cannoneers, and marches them to the point where the horses are to be discharged.

To prevent accidents, it is well to provide one or several movable bridges for discharging the horses, which are carried on the train. They are about sixteen feet long, a little wider than the car door, and are provided with hand-rails or ropes, movable at will. The bridge is supported at its upper extremity by a movable horse of a height corresponding to the sill of the door, and the cars are unloaded by passing them in succession in front of this bridge. The horse may be suppressed by fixing to the forepart of the bridge two strong flanges of iron which rest upon the floor of the car, and the bridge is applied in succession to each of the cars to be unloaded.

The non-commissioned officers in charge of the freight cars, immediately on arriving at the station, have the harness unloaded and placed in portions as originally divided in the cars. The material is discharged by the inverse means used to load it. As soon as the horse-cars reach the proper position, the men fix the movable bridges, open the doors, and bring the horses out in the inverse order in which they entered. If the horses have to be taken out of the same door they entered, the first two are backed out, and the rest follow after making a half turn. As soon as a rear team is disengaged it is taken to the place where the harness is deposited, and harnessed to a carriage which is conducted to the park, where the harnessing is completed.

CHAPTER XIII.

ON THE ATTACK AND DEFENSE OF FORTIFIED PLACES.

Sieges.—The siege of fortified places has always consisted in destroying the defenses from a distance, by means of large projectile machines, and effecting a breach in the wall for the purpose of entering the place.

In the early days of artillery, when defensive walls were very high, they were battered in breach with stone balls from 12 to 25 inches in diameter, after which the ditch was filled up and the place stormed. Mention is made of such balls at the siege of Metz, in the latter part of the fourteenth century.

The walls having been lessened in height, and supported behind with an earthen embankment, it was no longer possible to form the breach from a distance, and the use of stone balls having been abandoned, mines were resorted to for the purpose of blowing down the walls.

Cast-iron balls then coming into use, they were employed to make cuts in the masonry, instead of battering it down as formerly. This method, first invented by the Turks, was improved by Vauban, whose experience showed him that the breaches so formed were more regular and made more rapidly than by the use of mines.

Up to that period, the fire of the artillery, directed upon the pieces of the besieged and the interior edifices, expended its effect more upon the inclosed cities than upon the fortifications surround ing them. The ricochet fire was employed with full effect at this time in dismounting the enemy's pieces; and the war of sieges was carried to the highest perfection under its illustrious founder.

A besieging army is generally divided into two corps, one charged with the siege, the other with holding the surrounding country, for a short distance from the place. Sometimes the besieging army is covered by a corps of observation, in which case there is no necessity for its being so numerous as when it acts alone.

The artillery necessary for the attack of a place constitutes what is called a siege equipage or train. The strength of it depends upon the importance of the place to be attacked and the resources at hand. For besieging the strongest places, the French prescribe an equipage of—

40 24-pd. guns.	15 22ᶜ mortars (8.78 in.).
40 16 " (U. S. 18).	12 stone mortars.
40 22ᶜ howitzers (8.78 in,).	12 15ᶜ mortars (5.95 in.), and
15 27ᶜ mortars (10.79 in.).	60 rampart muskets.

A double attack would require about 200 cannon. The largest guns and howitzers, and smallest mortars, are provided with about 1,000 rounds of ammunition each. The large mortars with 750 rounds, and the stone mortars with 500 rounds.

In this country where the permanent works are so different from those in Europe, it is difficult to lay down any positive rules for the formation of siege equipages. The number and kind of pieces must be determined by the circumstances of each case, taking into consideration the strength, position, and condition of the place to be attacked. For the general principles on which estimates for a siege train are made, see Ordnance Manual, pp. 332–3.

An army, no matter how large, seldom carries along with it all the necessary means for a siege; and when fortified places obstruct the march, the attack is carried on with such pieces as can be obtained from the adjacent places in our possession; and generally the means made use of at sieges is far below the estimate which would be made were everything to be had close at hand.

Permanent works can be reduced only by the heaviest siege pieces, such as 18 and 24-pd. guns, 8-in. howitzers and mortars. The amount of ammunition will of course vary according to circumstances. If possible, the 18 and 24-pdrs. should be furnished with 1,000 rounds; the 12-pdrs. with a greater number. The 8-in. howitzers and mortars with 800, and the coehorn and 10-in. mortars with 600. Each *gun* should be provided, in addition, with 50 rounds of grape and canister, and 100 of spherical-case shot.

Military Reconnoissances.—On approaching a fortified place during a march, or in endeavoring to ascertain the position and force of the enemy, the resources of the country, &c., frequent reconnoissances have to be made. These are of three different kinds:

1st. The daily reconnoissances necessary for the safety of the camps, posts, &c. Their object is to discover the movements and preparations of the enemy and the disposition of his advanced posts. They are made by small detachments, and by patrols from the main guard. They should not be made at the same hour every day, nor by the same routes.

2d. Special reconnoissances designed to ascertain,—the topography of the country, and the means which it can furnish for attack and defense; the position of the enemy, and his strength at different points; and in fine, to furnish the information necessary for the determination of the method of operating, moving the different columns, &c. These reconnoissances are conducted in accordance with instructions from the general-in-chief, the commanders of separate corps, divisions, &c.

3d. Offensive reconnoissances, which ordinarily should be ordered only by the general-in-chief; for the purpose of ascertaining as accurately as possible, the position and strength of the enemy.

The result of every reconnoissance should be presented in a clear, simple, and positive report. The officer making it, must expressly distinguish between what he has seen himself, and what he has learned from others without being able himself to verify the accuracy of it. He adds to the report, the drawings necessary to represent the ground, the positions of the enemy, &c.

Troops making a reconnoissance, the object of which is simply to see and observe, should not become compromised or even seen; they should move with precaution, have an advance-guard and scouts out, and become engaged only when forced to it, in order to make prisoners when there is no other means of obtaining information, or when they encounter the enemy marching against their camp; in which case they give warning by preconcerted signals, such as a burning hay-stack, bonfires, &c.

To examine an enemy, choose the morning, when everything in his camp is in motion; observe the camp fires, the defensive works, the position of the parks, cavalry, &c. Take up a position on the flank of his column, and count the number of his battalions, squadrons, and batteries; note the space which they occupy, the time they take to march past, and the order of march.

For the topography, establish in advance, from the maps of the country, rough sketches upon a scale large enough to be able to figure in details at sight, and to delineate various distances and differences of level. Telescopes and compasses should be carried along with the party, which should also be furnished with good guides.

In making a reconnoissance without instruments, the survey is, of course, much less exact than when they are used. The distances are often measured by the pace, and many details inserted in the sketch, estimated simply by the eye.

A rough sketch is first made from the maps of the country. If the ground is open, a high level position is chosen, from which a great extent of country can be viewed; when practicable distances are measured with the chain, large triangles are formed in such a way as to obtain points near the center of the sketch, and these points are afterwards made use of to form smaller triangles, by intersections, offsets, &c. The intermediate details are inserted from the step and the eye.

In a wooded country, the method of examining with the eye and step is often the only practicable way of making the survey. To lessen the errors as much as possible, the principal lines of direction should be determined with care, as well as the points in which they intersect, and the sides of the inclosing perimeter multiplied with the compass, so as to verify the result.

The configuration of the ground is of essential importance in a military reconnoissance, in order to determine a proper distribution and employment of the different arms, and the positions for batteries, from a knowledge of the inclination of the slopes, the height of the hills, &c.

The principles laid down for making a regular survey should be followed as much as possible, in forming the map. If pressed for time, the conventional tints may be replaced by initial letters. Paths are represented by a single line, which should always be of the same size, and the thickness proportional to their importance. Roads are represented by two lines. Streams, which can be represented only by a single line, are marked in blue, which is increased in width, as the distance from the source increases.

When a more accurate survey is to be made, the position of the points, heights, distances, &c., are determined by ordinary surveying instruments, and the details are inserted in the map (of which a rough sketch is made as before), after personal observation.

If no maps or instruments for making the first sketch are at hand, a plane-table may be made with paste-board, or a plank, and used to determine the position of the most remarkable points. If these cannot be made, construct the sketch by triangles measured by the pace and by alignment.

A reconnoissance the object of which is to present a plan for occupying a position with fortified works, embraces, in general, a square surface of from 4,000 to 5,500 yards on a side.

A point is chosen near the center of the ground to be surveyed, which can be readily seen at a distance. Between this point and the limits of the survey, some 6 or 7 intermediate points are established, so that from each of them, the one on each side, as well as the central one, can be seen. There is thus formed a series of triangles around the central point, their bases constituting a circumscribing polygon. One of these bases is measured, either by the chain or in paces, and the bearings of the different triangles are taken with a compass.

Observe with a theodolite, or in the absence of one, with a sextant,* the inclination of the lines passing through the central

* The French have a simple little instrument, called the *inclination level*, or *èclimètre*, for determining angles in the field. It consists of two equal legs, *A C* and *B C* (see figure next page), kept apart by a metal arc, graduated in degrees in both directions from the middle point. From the point *C*, which is the center of the arc, a movable limb is suspended, furnished at the lower end with a vernier, by the aid of which the minutes are read. On this limb, and perpendicular to it, is fixed a small spirit level, *c f*, and

point, and through those to the right and left. As the length of these lines are known from the construction of the polygon, calculate the difference of level between the different points, and the results will verify each other.

The ground is then divided into sectors, which are surveyed in detail, the sides of the polygon serving as bases.

A sufficiently extended sketch, often accurate enough for military purposes, may be obtained in the following manner. Ascertain from the inhabitants of the country, the direction of the roads, rivers, and canals; the distances between towns, and to remarkable points, such as bridges, cross-roads, &c. Prepare in this way, the first rough draught, and correct it by means of the information which all foraging or scouting parties are directed to collect.

Among the methods employed to represent the configuration of a country, the one most used, consists in tracing upon the ground, level curves, and delineating upon the map the projections of their rectangular paths, or *lines of greatest inclination.*

If we determine the uniform distance between the curves, to the scale of the plan in such a way that it be represented by .04 in., the length of the normal lines for a slope of 45° will be .04 in., and

---

equal divisions traced on each side of the central point for adjusting the position of the bubble.

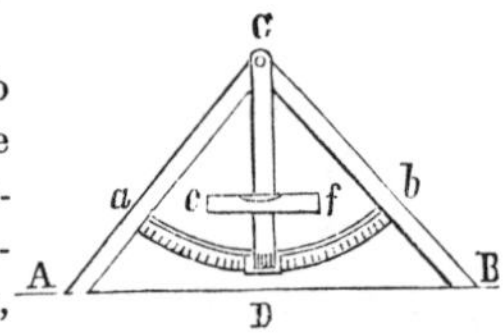

The instrument is made in such a way that the limb marks zero, when the base, *A B*, is horizontal, and the bubble in the center. To test its accuracy, the instrument is reversed in position. If the line *A B* is not horizontal, and the bubble is brought to the central position, the instrument will give a certain reading, say 3°, which should be the same on reversing it, if it is constructed properly. Suppose it is not so constructed, and in the reversed position it gives a reading of 1°: we conclude from this, that the sum of the two, 3°+1°, is double the inclination of *A B*, since in both positions, the limb has taken the same direction in regard to the vertical. The inclination is then 2° on the side of the first reading made. To facilitate the use of this instrument, it may be mounted on a tripod, by the aid of a ball and socket, which allows it to turn, and on the feet, *A* and *B*, two sighting vanes are fixed, to look through when getting the inclination. The instrument first used, had a graduated bar across, instead of the arc, and a plumb-line in place of the limb and vernier, and the one described has since been much improved and perfected, by adapting a telescope to the edge of a compass, with a vertical limb graduated to indicate the different inclinations of the telescope.

for one of 2°, 1.28 in., these being the extreme slopes so represented. The normal lines are placed a distance apart, equal to $\frac{1}{4}$ their length. If the ground to be represented has no very steep slopes, a better effect is produced by reducing the distance between the curves $\frac{1}{2}$ (.02 in.); and when very abrupt ground is met with, as the relation between the length of the normals and the spaces between them can no longer be observed, the lines are made a little thicker.

The most important slopes to be indicated on a map, are:

1st. Those of 60°, which are inaccessible to men, and have a base of 4 to a height of 7 units.

2d. Those of 45°, which are difficult for men, and have a base of 1 to a height of 1.

3d. Those of 30°, which are inaccessible to horses, and have a base of 7 to a height of 4.

4th. Those of 15°, easy accessible for carriages, and have a base of 12 to a height of 1.

The maximum inclination assumed by falling earth, in a mountainous country, is a slope of 100 base to a height of 71.

When great accuracy is necessary in representing the form of ground, the horizontal curves are traced with a spirit-level from yard to yard, or every two yards, each point being marked with a picket, and the ground afterwards surveyed with a plane-table or compass.

In ordinary surveys, the horizontal curves are determined with tolerable accuracy, by establishing a point at every five yards. For this purpose, suppose sections along the lines of greatest inclination to be made in the ground, and mark on the map the projection of several portions of these lines. Then measure with a sextant, or other instrument, the angles which these parts make with the horizontal. The angles and the projections being known, the height of the lines are deduced by making use of or table or particular scale. These projections are afterwards divided on the map, in such a manner that the points belonging to the horizontal curves are the desired distance apart.

The shading lines are made with a pen, in India ink. Masonry constructions are traced in red. Roads, paths, dikes, wooden bridges, isolated trees, wooden wind-mills, batteries, and entrenchments, in India ink. The different kinds of soil and cultivated ground, masses of houses, and water, are represented by conventional tints. The colors employed are, India ink, carmine, gamboge, indigo, and sepia. Masonry constructions are colored in carmine; rivers and streams, in blue; forests, in yellow, merging on a green, composed of gamboge and a very little indigo—the wet portions filled in with blue, like water; meadows in green, composed of indigo and gamboge. Marshy meadows are represented by breaking up the green, and filling in the open portions with a blue tint like water. Orchards have the same tint as meadows, and are regularly dotted over to represent the trees. Heath is made of a lighter blue than meadow land, and slightly diversified with red on another brush. Marshes are tinted like meadows, but with the wet portions filled in with blue. Untilled ground with a lighter blue than meadows, and slightly diversified with a color formed of gamboge and a little carmine, like sand. Vineyards with a violet color made of Indian ink, carmine, and indigo. Sand, with gamboge and a little carmine. Lines of troops are represented by short parallel lines filled in with blue.

The writing on the map is made in Indian ink. The altitude figures, the names of towns, cities, and dwellings, are written alongside, upon perpendiculars to the meridian, the north being placed above; those of rivers, streams, roads, and canals parallel to the directions in which these run. The roads going beyond the limits of the survey should be marked with the largest place to which they lead.

The names of cities are put in in upright capitals; those of market-towns in inclined capitals; wooden villages, in upright roman; hamlets and marshes in inclined roman; citadels, large rivers, and canals, in small inclined capitals; forts, small rivers, ponds, castles, and roads, in small upright roman capitals; mineral springs in italics; and farms, inns, manufactories, mills, paths, and small streams, in small italics.

In the field, if there is not time to form a topographical map, with pen-shading, &c., the ground may be figured with the stump or pencil, deeper tints being employed where the slopes are more rapid.

The principal mountain-chains which serve as a defense to a district of country should be distinctly marked, as well as the different branches which defend or favor the access to it. Note their direction with the compass, as also the relative heights of the different parts, if they are sufficiently extended to enable a plan of defense to be formed. Observe the communications to be preserved, the roads to be destroyed, and any other means of annoying the enemy; the proper positions for camps or entrenchments; the slopes, forests, rocks, &c.

A mountainous or hilly country, partly wooded, partly cultivated, is the most difficult to reconnoitre well. Commence, as much as possible, at the most elevated portions. Mark the commencement of the slopes; examine the ravines, water-courses, roads, and pathways. See if it is necessary to establish bridges, and if columns can follow the bottom of the valleys or the crest of the hills. Inspect the streams, as is directed in Chap. XIV; and canals in the same way, giving the communications which they establish, a description of the lockage, &c.

Give the position of springs and cisterns; the quality and quantity of water, &c.

The position, use, dimensions, construction, and solidity of all bridges should be given; the approaches, and water-way; the streets leading into towns and cities; the nature of the country in front of them; the means of fortifying them; of destroying them; of re-establishing a passage with the most advantage, regard being paid to the nature of the brooks, the current, width, embankment, fords, and communications.

Of ponds, marshes, &c., mention whether they are formed from springs, from inundations, or simply from wet ground; the best means of crossing them; at what seasons it is practicable to pass them, and with what kind of troops, whether they are unhealthy or covered with fogs. If there are causeways, state how they can be repaired; if not, how they can be made, and how defended. Sound the depth of any wagon-tracks. Prairies on which the grass is high and thick, or on which patches of yellowish-green moss occur, should be carefully examined, for in spite of their appearance, they are often impracticable even for infantry. Examine the ground around them.

Take the depth, extent, and level to which the water can be kept in inundations; the time required to raise the water to a given level; the working of the sluices, the means of taking possession of, or defending them; of opening or closing them.

On a coast, examine the shore, cliffs, rock, &c., which may make an approach more or less dangerous or quite impossible; extended and open portions, fit for making a landing; batteries and intrenchments established to defend the anchorages, channels, and accessible points, or positions where they may be placed in case of necessity. The adjacent islands which may be made use of for advanced works; the rise and fall of the tides, which affect more or less the approaches to the different points, and the variations which they produce in the course of rivers; and the means of passing them; the low-water marks; the creeks, bays, roadsteads, and ports; their advantages and inconveniences; the size of vessels they are capable of receiving, and the winds required to enter and leave them; the positions for camps, and the posts capable of covering the principal establishments in the interior of the country. Notice every thing characteristic of the accessible places; existing obstacles, and those which can be added to defend the approaches to them. State the condition of the forts, batteries, guards, and material. Analyze the system of defense adopted, and propose modifications, if necessary. Estimate the force which can be raised, in an emergency, either from the troops or the inhabitants, whilst waiting for the arrival of the regular troops at the point of attack.

In reconnoitering a wood or forest, state its position, extent, and thickness, as well as the height of the trees and undergrowth; the gaps existing between masses of the trees; their width; whether the trees on the right or left form a thicket, and whether they can be burned. The nature of the soil, the surrounding ground, &c.; the communications, the means of establishing, and their direction, in order that they shall not be taken in flank; the means of forming abattis. Go around the forest, noting the roads, streams, and ravines which lead to it, following them to their heads when not too long.

The kind, quantity, &c., of heath, underwood, and hedge, should be stated. Tall heath is usually practicable; low heath often marshy. Low and thick hedges form a very good defense.

Roads are surveyed either by the compass or the eye, and

their direction noted, as also the width, whether variable or constant. State whether railroads, paved or well-beaten, and if bordered by trees, hedges, or ditches; distances between the principal places, the ascents and descents, and estimate the time of marching them in hours; whether constantly practicable, or so only according to the weather and season. The country, streams, and towns along the road; the roads which cross it and where they lead. Heights which command them; whether in curves or zigzags crossing mountains. How formed, whether by excavation or embankment, and the length of the former; the dangerous points; repairs necessary for the passage of artillery. The width of the track, more especially in sunken roads, which should be avoided as much as possible, or filled up. An accident to a single carriage on such a road might stop a whole column. If but a single road exists in one direction, see if it is possible to open lateral ones for other columns. Trace the routes of these columns. Do not neglect the pathways or unfrequented roads, reported by the country people as impracticable, as they may often be repaired with little labor.

Examine passes, as to their being practicable for infantry, cavalry, and wagons; the communications between them; and if connected by the crests of the intervening hills; the means of guarding them; the time necessary to reach the summit by the established roads; the possibility of opening new routes.

In defiles, state the width and length of the gorge; positions to be occupied to protect a forward movement or cover a retreat. The nature of the ground at the entrance, and the troops which can be advantageously employed there in battle.

In forts, castles, and citadels, give their position and extent; their object, and the works connected with them. The protection which they afford to a city or the country. The obstacles or support offered by them to an enemy. The nature and condition of the fortification; whether ancient or modern, permanent or temporary; of great or small relief; revetted, wholly or in part; of masonry, brick, sod, or natural. The mines and galleries; surrounding grounds; the defense capable of being sustained by the works themselves, and those which may be added. The proper points of attack to be chosen.

For fortified cities, state their relation with each other, and with reference to the movements of armies. The positions of the 1st, 2d, lines, &c. Succors which they can give or receive. Means of directing these succors according to the direction of attack. Resources in provisions, &c., and the means of collecting them. Facilities for establishing depots, hospitals, &c. Nature and strength of the works, and of each front in detail. The surrounding ground, and the advantages which it offers for attack or defense. The positions to be occupied in the investment; communications to be established between the different quarters, and the works necessary for the safety of the lines.

With unfortified towns and villages, state the defense of which they are susceptible; the inclosing walls, towers, ditches, dry, wet, or full of water; the houses, whether against the walls or separated from them; the number of the gates. The surrounding grounds; gardens; the roads and paths adjacent to the place.

In reconnoitering any military position, three principal objects are to be considered. 1st. The ground itself of the position. 2d. The approaches to the place and means of debouching from them. And 3d. The communications, and the rear of the position.

A good position should be commanded neither on the front nor on the flanks.

Information to be collected in a catalogue. The names of cities, towns, and villages; opposite, those of the hamlets which are dependent upon them, with their distances from the chief places. Number of houses, united or isolated. Population. The number of men and horses which can be quartered. The quantity of grain, hay, straw, beeves, cows, calves, sheep, hogs, &c. The mills, ovens, wells, and fountains. The means of transportation, wagons, boats, horses, oxen, and mules. The number of farriers, wheelwrights, workmen in wood and iron, tailors, shoemakers, saddlers, &c. The taxes, revenues, commerce, and business of the places. The salubrity of the habitations, stables, air, and water. The quantity of grain which can be ground, and of rations which can be cooked in a given time. The combustibles. The iron, cloth, leather, wine, brandy, &c.

On arriving before the place to be besieged, the siege-park is established, as near as possible to the point of attack, generally between 2,500 and 3,000 yards from the advanced works, taking advantage of the form of the ground to cover it from the enemy's fire. The pieces are placed in the first line, at four yards apart from axis to axis. In rear of them the platforms, tools, and projectiles.

At 600 yards in rear of the park, the powder-magazines are established, in a line, and about 200 yards apart, each containing from 25,000 to 50,000 lbs. of powder in barrels. The magazines are covered with oil-cloth supported on a light frame-work. At 80 yards on the right and left of this line parks are established to serve as arsenals for distributing the powder.

The workshops for the artificers are 200 yards in rear of the magazines, and those where the fascines, &c., are made, 200 yards farther to the rear. They should be as near as possible to forests. Parks for the horses should be placed convenient to wood and water.

Batteries.—The points of attack are protected by a certain number of pieces placed together in position, when they constitute a *battery*, the term being also applied to the constructions necessary for the use of one or several pieces.

A battery consists of a covering mass A B C D, Fig. 284, called a parapet, designed to protect the men and pieces, and of one or two ditches to furnish the earth. Sometimes the parapet is made of earth brought to the position, and there is no ditch. The following are the usual dimensions of a battery:

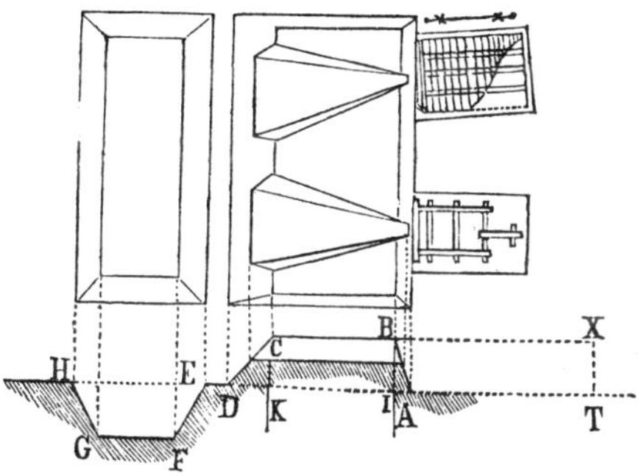

Fig. 284.

K I, the thickness of the parapet, is 18′, in order to protect the battery from pieces of large caliber. A T, the terre-plein of the battery, is ordinarily 26 feet wide, and has a slope of $\frac{1}{60}$ for the purpose of shedding water. B, the interior crest, is 7 ft. 6 in. high, and is so calculated that the enemy's line of fire B X, shall pass 6 ft. 6 in. above the rear of the terre-plein. A B, the interior slope, has a base equal to $\frac{2}{7}$ of the height I B. This slope is packed as hard as possible. C, the exterior crest is 6 ft. high.

B C, the top of the parapet, is given a slope to carry off the rain-water which falls upon it. C D, the exterior slope, has an inclination a little less than the natural slope of the earth, in order that the enemy's shot may not crumble it down, thus decreasing the height of the parapet. D E, the berm, is made from 2′ to 4′ wide, in order that the earth knocked down by the enemy's shot shall not fall into the ditch, which would render the repairing of the parapet more difficult. E F G H, is the ditch, the profile of which is calculated in such a way as to furnish earth sufficient to form the parapet. It is usually given a depth of 5 ft.; but sometimes the presence of water or rock renders it necessary to make it of less depth, and its width has then to be correspondingly increased. The scarp E F, and counter-scarp G H, are usually sloped so that their bases are equal to half their height.

Sometimes it becomes necessary, in order to protect the men and pieces from the flank fire of the enemy, to throw up épaulments, which join on to the ends of the parapet making a small angle with its direction. For the same purpose traverses are placed between the pieces. In consequence of the obliquity of the épaulment to the enemy's fire, it need not be thicker than 12 ft. The form is similar to that of the parapet, though the interior slope need not be so steep.

Guns and howitzers fire through embrasures. These are called *direct*, when the central line or line of fire is perpendicular to the parapet, which is usually the case. When this line is oblique to the parapet the embrasure is said to be *oblique*.

The embrasure is laid out so that the base or sole is 2 ft. wide at the neck or rear part, and at a distance of 5 ft. to the front is increased to 3 ft.

The solid part of the upper portion of a battery, between two embrasures, is called a *merlon*, and that part under the embrasures which is solid throughout, is called *the solid*.

That part of the parapet included between the base of the embrasure and the foot of the interior slope, is called the *genouillére*. Its height, for a piece mounted on the siege carriage, is 3 ft. 6 in. The base of the embrasure has a slope to the front, of ½ in. to 1 foot, to carry off the water.

The sole of the embrasure is traced out as soon as the parapet has been raised to that level, by laying off on the proper line of fire 5 feet from the interior slope, and at that point, on a perpendicular line, 1½ foot on each side. These two points and the sides of the neck are marked with short stakes; and lines drawn through them on each side and produced to the exterior slope, determine the splay of the cheeks of the embrasure.

In order to protect the cannoneers, the sides of the neck are sloped only enough to make the width at the interior crest 3 feet. But the slope of the cheeks is increased as they approach the exterior slope, so as to give a fire to the right and left, and prevent the cheeks from being blown away by the blast of the gun.

As steeply-sloped earth does not stand well, some kind of revetment has to be used. For this purpose, sod, gabions, saucissons, or hides, are used.

The *gabion* is a cylindrical basket with no bottom, 3 feet high and about 2 feet in diameter. They are set firmly in the ground and filled with earth. They are preferable to saucissons on account of requiring less wood, and being made and repaired more easily. They are generally employed for the cheeks of embrasures, for traverses, communications, &c., and are sometimes used in the same revetment with saucissons.

The *saucisson* is a cylindrical bundle of fagots, 1 foot in diameter and 18 or 20 feet long, bound together with withes. They are placed on top of each other with the proper slope, and secured to the parapet with stakes and withes, or iron wire.

When hides are used to protect the cheeks of an embrasure, it is generally in connection with some of the other revetting materials, and they are securely staked down over them.

In a sandy soil, sand-bags are used as a revetment, as was done at the attack on Vera Cruz; and in case of the scarcity of wood to make gabions, &c., common barrels, filled with earth, serve very well. These were made use of in the building of Fort Brown.

Common clay, mixed with chopped straw, makes a very good revetment, when well packed in layers of 1 foot thick.

When sand-bag revetments are used for the embrasures, the cheeks and sole should be covered with a double thickness of wicker-work to protect the bags from the blast of the guns. The bags are made to break joints, which is also the case with saucissons.

Sod revetments are seldom made use of in sieges, as it takes very long to construct them.

When the form of the ground permits it, the ditch is made on the inside of the parapet, and is 2½ feet deep, and from 18 to 20 feet wide. The men and pieces are placed in the ditch; and the parapet being formed in part of the solid ground, is more promptly made and more solid. A small ditch is cut in this case at the bottom of the exterior slope, to catch such of the enemy's shells as may roll along the ground.

Siege pieces should always be accompanied with their platforms, which are of especial importance with guns and howitzers, as these, to fire accurately, require that the trunnions should be horizontal. Without the platform, too, the ground would soon be worn into ruts, and in wet weather it would be almost impossible to manœuvre the guns.

The batteries are usually placed from 20 to 25 yards in front of the parallels, to which they are joined by trenches.

KIND.—Batteries are called, 1st, according to the kind of piece used in them; thus, *gun*, *howitzer*, and *mortar* batteries. Sometimes they are mixed batteries, and contain pieces of different kinds. 2d, from the nature of the firing, as *direct*, *ricochet*, and *breaching* batteries, according as the pieces fire directly upon the object, by ricochet, or for the purpose of making a breach; and 3d, from the direction of their fire. Thus, a *direct* battery is one which fires in a direction perpendicular to the face of the work to be struck, and the shot of which strikes the object without ricochet. An *oblique* battery is one whose line of fire is oblique to the work fired at, the firing being direct. A *reverse* battery strikes the interior of a face of the work, and under a small angle. An *enfilading* battery strikes the flank of a face, enfilading its length. Reverse and enfilading batteries generally fire ricochet shots. Ricochet and direct batteries dismount the enemy's artillery, and destroy his defenses. *Mortar* batteries render the communications between the different parts of the work difficult, especially for artillery; destroy the enemy's shelters, and set fire to or blow up the magazines. Under the protection of these different batteries, the works of the besiegers are pushed forward to the covered way, where the breaching batteries are established to open the scarp wall. The breaching batteries cannot be occupied until the fire of the work is almost silenced.

Gun and howitzer batteries may occupy four different positions in regard to the face attacked. 1st. They may be established in a position perpendicular to the prolongation of the face to be destroyed, (1) and as the projectiles have to pass over the face *A C*, Fig. 285, to fall upon the terreplein of *A B*, the pieces should fire in ricochet, and the battery is an enfilading ricochet battery. This is the best kind of a battery for dismounting the enemy's artillery. Usually, the first piece fires along the interior crest, the second parallel to it. The others are directed upon the middle of the adjoining face.

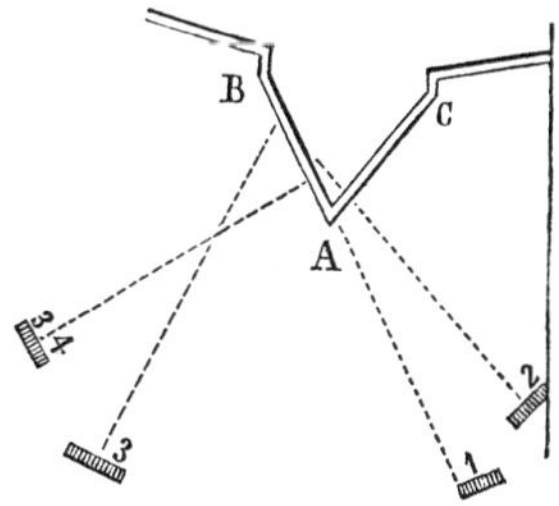

Fig. 285.

2d. If circumstances do not allow this position to be occupied the battery (2) is placed within the prolongation of the face, so as to strike the face on the interior, under a small angle. This battery fires in ricochet, and is called a reverse ricochet battery. Is becomes more effective as its distance from the prolongation of the face is lessened.

3d. The battery (3), may be established on the other side of the prolongation of the face, striking its exterior direct under a less angle than 90°. It then, by its oblique position, is not exposed to the fire from the face, *A B*, and taking in flank the embrasures of the place, demolishes them more easily. It is called an oblique battery.

4th. The locality sometimes renders it obligatory to make the battery a direct one, firing without ricochets, which is the least advantageous of all. (3 4.)

Tracing.—Batteries are laid out at night, the positions having been selected in the daytime. As this is a duty on which artillery officers may be ordered, although usually, in our service, performed by the engineers, the means required and method of tracing will be laid down. In the French service, the commanding officer of each battery traces his own.

With the assistance of some men, a dark lantern (to prevent the position from being discovered by the enemy, and to see by), some stakes, lines, and a measuring tape, two rectangles are laid out; one representing the base of the parapet, the other the upper opening of the ditch. These rectangles, if the platforms are to rest upon the natural ground, and the earth taken from the ditch in front, are placed about one yard apart to form the berm. But in case a *sunken* battery is to be made, that is, one in which the platforms will be below the natural level, and the earth for the parapet is taken from the rear of it, these rectangles are only 18 ins. apart. This sunken battery enables the men to be placed under cover sooner than the other; but it can only be employed when the trenches and other works in its front do not intercept the fire of its pieces.

Seven is, usually, the largest number of pieces placed in one battery, and generally not less than three. The number of guns being established, the length of the rectangles to be laid out follows from allowing 18 feet for each gun, and 6 feet for each splinter-proof gabion traverse, one of which is placed between every two guns when the battery is composed of more than three.

The width of the rectangles for a battery on the surface is, for the parapet, 26 feet; and for the ditch, about the same. For a sunken battery, the parapet rectangle is 24 feet wide, and that for the ditch is 22.

The rectangles for the épaulments and their ditches are laid out in the same way, allowing from 20 to 30 feet for the length of the épaulment, and 12 feet, or more, for its width, according to its obliquity to the line of the enemy's fire.

These rectangles, when laid out, are well marked with stakes and cords.

Forming.—In throwing up the battery, two men per yard of the parapet are allowed. They are placed one per yard in the ditch, one for every two yards on the berm, and one for every two yards on the parapet. The last two throw the earth toward the interior slope, and pack it. The men in the ditch are changed every two hours, and the working party relieved every twelve. They work ten hours, and rest two.

A man can throw earth with the shovel, 13 feet horizontally, or to a height of 5 feet.

When the excavation does not require the use of the pick, a tasked man can shovel and load on a wheelbarrow from 15 to 19 cubic yards of earth per day. When thrown horizontally, more than 6 feet, or less than 13, or to a height of 4 feet, or loaded on a cart, the number of cubic yards must be reduced to 10. An excavator by profession can remove with the shovel, and load in a cart, as much as 30 cubic yards of earth.

The nature of the ground is expressed by adding the entire or fractional number of pickers to the shovelers kept at work by them. Thus, one shoveler and one picker, ground for two men; one shoveler and two pickers, ground for three men; two shovelers and one picker, which is the same thing as one shoveler and one-half picker, ground for one man and one half.

In medium earth, one pick can keep two shovels at work; but that these shall not interfere with each other, they must be separated by a distance of from 5 to 6½ feet. For excavations, the estimate is generally one pick and two shovels for each space of from 3 to 4 yards long. The relieving shovelers are spaced horizontally at 4 yards, and vertically at about 5 feet. For each gang, one rammer and one leveler are required.

In excavating ditches, banquettes are left at a distance of 5 feet apart, and cut down so as not to interfere with the slopes. Those next the counterscarp are cut away as the work progresses, and those next the scarp when the parapet is about to be finished. The épaulment is formed of well-rammed horizontal layers from 8 to 10 inches thick, the edges projecting a little beyond the profile, and afterwards cut down with the proper slope.

Experiment proves that in siege-works, at the opening of the trenches, a workman from the line, in seven night hours and in ordinary ground will excavate about 2 cubic yards. In ten hours, this should be increased to very nearly 3 cubic yards.

In the day time, as many men as possible are placed in the ditch, and the rest employed in carrying forward supplies, and in

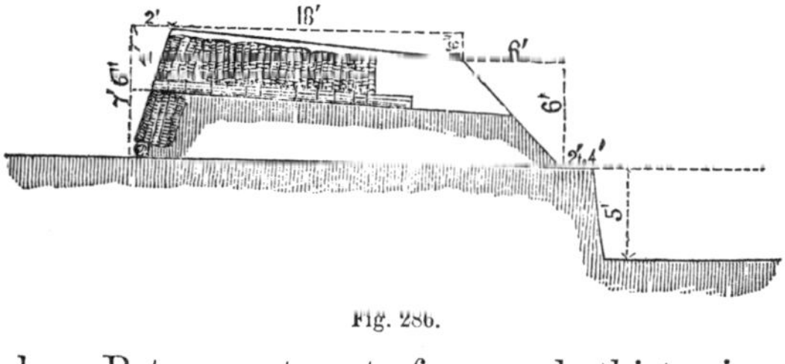

Fig. 286.

other work. Between twenty-four and thirty-six hours are required to form the battery, and only about half that time for a

sunken battery. The interior slope is riveted, Fig. 286, usually with gabions, the proper slope being given to them by placing underneath, along the foot of the slope, a row of fascines. When the first row of gabions has been filled, and the earth of the parapet reaches the same height, a second tier of fascines and gabions is placed on top, and the work continued, the requisite height being given either by placing sods on the top tier, or by earth alone.

Before the second tier gabions is placed in position, the embrasures are laid out. Their axes are 18 ft. apart, unless a traverse intervenes, when the distance is 24 ft., allowing 6 ft. for the width of the bottom of the traverse. The sides of the sole are at an inclination of $\frac{1}{10}$ with the axis. The top edges of the cheeks diverge more than the bottom. The cheeks are revetted for some distance from the neck, either with gabions secured as already described, the earth being compactly filled in around them; with fascines laid on top of each other, sloped outwards, and secured with pickets and withes; with sand-bags, which are laid in tiers, the layers breaking joints; or with sods, or hurdle-work.

The épaulments are not revetted, the interior slope being made as steep as the earth will stand.

Howitzer embrasures are sometimes made with a counter-slope, the sole receiving nearly the same inclination from the sill upwards as the least angle of elevation under which the piece will be required to fire.

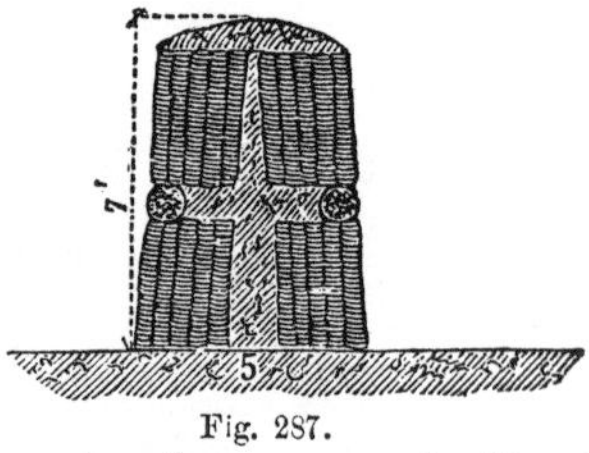

Fig. 287.

TRAVERSES.—The traverses, Fig. 287, are made only splinter-proof, to prevent the pieces of a bursting shell, for instance, from extending farther than among the cannoneers of two pieces. They are not designed to resist shot, and consist of only two tiers of gabions. The lower tier is made of two rows inclined towards each other at the top, filled and packed in between with earth. Two rows of fascines are placed on top of these, and the second tier of gabions on top of them, inclined like the first, filled and heaped up with earth. The traverse is from 15 to 18 ft. long, and has an interval between it and the parapet of 2 ft.

When traverses are used with sunken batteries, the distance

between guns separated by a traverse is increased to 26 ft. The

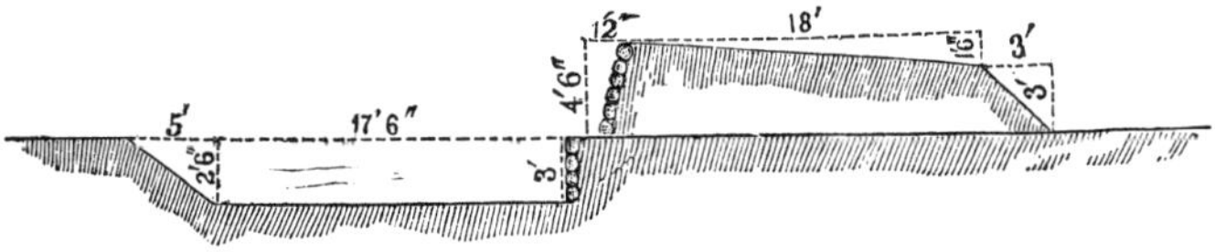

Fig. 288.

trench, Fig. 288, when first excavated is 16 ft. wide at the bottom, which is slightly inclined to the front, where the trench is 3 ft. deep. The front is cut down vertically in firm soil, and the rear receives a slope of $\frac{1}{2}$.

As the trench will furnish only sufficient earth for the parapet, that for the épaulment is obtained from a ditch 5′ deep outside of it, and that for the traverses from ditches cut outside the parapet and opposite the position of each traverse.

The front of the trench is cut down nearly vertically, and revetted with fascines, laid on top of each other, and pinned to the earth by stakes passing through them and obliquely downward. By this means, the guns can be run close up to the parapet, and their muzzles project well into the embrasures, which are made of the same form and in the same positions as prescribed for the other battery. When traverses are made in the sunken battery, a portion of the ground, 6 ft. wide, is left standing, and on this the traverse is built. This kind of traverse is made by first laying down on top of the ground left standing two rows of fascines, 5 ft. apart. On these two rows of gabions, inclined so as to touch at the top, are placed, filled with earth as before, and heaped up on top to the height of the interior crest. The sides of the solid part of the traverse are cut to a suitable slope, and revetted with fascines. A passage of 2 ft. is left between the interior slope and the gabionade of the traverse.

Several holes should be dug at the most convenient points in the trench of the battery, to collect the rain which may fall, and precautions should be taken to prevent water from running into the trench, either from the natural surface, or from the trenches leading to the battery.

In throwing up these batteries, a narrow ramp at the end o each épaulment is left, leading from the natural level to the bottom of the ditch, and serves for the convenience of the men whilst at work.

A trench of the ordinary dimensions (8′ wide at the bottom, and 3½′ feet deep one side, by 4′ the other), is made from each extremity of the battery to the parallel in its rear, Fig. 289.

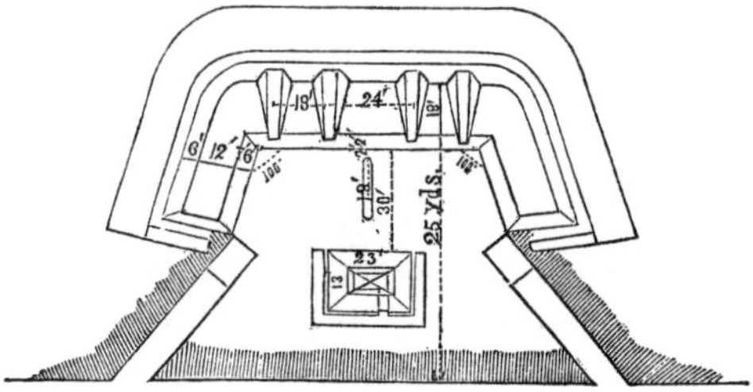

Fig. 289.

If the working parties are interrupted by sorties, the men are withdrawn into the trenches, care being taken to make them carry their tools with them, in order that the enemy may not make use of them in destroying the work.

The powder magazines should be at least 30′ in rear of the parapet, with its ceiling not more than a few inches above the natural level. The interior height need not exceed 5′, which will be about the depth of the excavation to be made. It should be 6′ × 12′ in the clear, and the sides may be formed of frames and sheeting boards, or of a row of gabions crowned with two courses of fascines. It is covered with splinter-proof timbers 6″ × 9″, over which is piled at least 3′ of earth, both on top and on the sides towards the parapet and épaulments.

The door is placed in rear, and is reached by one or two inclined trenches.

Mortar Batteries are of two kinds: those for mortars which throw shells, as the 8-in. and 10-in. siege; and those for the stone mortars. The coehorn mortar being small, is placed in any unoccupied corner of the trenches. The first kind are usually placed in front of the 1st and 2d parallel, and in such positions as to bring as large a portion of the place under their fire as possible. The stone-mortar batteries are used at shorter distances (in front of the 3d parallel), to annoy the covered way and adjacent parts.

The platforms may be laid on the natural surface, in which case the same form and dimensions are given to the parapet and épaulments as those laid down for gun-batteries; but as the mortars have to be set back far enough to enable the shell to clear the interior crest by about 3′, a revetment is unnecessary, and the interior slope may be made as steep as the earth will stand under firmly.

The front of the battery is estimated by allowing 15′ for each mortar, and 6′ for each splinter-proof traverse.

These batteries are, however, usually sunk beneath the natural surface, since several feet difference of level in the position of a mortar will have but little effect on the range. The trench is then made 13½′ wide at bottom; 3½′ deep in front, and 4′ in rear, with a reverse slope of ½. Height of parapet 4′; thickness 18′; berm 1′, and the front slope of the trench with a base of 2′.

The labor of throwing up batteries is performed by line soldiers. Eight artillerymen are assigned to each piece, and are relieved every 24 hours. They level the terrepleins, revet the slopes, and form the embrasures, in revetting which last, a mask is formed if necessary in front of the mouth. It is made of two ranks of gabions filled with earth, surmounted by a second rank filled with fascines, Fig. 290.

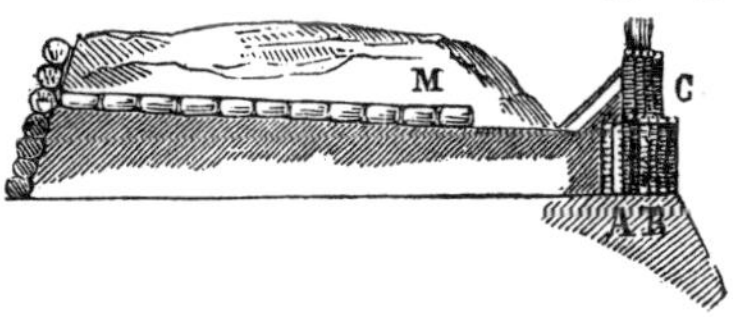

Fig. 290.

The platforms are laid whilst the battery is being finished. For mortars and ricochet-batteries they are laid level, and for direct firing have a counter-slope of 7 or 8 in. in order to diminish the recoil. The hurter should be placed perpendicular to the directrix.

Mortar-batteries are usually made without embrasures, but when, for the want of howitzers, the mortars are to be fired in ricochet, embrasures are made for them with genouillères 3½ ft. high, and an inclination of the sole of 9°. (Fig. 291.) The platform in this case is given a counter-slope in order that the mortar may fire under an angle sufficiently small, and to diminish the recoil.

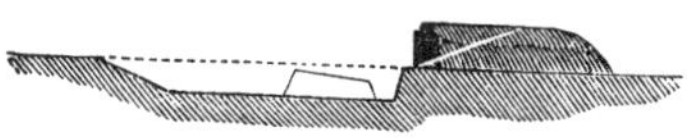
Fig. 291.

Enfilading and counter batteries are usually armed with 18-pds. and 24-pds. and 8-in. howitzers. The fire of the guns is mainly directed to destroy the enemy's artillery; that of the howitzers to sweep the covered ways and ditches, to destroy the palisadings, and injure the traverses by exploding shells in them.

All the batteries open their fire at the same time at a given signal, so as to divide the attention of the enemy.

Ricochet and mortar-batteries fire night and day; direct batteries in the daytime only.

Seven artillery-men are required for each gun; five for each howitzer and 10-in. mortar, and three for 8-in. mortars and stone-mortars. A portion of these may be detailed from the infantry. Each detail serves 24 hours at a time.

Breaching-Batteries.—Exposed revetments may be breached by heavy guns, at ranges from 400 to 600 yards; and batteries for such guns are in all respects the same as enfilading or counter-batteries. But when these revetments are covered from distant fire by the crest of the covered way, the breaching-batteries must be placed either on the glacis or on the terrepleins of the defenses, at points where no obstructions interfere with firing the guns low enough to form a breach practicable for the ascent of an assaulting column.

In either of these latter cases the batteries must be sunk low enough to subserve the object in view. The embrasures are usually cut out of the parapet, as an ordinary trench has generally to be first established, as a preparatory step. The form and dimensions of other sunken batteries, with such modifications as are demanded by the site of the battery and the position of the point to be attained, will apply in these cases.

Breaching-batteries, established either on the glacis or terre-plein of a work, will generally be exposed both on their flank and rear to the fire of dangerous commanding points, from which it will be necessary to cover them by traverses, the number and position of which will depend upon the command and position of the dangerous points. To cover from the flank fire, if the command of the dangerous point is considerable, it may be necessary to place a traverse at each interval between the guns. The traverses used in such cases receive a thickness of 14′ or seven gabions; their length depends upon the relative positions of the dangerous points and the exterior point of the battery to be covered.

When the reverse of these batteries is exposed, it is generally from the salient position of some comparatively distant point, from which a slanting fire may be brought to bear on this part of the battery. In this case it will generally be easy to cover the exposed part by running out from the reverse of the battery, an end of a trench, to form a wing traverse that shall intersect the lines of fire from the point upon every part exposed.

The guns of breaching-batteries should be placed so that the direction of their fire may be as nearly perpendicular to the line of the wall as possible; and when oblique, the angle should not exceed 45°, otherwise the effect of the shot will be greatly diminished, and the operation retarded.

Besides the breaching batteries, counter-batteries, in every respect similar to the others, are established on the glacis. Their object is to counter-batter and silence the artillery of the defense which may be brought to bear on the breaching batteries, or on the passage of the ditches. They are usually placed on the prolongation of the ditches.

The genouillère is equal in height to three ranks of saucissons, in order that the pieces may not tear away the revetment when they recoil. If these batteries cannot be sunk, they are formed of bags of earth brought up to the position. Breaching batteries are armed with from two to six 24 or 32 pds.

In these batteries, the necks of the embrasures are closed with a kind of oak shutter, bullet-proof, called embrasure-blinds, or mantlets, to protect the cannoneers from the fire of small arms. At the seige of Sebastopol, the Russians constructed their mantlets of several thicknesses of tightly twisted rope, securely bound together, and hung like a curtain from the top of the embrasure, an aperture being left in the lower part through which the muzzle of the gun was run. This opening was made large enough to allow sufficient field of fire to the piece; and the cannoneers were further protected by a circular mantlet made of the same material, which fit closely on the chase of the gun, between the wheels, a small opening on top being left for aiming through. These mantlets are perfectly bullet-proof, and must have been of great service in protecting the Russian gunners from the sharp shooters of their enemies in rifle-pits, &c.

Various obstacles may interfere with the construction of breaching batteries.

If the fire of the place is very deadly, or the nights very short, they must be thrown up from the interior. If it is impossible to sink them, the workmen must be covered by masks of stuffed gabions, earth-mounds, &c. Should the earth be stony, it must be screened, and the stones placed at the bottom of the parapet, so that the shot may not strike them and injure the men. If a battery has to be constructed on rock, or on marshy ground, the earth for the parapet must be carried to the position, or if possible the parapet be constructed of sand-bags, the workmen being covered by masks.

Marshy ground is consolidated by laying down beds of fascines, crossing each other at right angles, and fastened with hurdle-work. This is covered with earth mixed with straw. If a battery is made on the brow of a hill, or other position where the width of terreplein is limited, the rear of the platforms may be supported on trestle-work. This kind of floor is formed of beams 8″ square, placed about 2′ apart. A counter-hurter is fixed on it to prevent the piece from running off the platform to the rear. The platform planks are nailed in their positions. The work is protected by masks of gabions and fascines (Fig. 292), and these masks are increased in number, in order to divide the attention and fire of the enemy. Such a platform will not do to fire mortars from, the recoil being too directly transmitted to the supports.

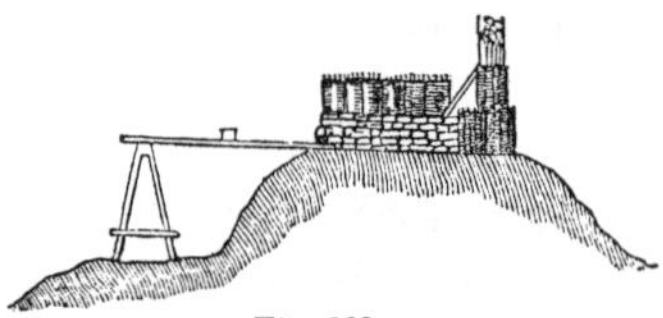

Fig. 292.

In order to increase the angle of fire, a platform broken into two steps is made use of. The lowest part, being in the rear, receives the trail, whilst the wheels rest on the higher part in front. Fig. 293 *a*. This kind of a battery platform is resorted to when the piece will not give, standing on its carriage in the ordinary way, an angle of fire sufficiently great for the object in view. The angle of fire is diminished by the reverse means, Fig. 293 *b*.

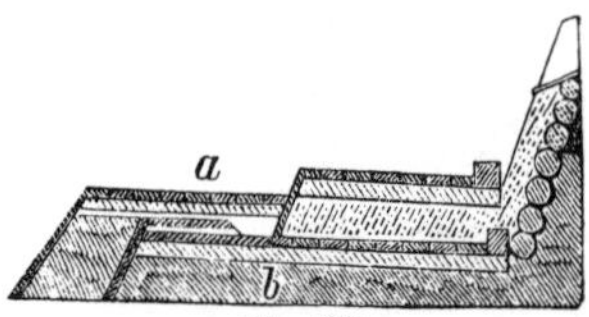

Fig. 293.

Floating batteries may be constructed on inundated ground, by making rafts, buoyed up with empty casks placed under them. A parapet is made in front, of woolsacks, saucissons, or oak timbers.

If the battery is commanded, it becomes necessary to raise the parapet or lower the terreplein in such a manner that the enemy's shot will pass 6 ft. 9 in. above the rear of the platform.

If it becomes necessary to establish a battery in a position very oblique, with regard to the object fired at, it is made in offsets, and the pieces placed as it were in échelon, in order to avoid oblique embrasures, which are objectionable.

One magazine should be provided for every two or three guns. They are sometimes placed at the sides of the battery, and sometimes in the parapet of the communications. For mortar and howitzer batteries small shelters are also required for the purpose of loading hollow projectiles. They are formed of a ditch 3 ft. wide, 3 ft. deep, and 6 ft. long, covered by a parapet and a blindage of wood and saucissons covered with earth. The entrance is to the rear, and approached by a ramp. Fig. 294.

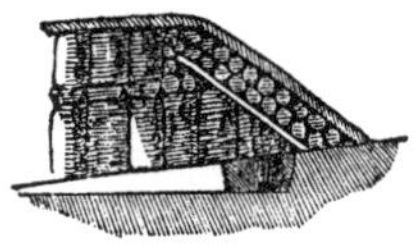

Fig. 294.

DEFENSE.—Fortifications should be armed in proportion to their importance and the military and other supplies which they contain. The pieces should be of medium calibers, to enable them to be promptly withdrawn from the fire of the besiegers, and replaced in more favorable positions. Some large pieces, however, are indispensably necessary to intimidate the enemy, and compel him to give his works the usual dimensions.

No definite rule can be laid down as to the armament of a fortification, more especially in regard to those in this country. Generally speaking, 18 and 24 pdrs. are large enough for any interior forts; whilst those on the sea-coast, where most of ours are situated, should be well provided with guns, howitzers, and mortars of the largest calibers, in order to cover the largest field of fire effectually, and prevent the approach of vessels of all kinds. They must, besides, possess a proportion of smaller calibers, including some field pieces for the land fronts, and to be used in interrupting the landing of troops.

The maximum amount of ammunition is 1,000 rounds per gun, 800 per howitzer, 500 for 10-inch mortars, and 700 for 8-inch; 2,000 rampart and 5,000 hand grenades; and the minimum, three fifths of this.

The artillery officers in a fort, and more especially the chief one, should be perfectly acquainted with all parts of the work, and all the adjacent ground within the sphere of attack, as on these will depend the proper disposition of the pieces. They should also have marks on the surrounding ground, so as to be able to estimate correctly the distances of the works and batteries of the besiegers.

There should be at least one skillful *gunner* for the service of each piece, and as only one third of the garrison are generally under arms at a time, there should be three times as many of these gunners as there are pieces in the place, including the reserve pieces. Three auxiliaries for each piece of small caliber, and four for the large, may be taken from the infantry. The workmen and artificers should number nearly one fifteenth of the gunners. In case of necessity, the services of citizens to assist in working the guns, may be made use of, or they may be employed as workmen and artificers.

Besides the battery-horses used in making sorties, a certain number are necessary to transport the pieces from place to place.

When the place is about to be attacked, the bastions are each armed with three pieces, placed in barbette; one 8-inch howitzer in the capital and one 18-pdr. on each face. The flanks are often armed with field guns to fire through the embrasures at reconnoissances and workmen. The front of attack is lighted up with pitched fascines, rampart-pots filled with composition, or with tar-barrels and fire-balls thrown forward towards the enemy.

The enemy should be closely watched, in order to ascertain the moment when he opens his trenches. For this purpose, the garrison should be in communication with persons outside, if possible, so as not to be taken by surprise. Some field pieces should then be run out in front of the glacis, the enemy's position lighted up with fire-balls, and grape-shot fired upon his workmen for two or three hours; after which, the enemy having sheltered himself from such projectiles, shot and shell are used, the latter with but little velocity, so that they may ricochet along the ground and enter the trenches.

So soon as the point of attack is satisfactorily determined, all the disposable artillery is placed in the most favorable positions for sweeping the ground in front. For this purpose, one 8-in. howitzer, firing along the capital, and five 18 or 24 pdrs. on each face, are placed in the bastion of attack. Seven or nine pieces of like caliber, and similarly disposed, may be placed in the cavalier of the bastion.

Eleven pieces are placed in each demi-lune of the attack, one an 8-in. howitzer, firing along the capital, the others 12 or 18 pdrs. Six on the face, bearing on the ground opposite the bastion of attack, the other three on the other face.

Five pieces of large caliber on the faces of the two collateral bastions which bear most directly on the trenches; and their flanks which bear on the bastion of attack, should each receive four pieces near the angle of the curtain.

In the collateral demi-lunes, six pieces are placed on the faces that bear on the trenches.

Sixteen 8-in. howitzers should be placed in the covered ways of the point of attack, and of the two collateral fronts; two being in each place of arms to fire in ricochet along the capitals. About twenty mortars should be distributed along the curtains and in the demi-lune redoubts.

The pieces on the faces which are enfiladed should be covered by gabionade traverses, one between every pair. The barbettes in the salients will be partly cut down, and the parapet raised to form embrasures for the guns. The traverses are 22 feet long, 9 feet high, and 12 feet thick at the base by 8 feet at top, and are revetted with gabions.

The pieces enumerated form a medium armament for the point of attack at this period of the defense; and whilst they are being placed in position, embrasures, platforms, and traverses are prepared at suitable positions, to place the artillery under shelter as soon as the enemy shall have established his enfilading batteries.

A continuous fire should be kept up on the parts of the trenches in progress, particularly when the enfilading batteries are commenced. The fire, instead of being scattered over all the batteries, should be concentrated on a few of the principal ones, as by delaying these, the others, if the besiegers act prudently, will not open their fire until all are ready.

As soon as the enfilading batteries begin to produce a marked effect, half the pieces on the faces of the defenses are withdrawn and held in reserve, those being retained in position which are best protected by the traverses, and keep up a steady fire on the heads of the approaches as they are advanced, redoubling tho ricochet firing whenever the workmen and guard of the trenches are relieved.

After the third parallel is constructed, the howitzers may be advantageously replaced by stone and coehorn mortars firing from the covered ways, and the redoubts of the re-entering places of arms. Guns will be placed in embrasures to fire in the direction of the ditches of the demi-lunes of attack against the crowning of their covered ways.

The defense at this time should be more than ever energetic. The converging and close fire of the artillery, combined with that of small arms, is capable of retarding very much the progress of the besiegers. Their works should now be well lit up at night with fire-balls, to prevent their rapid advancement under cover of the darkness. The embrasures should be protected with bullet-proof shutters, or masks. *Blinds*, or covers of timber, fascines and earth, under which guns can be secured from projectiles which would reach them at top or in flank, will now be very serviceable; but they must not be exposed to the direct fire of the enemy, and should allow free escape to the smoke, without which they would prove untenable. A few guns covered in this way, and placed in the salients of the collateral works, to take in reverse the trenches constructed on the glacis, will greatly retard the progress of the sappers.

As soon as the besiegers have arrived at within 30 yards of the covered way, hand-grenades are thrown into their works, stone-mortars are gathered in the adjacent portions of the work, and the establishment of the breaching-batteries retarded by a well-sustained fire of shot, shells, grenades, &c. The pieces are now withdrawn into the redoubts of the work abreast of the besiegers. The demi-lune redoubts receive 5 guns in the salient and 4 on the flank looking towards the bastion of attack. The redoubts themselves are not disarmed until the last extremity, when the pieces are, if possible, withdrawn into the body of the place.

As the moment for crowning the covered way approaches, the efforts of the besiegers to retard the works of attack will be redoubled. In addition to the measures already laid down, the flanks which bear on the point of attack will be armed with artillery; and oblique embrasures will be constructed in the curtains, to sweep the positions along the bastion-covered ways, where the besiegers are making the breaching and counter batteries.

The different breaches are defended with grape, hand and

rampart grenades, powder-bags, &c. Small mines, or bomb-fougasses, should be prepared at the summit of the breach, to be exploded as soon as the besiegers gain possession of it. The top of the breach is strewed with every possible obstacle that can retard the progress of the storming party; and grenades, thundering barrels, &c. will be rolled down on the troops as they ascend the slope. Coehorn-mortars and field-howitzers, loaded with grape, are placed in position, ready to fire upon the enemy as soon as he reaches the top.

SEA-COAST DEFENSES.—Artillery plays the most important part in sea-coast defenses; more especially now that the caliber has been so much increased, as by a single well-directed shot to endanger the safety of ships of the largest class. The fixed position of the land-battery, and the small surface which it presents, give it an immense advantage over vessels. It may be laid down as a principle, that a land-battery of 4 pieces is capable of contending advantageously with a ship of 120.

Sea-coast batteries are usually established near the entrances of ports, or at other points on the coast, for the defense of roadsteads, anchorages, small commercial ports, &c. They should be as far advanced as possible, as on islands, projecting points, &c., in order to maintain command over as great a distance as possible. A relief of from 30 to 45 ft. protects them from the ricochet fire of ships, whilst it gives the pieces the power of using that fire on the water up to 1,500 yards. Shot loses but little of its force by ricocheting over calm, still water. One of large caliber, which has ricocheted at 1,200 yards, has still sufficient force to penetrate the side of the largest-sized vessel. If the ground between the battery and the sea presents a slope favorable for the ricochet fire of an enemy, it should be cut into terraces, the rises of which will catch the shot.

The distance of the anchorage is determined by the depth of water at different points of the coast. The largest class of vessels require from 25 to 30 ft.; frigates, from 19 to 23 ft.; and sloops of war, from 16 to 18. It is, therefore, easy to ascertain the distance of an anchorage from a battery according to the kind of vessel.

The parapet of these batteries should be of earth, or at least covered with it to a depth of 2 ft. It is about 18 ft. thick, and 7 ft. 6 in. high. It is frequently revetted on the interior with masonry to a height of about 4 ft. 2 in. The width of the terre-plein is 18 ft. When subject to enfilade from the fire of shipping, returns are made at the extremities, and traverses placed at intervals. The ditch should be as deep as possible.

The distance between guns and howitzers is from 18 to 21 feet, and the height of the interior crest, above the platforms, about 5 feet, so that the pieces may fire over the parapet and follow vessels under way. The field of fire of each piece is 90°, or 45° on each side of the directrix, so that if a vessel passes a battery at a distance of 300 yards, it is under the fire of a piece for at least a distance of 600.

The armament of sea-coast batteries varies with their importance. The largest are armed with the largest guns, 32 and 42 pd. columbiads (now 64 and 128 pds.), and sea-coast howitzers. Some field-howitzers are sometimes also included. A certain quantity of light artillery (especially howitzers), is also necessary to accompany troops which may be detached to prevent an enemy from disembarking. But few pieces are kept mounted in our sea-coast defenses; and in case of a threatened war, this would be the first measure to be attended to.

In time of war a sentinel is posted night and day on these batteries, to give the earliest intelligence of the appearance of an enemy. Everything should then be prepared; charges, projectiles, &c. The platforms should be swept clean, and it should be ascertained beforehand that the chassis move freely on the traverse circles. The pieces are fired at the water-line of the vessels. If the shot falls short it will reach the vessel by ricochets; and the chances for producing a good effect are greater than when firing higher.

Hot shot may be most advantageously employed against vessels at anchor, and for combats of a certain duration, which allow time enough to heat the shot to the necessary degree, and the requisite careful and deliberate aiming.

Sea-coast batteries are provided with furnaces or grates for heating shot. One hour is required to set one of these furnaces going, but after that only from 30 to 35 minutes to heat a 32 or 42-pd. to a cherry red.

A cold shot makes, in the side of a vessel, simply a hole, which closes up in part by the elasticity of the wood, and is easily stopped with a large plug.

The use of shells with sea-coast artillery permits rapid firing, and enables us to seize the favorable occasions when the most injury can be inflicted on the enemy.

The fire of heavy 8 and 10 in. shell-guns is almost as accurate as that of guns, and takes effect at from 3,000 to 5,000 yards; but the most efficient range is from 1,700 to 1,900 yards.

10-inch mortars give ranges of over 4,000 yards, and are employed against distant anchorages. Their fire, although very uncertain, is of great moral effect, in the defense of sea-coasts, from the knowledge of the great havoc sometimes produced.

If the proximity of an anchorage enables an enemy to man his tops, and obtain a plunging fire into the battery defending it, field-pieces are established in the rear to fire upon the netting of the tops. Rockets and other incendiary fire-works would be fired against the sails and rigging to set them on fire.

Firing from ships is very uncertain, on account of the movement. As the range is shortened, the fire improves; but the ship being thereby brought within shorter range of the guns and howitzers of the battery, the contest becomes more than ever unequal. The bombardment, by the French, of the castle of St. Juan d'Ulloa, took place at 2,000 yards. Of 302 shells fired, 6 only fell in the fort, whilst some went 1,200 yards beyond.

In estimating the quantity of artillery necessary for the armament of a permanent fortification, considerable discrepancy of opinion exists among military writers. A resolute garrison might preserve a work of tolerable strength from a *coup de main* without the aid of cannon; and an idea of the maximum amount of artillery might be arrived at, by supposing each front armed with as many pieces as it can carry, with a suitable number of pieces in reserve to provide for casualties. But these are inadmissible extremes; and a medium estimate is to allow 3 heavy guns, and 1 heavy mortar, for each bastion or front, with 60 heavy guns, 20 heavy mortars, and 10 stone or light mortars for the armament of the point of attack.

An estimate of the allowance of ammunition is made by allowing 1,000 rounds for each gun, 800 for each mortar, 100 musket cartridges per day for each soldier on guard, and 300 lbs. of powder for each mine.

In estimating the quantity of artillery necessary for the

besiegers, the quantity employed by the besieged is taken as a basis, and as many guns and mortars allowed for the enfilading, counter, and mortar batteries as, is presumed, form the armament of the point of attack, with the addition of 40 or 50 heavy guns for the breaching batteries.

The medium allowance of ammunition for the artillery is 1,000 rounds for each gun, and 800 for each heavy mortar.

These estimates are of a very general character; and are introduced merely to give some idea of the relative proportions in question.

CHAPTER XIV.

THE PASSAGE OF RIVERS.

WATER-COURSES are generally the most important and usual obstacles offered to the march of an army; and consequently from the most remote periods armies are found carrying with them the materials to construct bridges or other means of crossing rivers.

The materials made use of are of various kinds. Sometimes boats made to be taken apart for transportation, sometimes trestle-work, and sometimes a flooring supported on boats, rafts, or inflated vessels.

In France copper pontoons were in use in the times of Louis XIV. and XV. Gribeauval substituted boats for the pontoons, which gradually disappeared.

Bridges constructed of the materials found on the spot are, however, much more frequently mentioned than those made from such as were already prepared and carried with armies.

Under Louis XIV. the bridge equipage was placed in charge of the artillery, where it still remains; it being considered that a corps which, like the French artillery, has charge of such a large number of workmen, teams, and supplies of different kinds, and which is itself so much interested in having stable bridges, should have the control of their construction. The pontooneers form now in the French service, a regiment of artillery. In the United States' service they are confined exclusively to the engineer corps.

A river flowing in front of an army forms a natural fortification, and compels an enemy to approach with a contracted front in order to cross it; whilst in rear it presents a grave obstacle to a successful retreat. Its importance however is much diminished in both views since the great improvements made in the construction of bridges. Skillful generals have usually overcome the obstacles of this operation, either by surprise or by forcing rapidly a passage.

In Surprising a Passage the attention of the enemy is distracted by false attacks, which, obliging him to divide his forces, allow the passage to be made at points deprived of troops, or badly defended. In the celebrated passage of the Rhine by Gen. Moreau, in 1793, five different attacks were made.

In Forcing a Passage, a commanding position is chosen at a point where a stream makes a sharp re-entering angle. Strong batteries are established there, crossing their fire on the salient made by the opposite bank. The fire of these batteries, being converging, is formidable against the enemy, whose fire is divergent, and whose artillery concentrated in a small space is more easily destroyed, or silenced and forced to retire. The troops are then advanced under cover of the fire of the artillery, and thrown across to the opposite bank.

When a river is from 400 to 600 yards wide it is very difficult to make the passage except by a surprise, unless there are islands the possession of which favors the construction of the bridge; for the enemy, placing himself at 400 yards from the end of the bridge, could destroy the attacking party by a fire of grape, without the latter being able to do much mischief at a distance of from 800 to 900 yards, however numerous his artillery.

The most favorable time to force the passage of a river, is just at the break of day. All the preparations having been made the preceding night without the knowledge of the enemy, an entire day remains in which to effect the possession of the opposite bank.

The passage of a river in retreat presents the greatest difficulties. Many instances might be cited where small streams have caused the ruin of considerable bodies of troops.

The most advantageous point for an army in retreat to pass a river, is where it forms a salient, and the bank to be abandoned is commanded by the other; for then the troops and artillery first passed over can cross their fire in front of the bridge, keep the enemy at a distance, and protect the movements of the troops.

The passage of a river is opposed by destroying the bridges and other means of crossing, which may exist; by establishing reserves in central positions, so as to be able to move them against the enemy at any required point; but a passage can be effectually opposed, only to an extent which is within a day's march for the troops. Upon the points menaced, batteries of large caliber are concentrated, and made to converge and cross their fire upon the enemy and his works. If time permits, this artillery should be protected by a parapet, 11 feet thick, and 2½ high. Solid shot and shell should be fired in ricochet against the bridge if established, and every attempt made to sink the boats of the enemy.

RECONNOISSANCE.—A careful reconnoissance of the river should precede all kinds of passage or navigation.

Data with regard to the course of the river, its bends, source, and mouth, should be collected, and the parts which flow through the enemy's country be discovered.

Ascertain if it is navigable, to what distance, and by what kind of boats; any dangerous places it may have in it; the channels which are to be followed, if there are more than one; whether bridged or not, and if so, the height of these above the water; the shoals and whirlpools.

Reefs or shallow water where boats are likely to ground, are recognized by a kind of ripple in the water.

Whirlpools are cavities in which the water is precipitated, turning and sinking below the level of the surrounding water. They are very dangerous in navigation.

The velocity of the water is not uniform all across the stream. It is the greatest in the *current*, where the water is the deepest, and diminishes towards each shore. The deepest part is called the navigable channel. The *current* is shown by an increased height in the water. Rivers are cited where the current is a yard higher than the water near the banks.

To determine the velocity of the river, a light body which will float is thrown in; a base is measured on the bank, and the time taken to pass over this distance is twice noted, and the mean taken. Dividing the length of the base line by the time reduced to seconds, gives the velocity.

If the current is far from the banks, two skiffs are anchored in it at a known distance apart, and the process gone through with.

The log line may also be made use of. For this purpose the float is attached to a small line which is unwound so rapidly as to prevent its arresting the movement of the float. The length of the cord run out at the end of a given time, is measured.

The velocity not being the same throughout the mass of the current, if the mean velocity is desired, a stick weighted so as nearly to touch the bottom, will be used as a float.

What is called a slight current shows a velocity of 1.6 ft. per second; an ordinary current, from 2.6 feet to 3.2 feet; a rapid current, from 4.9 feet to 6.6 feet; a very rapid current, from 6.6 feet to 9.9 feet; an impetuous current which nothing can resist, 9.9 feet, and beyond that.

Rivers are more rapid the straighter their banks become. The velocity is not the same at all seasons. The velocity of the current should be known for high, mean, and low water mark; as also the difference of level in the different stages of the water.

The influence and extent of the tides should be ascertained, as also the direction of the prevalent winds, the position of dams, sluices, and dikes, and their object. Whether the destruction of these will produce inundations, or fords by decreasing the depth of water.

Find whether the stream is liable to sudden freshets, and, if so, at what seasons. Freshets occasioned by the melting of snow occur first in March and April, and afterwards in July and August.

Freshets are shown by an increase of velocity in the water, which is disturbed from the bottom of the stream, throwing up mud, &c.

The right bank is on the right hand looking towards the mouth, and the left bank the opposite side. Sometimes, the winds from down the river acting against a freshet and then going down, a flood of water is created, which may cause accidents and frequently inundations.

Ascertain at what seasons the ice forms, and the strength of it. The breaking up of ice often causes sudden freshets, the water having been dammed up by it.

The more rapid the current, the heavier are the bodies which the stream will carry down. Heavy stones and pebbles are washed down in a mountainous country; and in a level country the velocity of the water is known to be so much the less as the sand deposited by the current becomes finer. The inspection, then, of the bed of a stream will determine to a certain extent the velocity of the current.

It should be ascertained whether the bottom is rocky, pebbly, or composed of angular stone, which would render it impracticable at the fords for horses and carriages; whether it is formed of gravel, earth, shifting sands, or reeds, which latter may be thick enough to interfere with the passage of boats.

Everything which tends to diminish the velocity of the water causes the precipitation of the earthy matter which it contains. The resistance offered by the waters of the sea to the current of a stream flowing into it, causes the formation of bars at the mouth, which interfere very much with navigation. Such bars are often formed at the mouths of the branches of a river.

The existence of islands in the river should be ascertained, as well as their extent, whether wooded or not, and the facilities they present for either attack or defense.

A stream is said to be *embanked* when the banks are considerbly above the level of the water.

The width of a stream is determined by stretching across it a light cord, which is graduated by tying knots along it. If this means, which is the most accurate, cannot be employed, and it is impossible to cross the river, the width is determined by the aid of two similar triangles, constructed by means of stakes and a common square, with a cord attached for determining the direction of the sides.

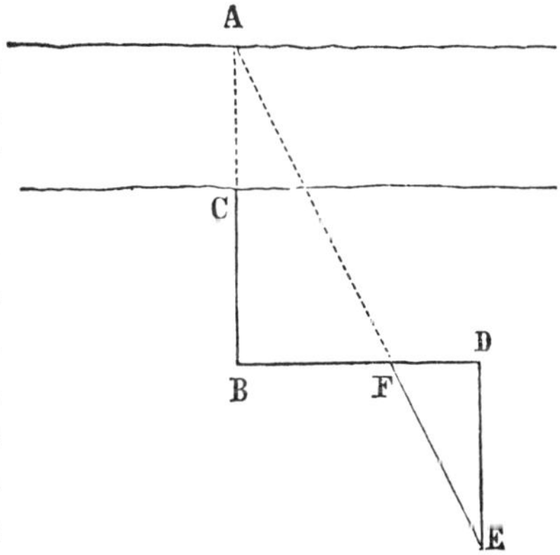

Fig 295.

For this purpose a point, A, Fig. 295, is selected on the opposite side of the stream, and a pole is planted at some point, B, on the observer's side. The small side of the square is then placed in the direction B C, and the direction of the long arm, B D, staked out to any convenient length. The square is now placed with the angle at D, the short arm on D B, and the perpendicular direction D E marked out. From E, the point A is sighted, and a man stationed on the line B D plants a pole at F, by direction of the observer at E.

The triangles A B F and F D E, give the proportion D F : D E : : B F : A B.

The first three lines being measured, the fourth results from the proportion, and from it the distance B C must be subtracted, to get the width of water.

The form of the bed of the stream is determined by sounding with a graduated pole or line at known distances from the shore.

The variations in the level of the water can be ascertained by observing the height on stakes driven into the bed near the shore.

It is noted whether the banks are steep, if they are very high or not, and if they present, within range, points advantageous for attack or defense.

Some streams, especially those subject to great swells, often change the direction of their waters.

Rivers with low banks often overflow to a considerable distance. The presence of dykes or levees indicate that a river is subject to floods, and the extent of the inundations should be ascertained.

It should be known whether the borders of a stream are marshy, and to what extent.

The branches of the river should be similarly examined, especially those which flow through the enemy's country.

Examine such bridges, ferries, fords, &c., which exist on the stream or its branches, and which may offer the means of crossing.

All these data serve to determine the manner in which the passage is to be effected, after considering the disposable resources and the time that can be spared.

Streams may be passed, 1st, by swimming; 2d, upon the ice; 3d, at a ford; 4th, on floating movable bodies; or 5th, on bridges.

The most perfect order should be maintained in these different operations. Care should be taken to give the banks a gentle slope at the crossing point, in order to render the approaches easy for the horses and carriages, and with the same idea a point is chosen where the stream is as easily accessible as possible.

Swimming.—1st. The passage by swimming can be successfully effected only when the troops have been instructed in the art. In France, the regulations prescribe swimming as one of the exercises, whilst in this country it is entirely neglected.

The experiments made with companies of swimmers, fully demonstrate the important results which could be obtained by a body of infantry well instructed in swimming.

Cavalry can pass by swimming, much more easily than infantry, as the horse swims naturally. All that is required, it not to oppose him too much in his movements, and to pass in a direction oblique enough to prevent the current getting too great a purchase upon him.

It is stated that an army once crossing the Rhine, found it fordable except for about 100 yards, and that several horsemen having lost their lives in attempting to swim it, the rest resorted to the expedient of crossing with a squadron front. The men mutually sustained each other, reached the opposite bank without loss, and dispersed the enemy. All the cavalry passed in the same way.

Sometimes the horses only are made to swim, that is, without their riders, and, being left to themselves, follow others which are led by men in boats. In disembarking horses, they are sometimes put overboard, and find their way to shore without further assistance.

On Ice.—This means of crossing a stream is very precarious, as the ice may be broken up by a change of temperature, and the communication between the two portions of an army interrupted. The passage must, besides, be made with the greatest precaution and order, to prevent terrible accidents which would result from the ice giving way, should it be overloaded.

Ice from 3 to 3.5 in. thick will sustain infantry when marching by file. With a thickness of 4.3 in., cavalry and light guns can pass over. With 6 in., heavy field-pieces; 8 in. will support 24-pdr. guns on sledges, with a weight not greater than 1,000 lbs to the square foot. Beyond this, the largest can be taken. For greater security, the wheels may be fixed on thick plank, with iron clamps, and the pieces drawn across by hand, the horses being unharnessed and led over.

The ice should always lie flat upon the water, otherwise the passage is not secure.

In very cold weather the thickness of the ice may be increased by covering it with a layer of straw or fascines, and throwing water over it.

If straw or fascines cannot be had, two rows of logs or beams are laid on the ice at a distance apart equal to the width of the passage, a layer of earth thrown in, and on top water is thrown. The operation repeated from time to time till a solid roadway is formed.

To prevent accidents, boards or straw are placed under the wheels and the horse's feet; an interval of twenty paces is left between the carriages; and, in case the ice is not strong enough to pass them over on their carriages, the heaviest pieces will be placed on sleds.

During the first part of the passage loud cracks will be heard, but they indicate no danger unless water should be seen coming through the fissures.

In the winter of 1794–5, the French army, passing over the frozen Zuyder Zee, captured the Dutch fleet with their light cavalry.

3d. Fording.—The fords on a river are very important points for crossing, in time of war, either a small detachment or a whole army.

The best are those with a firm solid bottom. In mountainous countries they are apt to be encumbered with large stones, which render them impracticable for carriages. In a flat country the bottom is frequently composed of mud or fine sand, which is worn away by the horses' feet, and it sometimes happens that after a passage is effected the ford is destroyed.

In very rapid streams, or those subject to sudden freshets, the fords are variable in position and depth, and can be but little depended on.

The ordinary depth for infantry to ford is about 39 inches, though when the current is not too strong, the men can pass at a depth of 51 inches. For cavalry, the ordinary depth is 51 inches, and for artillery, 27.5 in. when the ammunition boxes are at their usual height (34 in.), and not water-tight. If they are water-tight, or have been raised sufficiently, the carriages can cross with a depth of 40 inches.

When the water is deep and the current strong, great care is necessary, especially in passing over artillery. The person conducting a column over a direct ford, should keep his eyes steadily fixed on some object on the opposite bank, near the point where the road leaves the water. He should not look at the stream, which would deceive him by appearing to carry him down.

All those in rear should keep their eyes on those in front.

In order to resist the power of the stream, it is necessary to wade rather against it.

When the bottom, or the bank on the opposite side is bad, the leaders of the rear carriages are hitched to those in front, and an officer stationed at the entrance, and one on the opposite side. The first causes the drivers to observe the distances, and directs them how to drive across, while the second shows them how to leave the ford.

The horses should not be allowed to drink, halt, or trot, either in passing the ford or in leaving it, unless, there being no enemy in the front, the stream is a small one, neither deep nor rapid, and with no troops waiting in the rear to cross; in which case the horses may be allowed to drink while in the water.

The passage should be made with as large a front as possible, and the leading carriages, after reaching the opposite bank, should move a sufficient distance ahead so as not to interfere with the rear carriages.

A stream is sometimes found fordable all the way across except for a short distance, either from the natural depth of water at that place, or from the enemy having cut a ditch there. In this case it becomes necessary to throw in a filling of fascines loaded with stones, and laid up and down stream, or boxes or gabions filled with stones and sand, until the depth of water is sufficiently reduced, to a width of at least four or five yards.

The position of a ford is known by the information derived from the inhabitants in the vicinity; by the wheel-tracks on the bank and in the water; by the increased velocity of the current, or breadth of the stream; or by a double change in the direction of the road made in a short distance. In this last case, the ford runs diagonally from one bank to the other. They are sometimes found above bridges, above or below a bend, and at or near the mouths of streams.

The best way to discover and examine a ford, is to descend the stream in a small boat, to which is attached a sounding-line in such a way that it can go under water to a depth of 39 in. or 51 in., according as a ford for infantry or cavalry is sought for. The line should carry a float, which appears above the surface when the lead touches the bottom.

On the lead touching, the boat is stopped, and other fordable points sought for in every direction; the points touched by the lead are marked by stakes forced into the bed of the stream, the operation being continued all the way across. Two rows of stakes are established, showing the direction and breadth of the ford. To render the limits of the ford more apparent, cords are stretched along the stakes, which serve, besides, to assist those who lose their footing, or are carried down by the current. This is especially necessary in passing a ford at night, when torches or lanterns are often fastened to the stakes. If the water is shallow, the stakes and cords may be dispensed with.

A ford may also be discovered by using graduated poles for sounding. The Cossacks seek a ford in this way. They spread themselves along the bank of a stream, sounding it with their lances. As soon as one discovers the ford, the rest join him, and soon find out its breadth and direction.

When the situation of a ford has been discovered, and before the crossing is attempted, it should be explored by boatmen or good swimmers, and repaired if necessary. Should the enemy have dug ditches or *trous de loup* they should be filled up with gravel, or fascines loaded with stones. Should they have sunk obstacles in the water, they are removed with levers having hooks at the end of them. Should crows-feet or any thing of the kind have been spread along the bottom, they should be covered up with fascines or hurdles.

If the current is strong a rope should be stretched above the ford and supported on empty barrels. Cords are tied to this, and have attached to them buoys or pieces of light wood, which the men who lose their footing can catch hold of.

The passage is commenced by the infantry in columns of

platoons at nearly the full distance. The artillery next, and the cavalry last. This order is followed because the horses' feet hollow out and cut up the ford very much. The soldiers advance the upstream shoulder and carry the arm on that side. The cartridge-box should be placed on top of the knapsack to protect the ammunition from wet. The men should keep their eyes steadily fixed upon the opposite bank, and not upon the current, which would mislead them.

Sometimes cavalry is crossed along the upper side of the ford to break the force of the current, whilst more are passed over on the lower side to aid the men who may be carried down the stream. At times, especially in cold, wet weather, the foot soldiers are carried over behind the mounted men. In fine, the order in which the passage is made may be changed to suit the circumstances of the case.

In passing cavalry over, care should be taken to raise the bridle hands, in order that the horses may see the land; otherwise their sight may become disturbed by the current, and they allow themselves to be drawn down the stream. They should not be allowed to drink either, which would produce the same effect.

It is not prudent to attempt the passage of a ford during a swell, unless it is certain that all the troops can get across before the ford becomes impracticable.

A ford cannot be considered as a certain means of communication for an army, as a rain-storm or strong freshet may destroy it, or render it impracticable.

In retreating, fords are destroyed by cutting ditches across them, or sinking *trous-de-loup;* by spreading along the bottom crows-feet (four-pointed spikes, three of which stick in the ground whilst the fourth remains up); by fixing in the water, with strong pickets and ropes, common harrows, planks, or tables bristling with spikes with their points up; by tying together and sinking trees, the branches of which, turned towards the enemy, have been lopped off and pointed; by sinking cheveaux-de-frise: the best method of securing these last obstructions is to fix to them when in position, canvas bags filled with stone; and, last of all, the bank opposite the ford is cut down steep.

4th. FERRIES.—The passage in boats may be effected either in the presence of an enemy or by surprise, in order to protect the construction of a bridge; or, it may be effected by isolated bodies of greater or less strength, in which case the passage is performed much more rapidly than when it is necessary to construct a bridge, and then break it up again.

In the absence of a regular bridge-equipage, the common boats found on the river in the vicinity are seized, and taken to the point selected for the passage. This duty is done by boatmen, escorted by light-cavalry, and provided with ropes and the requisite tools. The movement should be well disguised and rapidly executed, as the enemy may sink the boats, or run them aground on his own side.

To float a submerged boat, two others are used, one on each side, with an interval between them greater than the width of the sunken one. They are lashed together by laying a couple of beams across their gunwales, near the prows and sterns, and firmly fastening them. The boats are kept in place by fixing anchors, both up and down stream; a rope or chain is passed under the prow of the sunken boat, and as far back as it will go. One end of this is fastened to one of the side boats, whilst the men in the other boat pull on the other end until the boat rises, when the rope is fastened to the nearest beam; and the same plan is pursued at the stern, until the boat, coming to the surface is bailed out with buckets, tubs, &c.

Another method is, to pass two ropes under the bottom of the sunken boat, and make them fast to those alongside. These are then partly filled with water, the ropes being tightened as they go down. The water is then bailed out of the side-boats, and they rise bringing the other with them. Repeat the operation if necessary, until the boat reaches the surface.

If the boats are simply stranded on the shore, the water and mud which they contain is bailed out.

If they are only partly sunk, with their gunwales above water, or only a little under, they are drawn into shallower water, and emptied as before.

Auger or bullet holes are stopped with conical plug. Those made with an ax or cannon-balls, with tow soaked in melted tallow, and kept in place by strips of boards nailed on the inside of the boat.

The boats having been collected at the designated point, the capacity of each one is determined, and if time permits, they are numbered.

The largest boats have the most stability, and are least liable to overset. The buoyancy of a boat, or its capacity to carry weight, is obtained by subtracting from its weight the weight of volume of water it displaces when the gunwales are level with the water.

To calculate the weight which a boat can carry, measure the height of the gunwale when afloat, above the water-line, and calculate the surface of a horizontal section of the boat taken at a distance below the gunwale equal to one half of the distance found. Multiplying the two expressed in feet together, will give the number of cubic feet of capacity, above the water-line. The weight of a cubic foot of water being taken as 62.4 lbs., if a capacity of 212 cubic feet has been found, we will have 13,228 lbs. as the weight which the boat is capable of carrying.

But as it would be dangerous to load a boat with its gunwale down to the water, the load must always be under that given by the calculation. Hence, the calculation can be made accurate enough for practical purposes, on the spot, and with the eye.

The boats should be accurately inspected with regard to their solidity.

The lower the center of gravity of the load, the less liable the boat is to upset.

The distance a boat is forced into the water by the load and its own weight, is called its draught of water. The draught is an important point of consideration on streams, the beds of which have reefs and shallow water, on or in which the boats may ground or upset.

The following data may serve to calculate the contents of boats.

A man completely armed weighs, say 160 lbs.; without arms, 145 lbs. Three men equipped, can stand in one square yard. A horse, by himself, weighs 1,000 lbs., and with his rider and equipments, 1,225 lbs. He occupies a space of 40 inches by 10 feet. Five men without equipments, occupy a space of one square yard, and weigh about 775 lbs.

The boats having been numbered, a statement will be drawn up, showing the proposed contents of each; and four boatmen and a pilot be assigned to each. To propel the boats, oars and boat-hooks are used; the latter are poles with an iron point and hook at the end. In the absence of a rudder, an oar is used for the purpose.

On the arrival of the troops, they are divided into platoons, and the chief of each platoon is informed of the number of his boat, its contents, and the time for embarking.

The men should not enter or leave the boats in close order, which is liable to upset them. The infantry enter the boats at the prow, the men seating themselves on the gunwale, or resting against the edge of the boat, commencing at the stern; the cartridge box is turned to the front, and the musket held between the legs. The rest of the men remain standing. Should there be but little water in the river, the boats will be pushed out to avoid their grounding, and the men, if it is necessary, wade a few steps in before entering the boat. This may be required also in disembarking. The men are ordered to remain perfectly silent and motionless; and in case of the boat striking aground, or against any object and leaning over, not to incline too suddenly to the opposite side, which might upset the boat. In case boatmen of the country have been employed, a strict watch is kept over them, to see that they perform their duty properly, and do not escape. They have been known, even in a friendly country, to jump overboard to escape the dangers of disembarking. Within range of the enemy, the men should have their arms loaded, and bayonets fixed, but should fire *only* when directed by their commander.

If possible, in shallow boats, rows of planks are laid in lengthways, in order to seat the men. This lowers the center of gravity, makes the men remain more quiet, and renders the navigation of the boats easier and more secure.

In the United States, no particular boat equipage has as yet been adopted; but the bridge-equipage boat of the French has been experimented upon in connection with our bridge equipage, and probably would be used were an emergency to arise, requiring a regular boat equipage to be fitted up. It is about 31 feet long, 6 feet wide at the widest part, and is capable of receiving 25 infantry soldiers, besides its equipment, and the men to work it, 5 in number. They can be employed on all kinds of water courses, however wide or rapid, and either as boats for transporting troops across rivers, or in disembarking, or as floating supports on which to build a bridge.

French Boats.—The French artillery supply any want of capacity in their boat equipage, and render the boats less liable to upset, by connecting together two, four, or even six of them; forming thus a *train* capable of carrying a much larger number of men than if the boats were separate. The four extreme knees of these boats are prolonged up above the gunwales. Against these, and across the gunwales of two boats lying side by side, are placed small beams to which these projecting knees are lashed. Four of these boats, which when single, carry only 100 men, are capable when connected together, as shown in the figure, 296, of carrying 180. The principle can be applied advantageously to all kinds of boats.

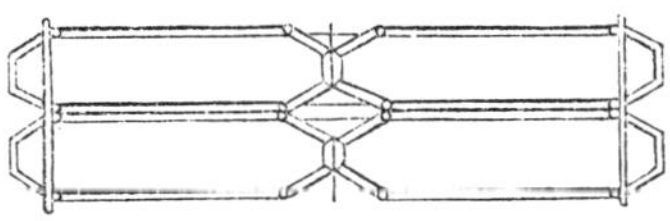

Fig. 296.

These *trains* do not travel as fast as single boats, and drift more. The men remain standing.

To pass cavalry over in these boats, six men are placed in each, holding by the rein their horses, which swim over, three on each side of the boats. If the current is very strong, but three are taken at a time, and these on the down-river side.

This method of passing a stream being very tedious and hurtful to the horses, especially in the winter, they are passed over in large boats, when they can be had. For this purpose a flooring of thick planks is laid in the bottom of the boat, to protect it from the feet of the horses.

They should be placed head and tail across the boat, each man holding his horse by the bridle near the bit. It is very dangerous to place them lengthways in the boat.

A ramp of thick planks is constructed at the bow of the boat to assist the horses in entering.

The Artillery is placed in dismounted, unless the form of the boats admits of the transportation of the guns on their carriages. Sometimes it is carried by uniting two or more boats by a platform. This renders the dismounting of the guns unnecessary, and the passage more rapid.

To prevent accidents, the river should be sounded at the place where it is to be crossed, in order to be certain that no reefs, shallow water, or whirlpools exist, which might endanger the safety of the boats. If a river is divided into several channels, the largest should be followed.

The heaviest objects should always be placed in the bottom of the boat. Such as would be damaged by water, as arms, ammunition and provisions, should be raised high enough above the bottom to keep them from getting wet, and covered with canvas or tarpaulins.

NAVIGATION.—If instead of simply crossing a river, the troops are to be transported some distance by water, the number of men in each boat is diminished, so that all may be comfortably seated, and a small boat should precede the rest to reconnoitre the stream.

A river with an inclination of $\frac{1}{4000}$ is easily navigable, and may be ascended with sails; but sails alone will not suffice with an inclination greater than $\frac{1}{2000}$, and towing becomes indispensable. It is impossible to ascend rivers having an inclination greater than $\frac{1}{500}$.

For difficult passages over rapid streams, it is necessary to employ pilots from the very neighborhood, or run great risk of disaster.

In descending currents, oars and rudders are used.

In ascending gentle currents, the boat-hooks alone may be made use of, but in more rapid ones the boats have to be towed by men or horses.

When the boats are united in a train, a mast is raised, the height of which depends upon the slope and height of the bank. It is supported by guys, and has fixed at the top the loop of a rope, to which is fastened the rope by which the boat is pulled along. The height of the mast should accord with the stiffness of the boat or boats, in order that the draft upon it may not pull it over.

To cross streams filled with floating ice, very strong boats, such as may be made out of a single tree, are required to resist the pressure; they may be dragged over such portions as are too solid to admit of canals being broken through them.

To open a communication through fixed ice, strong barges, if possible moved by screw propellers, are required. Their bows must be well protected with iron plates, and they should be provided with heavy beams to be raised by ropes and let fall upon the ice in front; saws may also be used, having heavy weights attached to the end under water, and the pieces of ice, when detached, may be hauled out of the water by ropes, or pushed under the solid ice.

Small barriers of ice which interrupt the navigation, or cause inundations, may be destroyed by turning streams of water upon certain points, so as to melt an opening, or by means of charges of powder in water-tight casks or bags, fixed underneath or lodged in holes bored in the ice, and fired simultaneously. A charge of 6 lbs. of powder, placed in the center of ice 2 feet thick, will break it up into small pieces throughout a circle of 20 feet in diameter.

Ice and snow, well rammed together, form temporary parapets, capable of even more resistance against shot than those of earth.

5. Rafts.—In the absence of boats, *rafts* may be used for passing rivers. They possess the advantage of not being sunk by the fire of the enemy, but they are very tedious to construct.

They are collections of the bodies of trees, ordinarily of light wood, not squared, but only roughly dressed and with the branches cut off in order to allow of their being connected together. These rafts, ordinarily rectangular in shape, are kept together by cross-pieces, fixed beneath by pins of iron or wood.

To give it the same thickness and stability throughout, the large and small ends of the logs are laid alternately on each side. Sometimes it is formed of two or three layers of logs crossing each other at right angles.

Rafts made of light timber have a greater floating capacity as they increase in size. If, for instance, the wood used is only 0.6 the weight of the water, each cubic yard is capable of supporting 675 lbs. Rafts which remain long in the water lose their buoyancy by becoming soaked.

In the absence of light wood, any that is at hand is used, and underneath are fixed a number of air-tight casks or barrels.

Rafts may be made of empty casks placed under a light frame covered with hurdles or the boughs of trees. Leather bags inflated are sometimes used in Europe in the same way; and in Spain Sir

Howard Douglass made such bags of the skins of the beeves killed for the subsistence of the troops. He had them cut in a circular shape, $5\frac{1}{2}$ feet in diameter, their edges joined, and firmly made fast to a piece of wood, and inflated this bag with a hand-bellows. The air was retained by a small piece of leather, which served as a valve, and was nailed on the end of the tube. This bag was capable of supporting a weight of 260 or 270 lbs. It remained very well inflated for 5 hours, and at the end of a day was still capable of supporting about 130 lbs.

The buoyant power of these bags, and of casks, is determined by estimating 62 lbs. for every cubic foot which they contain.

The dimensions of a raft can be easily calculated, knowing the weight it is to support and the material at hand to make it.

In loading a raft, great care must be taken that the troops go on in perfect order; for if they go on *en masse*, they will capsize it. The infantry, marching by a flank, should occupy the middle of the raft throughout its entire length. Two other parties are then placed, one in front the other in rear of the first command, and so on till the raft is loaded. The men should be at an order arms.

The same precautions are observed with cavalry. The horses must be placed across the raft, and alternately head and tail, the riders dismounted and holding them by the head, as laid down for the passage in boats. The same arrangement will be observed with regard to all horses.

With artillery, the heaviest carriages are placed in the middle of the raft, and the rest arranged so that the load will be spread uniformly, and the raft not inclined to either side.

Raft-navigation is advantageous upon rivers whose beds are unequal or rocky; but the passage with rafts is slower than with boats, the drift being much greater; and it is found necessary to depart from a higher point on the river to attain a given position. It is difficult to reach the exact point of disembarkation.

Fords being very often dangerous in cold and wet weather, attempts are usually made to construct means of making the passage.

A stream of little width may be passed by the aid of one or more trees, cut so as to fall across.

If a tree can be found on the bank large enough to reach the opposite side, it is cut down and placed in the water, with the largest end retained on shore. The other being put in up stream, is carried down by the current until it strikes the opposite bank, where the force of the current retains it. If the stream is too wide for this, a tree will be cut down on each side, and thrown into the water with their tops up stream, the feet being secured to the shore. Cords are attached to the tops of the trees, and as they float down with the current they are directed so that their branches will meet and interlace, forming a salient angle up stream. They are then tied together there, and any branches which obstruct the passage cleared away. If it were impossible to send any one to the opposite side, three trees would be cut down, two large and one small one. The two large ones are placed with their feet against the bank, and their heads forming an angle in the stream. The third is now passed out on these two, its foot rested on the others at the angle, and its head on the opposite bank. Men are then sent along this passage to clear away the branches.

Bridges.—Such means as the foregoing do very well for passing a stream in an emergency; but regularly constructed bridges form the only perfectly sure method, and the one generally used for crossing armies.

A military bridge is one which is thrown across a river temporarily for the purpose of crossing troops. It does not possess the same degree of solidity as a permanent one, and great care is necessary in preserving and using it.

It consists usually of a platform of heavy planks supported on boats, rafts, or trestles, and is named according to the kind of support used.

The velocity of the current, depth of water, and the resources at hand, determine the kind of support to be used; and it sometimes happens that all three kinds are used in the same bridge.

The construction of any kind of a bridge requires that both banks of the stream should be in possession of the party constructing it.

The banks, where too high, should be avoided, as they render the approaches to the bridge difficult, and require long ramps to be formed in order to approach and leave it. Marshy banks should also be avoided, as they require the construction of fascine causeways at the ends of the bridge. The ramps should not be inclined more than one sixth. The height of the bank above the water should not be greater than from 6½ to 8 ft., and should have at least a height of 3 ft. if possible.

To facilitate the construction of the bridge advantage should be taken of those branches of the stream under control, by floating the materials down them; and care should be taken to place it, if possible, above the mouths of those held by the enemy, who otherwise might destroy it by floating down heavy objects against it.

The place for the bridge should be sounded in order to ascertain the form of the bottom and the kind of supports to be used. If boats were employed in places where the water was too shallow they would probably be crushed and cause serious accidents. This is avoided by placing trestles in the parts too shallow for boats.

The bridge, if possible, should be made in a straight line, which best suits the form of the timbers. The body of the supports should be in the direction of the current, otherwise the tendency of the latter would be to turn them around and break up the bridge. The length of the bridge should be as near as possible perpendicular to the direction of the current.

A description of the regular pontoon or boat bridge, will not be given, for reasons before stated; but it may not be out of place to lay down some general observations upon the manner of making bridges from common boats picked upon a river, and the other materials which an army is able to lay its hands upon.

Ordinary boats being generally of various sizes and shapes, they must be placed under the bridge in such a way as to keep the flooring as near as possible in the same horizontal plane. The widest will be placed nearest the ends of the bridge, and the longest in the channel, as they are less acted on by the current. If any of the boats are too small, several of them will be joined together, side by side, by notched cross-pieces, upon which will be placed three beams lengthways of the boat, one about over each gunwale and the third over the axis of the boat. These three beams receive the timbers of the bridge flooring. Fig. 297.

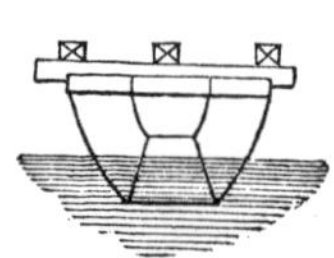

Fig. 297.

Should any of the boats be too low, a trestle or horse must be constructed in them in the direction of the axis of the boat, on which the timbers will rest. Fig. 298.

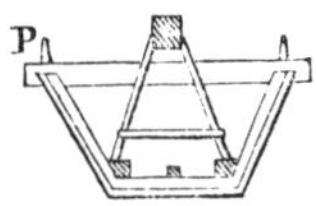

Fig. 298.

The. flooring timbers in these bridges, instead of being tied with ropes are fastened with iron clamps.

If no anchors can be had to hold the boats in their places, the anchoring-ropes may be fastened to boxes or baskets filled with stone or gravel, to mill-stones, or any other heavy body. The deeper the water the more distant should be these anchoring points, otherwise the great inclination of the ropes would tend to pull the boats down into the water.

If the river has floating ice in it, the anchoring ropes should be wound with twine or wire, or replaced in part by cables.

Should the flooring-beams not be long enough, they might be laid alternately in each row upon only three gunwales, or even upon one of each of two contiguous boats, taking care to fasten them securely to a shorter beam lying across both gunwales of the boats. Fig. 299.

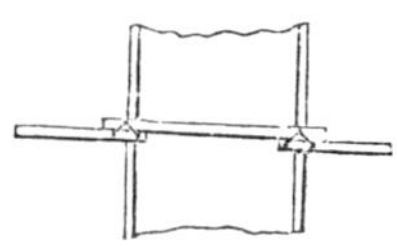
Fig 299.

RAFT-BRIDGES possess the advantage of not being liable to be sunk by the fire of the enemy; but they are very wide and require a great quantity of wood to make them. They offer too great a resistance to currents, which renders them hard to manage in the water and difficult to floor over. Experience has proved that raft-bridges cannot be established upon any but slow-running streams, with a velocity of current less than $6\frac{1}{2}$ ft. per second.

Light woods, such as pine, poplar, willow, birch, and all those whose specific gravity is less than that of water, are the only ones fit to construct rafts with.

The stability of a raft-bridge depends upon that of the rafts of which it is made; and these increase in stability as their length becomes greater. For this reason very long wood is employed in their construction. When it cannot be obtained long enough, the trees are joined together at the feet in couples.

The rafts present less resistance to the force of the current as they are increased in length; and besides, the distance between them becomes greater as they become narrower. If they are very wide and short, they must be close together; and the bridge, leaving thus but a small space open to the current, will experience a very great pressure.

Upon rivers which are sluggish, the bridge may be made of a continuous raft, the flooring being formed of faggots, fascines, and earth, mixed with straw.

The trees usually employed to make rafts are from 32 to 45 ft. long, and have a mean diameter of 1 ft. The number required is determined by the specific gravity of the wood, the size of the trees, and the weight to be supported.

The rafts are put together in the water, where the logs are more easily handled than on land.

Bridges and Rafts of Casks, Boxes, or Bags.—These bridges can only be used on small streams, far from any enemy, in consequence of the ease with which the rafts composing them may be sunk, either by accident or shot.

The casks or other floating materials made use of, are first guaged. They are then placed under a frame-work made from the timber torn from the neighboring houses, with their ends together, and bung-holes up, so that in case they leak they may be emptied with a hand-pump. Two rows of these casks at least are required, and in case they are small, more rows may be necessary.

Boxes well tarred and caulked could be easily decked over, and might be used to make rafts. In wine-producing countries, like France, Spain, and Portugal, the goats' skins used to put the wine in, have been frequently used as the buoyant material.

Trestle-Bridges.—These bridges are established upon streams whose bottoms are solid, depth of water not greater than 10 ft., and velocity of current not over 5 ft. a second. They have the advantage of requiring but little material for their construction, and are capable of being put up in almost any locality with the timber from the neighboring houses or woods.

The trestles or horses are a kind of tall frame, made of light wood in order to be more easily placed in position. They are placed four or five yards apart, and upon them the flooring timbers rest. They should not be too far apart, which would make them liable to be overset by the current.

The Horse, Fig. 300, is composed of a top-piece, A, about 15 ft. long and 8 in. square, designed to receive the bridge timbers, and

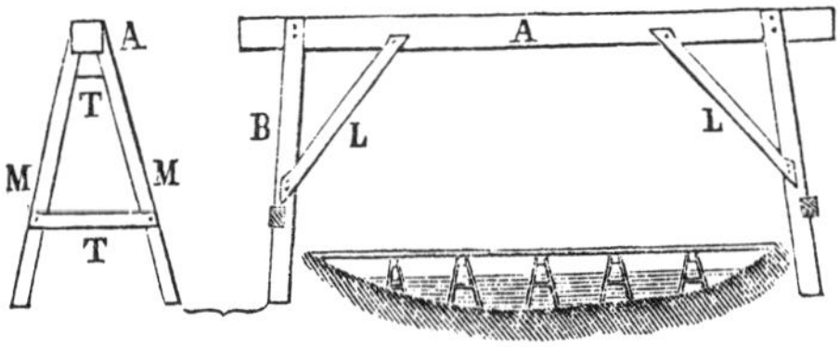

Fig. 300.

two double feet, B, 6 in. square, and of any required height, to support the top piece. The feet are composed, each of two inclined pieces, M and M, two crosspieces, T and T, and two braces, L and L. The distance between the feet of the uprights should not exceed one half the height of the horse. If they were not far enough apart the horse would not stand firmly, and would be easily overturned. If they were too far apart, there would not be sufficient stability, and the horses might be crushed under the load placed upon the bridge. The height of the horses will depend upon their positions. This requires the stream to be previously sounded at every point where a horse is to be placed. The feet should bear equally and solidly upon the bottom; and it becomes necessary sometimes to cut them to fit a rocky bottom, or nail thick plank on the bottom of the feet to keep them from running into the mud, if the bottom is formed of that.

If the timber is square, such as may be obtained from houses or fences, two good carpenters can make one of these horses in six hours, and by dividing the labor and tracing the adjustments by a pattern, in much less time.

If the stream is a shallow one, the men enter the water to place the horses in position. If deep, each horse is placed upon two small beams, resting on a roller placed in the rear, and run forward to its position by acting on the beams. This method requires the employment of a good deal of force; and an easier way is, to rest on the first horse established, two beams, the other ends of which rest on the bottom near where the second horse is to stand. The horse slides down these beams till it gains its position, when it is placed upright by pulling on it with ropes, and pushing against the top piece, or cap, with boat-hooks or poles; and so with the rest of the horses. Should any portion of the current be very rapid, two skiffs or boats of any kind joined together by beams, and anchored stem and stern, will be used for supports instead of horses.

Another method of placing these horses is to make use of a small raft, 21 ft. long by 6 ft. wide, carrying two forks, each of which consists of two vertical uprights, between which fits a beam, supported at one end on the cap of the last horse placed, and at the other on a pin which passes through holes bored in the uprights of the fork. The two beams form an inclined plane of any inclination, down which the horses are successively slid.

The string pieces lap each other over the caps of the horses, and are fastened to them by cords or clamps.

THE UNITED STATES' PONTOON BRIDGE.—During the Mexican war, a bridge-equipage was prepared and sent out to that country, for use with our armies. The supports used were cylindrical pontoons of *India-rubber*, which were inflated by means of bellows. Unfortunately, no opportunity occurred of testing this bridge to any extent in actual warfare: as the only stream where it could have been used on Genl. Taylor's line was crossed about the time of the passage of the law authorizing the equipment of a bridge-train; and the one afterwards sent to Genl. Scott's line was never found necessary, as no streams exist on that line of sufficient depth of water to require bridging, except at times, when little different from mountain torrents. It might, however, have been required by the army in its operations; and the great want of some such means to cross the Rio Grande at Matamoras, is sufficient argument in favor of the necessity of some organization of the kind in every properly constituted army.

Experiments have also been made by our engineer officers on the Birago trestle-bridge, of which the supports are, for a depth of water less than 12′, a peculiar kind of adjustable horse. For water deeper than this, pontoons made in several pieces joined together, or any large flat-bottomed boats, may be used.

The horse is formed of a movable cap sustained at each end by a single foot. The cap is of spruce or pine, 17′ long, 8¾″ deep, 6¼″ wide in the middle, and 8¼″ wide at the ends for a distance of 3′. A mortise 3½″ wide by 10½″ long is cut in each head of the cap for the passage of the foot, the bottom of the mortise being inclined 22°.

The feet, also of spruce, are of four different models, of the following dimensions:—

No. 1,..................8′ 3½″ long, by 4.8″ × 3½″.
" 2,..................12′5″ " " 5.8″ × 3½″.
" 3,..................16′7″ " " 4.8″ × 3½″.
" 4,..................20′9″ " " 4.8″ × 3½″.

Nos. 1 and 2 are placed single in the mortises; the remaining space being filled up by a false foot with two notches, 3′ 10″ long, 3½″ wide, a thickness of 3¾″ at the end, and 4¾″ at the middle, between the two notches.

Nos. 3 and 4 are often placed by pairs in the mortises, and pressed firmly together with wedges of hard wood.

To steady the feet and prevent them from being forced into the ground, they are armed at the bottom with wooden shoes; of which there are two models according to the nature of the ground, the larger being for very soft bottoms, and the smaller for hard bottoms and slopes. They have two holes for the points of the feet to pass through, and are provided with iron pins, fastened by chains, for securing the shoes to the feet.

The cap is supported in position by means of two iron suspension chains, with a ring at one end, with which the feet are capped. Below, the chains pass through suspension rings fixed by eye-bolts to the caps, and are secured there by iron keys fixed to chains attached to the caps.

The horses thus formed are retained in an upright position by the bridge-beams, which have, fastened at each end, a bracket of oak notched in the middle, forming catches to fit on top of the caps. These notches are placed 21′.9″ apart from center to center.

The Austrians have these beams made of such a length, and the distance between the notches such, that they will fit on the gunwales of the pontoons. They are particularly useful in fitting up the pontoons to form a raft, and in consolidating the abutment of a bridge by forming it of two mooring beams connected together by small beams.

The floor of this bridge is composed of planks 10′9″ long by 11½″ × 1½″, with a notch at each end ½″ deep and 18½″ long, to receive the balks at the side of the bridge. Half-planks, differing from the others only in the width, which is 6¼″, are used to complete the flooring in places where a plank would be too wide. They are also laid endwise and lashed against the ends of the planks, when these are placed perpendicular to the bridging-beams, to prevent them from sliding to one side.

In a stream with little current—

The horse No.	1	can be used in a depth of		3′.
"	" 2	"		5.′.
"	" 3	"		9′.
"	" 4	"		12½′.

But when the current is more rapid than 8½′ per second, the horse No. 3 should be used for a depth not greater than 8½, and No. 4 for a depth of only 11½′.

The pontoons can be used wherever the horses can. The highest point of the scaffold of a pontoon, under an unloaded bridge, is 37½″ above the surface of the water; and under a bridge supporting the heaviest load (a column of infantry), it is 18½″.

The pontoons are kept in place, according to their position, the force of the current, and that of the wind, by lines fastened on shore; anchors up or down stream; by cross lines; or by cables stretched across the river, usually above, sometimes below the bridge.

This bridge can be employed with a depth of water of nearly 13 feet, and therefore suffices for crossing the greater part of rivers of mean width, and may be used also advantageously in forming a considerable portion of military bridges across the widest and most rapid rivers.

It may be so laid as to form several passage-ways in the same bridge, by joining a larger number of the parts of pontoons together, and placing several rows of the horses. The Russians use in their bridge equipages a very light kind of pontoon, formed of a frame-work covered with canvas. The frame is composed of two side-frames, constructed of 4″ scantling, of the size and shape represented in Fig. 301.

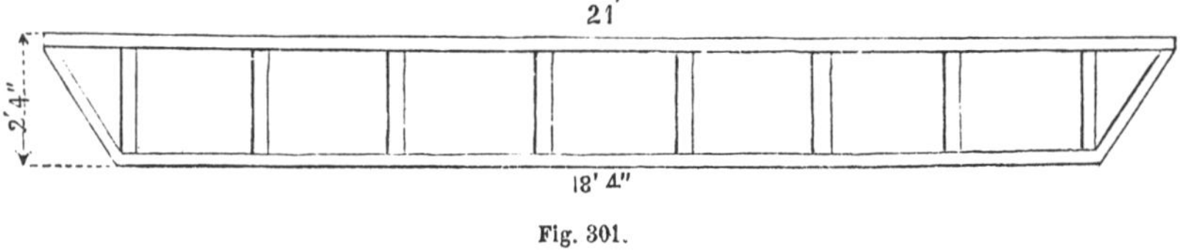

Fig. 301.

These two frames are connected below by movable transoms, with tenons at each end which fit into mortises in the bottom sills, and above by two transoms, one at each end, which are laid on the top string-pieces, about 2′ from the ends, and lashed to them. The canvas cover is stretched over the bottom, sides, and ends, brought over the ends, and lashed to the top transoms. It is secured along the sides of the top string-pieces by small nails passing through eyelet holes along the edges of the cloth. It is 10′ 8″ wide, 30′ long in the middle, 23′ long along the edges, and is painted black on both sides.

A plank is laid along the bottom for the pontoneers to stand on. The cables are attached to the top transoms. There are four balks for each boat, each one being 23′ 4″ long, 5″ deep, and 4″ wide; the side rails are 21′ long, and of 3″ scantling; four chesses are 12′ 2″ long, 1″.5 thick, and 18″.6 wide, the rest being of the same length and thickness, but only 9″.3 wide. The balks of the adjacent bais are connected by iron bolts and keys. There are special supports for the hand-ropes. The Birago trestle and abutments are used with these pontoons. Wagons with flat open bottoms, and a stanchion 3′ high at each angle, are used for transportation. In loading, the four broad chesses are laid on edge against the stanchions, two on each side, thus forming the sides of the wagon. The narrow chesses are laid on the bottom, then the balks, side-rails, pontoon-frames, oars, &c.; the anchor and cable on top of all. The canvas cover, rolled on a boat-hook, is hung to the stanchions on the right side of the wagon.

To pass a siege-train over this bridge, the boats are placed 8′ apart, from center to center, and six balks are used instead of four. In other cases, the distance between the centers of the supports varies from 11′ 8″ to 16′ 7½″, according to the method of construction, and the load to be crossed over.

The canvas pontoon, with its cover, complete, weighs 720 lbs.

Floatation of each ponton, 13,428 lbs.

Weight of flooring, &c., of one bay, 1,476 lbs.

Other materials packed on each wagon, vary from 144 to 378 lbs.

Total load of each wagon, from 2,340 to 2,574 lbs.

Weight of empty wagon, 1,206 lbs.

THE PRESERVATION OF BRIDGES.—In order to protect a bridge

against floating bodies launched by the enemy, or drifted down by the water, the following means are adopted:—

1. A guard of observation is established about 1,000 yards above the bridge, provided with boats, long cables, grappling-hooks, cramp-irons and hammers to drive them. The boats stationed at different distances, approach the floating bodies, attach the cables to them, carry the other end ashore, and by hauling on it, or fixing it ashore, they are run aground. If the current is very strong, and the stream very wide, the cable is fastened to a fixed anchor, and the floating body is thrown against the shore.

Sometimes boatmen might board the floating bodies, and take them ashore or aground. Ropes should be provided, terminated by a piece of chain and grappling-hooks, with which to grapple fire-ships.

Should any of these floating bodies escape, they should be directed, if possible, towards the *draw* in the bridge, which is constructed for security; and signals will be made to the sentinels at the bridge, to have it opened for the passage of the body.

2. Above the bridge a floating boom is placed, designed to stop the bodies carried down by the current. It is formed of large trunks of trees, of wood light enough to float it, joined by pieces of chain. It is placed obliquely across the stream so as to make an angle of 22° with the current, which gives little hold to floating bodies directed against the bridge, and throws them against the bank. It may, however, be carried away by large heavy bodies striking against it, and for this reason should be placed far enough above the bridge to prevent accidents in case of such an event.

3. The bridge may be composed entirely of draws, which may be rapidly removed out of the way of floating bodies.

4th. The bridge is protected from being carried away by sudden freshets, by making the ends, or shore-bays movable, so that they will rise and fall with the water.

Of the passage on bridges.—Military bridges being exposed to numerous accidents, it is necessary to multiply the number of them as much as possible. With only one, the least accident might lead to the ruin of an entire army. It is, therefore, usual to construct two or three from 200 to 300 yards apart. The instability of the floor requires the greatest care to prevent accidents resulting from an overload or too violent a rocking. The precautions used, are :

To post a sentinel at each end of the bridge, and others along it within sight and hail of each other, to cause the instructions given to be obeyed.

Infantry is marched by the flank, without music or noise of any kind, at the route-step, as the cadence step causes rocking, injurious to the bridge, and which may be even great enough to knock over the men who are on it. In case this rocking motion commences, the column should be at once halted and not allowed to proceed until the cessation of the motion.

In passing artillery, all but the rear driver should dismount and lead the horses. A distance of twenty yards should be kept between the carriages, which should travel as near as possible in the middle of the flooring. If the flooring is wet, great care should be taken to prevent the horses from slipping, battens being nailed across the floor if necessary. In some cases, it might be advisable to pass the horses and carriages over separately.

With cavalry, the men should dismount, each rider hold his horse by the bridle and prevent him from trotting. If the men do not dismount, the horses might become frightened, and jump off the bridge, or stop on it and cause disorder. If the horses move faster than a walk, they are liable to have their legs broken by getting them between the flooring planks.

Care should be taken not to overload a bridge, and not to allow fresh troops to go on it until those already on have finished the passage. Such an interval should be left between the different bodies of troops, that they will not have to wait after leaving the bridge.

Should the flooring crack under the weight of the carriages, they should not stop, but press on at a quick gait, as the elasticity of the wood might then prevent the rupture taking place.

If it becomes necessary to stop a carriage on the bridge, the team should be at once unhitched, the load taken out as quickly as possible, deposited inthe nearest boats, and the carriage thrown overboard.

A column of infantry and one of cavalry, or one of artillery and wagons, should not be allowed to occupy *any kind of a* bridge at the same time, but each separate column be made to pass by itself.

The troops should always stop at the command *halt*, from the sentinel, and resume the march only after receiving orders to do so.

With several bridges, some will be used for infantry and others for cavalry and carriages. If there is a ford, it will be used in preference for the cavalry.

If there are herds of cattle with the army, no more than six or eight head should be passed over at a time, as they instinctively flock together when frightened, and a greater number might prove disastrous. Cattle will swim across a stream very well; and this is a good way to get them over if they are in charge of intelligent men. After several have taken to the water and swam across, the rest follow without any difficulty.

No fire should be allowed to pass over the bridge; and, to guard against every chance of a conflagration or explosion, even lighted pipes and segars are not allowed.

The sentinels notice attentively the signals which may be made to them from up the river, communicate them at once to their commander, and in general give warning of every danger, halt the troops, or hasten them over, according to circumstances.

The greatest load which can pass a bridge, is a column of men without arms, marching in compact order. With the heaviest field batteries, all but two teams to each carriage, should be unhitched.

For keeping a bridge in repair a certain number of pontoneers are charged with opening the draws; tightening the ropes; bailing out the boats, and repairing them; closing the joints of the floor-planks; raising from time to time the anchors, to prevent them from burying themselves too deeply, which would prevent them being taken up easily when the bridge was broken up; and in general with repairing any damages which may show themselves.

If the bridge is thrown across a stream liable to be frozen, it should be taken up in time to prevent its being caught in the ice, and carried away by it when breaking up. If it is necessary for the bridge to remain, the ice should be broken every day around the supports.

A bridge may be taken up either, commencing at the bank abandoned, by removing in succession the floor and supports and taking them ashore, or by setting loose that end, and swinging the bridge around against the opposite shore, and then taking it apart.

FLYING-BRIDGES.—A flying-bridge is one which consists of a boat, draw (A), or raft Fig. 302, held by a rope, which prevents it from descending the stream. The bridge, of whatever make, is crossed from one bank to the other, by presenting its side obliquely to the action of the current by the use of a rudder. These bridges have the advantage of being easily established, and requiring but little material for their construction; but they do not furnish a continuous communication, and can be used only with a small body of troops.

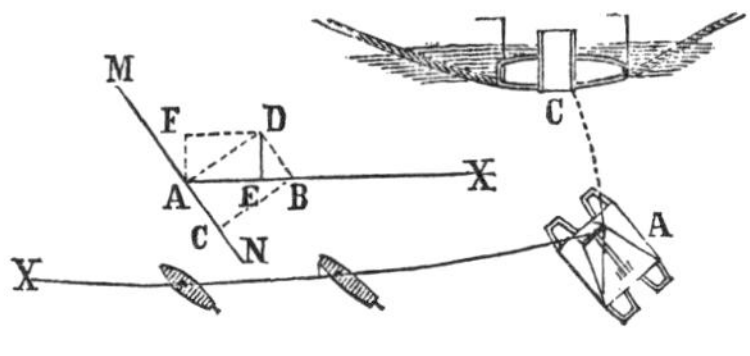

Fig. 302.

The bridge is made usually of two (Fig. 303), three, and sometimes six boats, connected together, and very solidly floored over, the beams being fastened to the gunwales of the boats with iron bolts or bands, and the flooring planks nailed down upon them. The floor is sometimes surrounded with a guard-rail. The most suitable boats are long, narrow and deep, with their sides nearly vertical, in order to offer greater resistance to the action of the current. At the end of the rope is fixed an anchor X, which is moored in the channel, if this is in the middle of the stream. If the channel is not in the middle, the anchor is placed a little on one side of it toward the most distant shore. By means of the rudder the bridge is turned in such a direction that it is struck obliquely by the current, and the force resulting from the decomposition of the action of the current makes it describe an arc of a circle around the anchor as a center, and this force acquires its maximum effect when the sides of the boats make an angle of about 55° with the direction of the current.

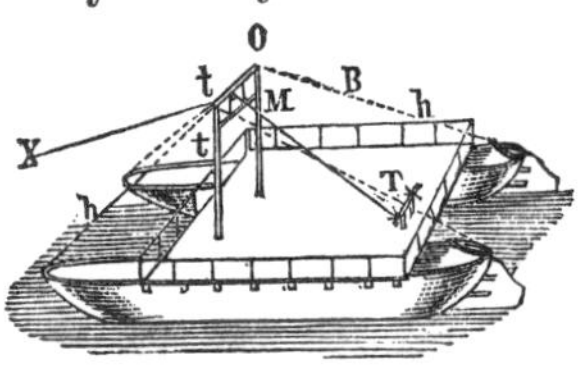

Fig. 303.

Suppose M N (Fig. 302) to represent the side of the boat, and A B the resultant of the forces of the current against it. The force A B will be decomposed into two forces; the one A C, will act in the direction M N as friction, and may be neglected, and the other A D will act perpendicularly to the side of the boat.

Were the boat free to move, and headed in the same direction, it would descend the river, at the same time crossing it. A D is then decomposed into two other forces, the one A E in the direction of the current, causing the boat to drift, the other A F, perpendicular to this, which pushes the boat across.

If the boat is now attached to a fixed point by the rope A X, the force A E will be neutralized, and all the effort of the current will be reduced to the force A F, which makes the boat revolve around the point X.

We see that should the velocity of the current be very small, the force A F will not be sufficient to move the bridge, and recourse must be had to boats drawn across by men or horses on the opposite banks, or by the men in the boats acting on ropes stretched across the river. Flying-bridges are, then, generally employed only on rapid streams.

The length of rope used should be once and a half or twice the width of the river. With a shorter rope the arc described by the bridge is too great, and it performs the ascending branch with difficulty; with a longer one, the rope becomes too heavy, sinks in the water, and fetters the movement. Generally, the arc described by the bridge should not be more than 90°.

To prevent the rope from dragging over the deck, which would interfere with the load, it is held up by an arrangement such as is indicated in Fig. 303, and buoyed out of the water nearly to the anchor by skiffs, empty casks, or other floating bodies.

When the stream to be crossed is not very wide, a flying-bridge may be made with two ropes, one fastened on each shore, the ropes being used alternately. If the stream, on the contrary, is very wide, several boats are fastened together, floored over, and anchored in the middle, and communication kept up with each shore by a flying-bridge, like the one already described.

In about one hour 36 men can construct a flying-bridge composed of 6 bridge-boats, and capable of carrying 250 infantry, or 2 pieces of artillery and 12 horses.

At least one spare anchor should always be carried on the bridge, to anchor it in case the rope should break or become detached; and oars, a small boat, and a long rope, should also be provided.

A flying-bridge may, in case of emergency, be made of any kind of boats with the means of fixing rudders to them. For want of an anchor, a large stone, mill-stone, or a bag or box of sand may be made use of.

A flying-bridge may be made of a raft, the best form being lozenge-shaped, with the front angle about 55°. It is attached to a rope stretched across the stream by three others with pulleys, which slide along the first rope, this being tightly stretched across and not allowed to hang in the water.

Buttresses constructed on boats or trestles, according to the means at hand, are formed on both sides of the river, at the points where the flying-bridge lands.

IRREGULAR BRIDGES are those established with the means found at any locality; and the bridges vary accordingly. The timber which is to serve as the beaming for the floor will determine the distance between the supports, which are themselves more or less irregular. For the want of boats of sufficient size, two or four small ones may be connected together to form a support sufficiently staple.

Within reach of woods, and over small streams, bridges of gabions may be made. The dimensions of the gabions should vary with the degree of strength which the bridge should possess. Those of from 51 to 59 in. in diameter, form a bridge capable of supporting the heaviest loads. The ordinary gabion of 20 in. in diameter forms a bridge strong enough to bear infantry.

These are stood on end, and filled with gravel or stone, and earth, mixed and well rammed, and capped by a piece of timber, on which the flooring beams rest. The height of the gabions will be determined by the height of the river banks, depth of water, &c.

The same kind of a structure is used in passing over marshy or overflowed ground, as, for instance, in the approaches to a bridge.

Field-carriages, military or common wagons, may all be made use of in making bridges for small streams, with slow current; and in the absence of these, carts can be used.

A bridge of carts is formed by establishing two rows of them

33

in pairs, each pair crossing their tongues in the air. The tongues are firmly bound together, and two opposite pairs form the supports for the ends of beams on which the flooring-beams rest. To increase the solidity of the structure, the axle-trees may be lashed together with ropes, or connected with beams. Boards are placed under the wheels to keep them from sinking too deep into the ground, and to chock them.

More solid bridges are made with four-wheeled vehicles placed parallel to the current, and connected by beams which support the cross-pieces on which the flooring timbers rest.

If timbers of sufficient length can be obtained, they are thrown across the stream from bank to bank, and covered with a flooring of plank; but if the timber is too short, frames, more or less complicated, must be constructed, which require skillful and intelligent workmen. For the want of square timber, common logs with the bark on may be used. Several ways of making such bridges are represented in Fig. 304.

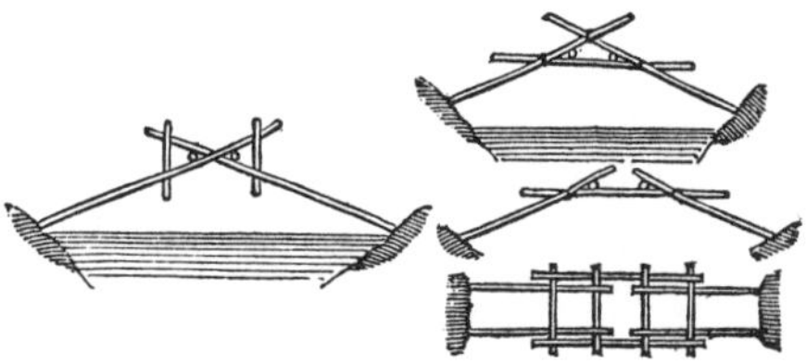

Fig. 304.

Suspension Bridges are sometimes formed over ravines, torrents, or narrow streams with steep, high banks, by suspending a flooring to ropes stretched across, and firmly supported at the ends. The structure is kept steady by stretching guys from the flooring to different points on the banks; and such a bridge can be made strong enough to bear field artillery.

Destroying Bridges.—A bridge which has to be abandoned, is destroyed either by burning it, blowing it up, or sinking it.

To burn a bridge, combustible materials, such as straw, dry fagots, tarred fascines, &c., are placed under the flooring, and fired. To hasten the destruction, fire must be communicated at several points. If time does not permit the burning of the bridge, a pile of logs, timber, &c., should be hastily raised on the floor, which, in the absence of anything better, would serve to arrest the progress of an enemy.

To blow up a bridge, barrels of powder, or loaded shells, are placed under the floor, and fired by means of slow fuzes, portfires, powder-trains, &c.

To sink a bridge of boats, holes are made in the bottoms of the boats, either by boring with augers, or knocking them through with axes; the cords are cut, and a portion of the floor thrown overboard. The sinking of a bridge may be much facilitated by boring the holes beforehand and inserting conical plugs projecting upwards. All that is then necessary to be done is to withdraw the plugs.

Masonry bridges are destroyed with powder, two or three of the arches being blown up, to render the passage more difficult.

If time permits and the necessary means are at hand, mines of from 100 to 120 lbs. of powder will be established in the piers of those arches designed to be destroyed. If the want of time does not allow this to be done, an excavation in the form of a cross will be made over the center of each arch, the arms being at least 3 yards long, care being taken to excavate down to the very masonry of the arch itself.

In each of the arms is placed about 150 lbs. of powder, which is well covered and rammed in; or if time does not allow this, the powder is covered over with planks, which are loaded down with stones and earth; and fire is communicated after taking the necessary precautions.

A more expeditious method is to suspend under the arch two or three barrels of powder, which are fired simultaneously. The bridge, in this case, offers much less resistance than when the powder has to crush the masonry by its explosion.

Enemy's Bridges.—The destruction of an enemy's bridges is an operation of the greatest importance, especially in repelling an attack in force. A portion of his army already passed over, may be compelled in this way to defend itself against the whole power of the assailed.

Boats or rafts heavily loaded to increase their mass and consequently their shock, may be floated down against their bridges. They should carry near the middle, a strongly supported mast, high enough to strike the bridge and shock it heavily.

Another method is by floating down fire-ships, which are boats filled with incendiary materials mixed with loaded shells and grenades in order to keep any one from approaching them. These boats have also a mast high enough to stop them under the floor of the bridge, and allow them to set fire to it. Fire is communicated by means of a gun-lock placed on the powder. The trigger or sear is joined by a chain to a lever which is made to turn on a pivot by the striking of the boat. A percussion-tube or match would serve better for the purpose.

Barrels or tarred boxes filled with powder, and provided with these explosive arrangements, form an excellent material for destroying bridges, as they easily escape the notice of the sentinels and can be floated down in numbers at a time. They are provided with wings on each side to prevent them from turning over, and to keep the bung or lock-lever up.

To be certain of success in the operation, floating bodies, fire and explosive boats should not be sent down in succession and separately, but a large number should be started in the current at once. Then a portion of them will run aground; others may be stopped by the parties placed on guard for the purpose; but some will probably reach the bridge, and either destroy or damage it. Night is the best time to perform these operations.

APPENDIX.

Principal Dimensions and Weights of Guns.

	IRON.							BRASS.		
	COLUMBIADS, NEW PATTERN.		SEA COAST.		SIEGE AND GARRISON.			FIELD.		
	128	64	42	32	24	18	12	12 NEW	12	6
	Inches.	Inches.	Inches.	Inches.	Inches.	Inches.	Inches.	Inches.	Inches.	Inches.
Diameter of the bore	10.000	8.000	7.000	6.400	5.820	5.300	4.620	4.620	4.620	3.670
True windage...............	.130	.120	0.160	0.150	0.140	0.130	0.100	0.100	0.100	0.090
Length of bore...............	101.170	108.320	110.000	107.600	108.000	109.000	103.400	63.600	74.000	57.500
Ditto.........in diameters	**10.717**	**13.54**	**15.71**	**16.78**	**18.56**	**20.56**	**22.38**	**13.76**	**16.00**	**15.67**
Length from rear of base ring to face of muzzle...........	116.000	116.000	117.000	114.000	114.000	114.000	108.000	66.000	78.000	60.000
Whole length of the piece.....	132.670	128.820	129.000	125.200	124.000	123.250	116.000	72.150	85.000	65.600
Semi-diameter of the base ring.	15.500	12.500	12.200	11.200	10.700	9.875	8.700	5.500	6.500	5.150
Semi-diameter of the swell of the muzzle................	9.500	7.500	8.400	7.700	7.793	6.935	5.932	4.250	5.170	4.125
Distance between these two semi-diameters.............	114.750	115.000	115.000	112.000	111.000	111.600	105.300	65.000	76.300	58.700
Natural angle of sight.........	2° 45′	2° 30′			1° 30′	1° 30′	1° 30′	1° 6′	1°	1°
Distance from rear of base ring to rear of trunnions	37.500	38.500	43.200	42.200	43.000	43.500	42.700	25.400	30.700	23.250
Diameter of the base ring......	31.000	25.000	24.400	22.400	21.400	19.750	17.400	11.000	13.000	10.300
Distance between the rimbases.	31.000	25.000	22.000	20.700	18.000	16.800	14.300	11.500	12.000	9.500
Length of the trunnions.......	9.000	6.500	6.500	6.000	5.000	4.750	4.500	3.250	3.500	2.800
Diameter of the trunnions.....	10.000	8.000	7.000	6.400	5.820	5.300	4.320	4.200	4.620	3.670
Distance from axis of trunnions to face of muzzle...........	73.500	73.500	70.300	68.600	68.090	67.850	63.390	38.500	44.990	34.910
Weight...............pounds	15,000	9,100	8,465	7,200	5,790	4,913	3,590	1,220	1,757	884
Preponderance ....pounds	950	740	440	466	255	200	200	105	60	33

Principal Dimensions and Weights of Columbiads and Howitzers.

	COLUMBIADS.		HOWITZERS.							
	OLD PATTERN.		IRON.				BRASS.			
	IRON.		Sea Coast.		Siege and Garrison.		Field.			Mountain
	10-inch.	8-inch.	10-inch.	8-inch.	8-inch.	24-pr.	32-pr.	24-pr.	12-pr.	12-pr.
	Inches.	Inches.	Inches.	Inches.	Inches.	Inches	Inches	Inches.	Inches.	Inches.
Diameter of the bore	10.000	8.000	10.000	8.000	8.000	5.820	6.400	5.820	4.620	4.620
True windage	0.120	0.120	0.120	0.130	0.130	0.140	0.150	0.140	0.100	0.100
Length of bore, exclusive of chamber	99.000	100.000	96.000	85.500	38.500	53.250	64.000	56.250	46.250	28.160
Ditto ... in diameters	**9 90**	**12.50**	**9.60**	**10.68**	**4.81**	**9.15**	**10.00**	**9.66**	**10.00**	**6.10**
Diameter of the chamber	8.000	6.400	7.000	6.400	4.620	4.620	4.620	4.620	3.670	3.340
Length of the chamber	12.000	11.000	9.500	7.500	8.000	4.750	7.000	4.750	4.250	2.750
Length from rear of base ring to face of muzzle	120.000	119.000	112.000	98.000	52.000	62.000	75.000	65.000	53.000	32.910
Whole length of the piece	126.000	124.000	124.250	109.000	61.500	69.000	82.000	71.200	58.600	37.210
Semi-diameter of base ring	16.000	13.000	13.250	11.100	9.125	6.900	6.900	6.000	5.000	3.800
Semi-diameter of swell of muzzle	10.750	8.500	10.125	8.250	8.225	5.850	5.600	4.875	4.100	3.450
Distance between these semi-diameters	117.500	117.000	109.500	96.000	51.500	61.800	74.750	64.800	52.850	32.910
Natural angle of sight	1°.21′	1° 23′			1°	1°	1°	1°	1°	0° 37′
Distance from rear of base ring to rear of trunnions	41.500	41.500	41.000	37.400	24.000	24.690	30.700	27.500	23.250	15.000
Diameter of base ring	32.000	26.000	26.500	22.200	18.250	13.800	13.800	13.800	12.000	10.000
Distance between the rimbases	31.000	25.000	25.000	20.700	18.000	12.800	12.000	11.500	9.500	6.900
Length of the trunnions	9.000	6.500	7.500	6.000	5.000	3.250	3.500	3.250	2.800	2.250
Diameter of the trunnions	10.000	8.000	8.000	6.400	5.820	4.620	4.620	4.200	3.670	2.700
Distance from axis of trunnions to face of muzzle	73.500	73.500	67.000	57.400	25.090	35.000	41.990	35.400	27.910	16.560
Weight ... pounds.	15,400	9,240	9,500	5,740	2,614	1,476	1,920	1,318	788	220
Preponderance ... pounds.	470	350	450	380	460	70	125	112	51	30

Principal Dimensions and Weights of Mortars.

	IRON.				BRASS.		IRON.
	HEAVY.		LIGHT.				
	13-inch.	10-inch.	10-inch.	8-inch.	Stone mortar.	Coehorn 24-pr.	Eprouvette.
	Inches.	Inches.	Inches.	Inches.	Inches.	Inches.	Inches.
Diameter of the bore	13.000	10.000	10.000	8.000	16.000	5.820	5.655
True windage	0.130	0.130	0.130	0.120		0.140	0.025
Length of the bore, exclusive of the chamber	26.000	25.000	15.000	12.000	19.800	8.820	11.500
Ditto...in diameters	**2.00**	**2.50**	**1.50**	**1.50**	**1.24**	**1.51**	**2 00**
Diameter of the chamber. Superior (at the bottom of the shell in iron mortars)	9.500	7.150	7.600	6.080	5.300	3.000	1.500
Diameter of the chamber. Inferior	7.250	5.640	5.000	4.000	3.000	2.000	1.500
Length of the chamber	13.000	10.000	5.000	4.000	6.750	4.250	1.350
Whole length of the mortar	53.000	46.000	28.000	22.500	31.550	16.320	
Distance from face of muzzle to front of trunnions	41.000	37.000	20.000	16.500	20.000	13.570	
Distance between the rimbases	36.000	27.500	20.500	16 250	18.000	7.500	
Length of the trunnions	8.500	6.500	5.000	4.000	6.000	2.500	
Diameter of the trunnions	12.000	9.000	8.000	6.000	8.000	2.750	
Weight...pounds	11,500	5,775	1,852	930	1,500	164	220

Principal Dimensions and Weights of Field-Gun Carriages and Limbers.

DIMENSIONS.	6-pdr. Gun and 12-pdr. Howitzer.	24-pdr. Howitzer	12-pdr. Gun and 32-pdr. Howitzer.
	In.	In.	In.
Distance between the inside of the trunnion plates	9.60	11.65	12.15
Diameter of the trunnion holes	3.70	4.25	4.65
Depth of the center of trunnion hole below the upper face of the trunnion plate	1.00	0.95	0.95
Distance of axis of trunnions in rear of axis of axle-tree, the piece being in battery on horizontal ground	0.50	1.00	0.80
Distance from axis of trunnions to axis of axle-tree	14.65	16.20	16.60
Height of axis of trunnions above the ground	43.10	44.80	45.20
Vertical field of fire, above the horizontal line, Gun	12°		13°
Vertical field of fire, above the horizontal line, Howitzer	13°	13°	12°
Vertical field of fire, below the horizontal line, Gun	8°		7°
Vertical field of fire, below the horizontal line, Howitzer	5°	8°	5°
Distance between the points of contact of trail and wheels with the ground line	74.40	79.80	79.80
Distance from front of wheels to end of trail, the piece being in battery	116.60	122.75	122.75
Distance of the muzzle of the piece in battery from the front of the wheels, Gun, in front of wheels	5.91		15.70
Distance of the muzzle of the piece in battery from the front of the wheels, Howitzer, front of wheels		5.90	12.70
Distance of the muzzle of the piece in battery from the front of the wheels, Howitzer, rear of wheels	1.09		
Length of gun-carriage without wheels	104.40	111.40	113.50
Length of limber without wheels	161.20	161.20	161.20
Length of limber without wheels or pole	52.85	52.85	52.85
Length of limber with wheels and pole	173.08	173.08	173.08
Distance between the centres of the axle-trees of gun carriage and limber	96.00	101.70	101.70
Length of the carriage limbered up	269.08	274.78	274.78
Distance from the muzzle of the piece, when limbered, to the front of pole. Gun	279.10		294.00
Distance from the muzzle of the piece, when limbered, to the front of pole. Howitzer	272.10	283.78	291.00
Whole length of the axle-tree	78.84	78.84	78.84
Track of the wheels	60.00	60.00	60.00
Height of wheel	57.00	57.00	57.00
Dish of finished wheel	1.50	1.50	1.50
	Lbs.	Lbs.	Lbs.
WEIGHTS. Gun-carriage, without wheels	540	736	783
Limber without wheels or ammunition chest	335	335	335
Ammunition-chest, without divisions	165	165	165
One wheel, Gun-carriage	180	196	196
One wheel, Limber	180	180	180
Gun-carriage complete, without implements	900	1128	1175
Limber complete, without implements	860	860	860
Gun-carriage and limber, without implements	1760	1988	2035

Principal Dimensions and Weights of Siege-Gun Carriages and Limbers.

DIMENSIONS.	12-pdr. Gun.	18-pdr. Gun.	24-pdr. Gun and 8-in. Howitzer.
	In.	In.	In.
Distance between the inside of the trunnion-plates	14.95	16.95	18.15
Diameter of the trunnion-holes	4.65	5.35	5.85
Depth of center of trunnion-hole below the upper face of trunnion-plate	1.10	1.20	1.40
Distance of axis of trunnions in rear of axis of axle-tree, the piece being in battery on horizontal ground	3.00	2.50	2.75
Distance from axis of trunnions to axis of axle-tree	22.45	22.85	23.25
Height of axis of trunnions above the ground	52.20	52.60	53.00
Vertical field of fire, above the horizontal line, Gun	13°	12°	12°
Vertical field of fire, above the horizontal line, Howitzer			15°
Vertical field of fire, below the horizontal line, Gun	4°	4°	4°
Vertical field of fire, below the horizontal line, Howitzer			10°
Distance between the points of contact of the wheels and trail with the ground line	100.00	101.00	101.00
Distance from the front of the wheels to the end of the trail, the piece being in battery	141.00	142.00	142.00
Distance of the muzzle of the piece in battery from the front of the wheels, Gun, in front of the wheels	30.74	35.35	35.34
Distance of the muzzle of the piece in battery from the front of the wheels, Howitzer, in rear of wheels			7.66
Length of gun-carriage, without wheels	130.00	133.00	133.60
Length of limber, without wheels	176.65	176.65	176.65
Length of limber, without wheels or pole	59.80	59.80	59.80
Length of limber, with wheels and pole	184.90	184.90	184.90
Distance between the centers of the axle-trees of gun-carriage and limber	94.00	96.00	96.00
Length of the carriage, limbered up	278.90	280.90	280.90
Distance from the muzzle of the gun, in its traveling position, to front end of pole	285.15	291.42	290.00
Whole length of the axle-tree	81.80	81.80	81.80
Track of the wheels	60.00	60.00	60.00
Height of wheel	60.00	60.00	60.00
Dish of finished wheel	2.00	2.00	2.00
	Lbs.	Lbs.	Lbs.
WEIGHTS. Gun-carriage, without wheels	1440	1542	1714
WEIGHTS. Limber, without wheels	585	585	585
WEIGHTS. One wheel	404	404	404
WEIGHTS. Gun-carriage, complete, without implements	2248	2350	2522
WEIGHTS. Limber, complete	1393	1393	1393
WEIGHTS. Gun-carriage and limber, without implements	3641	3743	3915

Field and Siege Wagons.

DIMENSIONS AND WEIGHTS.	Caisson.	Forge.	Battery Wagon.	Mortar Wagon.
	In.	In.	In.	In.
Length	125.50	130.00	154.00	143.60
Distance between the axle-trees of carriage and limber	92.00	97.80	112.93	102.95
Whole length, when limbered up	274.70	279.00	303.13	287.85
Height above the ground	58.75	70.50	73.55	60.00
	Lbs.	Lbs.	Lbs.	Lbs.
WEIGHT, Carriage-body, without wheels	432	997	910	984
WEIGHT, Limber, without wheels or chest	335	335	335	585
WEIGHT, One wheel	180	180	180	404
WEIGHT, Carriage and limber, complete, without implements or spare parts	1982	2217	2130	3185

INTERIOR DIMENSIONS.	Length.	Width.	Depth.	Weight.
	In.	In.	In.	Lbs.
Ammunition, or limber chest, without divisions	40.00	18.00	14.75	165
Traveling forge, Iron-room	40.00	32.00	7.50	
Traveling forge, Coal-box	31.00	13.00	17.00	100
Battery wagon, body	98.80	36.00	22.00	
Mortar wagon, floor	63.85	40.00		

Mortar Beds.

DIMENSIONS AND WEIGHT.	Siege.		Coehorn.	Eprouvette.
	8-inch.	10-inch.		
	In.	In.	In.	In.
Length	42.00	51.80	31.00	22.00
Exterior width, including maneuvering bolts	34.00	40.00	15.00	22.00
	Lbs.	Lbs.	Lbs.	Lbs.
Weight	920	1830	132	280

Weight of platform for siege-mortars, made of yellow pine, 837 lbs.

Principal Dimensions and Weights of Barbette Carriages.

DIMENSIONS.	12-pdr. Gun.	18-pdr. Gun.	24-pdr. Gun.	32-pdr. Gun, and 8-in. Howitzer.	42-pdr. Gun.
	In.	In.	In.	In.	In.
Distance between the inside of the trunnion plates	14.90	16.90	18.10	20.80	22.10
Diameter of the trunnion-holes	4.65	5.35	5.85	6.45	7.05
Depth of the center of trunnion-hole below upper face of trunnion-plate	1.00	1.00	1.00	1.00	1.00
Horizontal distance of axis of trunnions in rear of axis of axle-tree	3.90	3.90	4.00	4.10	4.20
Distance of axis of trunnions from axis of axle-tree	41.30	41.30	42.00	43.40	44.10
Height of axis of trunnions, in battery, above the traverse circle	71.76	71.76	73.45	75.77	77.47
Horizontal distance from axis of trunnions to axis of elevating screw — Gun	43.10	44.50	44.25	43.90	44.80
Horizontal distance from axis of trunnions to axis of elevating screw — Howitzer				39.00	
Vertical field of fire — above the horizontal line	11°	11°	11°	11°	11°
Vertical field of fire — below the horizontal line	5°	5°	5°	5°	5°
Length of gun-carriage, from front of wheels to rear of lunette	89.50	89.50	90.75	90.75	92.05
Whole length of the axle-tree	57.76	57.76	59.76	66.06	68.31
Distance between the exterior faces of the gun-carriage wheels	55.70	55.70	57.70	64.00	66.25
Inclination of the chassis in 100 inches	5.00	5.00	5.00	5.00	5.00
Whole length of the chassis	184.06	184.06	183.38	182.86	182.71
Width of the chassis between the outside of the rails	43.00	43.00	45.00	51.30	53.50
Horizontal distance from center of pintle to front end of rails	9.50	9.50	9.50	9.50	9.50
Horizontal distance from center of pintle to rear end of chassis	174.30	174.30	174.22	173.76	173.66
Horizontal distance from center of pintle to center of traverse-wheels	120.33	120.33	120.33	120.33	120.33
Horizontal distance from center of pintle to the face of the piece, in battery — Gun	59.00	63.15	63.29	63.70	65.30
Horizontal distance from center of pintle to the face of the piece, in battery — Howitzer				52.50	

Principal Dimensions and Weights of Barbette Carriages—Continued.

DIMENSIONS.	12-pdr. Gun.	18-pdr. Gun.	24-pdr. Gun.	32-pdr. Gun, and 8-in. Howitzer.	42-pdr. Gun.
	In.	In.	In.	In.	In.
Horizontal distance from center of pintle to axis of axle-tree, the piece being in battery	0.80	0.80	0.80	0.80	0.80
Horizontal distance from center of pintle to axis of axle-tree, the piece recoiled to counter-hurters	113.07	113.07	113.00	112.53	112.43
Diameter of gun-carriage roller	13.00	13.00	13.00	13.00	13.00
Diameter of gun-carriage wheel	43.50	43.50	43.50	43.50	43.50
Diameter of traverse-wheel of chassis	15.00	15.00	15.00	15.00	15.00
	Lbs.	Lbs.	Lbs.	Lbs.	Lbs.
WEIGHTS, Gun-carriage, without wheels or rollers	780	800	1073	1327	1400
One gun-carriage roller	135	135	135	146	146
One gun-carriage wheel	308	308	308	308	308
Chassis, without traverse-wheels or forks	1100	1100	1420	1836	2000
One traverse-wheel and fork	97	97	97	97	97
Pintle—new pattern	17	17	17	17	17
Gun-carriage complete, without implements	1666	1686	1959	2213	2308
Chassis complete, without pintle	1294	1294	1614	2030	2194

Principal Dimensions and Weights of Casemate Carriages.

DIMENSIONS.	24-pdr. Gun.	32-pdr. Gun.	42-pdr. Gun.	8-in. Columbiad.	24-pdr. Howitzer.
	In.	In.	In.	In.	In.
Distance between the inside of the trunnion-plates, measured in the axis of trunnion-holes	18.52	21.26	22.56	25.10	12.95
Diameter of the trunnion-holes	5.85	6.45	7.05	8.05	4.65
Depth of axis of trunnion-hole below the upper face of trunnion-plate	0.75	0.75	0.75	0.75	0.00
Horizontal distance of axis of trunnions in rear of axis of axle-tree	1.90	1.90	1.90	3.00	
Distance of axis of trunnions from axis of axle-tree	26.32	27.51	28.21	28.31	
Height of axis of trunnions, in battery, above the traverse circle	48.75	49.95	50.65	50.65	48.25
Vertical field of fire, above the horizontal line	9°	8°	8°	8°	
Vertical field of fire, below the horizontal line	4°	4°	4°	4°	
Length of gun-carriage, from front of cheeks to rear of trail-roller	67.35	67.35	67.35	67.20	48.25
Whole length of the axle-tree	48.50	52.00	55.05	57.60	
Distance between the exterior faces of gun-carriage trucks	40.00	44.50	46.55	49.10	
Inclination of the chassis in 100 inches	7.35	7.35	7.35	7.35	5.83
Whole length of chassis (including 3 inches for the tongue fork)	189.15	189.15	189.15	189.15	151.00
Width of chassis between the outsides of the rails	40.00	42.50	46.76	49.16	22.00
Length of rear transom of chassis	59.50	64.00	66.26	68.66	
Horizontal distance from center of pintle to front end of rails, upper side	49.94	49.94	49.94	49.94	7.00
Horizontal distance from center of pintle to middle of rear transom of chassis	193.40	193.40	193.40	193.40	
Horizontal distance from center of pintle to rear end of chassis	235.00	235.00	235.00	235.00	144.00

Principal Dimensions and Weights of Casemate Carriages—Continued.

DIMENSIONS.	24-pdr. Gun.	32-pdr. Gun.	42-pdr. Gun.	8-in. Columbiad.	24-pdr. Howitzer
	In.	In.	In.	In.	In.
Horizontal distance from center of pintle to center of rear traverse-wheels.	194.50	194.50	194.50	194.50	120.00
Horizontal distance from center of pintle to center of front traverse-wheels.	62.50	62.50	62.50	62.50	
Horizontal distance from center of pintle in rear of face of the piece, in battery	0.52	1.03	2.73	4.83	25.60
Horizontal distance from center of pintle to axis of axle-tree, piece in battery	65.67	65.67	65.67	65.67	
Horizontal distance from center of pintle to axis of axle-tree, piece recoiled to counter-hurters	177.13	177.13	177.13	177.13	
Diameter of gun-carriage truck-wheel	20.00	20.00	20.00	20.00	3.80
Diameter of rear traverse-wheels (mean)	15.90	15.90	15.90	15.90	6.00
Diameter of front traverse-wheels (mean)	7.84	7.84	7.84	7.84	
	Lbs.	Lbs.	Lbs.	Lbs.	Lbs.
WEIGHTS. Gun-carriage, without truck-wheels	908	1064	1120	1128	
WEIGHTS. One gun-carriage truck-wheel	223	223	223	223	
WEIGHTS. Chassis, without traverse-wheels	2120	2430	2575	2600	
WEIGHTS. One rear traverse-wheel	99	99	99	99	
WEIGHTS. One front traverse-wheel	30	30	30	30	
WEIGHTS. One pintle	92	92	92	92	50
WEIGHTS. Gun-carriage, complete, without implements	1354	1510	1566	1574	620
WEIGHTS. Chassis complete without pintle	2378	2688	2833	2858	660

Sling Carts.

	Large.	Hand.
	In.	In.
Length from rear of wheels to front end of pole	242.40	160.75
Length of axle-tree	92.00	75.50
Height of wheels	96.00	72.00
Distance between the wheels, on the ground	58.75	60.40
	Lbs.	Lbs.
WEIGHTS. One wheel	701	
WEIGHTS. Whole weight, without sling-chains	2282	1115
WEIGHTS. Trunnion chain and rings	23	
WEIGHTS. Sling-chain	84	

Gins.

	Field and Siege.	Garrison.	Casemate.
	In.	In.	In.
Length of legs	175.50	256.50	172.50
	Lbs.	Lbs.	Lbs.
WEIGHTS. Pry-pole	55	224	175
WEIGHTS. Gin, without blocks	455	823	642
WEIGHTS. Pulley-blocks, single	37		
WEIGHTS. " " double		65	65
WEIGHTS. " " triple		84	84

Dimensions, &c., of Siege Platforms.

NAMES OF PIECES.	GUNS AND HOWITZERS. No. of pieces.	Length.	Width.	Thickness.	Weight.	MORTARS. No. of pieces.	Length.	Width.	Thickness.	Weight.	Kind of timber used
		In.	In.	In.	Lbs.		In.	In.	In.	Lbs.	
Hurter	1	108	5.00	3.5	51						Yellow pine.
Sleepers	12	108	5.00	3.5	608	6	96	5.00	3.5	230	Yellow pine.
Deck planks	36	108	5.00	3.5	1854	18	108	5.00	3.5	927	Yellow pine.
Stakes (securing)	6	48	3.50	2.0	70	6	48	3.50	2.0	70	Yellow pine.
Stakes (implements)	4	32	2.00	1.0	10						Yellow pine.
Stakes (pointing)						4	48	1.00	1.0		Yellow pine.
Eye-bolts (iron)	4	14	0.75	r'nd		12	11	0.75	r'nd.		

Dimensions, &c., of the Rail Platform.

	SIEGE MORTARS.					Kind of timber used.
	No. of pieces.	Length.	Width.	Thickness.	Weight.	
		In.	In.	In.	Lbs.	
Sleepers	3	60	11.5	8.5		Yellow pine
Rails	2	84	10.0	10.0		
Stakes (securing)	14	48	3.5	3.0		
Platform complete					816	

Dimensions, &c., of the Ricochet Platform.

NAMES OF PIECES.	No. of pieces.	Length.	Width.	Thickness.	Weight.	Kind of timber used.
		In.	In.	In.	Lbs.	
Hurter	1	96	8.00	8.00	174	Yellow pine.
Sleepers	3	108	5.50	5.50	147	Yellow pine.
Planks	2	128	13.00	2.25	166	Beech, yellow pine, or oak.
Plank	1	84	13.00	2.25	60	Beech, yellow pine, or oak.
Pieces of plank	1	30	13.00	2.25	21	Beech, yellow pine, or oak.
Stakes	18	48	1.25	1.25	32	Hickory or oak.
Total weight		.....	......	......	600	

Ammunition carried in each Chest.

KIND.	NO.	WEIGHT.	REMARKS.
FOR 6-POUNDER GUN.		Lbs.	
Shot, fixed	25	190.00	
Spherical case, fixed	20	136.40	The distribution of this ammunition is not yet determined on.
Canister, fixed	5	42.00	
Spare cartridges, 1¼ lbs.	2	2.60	
Friction primers	75	0.97	In tube pouch, or in bundles.
Slow match, yds	1½	0.28	1½ yds slow match and 1 portfire are put in each packing box when sent from the arsenals.
Portfires	3	0.86	
		373.11	
FOR 12-POUNDER GUN.			
Shot, fixed	20	308.00	In left half, and in 4th division of right half.
Spherical case, ditto	8	108.00	In 1st and 2d divisions, right half.
Canister, ditto	4	67.64	In 3d division, right half.
Spare cartridges, 2½ lbs.	2	5.12	On the spherical case.
Friction primers	48	.63	In tube pouch or bundles.
Slow match, yds	2	.38	On the ammunition, in right half.
Portfires	4	1.15	
		490.92	
FOR 12-PDR. HOWITZER.			
Shells, fixed	15	157.50	In 2d, 3d, and 4th divisions, right half
Spherical case, fixed	20	250.00	In left half.
Canisters, fixed	4	47.40	In 1st division, right half.
Friction primers	60	1.20	In tube pouch, or in bundles, on the canisters, &c.
Slow match, yards	1½	0.38	On the canisters.
Port fires	3	0.87	
		457.35	
FOR 24-PDR. HOWITZER.			
Shells, strapped	12	225.60	In left half.
Spherical case, strapped	8	184.00	In front and middle divisions of right half.
Canisters	3	63.75	In rear divisions of right half.
Cartridges. Small charge.	23	53.82	12 in middle division, left half; 9 in middle division, right half; 2 on canisters.
Cartridges. Large charge.	2	5.40	On canisters.

Ammunition carried in each Chest.—Continued.

KIND.	NO.	WEIGHT.	REMARKS.
24-PDR.—*Continued.*		Lbs.	
Friction primers.........	35	0.45	As for 12-pounder howitzer.
Slow match, yards.......	2	0.38	
Port fires...............	4	1.15	
		534.55	
FOR 32-PDR. HOWITZER.			
Shells, strapped.........	8	196.80	Front and rear divisions of left half.
Spherical case, strapped..	6	186.00	Rear divisions, and right front division of right half.
Canister................	1	28.50	Left front division, right half.
Cartridges. Small charge.	15	46.50	1st division in each half.
Cartridges. Large charge.	1	3.88	
Friction primers.........	24	0.32	In tube pouch, or in the middle divisions.
Slow match, yards.......	6	1.15	In middle divisions.
Port fires...............	4	1.15	
		464.30	

Implements and Equipments for Field Carriages.

KIND.	NO.	WEIGHT.	PLACE.
FOR A GUN OR HOWITZER CARRIAGE.		Lbs.	
Sponges and rammers	2		12-pdr., 11.6 lbs.; 6-pdr., 9 lbs. On the gun carriage.
Sponge covers	2	0.28	
Worm and Staff	½	3.60	
Handspikes	2	14.50	
Sponge bucket	1	10.00	
Prolonge	1	18.00	
Tar bucket	1	7.00	On the limber.
Water buckets (leather)	2	16.00	
Gunner's havresacks	2	3.72	In the implement trays, or in other vacant spaces in the ammunition chest.
Tube pouch	2	1.90	
Vent punch	1	0.08	
Gunner's pincers	1	0.85	
Tow hook	1	0.60	
Pendulum hausse	1	0.65	
Thumb stalls	2	0.01	In the tube pouch.
Priming wire	1	0.08	
Lanyard for friction prim'rs	2	0.20	
Gunner's gimlet	1	0.08	
Fuse Wrench	1	0.30	
Tarpaulin, large	1	54.00	Strapped on the ammunition chest.
FOR A CAISSON.			
Felling axe	1	6.00	In the places provided for them on the caisson body.
Shovel, long handle	1	4.75	
Pick axe	1	6.50	
Spare handspike	1	7.25	
Spare pole	1	25.30	
Spare wheel	1	180.00	
Tow hooks	2	1.20	One in the limber chest, and one in a caisson chest.
Tar bucket	1	7.00	On the limber.
Watering buckets (leather)	2	16.00	
Tarpaulin, large	1	54.00	Strapped on the limber chest.

Two pairs of straps for the tarpaulins are fastened with screws to the edges of the lid of the limber chest, at 10 inches from the ends. The straps are 1.25 inch wide; the front straps, 24 inches long; the rear, 10 inches long, with buckles; each fastened with two 1-inch screws.

EQUIPMENT OF TRAVELING FORGES AND BATTERY WAGONS.

One forge and one battery wagon accompany each field battery. They are furnished with the tools and materials required for shoeing horses, and for ordinary repairs and preservation of carriages and harness.

Other forges and battery wagons, equipped for the general service of the army, accompany the field park which contains the general supplies of ordnance stores.

The forge for the field battery is designated by the letter A.
The forge for the field park " " " B.
The battery wagon for the field battery " " C.
The battery wagon for the field park " " D.

EQUIPMENT OF A FORGE FOR A FIELD BATTERY.

Interior arrangement of the Limber Chest.

The chest is marked, FORGE A.

There are *five boxes* for tools and stores; *one shoeing box*, and *one can* for oil.

The boxes are marked, A, Nos. 1, 2, 3, 4, 5.

They are made of white pine, 0.75 in. thick, with loose covers of the same thickness; the covers have three ¾ in. holes bored in each end, to lift them by.

Two handles of double leather are nailed on the inside of the ends of the boxes, so as not to interfere with the covers.

The sides and ends of all the boxes for the forges and battery wagons are dovetailed together, and fastened with 8*d. nails;* the covers are made with clamps on the ends.

Exterior Dimensions of the Boxes for FORGE A.

DESIGNATION.	Length.	Width.	Depth.	Weight.	REMARKS.
	In.	In.	In.	Lbs.	
A, Nos. 1 and 3.	17.8	13.25	7.5	8.25	
A, No. 2......	17.8	13.25	7.5	9.75	A parti'n at 4.5 in. from one end.
A, No. 4......	23.5	8.00	6.5	8.00	A partition for oil can, at 5.25 in. from one end.
A, No. 5......	39.8	9.80	6.5	14.50	
Shoeing box...	16.5	8.00	6.5	4.70	

The *oil can* is made of tin, to hold one quart; it is five inches square and four inches high, with a neck for a cork, one inch diameter and 0.5 in. high, near one corner. Weight 0.9 lb. It is marked, A, SPERM OIL.

Boxes Nos. 1, 2, and 3, are placed in the bottom of the chest; No. 1 against the left hand; No. 2 in the middle.

No. 4 is placed on top of Nos. 1 and 2, against the left end and the back of the chest; the division for the oil can on the left hand.

No. 5 is placed on top of Nos. 1, 2, and 2, against the front of the chest.

The shoeing box is placed on No. 3, against the right end and the back of the chest.

The tools and stores in all the boxes, and in the forges and battery wagons, are securely packed with tow.

Contents of the Limber Chest of Forge A.

SMITH'S TOOLS AND STORES.	NO.	WEIGHT.	PLACE.
		Lbs.	
Horse shoes, Nos. 2 and 3........lbs.	100	100.00	Box A, 1.
Horse shoes, Nos. 2 and 3........lbs.	100	100.00	Box A, 3.
Horse shoe nails, Nos. 2 and 3....lbs.	50	50.00	Box A, 2, large division.
Washers and nuts, No. 2............	30	5.25	In Box A, 2. 91.11 lbs.
Washers and nuts, No. 3............	10	3.20	
Washers and nuts, No. 4............	4	2.15	
Nails, No. 1, C..................lb.	1	1.00	
Nails, No. 2, C..................lb.	1	1.00	
Tire bolts........................	20	5.00	
Keys for ammunition chests.........	5	1.80	
Linch washers.....................	8	7.30	
Linch pins........................	12	8.37	
Chains, Nos. 1 and 2.............ft.	2	1.54	
Cold shut *S* links, No. 3............	50	2.50	
Cold shut *S* links, No. 5............	12	2.00	
Hand cold chisels..................	2	2.00	In Box A, 4. 28.52 lbs.
Hardie	1	0.75	
Files, assorted, with handles........	12	10.00	
Buttress	1	1.50	
Hand punches, round and square....	2	2.00	
Screw wrench......................	1	2.42	
Hand screw driver.................	1	0.32	
Hand vice.........................	1	1.00	
Pair smith's callipers..............	1	0.40	
Taps....... } Nos. 1, 2, 3, and 4....	4	1.50	
Pairs dies.. } Nos. 1, 2, 3, and 4....	4	1.83	
Wood screws, 1 in., No. 14....gross.	1	2.10	
Quart can of sperm oil.............	1	2.70	
Carried forward........	....	310.63	

Contents of Limber Chest.—Continued.

SMITH'S TOOLS AND STORES.	NO.	WEIGHT.	PLACE.
		Lbs.	
Brought forward	....	319.63	
Fire shovel	1	3.05	In Box A, 5. 80.05 lbs.
Poker	1	1.90	
Split broom	1	1.25	
Hand hammer	1	3.50	
Riveting hammer	1	1.05	
Nailing hammer	1	1.80	
Sledge hammer	1	10.50	
Chisels for hot iron	2	3.00	
Chisels for cold iron	2	3.00	
Smith's tongs	3	15.00	
Fore punch	1	1.00	
Creaser	1	1.00	
Fuller	1	2.40	
Nail claw	1	5.00	
Round punch	1	2.10	
Tap wrench	1	3.75	
Die stock	1	6.25	
Nave bands, developed	4	11.75	
Tire bands, developed	2	2.75	
Shoeing hammer	1	0.82	In shoeing box. 12.75 lbs.
Pair pincers	1	2.00	
Rasps (12 inches)	2	2.15	
Shoeing knife	1	0.33	
Toe knife	1	0.30	
Pritchel	1	0.85	
Nail punch	1	0.80	
Clinching iron	1	1.00	
Oil stone	1	1.50	
Leather aprons	2	3.00	
Iron square	1	2.00	Fastened on inside of the chest cover with two copper clamps.
Padlock	1	0.50	On the chest.
Tar bucket	1	7.00	On its hook.
Boxes	6	53.45	
Tow for packing	....	5.00	
Total	....	480.38	

Contents of the Forge Body, A.

Box A, 6, of the same dimensions as A, 1, is carried in the iron room. To put this box in, or take it out, loosen the thumb nuts and rai-e the rear of the bellows an inch.

TOOLS AND STORES.	NO.	WEIGHT.	PLACE.
		Lbs.	
Water bucket, wood	1	10.00	On its hook.
Anvil	1	100.00	On the fire place.
Vice	1	29.00	Fixed on the stock of the carriage.
Watering bucket, leather	1	8.00	On the vice.
Bituminous coal ....lbs.	250	250.00	In the coal box.
Coal shovel	1	4.75	(In the coal box.)
Padlock	1	0.50	On coal box.
Horse shoes, Nos. 2 and 3 ....lbs.	100	100.00	Box A, 6, in iron room.
Square iron, ½ and ⅝ in. ....lbs.	100	100.00	In the iron room. The bars not more than 3 feet long; the square iron in 2 bundles.
Flat iron, 1¼ in. × ⅝ in., 1 in. × ½ in. and 1½ in. × ¼ in. ....lbs.	50	50.00	(In the iron room.)
Round iron, ⅜ in. ....lbs.	50	50.00	(In the iron room.)
Cast steel, ⅝ in. square ....lbs.	5	5.00	(In the iron room.)
English blister steel ....lbs.	5	5.00	(In the iron room.)
Box	1	8.25	
Tow	....	2.00	
Total, exclusive of vice	....	693.50	

NOTE.—100 lbs. of horse shoes, assorted, contain 90 shoes.
1 lb. horse-shoe nails, No. 3, contains 140 nails.
1 lb. " " " 2, " 112 "

EQUIPMENT OF A BATTERY WAGON FOR A FIELD BATTERY.

Interior Arrangement of Limber Chest.

Tho chest is marked, BATTERY WAGON, C.

The tools and stores are carried in *four boxes*, marked C, Nos. 1, 2, 3, and 4, respectively, and in *one oil can*.

The *boxes* are made of white pine 0.75 in. thick, with leather handles inside, and loose covers, like those of the limber chest of Forge A.

The covers of Nos. 1 and 2 are 0.75 in. thick; those of Nos. 3 and 4 are 0.5 in. thick.

Exterior Dimensions of the Boxes.

DESIGNATION.	Length.	Width.	Depth.	Weight.
	In.	In.	In.	Lbs.
C, No. 1	17.8	13.25	7.50	8.25
C, No. 2	26.5	17.80	7.50	17.50
C, No. 3,	39.8	9.80	6.25	12.50
C, No. 4	39.8	8.00	6.25	11.00

No. 3 has a partition, at 5.25 in. from one end, for the oil can.

No. 4 has two partitions perpendicular to the sides, making three divisions 15.8 in., 10 in., and 11 in. long, respectively.

The *oil can* is like that for the limber chest of Forge A, and is marked: C, SPERM OIL.

Boxes Nos. 1 and 2 occupy the bottom of the chest; No. 1 against the left end.

Nos. 3 and 4 are placed on top of Nos. 1 and 2; No. 3 against the rear of the chest.

Contents of Limber Chest for Battery Wagon, C.

TOOLS AND STORES.	NO.	WEIGHT.	PLACE.
CARRIAGE MAKER'S TOOLS.		Lbs.	
Hand saws	2	4.00	Fastened to the inside of chest cover.
Tenon saw (14 in.)	1	1.50	
Jack plane	1	4.15	In Box C, 1. 17.20 lbs.
Smoothing plane	1	1.80	
Brace, with 24 bits	1	4.35	
Spoke shave	1	0.30	
Gauge	1	0.30	
Plane irons	2	1.05	
Saw set	1	0.25	
Rule (2 feet)	1	0.14	
Gimlets	12	0.95	
Compasses ... pair	1	0.18	
Chalk line	1	0.10	
Brad awls	2	0.17	
Scriber	1	0.15	
Saw files (4½ in.)	12	0.87	
Wood files (10 in.)	2	1.12	
Wood rasp (10 in.)	1	0.40	
Trying square (8 in.)	1	0.60	
Hand screw driver	1	0.32	
Carried forward	....	22.70	

Limber Chest for Battery Wagon, C.—Continued.

TOOLS AND STORES.	NO.	WEIGHT.	PLACE.
		Lbs.	
Brought forward......	....	22.70	
CARRIAGE MAKER'S TOOLS—*Cont'd.*			
Oil stone..........	1	1.50	In Box C, 2. 32.23 lbs.
Broad axe..........	1	6.00	
Hand axe..........	1	5.00	
Claw hatchet..........	1	2.00	
Claw hammer..........	1	1.50	
Pincers (small).......pair.	1	1.06	
Table vice..........	1	3.80	
Framing chisels (1 in. and 2 in.)	2	3.00	
Firmer chisels (¾ in. and 1½ in.).....	2	1.00	
Framing gouges (1 in. and 1½ in.)....	2	2.60	
Augers and handles (½ in., ⅝ in., ¾ in.).	3	2.35	
Screw wrench..........	1	2.42	
Felling axe } with handles.........	1	6.00	In Box C, 3. 23.25 lbs.
Adze..... } with handles	1	3.30	
Frame saw..........	1	4.50	
Quart can of sperm oil..........	1	2.70	
SADDLER'S TOOLS AND STORES.			
Mallet..........	1	1.75	
Clam..........	1	5.00	
Hammer..........	1	0.65	Box C, 4.
Shoe knife..........	1	0.09	
Half round knife..........	1	0.28	
Shears.......pair.	1	0.47	
Sand stone..........	1	1.54	
Rule (2 feet)..........	1	0.14	
Needles..........	100	0.08	
Awls and handles..........	12	0.75	
Punches..........	2	0.22	
Pincers.......pair	1	0.75	
Plyers.......pair.	1	0.22	
Claw tool..........	1	0.12	
Creaser..........	1	0.15	
Thimbles..........	4	0.06	
Strap awl..........	1	0.01	
Bees' wax.......lbs.	2	2.00	
Black wax.......lbs.	3	3.00	
Bristles.......oz.	8	0.50	
Shoe thread.......lbs.	5	5.00	
Patent thread.......lbs.	2	2.00	
Carried forward......	....	96.21	

Limber Chest for Battery Wagon, C—Continued.

TOOLS AND STORES.	NO.	WEIGHT.	PLACE.
		Lbs.	
Brought forward............	....	96.21	
SADDLER'S TOOLS AND STORES—*Cont'd.*			
Buckles (assorted, 0.75 in. to 1.5 in.)..doz.	3	1.00	In Box C, 4. 20.66 lbs.
Tacks..............................M.	3	0.75	
Gunner's callipers.....................	1	0.50	
Shoe knives	2	0.18	
Scissors.........................pairs.	2	0.20	
Padlock	1	0.50	
Tar bucket..........................	1	7.00	On its hook.
Boxes..........................	4	49.25	
Tow for packing................	....	7.00	
Total	....	162.59	

Interior Arrangement of Wagon Body, C.

A TILL, 9 in. wide and 9.5 in. deep, is placed at the back or right side of the wagon body.

AN AXE RACK extends along the whole length of the body, on the left side, 11 inches from the bottom; it is 2 in. deep and 1.5 in. wide, and is fastened to the side by the middle rivets of the side studs, and by 5 wood screws. The rack has notches, to hold three axes, a hatchet, and 3 hand-bills.

Four boxes, for stores, marked: C, Nos. 5, 6, 7, and 8.

One box, marked: C, CANDLES.

Exterior Dimensions of Boxes for Wagon Body, C.

DESIGNATION.	Length	Width.	Depth.	Weight.	REMARKS.	
	In.	In.	In.	Lbs.		
C, Nos. 5 and 6	23.0	18.50	11.25	17.50	No covers..	Of hard wood, 0.75 in. thick.
C, No. 7......	23.5	20.25	14.00	28.00	Loose cover	
C, No. 8......	13.0	13.00	5.00	6.00	Div. into four parts	White pine, 0.625 in. thick, with covers, hinges & locks.
Candle box...	11.0	6.50	5.50	2.85		

Seven tin cans, two marked C, NEATS' FOOT OIL; one marked C, LINSEED OIL; one C, TURPENTINE; two C, OLIVE PAINT; one C, BLACK PAINT.

Dimensions of Cans for Wagon Body, C.

KIND.	Capacity.	Diam.	Height.	Weight.	REMARKS.
		In.	In.	Lbs.	
For neatsfoot oil	2 gals.	8.00	11.50	2.20	Rounded tops and necks for corks.
For linseed oil and turpentine............	1 gal.	6.00	10.00	1.37	
For olive paint.......	25 lbs.	9.75	10.25	3.00	Flat tops; opening covered with a piece of tin, soldered on.
For black paint.......	5 lbs.	7.00	8.50	1.50	

Two kegs, for grease; exterior dimensions:

Diameter at the bilge....................10.5 inches.
Diameter at the heads................... 9.75 "
Height ....12.50 "
Weight..................................5 lbs.

Contents of the Wagon Body, C.

Box C, No. 5, is placed on the bottom of the wagon, next to the pile of harness which occupies the rear part of the body. Box No. 6 is on top of No. 5; No. 7 on the bottom of the wagon, in front of No. 5; No. 8 on top of No. 7. The candle box in No. 6.

TOOLS AND STORES.	NO.	WEIGHT.	PLACE.	
		Lbs.		
Linseed oil...........gal	1	9.17	In 1 tin can.	In Box C, 5. 80.44 lbs.
Spirits turpentine ....gal.	1	8.77	" 1 do.	
Olive paint ....lbs.	50	56.00	" 2 do.	
Black paint.............	5	6.50	" 1 do.	
Paint brushes..........	12	8.00	In candle box.	In Box C, 6. 28.73 lbs.
Sperm or wax candles, lbs.	5	7.85		
Rammer heads..........	4	2.90		
Sponge heads............	4	3.20		
Sponges................	12	3.00		
Priming wires...........	3	0.24		
Gunner's gimlets.........	3	0.24		
Lanyards for friction tubes	4	0.40		
Cannon spikes...........	6	0.30		
Dark lanterns...........	3	3.00		
Common lanterns........	4	4.60		
Carried forward....	....	109.17		

Contents of the Wagon Body, C—Continued.

TOOLS AND STORES.	NO.	WEIGHT.	PLACE.
		Lbs.	
Brought forward...	....	109.17	
Neats' foot oil.......gals.	4	32.80	In 2 tin cans. } In Box C, 7. 92.80 lbs.
Grease..............lbs.	50	60.00	" 2 kegs... } In Box C, 7. 92.80 lbs.
Nails (4d, 6d, 8d, 10d)..lbs.	20	20.00	Box C, 8.
Felling axes............	2	12.00	In the axe rack.
Claw hatchet...........	1	2.00	In the axe rack.
Hand bills.............	2	4.00	In the axe rack.
Caisson stock...........	1	35.00	Under the till, against the side and rear of the wagon.
Rammers and sponges....	3	13.50	On the caisson stock, against rear end.
Spokes...	40	72.00	On the bottom; piled lengthwise against the front end.
Fellies.................	24	160.00	On the spokes, crosswise.
Grindstone, 14 in. × 4 in..	1	50.00	On the fellies, against the left side of the wagon.
Arbor and crank for do...	1	6.50	On the fellies, against the left side of the wagon.
Screw jacks............	3	75.00	On the fellies, against the front and the till.
Wheel traces...........	10	47.50	In a pile occupying 30 inches at the rear end of the wagon, between the left side and the caisson stock, and up to the top of the till; the collars piled on each other, from the bottom.
Leading traces..........	10	57.50	
Collars................	6	27.50	
Girths.................	16	11.00	
Whips..................	16	8.00	
Bridles.................	6	18.00	
Halters................	6	21.00	
Halter chains...........	12	15.50	
Hame straps............	25	4.50	
Spare nose bags.........	12	13.50	On the harness.
Sash cord.........pieces.	6	10.00	On the harness.
Slow match..........yds.	50	6.00	On Box No. 7, to the left of No. 8.
Elevating screw.........	1	15.75	On the pile of harness.
Pole yoke..............	1	12.25	On the pile of harness.
Harness leather .....side.	1	25.00	Under till, in front of the pile of harness, and against the caisson stock.
Bridle leather........do.	2	22.00	Under till, in front of the pile of harness, and against the caisson stock.
Prolonge...............	1	18.00	On box No. 7, in front of No. 8.
Scythes................	4	9.00	In the till, against the front end.
Carried forward....	....	993.97	

Contents of the Wagon Body, C—Continued.

TOOLS AND STORES.	NO.	WEIGHT.	PLACE.
		Lbs.	
Brought forward....	....	993.97	
Scythe stones............	4	6.00	In the curve of the scythes.
Spades..................	6	30.00	In the till; the bits against the rear end.
Pickaxes and handles.....	2	13.00	Between the spade handles.
Corn sacks..............	24	20.00	On the scythes.
Tarpaulins, 5 feet square..	2	18.00	On the corn sacks, against front end.
Reaping hooks..........	4	3.85	Fastened to the ridge pole with a wooden clamp and a leather strap.
Scythe sneaths..........	4	12.00	Fastened to the ridge pole with two leather straps and buckles.
Spare stock for battery wagon................	1	90.00	In the spare stock stirrup.
Padlock................	1	0.50	
Watering bucket.........	1	8.00	Tied to the forage rack.
Forage..................			In the forage rack.
Boxes...........	4	69.00	
Tow	....	24.50	
Total......	....	1288.82	Exclusive of forage.

Dimensions and Weights of Gun-Carriage and Equipment for Mountain Howitzer.

DIMENSIONS.	
Gun-Carriage.	Inches.
Distance between the inside of trunnion-plates	7.00
Diameter of trunnion-holes	2.75
Depth of axis of trunnions below upper face of trunnion-plate	0.62
Distance of axis of trunnions in rear of axis of axle-tree, the piece being in battery, on horizontal ground	2.50
Distance from axis of trunnions to axis of axle tree	8.50
Height of axis of trunnions above the ground	27.00
Vertical field of fire, above the horizontal line	9°
Vertical field of fire, below the horizontal line	7°
Distance between the points of contact of wheels and trail with the ground line	43.70
Distance from front of wheels to end of trail, the piece being in battery.	71.80
Distance of the muzzle of the piece, in battery, in rear of wheels	2.44
Length of gun-carriage, without wheels	61.00
Length of thill	73.00
Whole length of the axle-tree	38.25
Track of the wheels	30.20
Height of wheel	38.00
Dish of finished wheel	2.00
Ammunition-Chest, or Carriage-maker's Tool-Chest.	
Interior length	32.80
" width	4.75
" depth	9.35
Forge-Chest, or Smith's Tool-Chest.	
Interior length	32.80
" width	8.00
" depth	16.25
WEIGHTS.	Pounds.
Howitzer	220.0
Gun-carriage, without wheels	157.0
One wheel	65.0
Handspike	5.0
Sponge and rammer	3.0
Gun-carriage complete, with implements	295.0
Thill	30.0
Bridle	3.0
Halter	3.5
Pack-saddle and harness	44.0
Lashing-girth and rope	3.0
Ammunition-chest, or carriage-maker's tool-chest, empty	20.0
Forge-chest, or smith's tool-chest, empty	42.0
Ammunition-chest, packed	112.0
Forge-chest, packed	115.0
Smith's tool-chest, packed	117.0
Coal-sack, filled with charcoal	25.0
Carriage-maker's tool-chests, A.	45.0
Carriage-maker's tool-chests, B.	45.0

Shot.

	13-in.	12-in.	10-in.	8-in.	42	32	24	18	12	9	6	4	3	1
Diameter....in.	12.87	11.87	9.87	7.88	6.84	6.25	5.68	5.17	4.52	4.10	3.58	3.12	2.84	1.95
Weight.....lbs.	294.00	231.00	128.00	65.00	42.70	32.60	24.40	18.50	12.30	9.25	6.10	4.07	3.05	1.00

Shells.

	For Columbiads and S. C. Howitzers.		For Mortars.			For Guns and Howitzers.				
	10-in.	8-in.	13-in.	10-in	8-in	42	32	24	18	12
	In.	In.	In.	In	In.	In.	In.	In.	In.	In.
Diameter	9.87	7.880	12.870	9.87	7.880	6.84	6.250	5.680	5.170	1.520
Thickness of sides and bottom, True	2.00	1.500	2.100	1.60	1.250	1.20	1.000	0.900	0.900	0.700
Thickness of sides and bottom, Greatest	2.10	1.580	2.250	1.70	1.330	1.25	1.050	0.950	0.940	0.740
Thickness of sides and bottom, Least	1.90	1.420	1.950	1.50	1.170	1.15	0.950	0.850	0.860	0.660
Thickness at fuze-hole	3.00	2.250	2.100	1.60	1.250	1.80	1.350	1.350	1.350	1.050
Diameter of fuze-hole, Exterior	1.45	1.338	1.800	1.75	1.300	1.00	0.900	0.900	0.900	0.900
Diameter of fuze-hole, Interior	1.00	1.000	1.483	1.51	1.113	0.73	0.698	0.698	0.698	0.743
Distance between ears	6.00	5.000	7.000	6.00	5.000					
Weight.....lbs.	101.00	50.500	197.000	87.50	44.500	31.00	22.500	17.000	13.400	8.400

The 8-inch mortar shell is used for the siege-howitzer.

Spherical Case Shot.

	8-in.	42	32	24	18	12	6
	In.	In.	In.	In.	In.	In.	In.
Diameter	7.880	6.840	6.250	5.680	5.170	4.520	3.580
Thickness of metal at the sides, True	0.700	0.650	0.600	0.550	0.500	0.450	0.360
Thickness of metal at the sides, Greatest	0.725	0.675	0.625	0.575	0.525	0.475	0.385
Thickness of metal at the sides, Least	0.675	0.625	0.575	0.525	0.475	0.425	0.335
Thickness of metal at the fuze-hole	1.600	1.500	1.500	1.100	1.100	0.750	0.750
Radius of reinforce at the fuze-hole	3.000	2.750	2.500	2.300	2.100	1.800	1.400
Diameter of fuze hole, Exterior	1.200	1.200	1.200	0.900	0.900	0.900	0.900
Diameter of fuze hole, Interior	0.960	9.075	0.975	0.735	0.735	0.788	0.788
Mean weight.....lbs.	30.000	20.320	16.000	11.800	8.700	6.100	3.000

The thickness of metal at the fuze-hole is supposed to be measured in the axis of the fuze-hole, between the spherical surfaces of the shell and of the reinforce.

The fuze-holes of shells and spherical-case shot taper 0.15 inch to 1 inch.

Diameters of Shot and Shell Gauges.

Diameter.	13-in.	10-in.	8-in.	42-pdr.	32-pdr.	24-pdr.	18-pdr.	12-pdr.	6-pdr.
Large	12.90	9.90	7.90	6.86	6.27	5.70	5.18	4.53	3.60
Small	12.80	9.80	7.80	6.76	6.18	5.61	5.10	4.46	3.54

Canister-Shot Gauges.

Diameter.	42-pdr.	32-pdr.	24-pdr. Gun, and 8-in. How.	18-pdr.	12-pdr. Gun, and 32-pdr. How.	9-pdr. Gun, and 24-pdr. How.	6-pdr.	12-pdr. How.
Large ...	2.26	2.06	1.87	1.70	1.49	1.35	1.17	1.08
Small ...	2.22	2.02	1.84	1.67	1.46	1.32	1.14	1.05

Grape-Shot Gauges.

Diameter.	42-pdr.	32-pdr.	24-pdr.	18-pdr.	12-pdr.
Large......	3.17	2.90	2.64	2.40	2.06
Small	3.13	2.86	2.60	2.36	2.02

Dimensions of Cylinder Gauges.

	10-in.	8-in.	42-pdr.	32-pdr.	24-pdr.	18-pdr.	12-pdr.	6-pdr.
a........	50.00	40.00	35.00	32.00	29.00	26.50	23.00	18.50
b........	9.90	7.90	6.86	6.27	5.70	5.18	4.53	3.60
c........	10.90	8.90	7.86	7.27	6.50	6.00	5.35	4.40
d........	11.15	9.15	8.10	7.50	6.70	6.20	5.55	4.60
e........	2.50	2.00	2.00	2.00	1.75	1.75	1.50	1.50

Dimensions and Weights of Canisters.

DIMENSIONS.	FOR SIEGE AND GARRISON GUNS.					FOR 8-IN. HOWITZERS.	
	42	32	24	18	12	Siege.	Sea Coast.
	IN.	IN.	IN.	IN.	IN.	IN.	IN.
Length of tin for cylinder, developed.	21.50	20.00	18.30	16.70	14.40	25.10	25.10
Height of ditto..................	9.60	9.00	8.25	7.70	6.75	8.60	8.60
Interior diameter of cylinder.......	6.78	6.19	5.63	5.12	4.47	7.80	7.80
Diameter of plates................	6.73	6.14	5.58	5.07	4.42	7.75	7.75
Height of finished canister	8.70	8.10	7.35	6.80	6.00	12.03	12.35
Number of tiers of shot	4	4	4	4	4	4	4
Number of shot in each of the three lower tiers....................	7	7	7	7	7	12	12
Number of shot in 4th tier	6	6	6	6	6	12	12
Whole number of shot	27	27	27	27	27	48	48
Weight, finished canister.......lbs.	48.00	37.00	29.00	23.00	15.00	53.50	54.50

Grape-shot.

	8-in.	42	32	24	18	12
	IN.	IN.	IN.	IN.	IN.	IN.
Diameter of large gauge	3.60	3.17	2.90	2.64	2.40	2.06
Diameter of small gauge	3.54	3.13	2.86	2.60	2.36	2.02
Mean weight ....lbs.	6.10	4.20	3.15	2.40	1.80	1.14

Canister-shot.

	NATURE OF ORDNANCE.								
	42-pdr Gun.	32-pdr. Gun.	24-pdr. Gun, and 8-in Siege-How-itzer.	18-pdr. Gun.	12-pdr. Gun, and 32-pdr. Howit-zer.	24-pdr. Howitzer.	6-pdr. Gun.	12-pdr. How-itzer. Field.	12-pdr. How-itzer. Mountain.
	IN.	IN.	IN.	IN.	IN.	IN.	IN.	IN.	
Diameter of large gauge.....	2.26	2.06	1.87	1.70	1.49	1.35	1.17	1.08	Mus-ket ball.
Diameter of small gauge.....	2.22	2.02	1.84	1.67	1.46	1.32	1.14	1.05	
Mean weight..........lbs.	1.50	1.14	0.86	0.64	0.43	0.32	0.21	0.16	

Grape-shot Stands.

DIMENSIONS.	8-in.	42	32	24	18	12
	IN.	IN.	IN.	IN.	IN.	IN.
Diameter of plates	7.85	6.83	6.24	5.68	5.17	4.52
Thickness of plates	0.60	0.60	0.50	0.50	0.40	0.40
Interior diameter of rings	6.55	5.73	5.16	4.75	4.26	3.80
Diameter of round iron for rings and pins.	0.60	0.50	0.50	0.38	0.38	0.32
Length of pin, including tapped part	14.70	9.25	8.70	7.88	7.18	6.12
Height of stand between the outsides of the plates	9.85	8.75	8.20	7.50	6.80	5.80
WEIGHTS.	LBS.	LBS.	LBS.	LBS.	LBS.	LBS.
Plates	13.60	10.20	8.00	6.75	4.56	3.44
Pin, nut, and rings	4.75	2.80	2.50	1.81	1.12	0.69
Stand, complete	74.50	51.25	39.75	30.61	22.15	14.84

Charges for Mortar Shells.

	13-in.		10-in.		8-in.		Cohorn.	
	LBS.	OZ.	LBS.	OZ.	LBS.	OZ.	LBS.	OZ.
CHARGE, Of the shell filled with powder	11	0	5	0	2	9	1	0
CHARGE, To burst the shell	6	0	2	0	1	0	0	8
CHARGE, To blow out the fuze	0	6	0	5	0	4	0	2
ORDINARY SERVICE-CHARGE, Cannon-powder.	7	0	3	0	1	12		
ORDINARY SERVICE-CHARGE, Incend'ry match, or other comp.	0	8	0	6	0	6		

Dimensions for Field Cartridge Bags.

DIMENSIONS.	12-pdr. gun; 32 and 24-pdr. howitzers.	6-pdr. gun, and 12-pdr. howitzer.	REMARKS.
	In.	In.	
Length of rectangle (cylinder developed)	14.20	11.40	1 in. allowed for seam.
Height	10.00	7.25	0.5 in. do.
Diameter of bottom	5.25	4.37	1 in. do.

POWDER MEASURES.

Made of sheet copper; those for use in the park should be made without handles, for the convenience of putting them up in a nest; their form is cylindrical, the interior diameter and height being equal.

To find the diameter and height of a cylinder to contain a given quantity of gunpowder, multiply the weight in pounds by

38.2 for cannon powder....... } of medium density,
39.4 for musket or rifle powder. }

and take the cube root of the product.

Dimensions of Powder Measures.

Weight of powder.		Diameter and height.	Weight of powder.		Diameter and height
lbs.	oz.	In.	lbs.	oz.	In.
0	1	1.337	2	0	4.240
0	2	1.685	2	8	4.571
0	4	2.122	3	0	4.857
0	8	2.673	4	0	5.346
1	0	3.368	4	8	5.560
1	4	3.628	6	0	6.120
1	8	3.855	8	0	6.736

Charges for Field Pieces.

KIND.	FOR GUNS.		FOR HOWITZERS.		
	12-pdr.	6-pdr	32-pdr.	24-pdr.	12-pdr.
	Lbs.	Lbs.	Lbs.	Lbs.	Lbs.
For shot...............................	2.5	1.25			
For spherical case or canister............	2.0	1.00	2.50	2.0	1.0
For shells. Small charge..............			2.50	2.0	1.0
For shells. Large charge..............			3.25	2.5	1.0

Dimensions for Field Howitzer Cartridge Blocks.

DIMENSIONS.	32-PDR. HOWITZER.		24-PDR. HOWITZER.	
	Small charge.	Large charge.	Small charge.	Large charge.
	In.	In.	In.	In.
Diameter	4.15	4.150	4.15	4.15
Height	2.00	0.750	1.00	0.50
Distance from middle of groove to bottom of block	0.40	0.375	0.40	0.25
Width of groove	0.30	0.300	0.30	0.30
Depth of groove	0.15	0.150	0.15	0.15

Stamps for Blank Field Cartridges.

			In.	In.
Stamps,	for cutting,	Width	7.6	6.0
		Length, including semicircular ends	10.5	8.5
	for sewing,	Width	6.6	5.2
		Length, including semicircular ends	10.0	8.0

Dimensions for Formers for making Paper Cylinders and Caps.

DIMENSIONS.		FOR GUNS.		FOR HOWITZERS.		
		12	6	32	24	12
		In.	In.	In.	In.	In.
Paper for a cylinder and a cap.	Length, developed	14.40	11.60	14.40	14.40	11.6
	Height	12.50	11.50	12.00	10.00	8.0
Height of cylinder.	For large charge	5.00	4.00	5.25	3.50	
	For small charge	4.00	3.50	5.00	3.00	3.0
Formers for cylinders and caps.	Length (exclusive of handle)	15.00	13.00	Same as 12-pdr. gun.		Same as 6-pdr. gun.
	Width at upper end.	6.71	5.25			
	Width at lower end.	6.60	5.17			
	Thickness	0.15	0.15			
Cylindrical formers for choking caps.	Length	10.00	10.00			
	Diameter	4.30	3.30			

Dimensions of Cartridge Bags for Siege and Garrison Ammunition.

	GUNS.					COLUMBIADS.		HOWITZERS.		
									Sea Coast.	
	42-pdr.	32-pdr.	24-pdr.	18-pdr.	12-pdr.	10-in.	8-in.	Siege, 8-in.	10-in.	8 in.
Charge of powder.....lbs.	10.50	8.00	8.00	6.00	4.00	20.00	12.00	4.00	12.00	8.00
	In.	In.	In.	In.	In.	In.	In.	In.	In.	In.
Diameter of chamber.....	7.00	6.40	5.82	5.30	4.62	8.00	6.40	4.62	7.00	6.40
Length of chamber.......						12.00	11.00	8.00	9.50	7.50
Diameter of cartridge....	6.00	5.50	5.00	4.60	4.20	7.50	6.00	4.20	6.50	6.00
Length of 1 lb. of powder in a cartridge..........	0.98	1.16	1.45	1.75	2.00	0.63	0.98	2.00	0.83	0.98
Width of cutting stamp...	10.35	9.55	8.75	8.15	7.60	12.70	10.35	7.60	11.15	10.35
Width of sewing stamp, and of the finished bag..	9.35	8.55	7.75	7.15	6.60	11.70	9.35	6.60	10.15	9.35
Whole length of bag. cut.	18.00	18.00	18.00	17.00	14.00	24.00	20.00	14.00	18.00	15.00
Length of cartridge filled.	11.00	10.50	12.00	11.00	9.00	14.00	12.50	9.00	11.00	9.00
Quantity, 5-4 stuff for 100 bags...............yds.	30	27	25	23	14	36	30	14	31	20

Dimensions of Sabots for Siege and Garrison Ammunition.

	SIEGE AND GARRISON GUNS.					S. C. H.		COLUMBIADS	
	42	32	24	18	12	10-in.	8-in.	10-in.	8-in.
	In.	In.	In.	In.	In.	In.	In.	In.	In.
Whole height................	2.00	1.50	1.50	1.50	1.50	2.00	2.00	2.00	2.00
Greatest diameter............ }	6.58	6.00	5.43	4.92	4.35 }	7.75	6.79	8.41	6.79
Diameter of bottom........... }						6.75	6.15	7.75	6.15
Cavity for { Depth..............	1.00	0.75	0.75	0.75	1.00	1.00	1.00	1.00	1.00
the ball. { Radius of curv......	3.42	3.12	2.84	2.58	2.26	4.93	3.93	4.93	3.93
STRAPS. { Width...............	0.65	0.60	0.55	0.50	0.45	1.00	0.75	1.00	0.75
STRAPS. { Length...............	21.00	19.00	17.50	16.00	14.00	29.00	23.50	29.00	23.50

Dimensions of 8-inch Canister Sabots.

	SIEGE.	SEA COAST.	
	In.	In.	
Whole height.......................	4.68	5.00	
Greatest diameter..................	7.85	7.85	
Diameter at the bottom.............	*	6.40	* Bottom hemispherical.
Diameter of cylinder for the tin.......	7.80	7.80	
Height " "	0.75	0.75	

Dimensions of Sabots, Field Guns, and Howitzers.

DIMENSIONS OF SABOTS.	12-PDR. GUN.		6-PDR. GUN.		32-PDR. HOWITZER.		24-PDR. HOWITZER.		12-PDR. HOWITZER.	
	Shot and spherical case	Canister.	Shot and spherical case.	Canister.	Shells and spherical case.	Canister.	Shells and spherical case.	Canister.	Shells and spherical case.	Canister.
	In.	In.	In.	In.	In.	In.	In.	In.	In.	In.
Whole height	2.00	2.25	1.55	2.25	2.40	4.75	2.40	4.45	3.20	4.45
Greatest diameter	4.85	4.52	3.35	3.58	5.60	6.24	5.30	5.68	4.27	4.52
Diameter of bottom of conical part					4.50	4.50	4.60	4.60	3.60	3.60
Height of conical part					2.40	4.00	2.40	3.75	2.00	2.75
Diameter at bottom of sabot	4.15	4.15	3.20	3.20	4.50	4.50	4.60	4.60	3.20	3.20
Cavity for ball { Depth	1.50		1.00		1.50		1.50		1.30	
Cavity for ball { Radius of curvature	2.26		1.80		3.12		2.84		2.26	
Height of cylinder for tin		0.50		0.50		0.75		0.70		0.50
Diameter of " "		4.47		3.53		6.19		5.63		4.47
Distance from middle of lower groove to bottom of sabot	0.40	0.40	0.40	0.40					0.40	0.40
Distance between centers of grooves		0.80		0.80					0.50	0.50
Distance between holes for handles					1.50	2.30	1.50	2.30		
Length of cord for handles					12.00	20.00	12.00	19.00		

Dimensions of Fuze Plugs.

DIMENSIONS.		For 32-pdr. spherical case.	For other shells and spherical case.	REMARKS.
		In.	In.	
Exterior diameter,	at top	1.25	0.95	Exterior taper 0.15 in. to 1 in.
	at bottom	1.025	0.75	
Interior diameter,	at top	0.50	0.50	Interior taper 0.05 in. to 1 in.
	at bottom	0.425	0.4325	
Height		1.50	1.35	

Charges for Field Shells.

		32-PDR.	24-PDR.	12-PDR.	REMARKS.
		lbs. oz.	lbs. oz.	lbs. oz.	
Powder required	To fill the shell............	1 5	1 0	0 8	Rifle or musket powder is used in preference to cannon.
	To burst "	0 11	0 8	0 5	
	To blow out the fuze plug..	0 2	0 2	0 1	
	For service charge.........	1 0	0 12	0 7	

Charges for Spherical Case Shot.

CHARGE.	8-in.	42	32	24	18	12	6
Number of musket balls.......	486	306	225	175	120	78	38
Bursting charge of powder..oz.	15	9	8	6	5	4.5	2.5
Weight of shot loaded.....lbs.	59.5	39.00	30.13	22.75	16.30	11.00	5.5

Dimensions of Canisters.

DIMENSIONS OF CANISTERS.	FOR GUNS.		FOR HOWITZERS.		
	12	6	32	24	12
	In.	In.	In.	In.	In.
Length of tin for cylinder (developed).....	14.40	11.50	20.00	18.30	14.40
Height of ditto......................	6.65	5.40	7.10	6.30	5.20
Interior diameter of cylinder............	4.45	3.53	6.19	5.63	4.45
Diameter of plates for bottom and cover...	4.40	3.48	6.14	5.58	4.40
Thickness of bottom plate................	0.25	0.25	0.25	0.25	0.25
Thickness of sheet iron cover....	0.07	0.07	0.10	0.10	0.07
Height of finished canister, including sabot.	8.00	6.75	10.50	9.55	8.75
Number of tiers of shot..................	4	4	4	4	4
Number of shot in each of 3 lower tiers...	7	7	12	12	12
Number of shot in 4th tier...............	6	6	12	12	12
Whole number of shot....................	27	27	48	48	48
	LBS.	LBS	LBS.	LBS.	LBS.
Weight of finished canister..............	14.80	7.32	28.50	21.25	10.80

Dimensions and Weights of Fixed Ammunition.

DIMENSIONS.		FOR GUNS.		FOR HOWITZERS.		
		12	6	32	24	12
		In.	In.	In.	In.	In.
Height of charge of powder, including cartridge blocks for 32 and 24-pdr. howitzers	Large charge	5.00	4.00	7.40	5.90	3.25
	Small charge	4.00	3.25	7.40	5.40	
Height of strapped shot or shell		5.02	4.13	7.14	6.58	6.42
Height of canister with sabot		8.00	6.75	10.50	9.55	8.75
Height of a round of fixed ammunition, with cap	Shot	10.40	8.43			
	Shell					10.00
	Spherical case	9.50	7.80			10.00
	Canister	12.40	10.30			12.30
WEIGHTS.		LBS.	LBS.	LBS.	LBS.	LBS.
Cartridge, including cartridge block	Large charge	2.56	1.30	3.88	2.70	
	Small charge	2.06	1.05	3.10	2.34	1.05
Shot, strapped		12.75	6.28			
Shell, strapped and charged				24.60	18.80	9.35
Spherical case, strapped and charged		11.43	5.75	31.00	23.00	11.30
Canister with sabot		14.80	7.32	28.50	21.25	10.80
Round of ammunition, complete	Shot	15.40	7.60			
	Shell, with small charge			27.70	21.15	10.50
	Spherical case	13.50	6.82	34.10	25.34	12.50
	Canister	16.91	8.40	31.60	23.60	11.85

Charges for Shells for Columbiads and Heavy Guns.

CHARGE OF POWDER.	COLUMBIADS.				GUNS.									
	10-in.		8-in.		42		32		24		18		12	
	Lbs.	oz.	Lbs.	oz.	Lbs.	oz.	Lbs.	oz.	Lbs.	oz	Lbs.	oz	Lbs.	oz.
To fill the shell	3	4	1	12	1	8	1	5	1	0	0	11	0	8
To burst the shell	1	6	1	0	0	12	0	11	0	8	0	7	0	5
To blow out the fuze plug	0	10	0	8	0	6	0	2	0	2	0	1½	0	1
For ordinary service	3	0	1	8	1	4	1	0	0	12	0	10	0	7

Fuzes for Mortar Shells.

DIMENSIONS AND WEIGHTS.	13-IN.	10-IN.	8-IN.
	IN.	IN.	IN.
Diameter of fuze. at upper end	1.85	1.70	1.25
Diameter of fuze. at lower end of first cone	1.65	1.55	1.15
Diameter of fuze. lower end of fuze	1.25	1.00	0.90
Diameter of cup. at the top	1.25	1.00	0.75
Diameter of cup. at the bottom	0.90	0.80	0.60
Diameter of the bore	0.40	0.30	0.30
Length of first cone	2.80	2.25	1.25
Depth of the cup	0.60	0.50	0.40
Thickness of wood at the bottom of the fuze	1.20	0.90	0.90
Length of composition	9.00	8.00	5.00
Whole length of fuze	10.80	9.40	6.30
Length of 1st drift, exclusive of the handles	9.00	8.00	8.00
Length of 2nd drift, exclusive of the handles	4.50	4.00	4.00
Diameter of drifts	0.36	0.27	0.27
	LBS.	LBS.	LBS.
Weight of composition for 100 fuzes	8	4	2½
Weight of 100 fuzes, complete	54	33	16

Composition for Mortar Fuzes.

No.	Nitre.	Sulphur.	Mealed powder.	Time of burning 1 inch.	REMARKS.
1....	2	1	3	3.8 sec.	For 10-inch and 8-inch mortars, light.
2....	2	1	2¼	5.0 sec.	For 13-inch and 10 inch mortars, heavy.
3....			1	2.2 sec.	For 8-inch howitzers.

Composition for Paper Fuzes for the Field Service.

	Mealed powder.	Sulphur.	Time of burning 1 inch.	Color of Fuze.
1................	1	0.00	2 seconds.	Black.
2................	8	3.00	3 "	Red.
3................	8	3.50	4 "	Green.
4................	8	4.00	5 "	Yellow.

Composition for Paper Fuzes for Heavy Guns, Howitzers, and Columbiads.

No.	Nitre.	Sulphur.	Mealed powder.	Time of burning for 1 inch.	
1....	26	9	14	10 sec.	
2....	26	9	12	14 sec.	
3....	26	9	10	20 sec.	

Portfire Composition.

The English use .... Nitre.................. 60 parts.
Sulphur .... 40 "
Mealed Powder......... 20 "

The French use several kinds, one of which is
Nitre.................. 16 parts.
Sulphur .... 7 "
Mealed Powder......... 6 "

Sheet Metals and Wire.

Sheet metals and metallic wires are designated by *Nos.* corresponding with the thickness, as indicated by the ordinary iron-wire gauge. These gauges do not exactly agree with each other, but the following table shows very nearly the thickness corresponding with the several Nos.

No.	Thickness.	No.	Thickness.	No.	Thickness.	No.	Thickness.
	IN.		IN.		IN.		IN.
1	0.313	9	0.161	17	0.057	25	0.023
2	.292	10	.143	18	.052	26	.020
3	.271	11	.125	19	.047	27	.018
4	.250	12	.113	20	.042	28	.016
5	.233	13	.100	21	.037	29	.014
6	.216	14	.088	22	.031	30	.012
7	.200	15	.075	23	.028		
8	.180	16	.063	24	.025		

Making Cartridges for Small Arms.

DIMENSIONS OF PAPER FOR CARTRIDGES.		SHEETS.		TRAPEZOIDS.			
		Length.	Breadth.	Height.	Long side.	Short side.	Number in one sheet.
		IN.	IN.	IN.	IN.	IN.	
MUSKET.	Single ball, or ball and buck-shot	16.5	13	4.33	5.25	3.00	12
	Blank	20.0	15	4.00	4.75	2.75	20
	12 buckshot	16.5	13	5.50	5.00	3.00	9
RIFLE...	Ball	16.5	13	4.00	4.25	2.25	16
	Blank	20.0	15	3.00	4.25	2.25	30
PISTOL..	Ball	16.5	13	2.75	4.25	2.25	24
	Blank	20.0	15	2.50	4.25	2.25	36

Table of Dimensions for Formers for making Cartridges with elongated expanding balls. (The dimensions are referred to the plate by means of the letters placed opposite to them.)

	Altered musket.	New Rifle musket.	Pistol carbine.	
	Inches.	Inches.	Inches.	
a	3.50	3.50	3.50	Outer wrapper.
d	2.50	2.25	2.25	
c	5.25	4.25	4.25	
a	1.10	1.00	0.80	Cylinder case.
e	2.75	2.00	2.00	
f	1.50	1.30	1.10	Cylinder wrapper.
g	2.75	2.20	2.20	
h	3.75	3.00	3.00	

Ranges of Field Guns and Howitzers.

The range of a shot or shell in this table is the first graze of the ball on horizontal ground, the piece being mounted on its appropriate field carriage.

The range of a spherical case shot is the distance at which the shot bursts near the ground, in the time given; thus showing the elevation and the length of fuze required for certain distances.

KIND OF ORDNANCE.	Powder.	Ball.	Elevation.	Range.	REMARKS.
	LBS.		° ′	YARDS.	
6-PDR. FIELD GUN.	1.25	Shot	0 00	318	
		"	1 00	674	
		"	2 00	867	
		"	3 00	1138	
		"	4 00	1256	
		"	5 00	1523	
	1.00	Sph. case shot.	2 00	650	Time of flight 2 secs.
			2 30	840	" " 3 "
		"	3 00	1050	" " 4 "
12-PDR. FIELD GUN.	2.50	Shot	0 00	347	
		"	1 00	662	
		"	1 30	785	
		"	2 00	909	
		"	3 00	1269	
		"	4 00	1455	
		"	5 00	1663	
	1.50	Sph.case.	1 00	670	Time 2 seconds.
		"	1 45	950	" 3 "
		"	2 30	1250	" 4 "

Ranges of Field Guns and Howitzers.—Continued.

KIND OF ORDNANCE.	Powder.	Ball.	Elevation.	Range.	REMARKS.
	LBS.		° ′	YARDS.	
NEW 12-PDR. GUN.	2.50	Shot	0 00	325	
		"	1 00	615	
		"	1 30	784	
		"	2 00	876	
		"	3 00	1201	
		"	4 00	1322	
		"	5 00	1619	
	2.00	Shell.	2 00	787	
		"	2 30	926	
		"	3 00	1079	
		"	3 75	1300	
	2.50	Sph.case.	0 30	304	
		"	1 00	574	
		"	1 30	633	
		"	2 00	731	
		"	3 00	960	
		"	3 30	1080	
		"	3 75	1135	
12-PDR. FIELD HOWITZER.	1.00	Shell	0 00	195	
		"	1 00	539	
		"	2 00	640	
		"	3 00	847	
		"	4 00	975	
		"	5 00	1072	
	0.75	Sph.case.	2 15	485	Time 2 seconds.
		"	3 15	715	" 3 "
		"	3 45	1050	" 4 "

Carried over.

Ranges of Field Guns and Howitzers.—Continued.

KIND OF ORDNANCE.	Powder.	Ball.	Elevation.	Range.	REMARKS.
	LBS.			YARDS.	
24-PDR. FIELD HOWITZER.	2.00	Shell.	0 00	295	
		"	1 00	516	
		"	2 00	793	
		"	3 00	976	
		"	4 00	1272	
		"	5 00	1322	
	1.75	Sph. case.	2 00	600	Time 2 seconds.
		"	3 00	800	" 3 "
		"	5 30	1050	" 4 "
	2.00	"	3 30	880	" 3 "
32-PDR. FIELD HOWITZER.	2.50	Shell	0 00	290	
		"	1 00	531	
		"	2 00	779	
		"	3 00	1029	
		"	4 00	1203	
		"	5 00	1504	
	2.50	Sph. case.	3 00	800	Time 2.75 seconds.

Ranges of Mountain Howitzer.

Charge.	Ball.	Elevation.	Range.	REMARKS.
LBS.			YARDS.	
0.5	Shell	0 00	170	
	"	1 00	300	
	"	2 00	392	
	"	2 30	500	Time 2 seconds.
	"	3 00	637	
	"	4 00	785	Time 3 seconds.
	"	5 00	1005	
0.5	Sph. case	0 00	150	
	"	2 30	450	Time 2 seconds.
	"	3 00	500	
	"	4 00	700	Time 2.7 seconds.
	"	4 30	800	Time 3 seconds.
0.5	Canister	4° to 5°	250	

Ranges of Heavy Ordnance.

The *range* of a gun or howitzer in this table is the first graze of the ball on the horizontal plane on which the carriage stands.

KIND OF ORDNANCE.	Powder.	Ball.	Elevation.	Range.	REMARKS.
	LBS.		° ′	YARDS	
18-PDR. SIEGE AND GARRISON GUN. On barbette carriage.	4.50	Shot	1 00	641	
		"	2 00	950	
		"	3 00	1256	
		"	4 00	1450	
		"	5 00	1592	
24-PDR. SIEGE AND GARRISON GUN. On siege carriage.	6.00	Shot	0 00	412	
		"	1 00	842	
		"	1 30	953	
		"	2 00	1147	
		"	3 00	1417	
		"	4 00	1666	
		"	5 00	1901	
	8.00	"	1 00	883	
		"	2 00	1170	
		"	3 00	1454	
		"	4 00	1639	
		"	5 00	1834	
32-PDR. SEA-COAST GUN. On barbette carriage.	6.00	Shot	1 45	900	
	8.00	"	1 00	713	
		"	1 30	800	
		"	1 35	900	
		"	2 00	1100	
		"	3 00	1433	
		"	4 00	1684	
		"	5 00	1922	
	10.67	"	1 00	780	
		"	2 00	1155	
		"	3 00	1517	
42-PDR. SEA-COAST GUN. On barbette carriage.	10.50	Shot	1 00	775	
		"	2 00	1010	
		"	3 00	1300	
		"	4 00	1600	
		"	5 00	1955	
	14.00	"	1 00	770	
		"	2 00	1128	
		"	3 00	1380	
		"	4 00	1687	
		"	5 00	1915	

Ranges of Heavy Ordinance.—Continued.

KIND OF ORDNANCE.	Powder.	Ball.	Elevation.	Range.	REMARKS.
	LBS.	Shell	° ′	YARDS.	
8-INCH SIEGE HOWITZER.	4.00	45 lbs.	0 00	251	
On siege carriage.		"	1 00	435	
		"	2 00	618	
		"	3 00	720	
		"	4 00	992	
		"	5 00	1241	
		"	12 30	2280	
		Shell			
8-INCH SEA COAST HOWITZER.	4.00	45 lbs.	1 00	405	
On barbette carriage.		"	2 00	652	
		"	3 00	875	
		"	4 00	1110	
		"	5 00	1300	
	6.00	"	1 00	572	
		"	2 00	828	
		"	3 00	947	
		"	4 00	1168	
		"	5 00	1463	
	8.00	"	1 00	646	
		"	2 00	909	
		"	3 00	1190	
		"	4 00	1532	
		"	5 00	1800	
		Shell			
10-INCH SEA-COAST HOWITZER.	12.00	90 lbs.	1 00	580	
On barbette carriage.		"	2 00	891	Time, flight 3.00 sec.
		"	3 00	1185	" " 4.00 "
		"	3 30	1300	
		"	4 00	1426	" " 5.25 "
		"	5 00	1650	" " 6.00 "
		Shot			
8-INCH COLUMBIAD.	10.00	65 lbs.	1 00	932	Axis of gun 16 feet above the water.
On barbette carriage.		"	2 00	1116	
		"	3 00	1402	
		"	4 00	1608	
		"	5 00	1847	
		"	6 00	2010	
		"	8 00	2397	Shot ceased to ricochet on the water.
		"	10 00	2834	
		"	15 00	3583	
		"	20 00	4322	
		"	25 00	4875	
		"	27 00	4481	
	15.00	"	27 30	4812	

Ranges of Heavy Ordnance.—Continued.

KIND OF ORDNANCE.	Powder.	Ball.	Elevation.	Range.	REMARKS.
	LBS.		° ′	YARDS.	
8-INCH COLUMBIAD—*Continued.*	10.00	Shell 50 lbs.	1 00	919	
		"	2 00	1209	
		"	3 00	1409	
		"	4 00	1697	
		"	5 00	1813	
		"	6 00	1985	
		"	8 00	2203	
		"	10 00	2657	
		"	15 00	3556	
		"	20 00	3716	
		"	25 00	4387	
		"	27 00	4171	
	15.00	"	27 30	4468	
10-INCH COLUMBIAD. On barbette carriage.	18.00	Shot	0 00	394	Axis of gun 16 feet above the water.
		128 lbs.	1 00	752	
		"	2 00	1002	
		"	3 00	1230	
		"	4 00	1570	
		"	5 00	1814	
		"	6 00	2037	Shot ceased to ricochet on water.
		"	8 00	2519	
		"	10 00	2777	
		"	15 00	3525	
		"	20 00	4020	
		"	25 00	4304	
		"	30 00	4761	
		"	35 00	5433	
	20.00	"	39 15	5654	
	12.00	Shell	1 00	800	
		100 lbs.	2 00	1012	
		"	3 00	1184	
		"	4 00	1443	
		"	5 00	1604	
	18.00	"	0 00	448	
		"	1 00	747	
		"	2 00	1100	
		"	3 00	1239	
		"	4 00	1611	
		"	5 00	1865	
		"	6 00	2209	
		"	8 00	2489	
		"	10 00	2848	
		"	15 00	3200	
		"	20 00	3885	
		"	25 00	4150	
		"	30 00	4651	
		"	35 00	4828	Time of flight 35 sec.

Ranges of Heavy Ordnance.—Continued.

KIND OF ORDNANCE.	Powder.	Ball.	Elevation.	Range.	REMARKS.
	LBS.		° ′	YARDS.	
12-INCH COLUMBIAD.	20.00	Shell 172 lbs.	10	2770	Time of flight 11 sec.
		"	15	3731	" " 16 "
		"	22	4280	" " 20 "
		"	25	4718	" " 26 "
		"	30	5004	
		"	35	5339	" " 32 "
		"	37	5266	" " 31 "
		"	39	5064	
	25.00	"	10	2881	" " 11.5 "
		"	15	3542	" " 15 "
		"	30	5102	
		"	35	5409	" " 32 "
		"	37	5373	" " 32 "
		"	39	5506	" " 36 "
		Shell	35	5644	
		180 lbs.	39	5615	
	28.00	"	35	5671	
		"	39	5761	3⅛ miles. Time 36 sec.
13-IN. SEA-COAST MORTAR.	20.00	Shell 200 lbs.	45	4325	
12-IN. SEA-COAST MORTAR.	20.00	Shell 200 lbs.	45	4625	Experimental.
10 IN. SEA-COAST MORTAR.	10.00	Shell 98 lbs.	45	4250	Time of flight 36 sec.
10-INCH SIEGE MORTAR.	1.00	Shell	45	300	Time of flight 6.5 sec.
	1.50	90 lbs.	45	700	" " 12 "
	2.00	"	45	1000	" " 14 "
	2.50	"	45	1300	" " 16 "
	3.00	"	45	1600	" " 18 "
	3.50	"	45	1800	" " 19 "
	4.00	"	45	2100	" " 21 "
	lbs. oz.				
8-INCH SIEGE MORTAR. (From Griffith's Artillerist's Manual.)	0 10¾	Shell 46 lbs.	45	500	Time of flight 10 sec.
	13¾	"	45	600	" " 11 "
	1	"	45	750	" " 12¼ "
	1 02	"	45	900	" " 13 "
	1 03½	"	45	1000	" " 13½ "
	1 04¾	"	45	1100	" " 14 "
	1 06	"	45	1200	" " 14½ "

Ranges of Heavy Ordnance.—Continued.

KIND OF ORDNANCE.	Powder.	Ball.	Elevation	Range	REMARKS.
	oz.		° ′	YARDS.	
24-POUNDER COEHORN MORTAR.	0.50	Shell 17 lbs.	45	25	
	1.00	"	45	68	
	1.50	"	45	104	
	1.75	"	45	143	
	2.00	"	45	165	
	2.75	"	45	260	
	4.00	"	45	422	
	6.00	"	45	900	
	8.00	"	45	1200	

Ranges of Hale's War Rockets.

ELEVATION.	RANGE (FIRST GRAZE.)		REMARKS.
	2¼-inch.	3¼-inch.	
	Yards.	Yards.	
4° to 5°	500 to 600	500 to 600	The rockets were fired from a trough 10 feet long.
8°	700	800 to 1000	
10°	800 to 900	1000 to 1200	
15°	1200	1200 to 1400	Weight of 2¼-inch rocket.... 6 lbs.
47°	1760	2200	" 3¼-inch " ....16 lbs.

Initial Velocities of Cannon Balls.

From experiments made with the ballistic pendulum, at Washington Arsenal.

KIND OF ORDNANCE.	PROJECTILE.		Charge of Powder.	Initial Velocity.
	Kind.	Weight.		
		Lbs.	Lbs.	Feet.
6-pdr. field gun	Shot	6.15	1.25	1130
			1.50	1563
			2.00	1741
	Sph. case	5.50	1.00	1357
	Canister........	6.80	1.00	1230
12-pdr. field gun	Shot	12.30	2.50	1486
			3.00	1597
			4.00	1826
	Sph. case	11.00	2.00	1392
	Canister........	13.50	2.00	1262

Initial Velocities—Continued.

KIND OF ORDNANCE.	PROJECTILE. Kind.	PROJECTILE. Weight.	Charge of Powder.	Initial Velocity.
		Lbs.	Lbs.	Feet.
12-pdr. field howitzer	Shell	8.90	1.00 1.25	1054 1178
	Sph. case	11.00	1.00	953
	Canister	9.64	1.00	1015
12-pdr. siege and garrison gun	Shot	12.30	2.00 3.00 4.00	1378 1674 1906
	Shell	8.90	2.00 3.00	1611 1929
12-pdr. gun, 25 cal. long	Shot	12.30	2.00 3.00 4.00 5.00 6.00 7.00 8.00	1411 1734 1933 2098 2239 2300 2324
24-pdr. siege and garrison gun	Shot	24.25	3.00 4.00 6.00 8.00	1240 1440 1680 1870
	Shell	17.00	3.00 4.00	1470 1670
	Canister	29.00	3.00 4.00	1135 1303
	Grape	30.60	3.00 4.00	1108 1272
32-pdr. sea-coast gun	Shot	32.30	4.00 5.33 8.00 10.67	1250 1430 1640 1780
	Shell	23.40	4.00 5.33	1450 1657
	Canister	37.00	4.00 5.33	1172 1342
	Grape	39.75	4.00 5.33	1133 1297

Initial Velocities of Elongated Expanding Balls,

As determined by the Musket Ballistic Pendulum at the Washington Arsenal.

No. of fire.	Kind of Gun.	Weight of powder.	Weight of balls	Point of impact.	Vibration of Pendulum.		Velocity.	Date, &c.
		Grs.	Grs.	Inches.	°		Feet.	
1	Altered Harper's	60	510	79.00	5	42	826	July 11, 1856.
2	Ferry rifle, 1856.	60	510	78.75	6	10	896	
3	" "	60	510	78.50	6	26	937	
4	" "	60	510	78.75	6	14	905	
5	" "	60	510	79.00	6	35	951	
6	" "	60	510	78.75	6	20	920	
7	" "	60	510	79.00	6	26	932	
8	" "	60	510	78.87	6	24	929	
9	" "	60	510	79.00	6	28	936	
	Mean velocity or feet per second....						914	
1	New rifle-musket,	60	510	80.00	6	00	858	July 24.
2	1856.	60	510	78.87	6	36	957	
3	" "	60	510	79.00	6	42	970	
4	" "	60	510	79.12	6	42	967	
5	" "	60	510	79.00	6	49	989	
6	" "	60	510	79.00	6	49	989	
7	" "	60	510	79.00	6	33	945	
8	" "	60	510	79.00	6	42	970	
9	" "	60	510	79.00	6	52	994	
10	" "	60	510	79.00	6	50	989	
	Mean velocity						963	
1	Altered rifle-musket Harper's Ferry, 1856	70	740	79.75	8	8	809	July 11.
2	" "	70	740	79.25	9	00	901	July 11.
3	" "	70	740	79.25	8	46	878	July 11.
6	" "	70	740	79.25	8	51	884	July 23; gun cool.
7	" "	70	740	79.50	8	28	850	" "
8	" "	70	740	79.00	8	43	873	July 23.
9	" "	70	740	79.00	9	15	927	July 23.
10	" "	70	740	79.00	9	24	943	July 23.
	Mean velocity....................						883	
1	Altered rifle-musket Springfield, 1856.	70	740	79.75	8	17	822	July 11.
2	" "	70	740	79.62	8	18	827	
3	" "	70	740	79.00	8	41	870	
4	" "	70	740	79.62	8	58	900	
5	" "	70	740	79.00	8	56	897	
6	" "	70	740	79.00	9	10	920	
7	" "	70	740	79.00	9	06	914	
	Mean velocity....................						879	

Initial Velocities.—Continued.

No. of fire.	Kind of Gun.	Weight of powder.	Weight of ball.	Point of impact.	Vibration of pendulum.		Velocity.	Date, &c.
		Grs.	Grs.	Inches.	°		Feet.	
8	Smooth-bored percussion musket.	70	740	79.62	9	33	949	July 23; gun warm, had been fired seven times previous to the commencement of this series of fires.
9	" "	70	740	79.37	9	42	969	
10	" "	70	740	79.50	9	18	928	
11	" "	70	740	79.50	9	33	951	
12	" "	70	740	79.50	9	45	971	
	Mean velocity						954	
1	Pistol-carbine, 1856.	40	468	79.37	3	23	533	July 24.
2	" "	40	468	78.62	3	25	544	July 24.
3	" "	40	468	78.75	3	27	548	July 24.
4	" "	40	468	78.37	3	46	598	July 24.
5	" "	40	468	78.37	3	40	582	July 24.
6	" "	40	468	78.87	3	55	616	July 24.
7	" "	40	468	78.75	4	28	706	July 24.
8	" "	40	468	79.37	4	20	680	Ball struck wooden block sidewise, and fell to the ground.
9	" "	40	468	78.62	3	53	617	July 24.
10	" "	40	468	79.00	3	51	610	July 24.
	Mean velocity						603	

PENETRATIONS.

Table of penetrations in a target made of seasoned white pine plank one inch thick, and placed one and a half inches apart.

ARM.	Weight of ball.	Weight of powder.	Diameter of ball.	Planks penetrated	Distance.
	Grains.	Grains.	Inch.	Number.	Yards.
Altered rifle..................	500	60	0.5775	$9\frac{1}{3}$	200
Altered musket..............	730	70	0.6850	$10\frac{1}{2}$	200
New rifle-musket............	500	60	0.5775	11	200
Pistol-carbine................	450	40	0.5775	$5\frac{3}{4}$	200
Altered rifle..................	500	60	0.5775	$5\frac{2}{3}$	600
Altered musket..............	730	70	0.6850	$6\frac{1}{3}$	600
New rifle-musket............	500	60	0.5775	$6\frac{1}{3}$	600
Pistol-carbine................	450	40	0.5775	3	500
Altered musket..............	730	70	0.6850	$3\frac{1}{2}$	1000
Altered rifle..................	500	60	0.5775	3	1000
New rifle-musket............	500	60	0.5775	$3\frac{1}{4}$	1000

At 1,000 yards a ball from the new rifle-musket passed completely through the frame of the target, which was made of solid white pine three inches thick.

Interior Dimensions of Cylindrical Powder Measures.

Contents.		Diameter and height.	Contents.		Diameter and height.
Lbs.	oz.	In.	Lbs.	oz.	In.
0	1	1.337	2	0	4.240
0	2	1.685	2	8	4.571
0	4	2.122	3	0	4.857
0	8	2.673	4	0	5.316
1	0	3.368	4	8	5.560
1	4	3.628	6	0	6.120
1	8	3.855	8	0	6.736

Composition for Filling Rockets.

COMPOSITIONS.	Nitre.	Sulphur.	Charcoal.	Steel Filings.
1	16	4	6	4
2	10	2	3	
3	8	2	3	

Composition for Stars.

COMPOSITIONS.	Nitre.	Sulphur.	Mealed Powder.	Antimony.
1	16	8	3	
2	16	7	4	
3	16	8	4	*1.5

* With the nitre, produces a whitish flame.

Composition for Serpents.

COMPOSITIONS.	Nitre.	Sulphur.	Mealed Powder.	Charcoal.	Steel Filings.
1	3	2	16	½	
2	15	4		2½	
3	16	2	4	6	
4	16	4	4	2	6

Composition for Gold Rain.

COMPOSITIONS	Nitre.	Sulphur.	Mealed Powder.	Charcoal.	Pulverized Soot.	German Black.	Dissolved Gum.
1	½	1½	8		½	½	½
2	16	10	6	4		2	
3	16	8	8	2		2	

Composition for Fire and Light Balls.

SOFT COMPOSITION.	13-in.	10-in.	8-in.
	LBS.	LBS.	LBS.
Rosin	8.00	5.50	2.75
Pitch	4.00	2.75	1.25
Mutton tallow	1.50	1.00	0.50
Spirits of turpentine	1.00	0.66	0.33
Linseed oil	1.00	0.66	0.33
Gunpowder	12.00	8.00	4.00
Dry composition	10.00	6.66	3.33
Chopped tow	1.00	0.66	0.33
Dry composition, additional	2.00	1.33	0.66
	IN.	IN.	IN.
Height of composition. Before inserting the tarred link and above the shell	1.00	1.00	0.50
Height of composition. Whole height	12.00	10.00	8.00
Tarred link. Exterior diameter	6.50	6.00	4.75
Tarred link. Thickness	2.00	1.75	1.50
Size of shell to be inserted	32-pd.	24-pd	12-pd.

The dry composition consists of:

Beeswax	0.66
Nitre	16.00
Flowers of sulphur	6.00
Inflammable sawdust	1.08
Regulus of antimony	2.66
Gunpowder	1.00

Melt the beeswax over the fire and add the nitre to it; when the mixture is about to melt, take it off from the fire and stir in the sulphur; then add the sawdust, the antimony, and lastly, the powder, and mix them with the hands. Work with great caution against their taking fire.

To prepare the inflammable sawdust: Boil the sawdust in a solution of half its weight of nitre dissolved in an equal quantity of water—crude nitre or damaged gunpowder may be used for this purpose. Evaporate to dryness, stirring frequently; then spread out the sawdust, to become perfectly dry before being used.

Kit.

Composition: 9 rosin, 6 pitch, 6 beeswax, 1 tallow. To be melted together and poured into water; then worked with the hands until it becomes soft and pliable.

PHYSICAL DATA.

Working Power of Men and Horses.

Men. *A foot soldier* travels in 1 minute,

In common time, 90 steps = 70 yards.
In quick time, 110 " = 86 "
In double quick, 140 " = 109 "

He occupies in the ranks a front of 20 in., and a depth of 13 in., without the knapsack; the interval between the ranks is 13 in. 5 men can stand in a space of 1 square yard. Average weight of men, 150 lbs. each.

A *man* travels, without a load, on level ground, during 8½ hours a day at the rate of 3.7 miles an hour, or 31¼ miles a day. He can carry 111 lbs. 11 miles, in a day. A porter going short distances and returning unloaded, carries 135 lbs. 7 miles a day. He can carry, in a wheelbarrow, 150 lbs. 10 miles a day.

The maximum power of a strong man, exerted for 2½ minutes, may be stated at 18,000 lbs. raised 1 foot in a minute.—*Mr. Field's Experiments* 1838.

A man of ordinary strength exerts a force of 30 lbs. for 10 hours a day, with a velocity of 2½ feet in a second = 4500 lbs. raised 1 foot in a minute = *one-fifth* the work of a horse.

Daily allowance of water for a man, 1 gallon, for all purposes.

Horses.—*A horse* travels the distance of 400 yards, at a walk, in 4½ minutes; at a trot, in 2 minutes; at a gallop, in 1 minute.

He occupies in the ranks a front of 40 in., a depth of 10 feet; in a stall, from 3½ to 4½ feet front; at picket, 3 feet by 9. Average weight of horses, 1000 lbs. each.

A horse carrying a soldier and his equipments (say 225 lbs.), travels 25 miles in a day (8 hours).

A *pack horse* can carry 250 to 300 lbs., 20 miles a day.

A *draught horse* can draw 1600 lbs. 23 miles a day; weight of carriage included.

Artillery horses should not be made to draw more than 700 lbs. each, the weight of the carriage included.

The ordinary work of a horse for 8 hours a day, may be stated at 22,500 lbs., raised 1 foot in a minute.

In a horse mill, the horse moves at the rate of 3 feet in a second. The diameter of the path should not be less than 25 or 30 feet.

Daily allowance of water for a horse, 4 gallons.

Forage.—Hay, pressed in bundles, 11 lbs. to the cubic foot.
Oats, 40 lbs. to the bushel, or 32.14 lbs. to the cubic foot.
Wheat, 60 lbs. to the bushel, or 48.21 lbs. to the cubic foot.

A horse-power in steam engines, is estimated at 33,000 lbs. raised 1 foot in a minute; but as a horse can exert that force but 6 hours a day, one steam horse-power is equivalent to that of 4 horses.

Table of Lengths.

Millimètre	=	0.03937 inches.		
Centimètre	=	0.39371 "		
Décimètre	=	3.93710 "		
Mètre	=	39.37100 "	=	1.094 yards.
Décamètre	=	393.71000 "	=	10.940 "
Hectomètre	=	3937.10000 "	=	109.400 "
Kilomètre	=	39371.00000 "	=	1094.000 "
Myriomètre	=	393710.00000 "	=	10940.000 "

The basis of the French measures is the measure of a meridian of the the earth, a quadrant of which is 10,000,000 mètres, measured at the temperature of 32° F.

Table of Weights.

Millogramme	=	0.0154 grains troy.	
Centigramme	=	0.1544 "	
Décigramme	=	1.5444 "	
Gramme	=	15.4440 "	= 0.035 oz. avoir.
Décagramme	=	154.4402 "	
Hectogramme	=	1544.4023 "	
Kilogramme	=	15444.0234 "	= 2.206 lbs. avoir.
Myriogramme	=	154440.2344 "	

The basis of the French system of weights is the weight in vacuo of a litre, or a cubic décimètre, of distilled water at the temperature of 39.2° F. $\frac{1}{1000}$ part of this weight is a gramme.

1 lb. avoir.=0.4535685 kilogramme.

Heat.

	Fahrenheit.	Reaumer.	Centrigrade.
Freezing point of water.......	32°	0°	0°
Boiling " "	212°	80°	100

For converting from one scale to the other,

$$F^\circ = \frac{C^\circ \times 9}{5} + 32^\circ = \frac{R^\circ \times 9}{4} + 32^\circ$$

MEASURES OF WEIGHT.

Avoirdupois Weight.

Drams.	Ounces.	Pounds.	Quarters.	Cwt.	Ton.
16	1				
256	16	1			
7168	448	28	1		
28672	1792	112	4	1	
573440	35840	2240	80	20	1

The standard avoirdupois pound of the United States, as determined by Mr. Hassler, is the weight of 27.7015 cubic inches of distilled water weighed in air, at the temperature of the maximum density (39°.83), the barometer being at 30 inches.

Troy Weight.

Grains.	Dwt.	Ounce.	Pound.
24	1		
480	20	1	
5760	240	12	1

The pound, ounce, and grain are the same in Apothecaries' and Troy weight; in the former, the ounce is divided into 8 drachms, the drachm into 3 scruples, and the scruple into 20 grains.

7000	Troy grains	=	1 lb.	avoirdupois.
175	Troy pounds	=	144 lbs.	"
175	Troy ounces	=	192 oz.	"
$437\frac{1}{2}$	Troy grains	=	1 oz.	"

7000 Troy grains to the pound are used in estimating weights in artillery.

The common musket ball is $\frac{1}{17}$ of a pound, and consequently, equal to $411\frac{3}{4}$ grains.

INDEX.

A.

B.

C.

D.

E.

F.

G.

H.

I.

K.

L.

M.

N.

O.

P.

Q.

R.

S.

T.

V.

W.

PLATES